Mathematical
Economics and
Operations Research

ECONOMICS INFORMATION GUIDE SERIES

Series Editor: Robert W. Haseltine, Associate Professor of Economics, State University College of Arts and Science at Geneseo, Geneseo, New York

Also in this series:

AMERICAN ECONOMIC HISTORY—*Edited by William K. Hutchinson**

ECONOMIC DEVELOPMENT—*Edited by Thomas A. Bieler**

ECONOMIC EDUCATION—*Edited by Catherine A. Hughes*

EAST ASIAN ECONOMIES—*Edited by Molly K.S.C. Lee**

ECONOMIC HISTORY OF CANADA—*Edited by Trevor J.O. Dick*

ECONOMICS OF EDUCATION—*Edited by William Ganley**

ECONOMICS OF MINORITIES—*Edited by Kenneth L. Gagala*

HEALTH AND MEDICAL ECONOMICS—*Edited by Ted J. Ackroyd*

HISTORY OF ECONOMIC ANALYSIS—*Edited by William K. Hutchinson*

INTERNATIONAL TRADE—*Edited by Ahmed M. El-Dersh**

LABOR ECONOMICS—*Edited by Ross E. Azevedo*

MONEY, BANKING, AND MACROECONOMICS—*Edited by James M. Rock*

PUBLIC POLICY—*Edited by Michael Joshua**

REGIONAL ECONOMICS—*Edited by Jean Shackleford**

RUSSIAN ECONOMIC HISTORY—*Edited by Daniel R. Kazmer and Vera Kazmer*

SOVIET-TYPE ECONOMIC SYSTEMS—*Edited by Z. Edward O'Relley*

STATISTICS AND ECONOMETRICS—*Edited by Joseph Zaremba**

TRANSPORTATION ECONOMICS—*Edited by James P. Rakowski*

URBAN ECONOMICS—*Edited by Jean Shackleford**

*in preparation

The above series is part of the

GALE INFORMATION GUIDE LIBRARY

The Library consists of a number of separate series of guides covering major areas in the social sciences, humanities, and current affairs.

General Editor: Paul Wasserman, Professor and former Dean, School of Library and Information Services, University of Maryland

Managing Editor: Denise Allard Adzigian, Gale Research Company

Mathematical Economics and Operations Research

A GUIDE TO INFORMATION SOURCES

Volume 10 in the Economics Information Guide Series

Joseph Zaremba

Professor of Economics
State University
College of Arts and Science at Geneseo
Geneseo, New York

Gale Research Company
Book Tower, Detroit, Michigan 48226

Library of Congress Cataloging in Publication Data

Zaremba, Joseph.
    Mathematical economics and operations research.

    (Economics information guide series ; v. 10) (Gale
information guide library)
    Includes indexes.
    1. Economics, Mathematical—Bibliography. 2. Linear
programming—Bibliography. 3. Economics—Bibliography.
I. Title.
Z7164.E2Z37   [HB135]   016.33   73-17586
ISBN 0-8103-1298-0

VITA

Joseph M. Zaremba is professor of economics at State University College of Arts and Science, Geneseo, New York. He has degrees in forestry, economics, and mathematics, including a Ph.D. in economics from Harvard University. Zaremba began his teaching career on the faculty of the Yale School of Forestry and has taught at the State University of New York College of Forestry and Fordham University. He is the author of ECONOMICS OF THE AMERICAN LUMBER INDUSTRY and is currently working on ECONOMETRICS AND ECONOMIC STATISTICS, which will be published as part of the Economics Information Guide Series of the Gale Information Guide Library.

CONTENTS

Contents

PREFACE

This guide is an annotated bibliography of more than 1,600 books which have been published in the English language, including English translations of foreign titles, prior to and including 1974, and some books in 1975. It is divided into eighteen categories (sixteen chapters and two appendixes) on the basis of subject matter. Within each category the books are arranged according to the author's last name, or the last name of the first author in the case of joint authorship. The relatively few books in which no author is listed are arranged alphabetically by title. The book's content, as conveyed in its title, generally serves as the main basis for classification into categories. However, in some cases a book's main analytical technique was used for classification purposes. For example, the book, NATURAL GAS AND NATIONAL POLICY: A LINEAR PROGRAMMING MODEL OF NORTH AMERICAN GAS FLOW, by L. Waveman, was placed in chapter 10, "Mathematical Programming."

The selection of books is more complete and thorough for some categories than for others. For example, the chapters on mathematical programming and input-output analysis are relatively complete and thorough. The coverage of books is also more thorough in the case of American publishers than foreign publishers.

In some cases American distributors of foreign titles are listed. For example, most books published by North-Holland are distributed by American Elsevier, and the distributor's name follows the title. For American books, the location or city of the publisher's headquarters is given, rather than the location of a branch office of the publisher.

Since the fields of mathematical economics and operations research fall into the broader fields of economics, business, and mathematics, some guidelines were established for book selection. To qualify, a book must have a substantial portion of its analysis in terms of a well-defined branch of mathematics, such as differential calculus, matrix algebra, and set theory, or its analysis must be undertaken in terms of established techniques of analysis, such as linear programming, dynamic programming, or classical optimization. The main exceptions to this rule are the books on methodology and philosophical issues in mathematical economics and operations research, which are included even

Preface

though no mathematical formulas may be used. Most of the books in economics which were published until very recently are excluded because their content is not sufficiently mathematical. Similarly, numerous books in production, marketing, and other business fields have been excluded. However, an attempt was made to include all books in such traditional areas as mathematical programming, inventory theory, queueing theory, and mathematics for economists.

Since many branches of pure and applied mathematics--calculus, mathematical analysis, differential equations, linear algebra, to mention a few--contribute to the development of economics and management science, it was decided to include selected mathematics books in this guide, particularly for chapters 1 and 12. To qualify, a mathematics book had to have applications of mathematical concepts illustrated with examples from economics, operations research, business, or some other field of the social sciences as well as examples from the physical, biological, or natural sciences. Thus many books on matrix algebra, linear algebra, finite mathematics, and calculus are included because they are application oriented. Moreover, many books in pure mathematics are included since they provide an important prerequisite for applications in economics and management science. For example, most of the books available on the calculus of variations are listed in chapter 12, "Optimization and Control." A few books on set theory, functional analysis, and other branches of pure mathematics are also included. Because of the importance of stochastic processes in economics and management science, a number of books in the theory of probability and stochastic processes are listed in chapter 14, "Stochastic Processes."

An attempt was made to include the following types of information in the annotations where available: (1) type of book, that is, monograph, textbook, or reference; (2) the audience that the book is aimed at, that is, student, practitioner, applied mathematician, etc.; (3) a brief summary of the content of the book, based on topics and applications covered; (4) the mathematical prerequisites for comprehension and the extent of mathematical reviews provided, if any; (5) availability of exercises, problems, references, and summaries; and (6) the style of presentation and the mathematical rigor employed. In some cases, this information is conveyed implicitly in the book's title and length, and in the titles of the appendixes.

The annotations are formulated on the basis of the author's statements, views, and opinions as conveyed in the preface, table of contents, summaries, and introduction, and only rarely was a subjective opinion about the book offered. For example, no attempt was made to determine whether or not a given book succeeded in light of the author's expectations.

Joseph Zaremba
Geneseo, New York

LIST OF ABBREVIATIONS

The following abbreviations have been used in this guide:

ABC   Analysis of Bar Charting

CPM   Critical Path Method

DOD   Department of Defense

EOQ   Economic Order Quantity

ITC   International Textbook Company

NASA   National Aeronautic and Space Administration

NBER   National Bureau of Economic Research

OBE   Office of Business Economics

OR   Operations Research

PERT   Program Evaluation and Review Techniques

RPMS   Resource Planning and Management System

SIAM   Society for Industrial and Applied Mathematics

SRA   Science Research Associates

SSRC   Social Science Research Council

Part I

MATHEMATICS

Chapter 1

MATHEMATICS IN ECONOMICS,

OPERATIONS RESEARCH, AND BUSINESS

Adams, Lovincy Joseph. MODERN BUSINESS MATHEMATICS. New York: Holt, Rinehart and Winston, 1963. x, 348 p. Tables, pp. 304-26. Answers to Odd-numbered Problems, pp. 327-43. Index, pp. 345-48.

A text in business mathematics for students with limited mathematical background. Topics include algebra, percents, simple and compound interest and annuities, linear programming, break-even point analysis, inventory control, and matrices. The sections of the book conclude with exercises, and the chapters conclude with miscellaneous exercises.

Adams, William J. CALCULUS FOR BUSINESS AND SOCIAL SCIENCE. Lexington, Mass.: Xerox College Publishing, 1975. xiii, 226 p. Appendixes, pp. 200-216. Answers to Selected Exercises, pp. 217-22. Index, pp. 223-26.

A text for a one-semester course in calculus for students of business and social science who have knowledge of high school intermediate algebra or its equivalent. Topics include functions and limits, differential calculus and optimization problems, integral calculus, and topics in multivariable calculus. Applications include marginal revenue and cost, elasticity, profit-maximization behavior of a firm, and the Cobb-Douglas production function. The appendixes contain a review of algebra, the mathematics of a straight line, and the mathematics of finance. The sections of the book contain numerous problems for assignment.

_____. FINITE MATHEMATICS FOR BUSINESS AND SOCIAL SCIENCES. Lexington, Mass.: Xerox College Publishing, 1974. xiv, 354 p. Appendixes, pp. 327-38. Answers to Selected Exercises, pp. 339-47. Index, pp. 349-54.

A text for a course in finite mathematics for business and social science majors who have a good high school algebra background. It is divided into the following five parts: linear programming, probability, theory of games, matrices, and mathematics of finance. Numerous examples with detailed solutions are provided. The sections conclude with exercises, and a study guide is available which

3

includes answers with detailed solutions to all exercises. The four appendixes take up the straight line, sets, a probability function, and systems of linear equations.

Allen, Clark Lee. ELEMENTARY MATHEMATICS OF PRICE THEORY. Belmont, Calif.: Wadsworth, 1962. xv, 155 p. Bibliography, pp. xiii–xv. Appendixes, pp. 123–45. Answers to Odd-numbered Problems, pp. 146–50. Index, pp. 151–55.

A supplementary text covering the basic mathematics needed in a mathematically oriented price theory course. It is divided into the following three parts: functions and graphs; calculus, including elasticity and homogeneous functions; and geometry, including linear programming. The appendixes are devoted to reviews of algebra, nomographic solutions of economic problems, and notes on general equilibrium. Exercises follow the sections of the book, and the chapters conclude with bibliographical notes.

Allen, R.G.D. BASIC MATHEMATICS. New York: St. Martin's Press, 1962. xii, 512 p. Appendix: Formulae of Elementary Algebra and Trigonometry, pp. 489–501. Exercises: Solutions, pp. 502–5. Index, pp. 507–12.

An elementary treatment of modern mathematics, including groups and fields, calculus, and linear algebra. It is suitable as a text for college courses in mathematics for liberal arts students, or as a reference, and requires no background in college mathematics. Each chapter concludes with about two dozen exercises.

_____. MATHEMATICAL ANALYSIS FOR ECONOMISTS. New York: St. Martin's Press, 1938. xvi, 548 p. A Short Bibliography, pp. xiv–xv. Greek Letters, p. xvi. Index, pp. 543–546. Economic Applications Index, pp. 546–48. Authors, p. 548.

A classic work on the mathematics most useful to students of economics with illustrative examples from economics. Topics include functions and limits; differentiation and integration, including partial differentiation; differentials; differential equations; series; determinants; and an introduction to the calculus of variations. Each chapter concludes with a lengthy set of problems stressing economic applications. This book has gone through numerous printings, and it remains a valuable reference and supplemental text.

_____. MATHEMATICAL ECONOMICS. 2d ed., rev. New York: St. Martin's Press, 1960. xviii, 812 p. Appendix A: The Algebra of Operators and Linear Systems, pp. 725–39. Appendix B: The Algebra of Sets, Groups and Vector Spaces, pp. 740–80. Appendix C: Exercises: Solutions and Hints, pp. 781–803. Index, pp. 805–12.

An advanced text on the newer developments of mathematical economics arising from about 1938 to 1958. Topics include the cobweb and other simple dynamic models; multiplier and accelerator; complex variables; linear differential and difference equations; trade cycle theory: Samuelson-Hicks, Goodwin, Kalecki, and Philips; equilibrium; interindustry relations; matrix algebra; games and linear programming; allocation of resources; the theories of the firm and of value; and the aggregation problem. Assumes knowledge of calculus, differential equations, and linear algebra. Exercises follow the sections of the book, and the chapters conclude with references.

Almon, Clopper, Jr. MATRIX METHODS IN ECONOMICS. Addison-Wesley Series in Behavioral Sciences: Quantitative Methods. Reading, Mass.: Addison-Wesley, 1967. x, 164 p. Supplementary Readings, pp. 153-54. Appendix on Determinants, pp. 155-61. Index, pp. 163-64.

An exposition of mathematical methods in economics beyond calculus. It is suitable for a text in quantitative methods for graduates and undergraduates who have knowledge of calculus and statistics. Matrix algebra applications include input-output analysis, regression, linear programming, and related areas. Emphasis is on computational procedures, and many exercises are included in each chapter.

Anderson, Chaney, and Pierce, R.C., Jr. ELEMENTARY CALCULUS FOR BUSINESS, ECONOMICS, AND SOCIAL SCIENCE. Boston: Houghton Mifflin, 1975. xi, 251 p. Table of Symbols, p. xi. Appendix One: Review of Exponents, pp. 199-203. Appendix Two: Introduction to Logarithms, pp. 204-11. Appendix Three: Table of Exponents, p. 212. Answers to Selected Exercises, pp. 213-41. Answers to Self-tests, pp. 243-48. Index, pp. 249-51.

A text for a one-semester course in calculus for students of business and the liberal arts who have a good high school algebra background. Each topic is illustrated with examples and non-examples, as well as applications, and logarithms are treated completely. The number 'e' and its applications are introduced early, and problems requiring maximums and minimums are presented in three different chapters, and include the use of functions of several variables. The sections conclude with exercises, and the chapters conclude with self-tests.

Archibald, George C., and Lipsey, Richard G. AN INTRODUCTION TO A MATHEMATICAL TREATMENT OF ECONOMICS. London: Weidenfeld and Nicolson, 1967. 399 p. Answers to Exercises, pp. 367-91. Index, pp. 393-99.

A text for an introductory course in the use of mathematics in economics for students with a minimal mathematics background. Calculus is introduced as needed, and the emphasis is on applications. Topics include functions and graphs, simple linear models, calcu-

lus and applications of derivatives, maxima and minima, and
economic dynamics. The chapters conclude with exercises.

Ayres, Frank, Jr. MATHEMATICS OF FINANCE, INCLUDING 500 SOLVED
PROBLEMS. Schaum's Outline Series. New York: McGraw-Hill Book Co.,
1963. 535 p. Review Problems. Tables. Index.

Presents an introduction to the mathematics of finance for use as
a supplement in finance courses and as a reference. Each chapter begins with statements of relevant definitions, theorems, and
principles followed by a set of solved problems. Topics include
operations with numbers, exponents and logarithms, progressions,
simple and compound interest and discount, ordinary and deferred
annuities, amortization and sinking funds, bonds, probability and
the mortality table, and life insurance.

_____. MATRICES, INCLUDING 340 SOLVED PROBLEMS. Schaum's Out-
line Series. New York: McGraw-Hill Book Co., 1962. 219 p. Index,
pp. 215-18. Index of Symbols, p. 219.

A supplementary text for courses in matrix algebra and a reference
for all applied fields. Topics include types of matrices, determinants, adjoint and inverse, fields, linear dependence of vectors
and forms, vector spaces and linear transformations, vectors over
real and complex fields, congruence, bilinear forms, quadratic
and Hermitian forms, characteristic roots of a matrix, similarity,
polynomials over a field, lambda matrices, Smith normal forms,
minimum polynomial of a matrix, and canonical forms under similarity.

Baggaley, Andrew R. MATHEMATICS FOR INTRODUCTORY STATISTICS: A
PROGRAMMED REVIEW. New York: Wiley, 1969. xiv, 171 p. Instructions to
the Teacher, pp. ix-x. Instructions to the Reader, p. xi. Answers to Review
Tests, pp. 159-69. Index, p. 171.

A self-study review of the mathematics required for understanding
basic statistics. It is divided into the following four sections:
algebra, plotting points on graphs and the concepts of intercept
and slope, extraction of the square root, summation operator and
summation laws. About 190 minutes are required for the average
reader to work through all four sections.

Bailey, Norman T. THE MATHEMATICAL APPROACH TO BIOLOGY AND
MEDICINE. Series in Quantitative Methods for Biologists and Medical Scien-
tists. New York: Wiley, 1967. xiii, 296 p. References, pp. 284-86. Author
Index, pp. 287-88. Subject Index, pp. 289-96.

A presentation of the mathematical approach to biology and medicine, with emphasis on applications. Part 1 requires little mathematics preparation and deals with general principles and the basic

philosophy of biomathematics, including mathematics, statistics, scientific research, operational research, automatic electronic computing, and the organization of biomathematics. Part 2 contains a wide range of applications to biological and medical subjects such as taxonomy, ecology, epidemics, genetics, medical diagnosis, and operational research in medicine. Assumes knowledge of elementary differential and integral calculus.

Bak, Thor A., and Lichenberg, Jonas. MATHEMATICS FOR SCIENTISTS. New York: Benjamin, 1966. xiv, 487 p. Index, pp. a–g.

A text for a one-year course in mathematics for scientists in physical chemistry and biophysics. It is a revised and expanded edition of VIDEREGAENDE MATEMATIK, which appeared in Denmark in 1960. The mathematical results are formulated in a precise manner even where proofs are omitted. Part 1 deals with vectors and matrices, tensors, and groups; part 2 is devoted to functions of one and several real variables and includes differential and integral calculus; part 3 contains material normally included in books on advanced calculus: series, differential equations, functions of a complex variable, and numerical methods. Extensive use is made of examples, and the sections of the book conclude with exercises and answers. This book is useful to those in the applied fields desiring an introduction to a wide coverage of topics in undergraduate mathematics.

Bashaw, W. Louis. MATHEMATICS FOR STATISTICS. New York: Wiley, 1969. xvi, 326 p. Appendix A: Table of Squares, pp. 282–83. Appendix B: Table of Common Logarithms, pp. 284–85. Appendix C: Table of Natural or Naperian Logarithms, pp. 286–87. Appendix D: Answers to Problems, pp. 288–322. Index, pp. 323–26.

A presentation of the arithmetic and mathematics background that is essential to the learning of introductory applied statistics. It is a self-study book addressed to students who need a review of mathematics but do not wish to enroll in a mathematics course. It is divided into the following five parts: basic arithmetic refresher, basic algebra refresher, basic matrix algebra, set algebra and probability, and miscellaneous skills, including graphing, logarithms, and computational accuracy. The sections of the book conclude with problems, and the chapters conclude with review problems.

Batschelet, Edward. INTRODUCTION TO MATHEMATICS FOR LIFE SCIENTISTS. Biomathematics, vol. 2. New York: Springer-Verlag, 1970. xiv, 495 p. Index of Symbols, pp. xiii–xiv. Answers to Problems, pp. 459–67. References, pp. 468–76. Author and Subject Index, pp. 477–95.

A text for an introductory course in mathematics for biology and medical students with a background in high school mathematics.

Reliance is placed on the intuitive approach to concepts, and examples selected from a wide area of research in the life sciences are provided. Topics include real numbers, sets and symbolic logic, relations and functions, the power function, exponential and logarithmic functions, limits of sequences and functions, probability, matrices, and complex numbers. The chapters conclude with problems.

Baxter, Willard E., and Sloyer, Clifford W. CALCULUS WITH PROBABILITY, FOR THE LIFE AND MANAGEMENT SCIENCES. Addison-Wesley Series in Management. Reading, Mass.: Addison-Wesley, 1973. xiii, 648 p. Answers to Odd-numbered Problems, pp. 605-18. Comprehensive Problem Sets, pp. 619-41. Index, pp. 645-48.

A text for a one-year course in college mathematics for students in the life and management sciences, including economics, who have only high school mathematics. Emphasis is on mathematics, rather than on applications, although the last chapter applies difference equations to the cobweb phenomenon in economics and the spread of a disease. Exercises follow the sections of the book. Topics include differentiation and integration, optimization with several variables, multiple integrals, differential equations, and difference equations.

Beighey, D. Clyde, and Borchardt, Gordon C. MATHEMATICS FOR BUSINESS; COLLEGE COURSE. 5th ed. New York: McGraw-Hill Book Co., 1974. 250 p.

A text-workbook designed to develop mathematical and computational skills needed by college students who are preparing for a business career. It offers business-oriented exercises and realistic applications. Topics include fractions and decimals, percents, mathematics of finance with consumer applications, payroll and depreciation, insurance and taxes, stocks, bonds, annuities, and mathematics of business management.

Bellman, Richard E. INTRODUCTION TO MATRIX ANALYSIS. 2d ed., rev. New York: McGraw-Hill Book Co., 1970. xxiii, 403 p. Appendix A: Linear Equations and Rank, pp. 371-78. Appendix B: The Quadratic Form of Selberg, pp. 379-82. Appendix C: A Method of Hermite, pp. 383-84. Appendix D: Moments and Quadratic Forms, pp. 385-89. Index, pp. 391-403.

An introduction to the methods of modern matrix theory for analysts, engineers, physicists, and mathematical economists. Topics include reduction of symmetric matrices to diagonal form, functions of matrices, characteristic roots, inequalities, dynamic programming, matrices and differential equations, stability theory, control processes, and stochastic matrices. Alternative proofs of theorems are given in order to present a variety of fundamental methods. Computational treatment of matrices is omitted. Sections contain exercises, and the chapters conclude with miscellaneous exercises, bibliographies, and discussions.

_____. METHODS OF NONLINEAR ANALYSIS. VOLUME I. Mathematics in Science and Engineering, Vol. 61-1. New York: Academic Press, 1970. xx, 340 p. Author Index, pp. 331-35. Subject Index, pp. 337-40.

An introduction to the basic concepts of stability and variational analysis of nonlinear systems and the analytic techniques required for solving such systems. These techniques employ ordinary differential equations and matrix analysis and are illustrated easily with numerical examples. The first four chapters consist of a view of first- and second-order differential equations, matrix theory, and stability of solutions of equations. The remaining four chapters take up the following methods of solutions: Bubnov-Galerkin method, differential approximation, and the Rayleigh-Ritz method. The mathematical results are stated without proof, and the emphasis throughout is on the development and applications of techniques of solutions. A unique feature of this book is the large number of exercises that serve to apply and extend the topics. Each section concludes with a large number of miscellaneous exercises (one chapter has fifty-two miscellaneous exercises), some with references to journal articles where the topics are treated in greater detail, and three to four pages of bibliographies and comments. Assumes knowledge of ordinary differential equations, linear algebra, and the calculus of variations.

_____. METHODS OF NONLINEAR ANALYSIS. VOLUME II. Mathematics in Science and Engineering, Vol. 61-2. New York: Academic Press, 1973. xvii, 261 p. Contents of Volume I, p. xvii. Author Index, pp. 255-58. Subject Index, pp. 259-61.

The topics of volume II (chapters 9 to 16) include linear and nonlinear partial differential equations; the following methods of solutions: duality, differential inequalities, quasilinearization, dynamic programming, and invariant embedding; and the following methods requiring the use of the computer: iteration, infinite systems of differential equations, and differential and integral quadratures. The large number of exercises in both the sections and at the ends of the chapters are designed to apply and extend the mathematical results which are presented without proof. Assumes knowledge of partial differential equations, linear algebra, and familiarity with functional analysis. Each chapter concludes with three or four pages of bibliography and comments.

_____. MODERN ELEMENTARY DIFFERENTIAL EQUATIONS. Reading, Mass.: Addison-Wesley, 1968. xii, 196 p. Author Index, p. 193. Subject Index, pp. 195-96.

A text for a course in differential equations for students having knowledge of calculus and the theory of power series. Topics include second-order differential equations, power-series and numerical solutions of differential equations, linear systems of nth-order differential equations, and existence and uniqueness theorems. Exercises follow the sections of the book, and the chapters conclude with bibliographies and comments.

Bellman, Richard E.; Kagiwada, Harriet H.; Kalaba, Robert E.; and Prestrud, Marcia C. INVARIANT IMBEDDING AND TIME-DEPENDENT TRANSPORT PROCESSES. Modern Analytic and Computational Methods in Science and Mathematics, vol. 2. New York: American Elsevier, for the Rand Corporation, 1964. x, 256 p. References, pp. 50-53. Appendixes, pp. 54-254. Index, pp. 255-56.

A presentation of a new technique, called the numerical inversion of the Laplace transform, for obtaining a numerical solution for time-dependent equations of radiative transfer. The appendixes consist of tables of reflection functions for time-dependent isotropic scattering from a slab, coefficients of the first fifteen shifted Legendre polynomials in order of lowest to highest power, roots of the shifted Legendre polynomials and corresponding weights, coefficients of these polynomials, derivatives of shifted Legendre polynomials evaluated at the roots, the elements of the inverse matrix without division by weights, angles in degrees whose cosines are roots of shifted Legendre polynomials, and FORTRAN programs. Assumes knowledge of advanced calculus and matrix algebra.

Bellman, Richard E., and Kalaba, Robert E. QUASILINEARIZATION AND NONLINEAR BOUNDARY-VALUE PROBLEMS. Modern Analytic and Computational Methods in Science and Mathematics, vol. 3. New York: American Elsevier, for the Rand Corporation, 1965. 206 p. Appendix One: Minimum Time Program, pp. 163-66. Appendix Two: Design and Control Program, pp. 167-72. Appendix Three: Program for Radiative Transfer-inverse Problem for Two Slabs, Three Constants, pp. 173-83. Appendix Four: Van der Pol Equation Program, pp. 185-88. Appendix Five: Orbit Determination Program, pp. 189-92. Appendix Six: Cardiology Program, pp. 193-202. Index, pp. 203-6.

An introduction to quasilinearization which is a technique for solving nonlinear ordinary differential equations by transforming them into initial value problems for ordinary differential equations using linear approximation techniques and the capabilities of the digital computer. Topics include quasilinerization and the Riccati equation; two-point boundary value problems for second-order differential equations; monotone convergence of linear differential inequalities; monotonicity for higher-order differential equations and for linear systems; application of quasilinearization to the solution of nonlinear elliptic partial differential equations; and applications of quasilinearization to treat descriptive and variational processes arising in design and control, in orbit determination, in the detection of periodicities, and in the identification of systems (including inverse problems in radiative transfer and cardiology). The final chapter discusses the relationship between dynamic programming and the theories of positive operators and quasilinearization. The appendixes contain FORTRAN programs for problem solving. Numerical examples in physics, engineering, and biology are included. Assumes knowledge of partial differential equations. Useful to anyone interested in solving systems of ordinary and partial differential equations. Chapters conclude with comments and bibliographies.

Bellman, Richard E.; Kalaba, Robert E.; and Lockett, Jo Ann. NUMERICAL INVERSION OF THE LAPLACE TRANSFORM: APPLICATIONS TO BIOLOGY, ECONOMICS, ENGINEERING, AND PHYSICS. New York: American Elsevier, for the Rand Corporation, 1966. viii, 249 p. Appendixes, pp. 175-246. Index, pp. 247-49.

Presents methods for obtaining numerical solutions of a large class of functional equations that occur repeatedly in the description of scientific problems by slide rule, desk computer, and numerical integration of a system of ordinary differential equations or of a system of linear algebric equations. The chapter titles are: "Elementary Properties of the Laplace Transform," "Numerical Inversion of the Laplace Transform," "Linear Functional Equations," "Nonlinear Equations," "Dynamic Programming," and "Ill-Considered Systems." The appendixes contain FORTRAN programs for numerical computations, roots of shifted Legendre polynomials, coefficients of certain polynomials, and derivatives of shifted Legendre polynomials evaluated at the roots. The chapters conclude with bibliographies and comments.

Bellman, Richard E., and Wing, G.M. AN INTRODUCTION TO INVARIANT EMBEDDING. Pure and Applied Mathematics Series. New York: Wiley-Interscience, 1975. xv, 250 p. Author Index, pp. 247-48. Subject Index, pp. 249-50.

An introduction to invariant imbedding with emphasis on methodology rather than on the particular physical problems to which it can be applied. The presentation is historical with regard to concepts, and applications, which are of interest to applied mathematicians, are included. Topics include existence and uniqueness problems, random walk, wave propagation, time-dependent problems, calculation of eigenvalues for Sturm-Lioville type systems, Schrodinger-like equations, transport theory and radiative transfer, and integral equations. Assumes knowledge of advanced calculus, functional analysis, and Laplace transforms. Mathematical derivations are stressed, but rigorous proofs are omitted. The chapters conclude with problems and references.

Bello, Ignacio. CONTEMPORARY BUSINESS MATHEMATICS. Philadelphia: W.B. Saunders, 1975. xv, 572 p. Appendixes, pp. 481-518. Answers: Exercises and Self-tests, pp. 519-65. Index, pp. 567-72.

A text for a first course in mathematics for business students in an

associate degree program in a junior college or university. Part 1
discusses the essentials of business mathematics and includes a review
of arithmetic operations, fractions, and decimals, as well as the
applications of these skills. Part 2 introduces the topics of ratio,
proportions, and percent with applications in retailing, accounting,
and related fields. Part 3 is concerned with simple and compound
interest. Part 4 deals with the mathematics used by investors in
the areas of stocks, bonds, and insurance. Part 5 consists of an
introduction to statistics and probability with emphasis on decision
making. Each section contains many worked examples, progress
tests, and exercises. The chapters conclude with self-tests, and
the appendixes take up the metric system and the use of hand
calculators, and interest and other tables.

Benavie, Arthur. MATHEMATICAL TECHNIQUES FOR ECONOMIC ANALYSIS.
Prentice-Hall Series in Mathematical Economics. Englewood Cliffs, N.J.:
Prentice-Hall, 1972. xii, 287 p. Index, pp. 282-87.

A text for a course in mathematical economics and economic theory
for students who are familiar with differential and integral calculus, matrix algebra, and differential and difference equations.
With this background, the book is self-contained, and the few
theorems that are not proved are accessible elsewhere. Chapter 1
deals with static economic analysis. Chapter 2 is concerned with
constrained and unconstrained optimization, the Lagrange-Kuhn
theorem, and applications to comparative static theorems. Chapter
3 presents two solution techniques for linear differential and difference equation systems of any order. Chapter 4 states and proves
the Liapunov theorems pertaining to the stability of linear and nonlinear systems of differential and difference equation. Chapter 5
consists of an introduction to optimal control and contains a proof
of the maximum principle of Pontryagin, and its economic interpretation. The development of each economic technique is followed by exercises and examples.

Ben-Israel, Adi, and Greville, Thomas N.E. GENERALIZED INVERSES: THEORY
AND APPLICATIONS. Pure and Applied Mathematics Series. New York: Wiley-
Interscience, 1974. xi, 395 p. Bibliography, pp. 359-81. Glossary of sym-
bols, pp. 383-85. Author Index, pp. 387-91. Subject Index, pp. 393-95.

A survey of generalized inverses from a unified point of view, illustrating the theory with applications taken from many areas. The
book consists of an introduction and eight chapters, seven of which
treat generalized inverses of finite matrices, while the eighth introduces generalized inverses of operators between Hilbert spaces.
Numerical methods are also treated. Assumes knowledge of linear
algebra. Included are more than 450 exercises at different levels
of difficulty, many of which contain detailed solutions, making
this book suitable for either self-study or classroom use.

12

Berge, Claude. THE THEORY OF GRAPHS AND ITS APPLICATION. Translated by A. Doig. New York: Wiley, 1962. vii, 247 p. Appendix 1: Note on the General Theory of Games, pp. 220-26. Appendix 2: Note on Transport Problems, pp. 227-37. Bibliography, pp. 238-44. List of Symbols, p. 245. Index of Terms Used, pp. 246-47.

> Originally published in French (Paris: Dunod, 1958), this book has extensive applications of graph theory in many diverse fields. A knowledge of matrices and determinants is assumed in only three chapters. Topics include descendance relations, the ordinal function on an infinite graph, the fundamental numbers of the theory of graphs, application to information theory, kernels of graphs (application to game theory), games on a graph, the problem of the shortest route, transport networks, the theorem of the demi-degrees, the assignment problem, centers of a graph, incidence matrices, trees and aborescences, Euler's problem, semifactors, the connectivity of a graph, and planar graphs. Assumes knowledge of elementary set theory.

Berkeley, Edmund C. A GUIDE TO MATHEMATICS FOR THE INTELLIGENT NONMATHEMATICIAN. New York: Simon & Schuster, 1966. 352 p. Appendixes: 1, A Short Introduction to Algebra; 2, Studying by Yourself; 3, A Short Bibliography, pp. 311-39. Index, pp. 340-52.

> A guide to mathematics expressed in terms of ideas that are familiar to the layman. It attempts to demonstrate three main propositions: everyone is a mathematician without being fully aware of it; important mathematical ideas; and the process of understanding. It is divided into the following four parts: a way in; roughly and approximately; a dozen powerful ideas made easy to understand; and mathematics and the environment. The derivative and the integral are among the calculus concepts which the author explains. This book is useful to anyone desiring an intuitive understanding of basic concepts and ideas of mathematics.

Berston, Hyman Maxwell, and Fisher, Paul. COLLEGIATE BUSINESS MATHEMATICS. Homewood, Ill.: Richard D. Irwin, 1973. xii, 328 p. Answers to Odd-numbered Problems, pp. 299-313. Index, pp. 315-28.

> One third of this text deals with a review of basic arithmetic, and two thirds covers basic applications to business problems. Sixty-seven two-page assignments are included at the ends of the chapters, with ample space in the book for solutions.

Billingsley, Patrick. ERGODIC THEORY AND INFORMATION. Foreword by D. Kendall. Tracts on Probability Theory and Statistics. New York: Wiley, 1965. xiii, 193 p. Bibliography, pp. 181-85. Index of Examples, pp. 187-89. General Index, pp. 191-93.

> A brief survey of ergodic theory and information theory designed to follow a brief course in measure-theoretic probability. Chap-

ter 1 deals with the ergodic theorem, and chapter 2 treats the Kolmogorov-Sinai application of Shannon's theory. Chapters 4 and 5 apply ergodic theory to information theory and coding. Chapter 3 reviews conditional probability and expectation.

Bingham, Robert C. ECONOMICS, MATHEMATICALLY SPEAKING. New York: McGraw-Hill Book Co., 1972. xii, 424 p. Answers to Exercises, pp. 411-24. No index.

Mainly a supplement to a one-year course in economic principles covering both microeconomics and macroeconomics. It translates the economic theory found in a principles text into the language of mathematics utilizing only college algebra. No mathematics is included, and subjects which do not lend themselves to mathematical treatment are omitted. There are forty-six sets of exercises involving both mathematics and numerical applications.

Birkhoff, Garrett, and MacLane, Saunders. A SURVEY OF MODERN ALGEBRA. 3d ed., rev. New York: Macmillan, 1965. x, 437 p. Bibliography, pp. 423-25. List of Special Symbols, pp. 426-28. Index, pp. 429-37.

A text for a course in modern (abstract) algebra requiring a prerequisite of high school algebra, and a reference for the applied fields. Topics include polynomials, real and complex numbers, groups, rings and ideals, vector spaces, canonical forms, Boolean algebras and lattices, algebraic number fields, and Galios theory. Sections of the book conclude with exercises.

Bishir, John W., and Drewes, Donald W. MATHEMATICS IN THE BEHAVIORAL AND SOCIAL SCIENCES. New York: Harcourt, Brace and World, 1970. xiii, 714 p. Answers to Selected Exercises, pp. 691-706. Author Index, pp. 707-8. Subject Index, pp. 709-14.

A text designed to introduce the mathematical models most frequently used in the behavioral and social sciences. Part 1 considers finite mathematics; matrices and linear algebra are discussed in part 2; part 3 introduces differential and integral calculus, and difference equations; and part 4 deals with probability theory and its application to the development of random models in the social and behavioral sciences. Over 1,500 examples and problems are included, and the chapters conclude with reading lists. Assumes knowledge of college algebra.

Black, John M.A., and Bradley, James F. ESSENTIAL MATHEMATICS FOR ECONOMISTS. New York: Wiley, 1973. x, 268 p. Appendix 1: The Derivative of $y=x^n$, pp. 253-54. Appendix 2: Euler's Theorem, p. 255. Appendix 3: Trigonometric functions, pp. 256-59. Appendix 4: De Moivre's Theorem, pp. 260-61. Appendix 5: The Stability of Second-order Difference Equations, pp. 262-64. Index, pp. 265-68.

A text for a one-semester undergraduate course in mathematical economics for students who have a good high school algebra background. The chapter titles are: "Linear Functions," "Slope and Elasticity," "Simultaneous Linear Equations," "Curvilinear Functions," "Differentiation," "Maxima and Minima," "Partial Differentiation," "Maxima and Minima Subject to Constraints," "Integration," "Difference Equations," "Exponential Functions," "Second-order Difference Equations," "Complex Numbers," "Difference Equations in Two Variables," and "Second-order Differential Equations." The mathematics is used to illustrate economic concepts, and the examples are relegated to exercises for which detailed solutions are provided. No theorems are proved.

Bochner, Salomon. THE ROLE OF MATHEMATICS IN THE RISE OF SCIENCE. Princeton, N.J.: Princeton University Press, 1966. x, 386 p. Index, pp. 373-86.

Part 1 consists of six essays previously published on the role of mathematics in a variety of fields and settings. Part 2 consists of biographical sketches.

Boot, John C.G. MATHEMATICAL REASONING IN ECONOMICS AND MANAGEMENT SCIENCE: TWELVE TOPICS. Prentice-Hall International Series in Management. Englewood Cliffs, N.J.: Prentice-Hall, 1967. xii, 178 p. Index, pp. 175-78.

A selection of twelve topics which illustrates the use of mathematics in the formulation and/or solution of problems. The topics fall into three groups. The first group consists of mathematical topics such as difference equations, Markov chains, characteristic-value problems, and probability, with applications. The second group consists of such topics as decision theory, game theory, strategies, and dynamic programming, with emphasis on the philosophy of approach. The third group consists of sensitivity analysis, inventory models, input-output models, and growth models. Assumes knowledge of calculus and matrix algebra. The chapters conclude with postscripts consisting of only a reference, elaboration of the mathematics, or a more detailed discussion of topics only tangentially considered in the text.

Bowen, Earl K. MATHEMATICS WITH APPLICATIONS IN MANAGEMENT AND ECONOMICS. 3d ed., rev. Irwin Series in Quantitative Analysis for Business. Homewood, Ill.: Richard D. Irwin, 1972. xv, 687 p. Appendix 1: Elements of Algebra, pp. 580-627. Appendix 2: Formulas, Equations, and Graphs, pp. 628-70. Tables, pp. 673-82. Index, pp. 683-87.

A text for a one-semester finite mathematics course and a one-semester calculus course for students in economics and business who have a good background in high school algebra. The basic explanation-example-exercise approach makes this book semiprogrammed and suitable for self-study. Chapters 1 to 6 cover linear systems and matrices in which linear programming forms the main application. Chapter 7 takes up logarithms, and chapter 8 discusses the mathematics of finance. Chapters 9 to 15 are devoted to differential and integral calculus and probability. The sections conclude with problems with answers, and the chapters conclude with review problems. Almost every page contains one or more exercises.

Bowley, Arthur Lyon. THE MATHEMATICAL GROUNDWORK OF ECONOMICS: AN INTRODUCTORY TREATISE. Reprints of Economic Classics. 1924. Reprint. New York: Augustus M. Kelley, 1960. viii, 98 p. Appendix: Summary of the Mathematical Ideas and Formulae Used, pp. 78-96. Index, pp. 97-98.

Presents the main mathematical methods used by Cournot, Jevons, Pareto, Edgeworth, Marshall, Pigou, and Johnson. It is one of the first books to appear in mathematical economics and treats the following topics: simple and multiple exchange, production, supply and demand of the factors of production, general equations of supply and demand in a stationary population with applications, and surplus value, rent, and taxation.

Bowman, Frank. AN INTRODUCTION TO DETERMINANTS AND MATRICES. Applied Mathematics Series. London: English Universities Press, 1962. ix, 163 p. Answers to Exercises, pp. 156-59. Brief bibliography, p. 160. Index, pp. 161-63.

An introduction to the more advanced books, theoretical and numerical, on the algebra of matrices and their applications. Topics include determinants, rank, quadratic forms, matrices, differential equations, and approximations. Sections of the book conclude with exercises.

Brand, Heinz W. THE FECUNDITY OF MATHEMATICAL METHODS IN ECONOMIC THEORY. Preface by H. Rittershausen. Translated by E. Holmstrom. Boston: Reidel, 1961. viii, 56 p. Bibliography, pp. 50-53. Notes, pp. 54-56.

First published in German, this book is a nonmathematical discussion of the role of mathematics in economic theory. The author distinguishes between two types of methods: mathematical presentation and mathematical operations. An example of the former type is the consumption function which expresses a relationship between two economic variables, income and consumption; an example of the latter type is the influence of autonomous investment on national income and on the balance of payments between two countries. Other problems, such as the significance of psychophysical laws on the exactness of mathematical methods, are discussed.

Brent, Richard P. ALGORITHMS FOR MINIMIZATION WITHOUT DERIVATIVES. Prentice-Hall Series in Automatic Computation. Englewood Cliffs, N.J.: Prentice-Hall, 1973. xii, 195 p. Bibliography, pp. 169-85. Appendix: FORTRAN Program Subroutines, pp. 187-91. Index, pp. 193-95.

A monograph which describes improvements in existing algorithms for minimization without derivatives and extends the mathematical theory underlying these algorithms. Some new algorithms for appoximating local and global minima are also given. Four of the seven chapters contain ALGO procedures for use on IBM 360/67 and 360/91 computers, and the appendix contains FORTRAN subroutines for the one-dimensional zero-finding and minimization algorithms. The presentation is primarily descriptive and oriented towards illustration of the algorithms. This book is useful for graduate students and research workers in numerical analysis, computer science, and operations research.

Bressler, Barry. A UNIFIED INTRODUCTION TO MATHEMATICAL ECONOMICS. New York: Harper and Row, 1975. xviii, 667 p. References for Further Study, pp. 621-23. Tables, pp. 624-34. Answers to Selected Exercises, pp. 635-57. Index, pp. 659-67.

A text for an undergraduate course in mathematical economics for students who have only a high school algebra background. Mathematics and economics are integrated throughout, with the necessary mathematics introduced as needed. Topics include set theory, the number system, matrix algebra and systems of equations, differential and integral calculus with applications to maxima and minima problems in economics, functions of several variables and partial derivatives, the theory of consumer choice, the theory of the firm and its decisions, monopoly and oligopoly equilibrium, income and employment theory, integration, and difference and differential equations. The omitted topics include linear programming, input-output analysis, and trigonometric functions. Proofs of mathematical theorems are included wherever they are instructive for future work, and the more difficult proofs and formulations are relegated to appendixes at the end of several chapters. The sections of the book conclude with exercises.

Burford, Roger L. INTRODUCTION TO FINITE PROBABILITY. Merrill's Mathematics and Quantitative Methods Series. Columbus: Charles E. Merrill, 1967. xiv, 145 p. Appendix, pp. 135-42. Index, pp. 143-45.

An introduction to the theory of probability in finite sample spaces for students not majoring in mathematics or the sciences. One or more examples follow each theoretical argument and derivation. Assumes knowledge of symbolic logic, Boolean algebra, and set theory. The applications are in inventory control, waiting lines, equipment replacement, quality control, and statistics. The chapters conclude with exercises and references.

Burington, Richard S. HANDBOOK OF MATHEMATICAL TABLES AND FORMULAS. 5th ed. New York: McGraw-Hill Book Co., 1973. xi, 500 p. References, pp. 460-61. Glossary of Symbols, pp. 462-65. Index of Numerical Tables, pp. 466-68. Subject Index, pp. 469-500.

A reference for those in academic, professional, scientific, engineering, and business fields in which mathematical reasoning, processes, or computations are required. It contains information of a more traditional nature as well as definitions, theorems, formulas, and tables needed for contemporary applications. The first part contains sections on the following topics: sets, logic, number systems, elementary statistics, tables of integrals and derivatives, annuities, linear algebra, linear vector spaces, numerical analysis, solutions of linear and nonlinear systems of equations, approximate differentiation and integration, calculus of finite differences, partial differential equations, Legendre and Bessel functions, Fourier analysis, Laplace transforms, and complex variables. The second part of the book contains the following tables: five-place logarithmic and trigonometric tables; tables of natural logarithms, and exponential and hyperbolic functions; tables of squares, cubes, square roots and cube roots; reciprocals and other numerical quantities; normal curve tables; and tables of chi-square and Poisson distributions. This book is designed to serve as a companion to HANDBOOK OF PROBABILITY AND STATISTICS WITH TABLES, by R.S. Burington and D.C. May (New York: McGraw-Hill Book Co., 1970).

Busacker, Robert C., and Saaty, Thomas L. FINITE GRAPHS AND NETWORKS: AN INTRODUCTION WITH APPLICATIONS. International Series in Pure and Applied Mathematics. New York: McGraw-Hill Book Co., 1965. xiv, 294 p. Glossary, pp. 267-70. Answers to Exercises, pp. 271-83. Index, pp. 285-94.

A text for a course in graph theory for undergraduates and graduates in mathematics, physical and social sciences, engineering, economics, and operations research who are familiar with the basic concepts of set theory, matrices, and vector spaces. It is divided into the following two parts: basic theory and applications. The theory part includes concepts of undirected and directed graphs, partitioning and measuring distances on graphs, planar graphs, and geometric and algebraic considerations, including the role of matrices in the analysis of graphs. The applications are in the areas of combinatorial problems, puzzles and games, matchings, physical sciences, engineering, and network flows. A brief introduction to stochastic network flows with applications to queueing theory is also given. Although proofs are stressed, the concepts are illustrated by graphs and examples. Exercises are dispersed throughout the chapters, and the chapters conclude with references.

Bush, Grace, and Young, John E. FOUNDATIONS OF MATHEMATICS: WITH APPLICATIONS TO THE SOCIAL AND MANAGEMENT SCIENCES. 2d ed., rev. New York: McGraw-Hill Book Co., 1973. xiv, 536 p. Appendix A: Logarithms, pp. 455-63. Appendix B: Tables, pp. 465-92. Answers to Selected Exercises, pp. 493-529. Index, pp. 521-36.

A text for liberal arts students who have a background in at least two years of high school mathematics. Topics include sets, relations, functions, probability, linear systems and matrices, linear programming and the simplex method, differential and integral calculus, and basic statistics. Numerous examples and applications are provided, and exercises follow the sections.

Bushaw, D.W., and Clower, R.W. INTRODUCTION TO MATHEMATICAL ECONOMICS. The Irwin Series in Economics. Homewood, Ill.: Richard D. Irwin, 1957. xii, 345 p. List of Symbols, p. 335. Index, pp. 336-45.

An introduction to mathematical methods of economic analysis for

students with a limited mathematics background. Part I (chapters
1 to 7) applies mathematics to economic analysis, and part II
(chapters 8 to 12) consists of a review of the necessary mathematics. The sections conclude with exercises, and the chapters conclude with suggestions for further reading.

Campbell, Hugh G. AN INTRODUCTION TO MATRICES, VECTORS, AND
LINEAR PROGRAMMING. The Appleton-Century Mathematics Series. New
York: Appleton-Century-Crofts, 1965. xiv, 244 p. Answers to Odd-numbered
Exercises, pp. 217-38. Index, pp. 239-44.

A text for students of social science, engineering, or business administration who have knowledge of college algebra and geometry.
It is more elementary than most books on this subject, and it contains a chapter on mathematical systems and one on linear programming. Other topics include special matrices, determinants, the
inverse matrix, matrix transformations, systems of linear equations,
vector spaces, linear transformations, and convex sets. The sections
conclude with exercises, and the chapters conclude with new vocabularies.

_____. LINEAR ALGEBRA WITH APPLICATIONS INCLUDING LINEAR PROGRAM-
MING. The Appleton-Century Mathematics Series. New York: Appleton-Century-
Crofts, 1971. xiii, 441 p. References, pp. 395-96. Appendixes, pp. A1-A4.
Answers to Selected Odd-numbered Exercises, pp. A9-A37. Index, pp. A39-A45.

A text for a linear algebra course containing numerous examples
to motivate nonmathematics majors. Most sections of the chapters
conclude with applications and exercises, and the chapters conclude with new vocabularies and special projects.

Cangelosi, Vincent E. COMPOUND STATEMENTS AND MATHEMATICAL LOGIC.
Columbus: Charles E. Merrill, 1967. xii, 114 p. Index, pp. 113-14.

A text for a basic mathematics course for students planning careers
in fields other than mathematics, especially business and economics.
It introduces the student to the process of deductive logic and syl-
logistic reasoning, the use of symbolic notation, and the abstract
nature of mathematical science. The propositions are presented,
explained, and illustrated, and the problems at the end of the chapters are designed to reinforce the material.

Casson, Mark. INTRODUCTION TO MATHEMATICAL ECONOMICS. New
York: Crane, Russak, 1973. xi, 402 p. Bibliography. Index.

A text for an undergraduate course in mathematical economics covering the following topics: calculus, including maxima and minima,
differentiation, constrained maximization and minimization, integration, and applications; matrix algebra, including simultaneous linear
equations, the matrix multiplier, characteristic values, nonnegative
matrices, and symmetric matrices; and differential and difference
equations with applications.

Casti, John, and Kalaba, Robert E. IMBEDDING METHODS IN APPLIED MATHEMATICS. Applied Mathematics and Computation Series. Reading, Mass.: Addison-Wesley, 1973. xiv, 306 p. Appendix A: General Computer Programs for Integral Equations with Displacement Kernels, pp. 285-303. Index, pp. 305-6.

> This book shows how the theory of invariant imbedding may be used to recast broad classes of boundary value problems as initial value problems which can be solved efficiently by computer programs. Topics include: linear and nonlinear two-point boundary value problems; Fredholm integral equations; the use of Euler equations; dynamic programming, and the maximum principle in solving constrained and unconstrained optimization problems; and applications of invariant embedding to problems arising in radiative transfer, analytical mechanics, optimal filtering theory, and nonlocal wave interaction. This book is useful to anyone interested in developing efficent solution techniques for optimal control problems. Assumes knowledge of advanced calculus and elementary differential equations. The chapters conclude with notes and references.

Chiang, Alpha C. FUNDAMENTAL METHODS OF MATHEMATICAL ECONOMICS. 2d ed., rev. New York: McGraw-Hill Book Co., 1974. xv, 784 p. The Greek Alphabet, p. 772. Short Reading List, pp. 773-76. Index, pp. 777-84.

> A text for a one-year course in graduate and undergraduate mathemathical economics. It stresses equally mathematics and applications to static, comparative static, and dynamic economics. The book includes a detailed development of classical constrained and unconstrained optimization techniques, linear and nonlinear programming, and game theory. Topics that are new in the second edition include the implicit function theorem, characteristic roots of a matrix, quasiconcave functions, L'Hospital's rule, and integration by parts. Assumes knowledge of college algebra. The sections conclude with exercises.

Childress, Robert L. CALCULUS FOR BUSINESS AND ECONOMICS. Englewood Cliffs, N.J.: Prentice-Hall, 1972. xii, 267 p. Appendix A: Establishing Functions, pp. 217-25. Appendix B: Logarithms: Laws of Exponents, pp. 226-32. Tables, pp. 234-53. Selected Answers to Odd-numbered Questions, pp. 254-67.

> A text for a one-semester calculus course for students in economics, business, and finance. The presentation relies on the explanation-example approach, and about one-half of the book is devoted to concepts and methodology and the remaining half to applications. Topics include differential and integral calculus, optimization, single-variable models, multivariate and transcendental functions, and multivariate and exponential business models. Assumes knowledge of college algebra. The chapters conclude with problems.

_____ . MATHEMATICS FOR MANAGERIAL DECISIONS. Englewood Cliffs, N.J.: Prentice-Hall, 1974. xiv, 689 p. Appendix A: Establishing Functions, pp. 616-24. Appendix B: Logarithms: Laws of Exponents, pp. 625-31. Tables, pp. 632-64. Selected Answers to Odd-numbered Questions, pp. 665-80. Index, pp. 681-89.

A text for a one-semester course in mathematics for students of business and economics who possess a background in high school algebra. Emphasis is on introducing the quantitative tools of sets, matrices, linear programming, calculus, and probability in a manner that is easy to grasp for the nonmathematically inclined student, and providing numerous examples of the applicability of quantitative techniques in the administration of an enterprise. Topics include systems of equations; matrix representation of systems of equations with applications to input-output analysis; simplex method for solving linear programs; duality and sensitivity analysis; transportation and assignment problems; optimization using calculus; multivariate and exponential business models, including inventory models and the method of least squares; Bayes' theorem; random variables and the commonly used discrete and continous probability distributions; and growth rate functions. Over 500 problems are presented at the ends of the chapters. Some of the chapters conclude with suggested references.

_____ . SETS, MATRICES, AND LINEAR PROGRAMMING. Englewood Cliffs, N.J.: Prentice-Hall, 1974. xii, 356 p. Selected Answers to Odd-numbered Questions, pp. 341-51. Index, pp. 353-56.

A text for a one-semester introductory course in linear programming for undergraduates and a supplemental text for courses in operations research and quantitative methods. Topics include sets, functions, matrix algebra, linear programming and the simplex method, duality and sensitivity analysis, the transportation and assignment problems, and integer programming. The emphasis is on examples of the applicability of quantitative techniques in the administration of an enterprise. Assumes knowledge of only high school algebra. The chapters conclude with problems and selected references.

Cissell, Robert, and Bruggeman, Thomas J. MATHEMATICS FOR BUSINESS AND ECONOMICS. Boston: Houghton Mifflin, 1962. ix, 229 p. Tables, pp. 195-225. Index, pp. 226-29.

A text for a one-semester course in business mathematics for students with only a high school algebra background. Topics include the following functions: linear, rational, exponential and logarithmic, quadratic and higher degree polynomials, systems of linear functions, functions of more than one variable, step functions and periodic functions. This book provides some of the basic mathematical tools for business and economic applications. The sections conclude with exercises.

Cissell, Robert, and Cissell, Helen. MATHEMATICS OF FINANCE. 4th ed., rev. Boston: Houghton Mifflin, 1973. x, 549 p. Appendix: Computational Methods, pp. 415-39. Index, pp. 441-46. Mathematical Tables, pp. 1-103.

A text for a one-semester course in the mathematics of finance for students with only a high school algebra background. Topics include simple and compound interest, annuities, amortization and sinking funds, capital budgeting, life insurance, stocks, and continuous compounding. Practical problems are used to illustrate the application of the formulas and tables. The sections conclude with exercises, with answers given at the end of each exercise. The appendix is devoted to computational methods for rapid arithmetic, logarithms, and computer programs.

Coughlin, Raymond F. ELEMENTARY APPLIED CALCULUS: A SHORT COURSE. Boston: Allyn and Bacon, 1974. viii, 280 p. Appendix A: Trigonometric Functions, pp. 225-46. Appendix B: Exponential and Logarithmic Functions, pp. 247-50. Appendix C: Basic Integration Formulas, pp. 251-53. Appendix D: Tables of Natural Logarithms and Exponential Function, pp. 254-61. Appendix E: Mathematical Induction, pp. 262-63. Solutions to Selected Exercises, pp. 264-76. Selected Bibliography, pp. 277-78. Index, pp. 279-80.

A text for a one- or two-semester course in calculus for students in the liberal arts or business who have a high school algebra background. It includes actual mathematical models cited in texts and papers on various subjects such as biology, psychology, ecology, and sociology. Proofs of theorems are included if the mathematical result is not intuitively obvious and if the details of the proof are within the reach of the student. Topics include set theory; limits and the derivative; the integral; applications of the derivative and the integral; differential equations; functions of several variables; and Lagrange multipliers. Sections of the book conclude with exercises.

Coyle, R.G. MATHEMATICS FOR BUSINESS DECISIONS. London: Thomas Nelson, 1971. vii, 309 p. Appendix, pp. 257-62. Solutions to Problems, pp. 263-305. Index, pp. 307-9.

A self-study text for students of business and economics and for practicing managers who have little formal mathematics background. The approach is by way of example and illustration, rather than by theorem and proofs, and each chapter includes examples with fully worked solutions. Topics include advanced optimization by calculus, linear programming, basic statistics, regression analysis, decision theory, and Markov processes. The last chapter contains three extensive real life case problems in manufacturing and mining. The appendix consists of statistical tables, and the chapters conclude with exercises.

Crowdis, David G.; Shelley, Susanne M.; and Wheeler, Brandon W. CALCULUS FOR BUSINESS, BIOLOGY, AND THE SOCIAL SCIENCES. Beverly Hills, Calif.: Benziger, Bruce, & Glencoe, 1972. xii, 547 p. Appendix: Tables, pp. 475-82. Answers to Odd-numbered Exercises, pp. 485-542. Index, pp. 545-47.

A text for a course in calculus for students who have four years of high school mathematics or its college equivalent. The approach is intuitive, rather than formal, and the first six chapters parallel the introductory calculus course approved by the Committee on Undergraduate Program in Mathematics, while the remaining three chapters introduce concepts of multiple-variable calculus and infinite series. Wherever possible, applications are taken from the social, management, and biological sciences rather than from engineering and physics. Exercises follow the sections, and the chapters also conclude with review exercises.

Cullen, Charles G. MATRICES AND LINEAR TRANSFORMATIONS. Addison-Wesley Series in Mathematics. Reading, Mass.: Addison-Wesley, 1966. 227 p. Appendix: Greek Alphabet, p. 206. Answers to Selected Exercises, pp. 209-22. Glossary of Mathematical Symbols, pp. 223-24. Index, pp. 225-27.

A text for a one-semester course in linear algebra and matrix theory for a mixed group of students, including students from the social sciences, who have knowledge of calculus and analytic geometry. Topics include vector spaces, determinants, linear transformations, polynomial matrices, matrix analysis, and numerical methods. The sections of the book conclude with exercises.

Cuming, Henry George, and Anson, Cyril Joseph. MATHEMATICS AND STATISTICS FOR TECHNOLOGISTS. New York: Chemical, 1966. vii, 490 p. Answers to Exercises, pp. 467-86. Index, pp. 489-90.

A text for a one-semester course in mathematics and statistics for a varied audience in the applied disciplines. Topics include a review of basic algebra, geometry, and trigonometry, differentiation and integration, ordinary differential equations, basic statistics, control charts, analysis of variance, and regression. Many chapters are virtually self-contained, and the large number of worked examples are designed to assist the reader who is attempting to teach himself. The sections of the book conclude with exercises.

Curry, Othel Jackson, and Pearson, John E. BASIC MATHEMATICS FOR BUSINESS ANALYSIS. Rev. ed. Homewood, Ill.: Richard D. Irwin, 1970. viii, 290 p.

A text for a course in business mathematics for students who have a high school algebra background, although college algebra is helpful for some sections. Topics include algebraic expression and operators, problems in marketing, wages and taxes, interest, borrowing and lending, financial statement analysis, basic statistics, and data processing and computer languages. The emphasis is on understanding mathematical concepts, rather than on applications to practical problems.

Curtis, Alan R. PRACTICAL MATH FOR BUSINESS. Instructor's Edition. Boston: Houghton Mifflin, 1973. xii, 397 p. Transparency Masters, pp. 299-318. Examins, pp. 319-85. Answers to Examins, pp. 385-97. No Index.

A text for a course in business mathematics for junior colleges.

It consists of thirteen chapters, with each chapter beginning with
a summary of key points followed by a short minicase with illustrations. Each chapter contains several sections with one or more
projects consisting of problems and applications. Topics include
arithmetic and basic algebraic operations, checking accounts and
payrolls, interest with applications, balance sheet and income
statements, retailing, mathematics, inventory, depreciation, and
stocks and bonds.

Daniel, James W. THE APPROXIMATE MINIMIZATION OF FUNCTIONALS.
Prentice-Hall Series in Automatic Computation. Englewood Cliffs, N.J.:
Prentice-Hall, 1971. xi, 228 p. Epilogue, pp. 211-12. References,
pp. 213-23. Index, pp. 225-28.

Presents the problems encountered in the approximate minimization
of functionals. The main topics include the existence of solutions,
the convergence of approximate solutions of discretized variational
problems in general and in specific cases, the theory of gradient
methods of minimization in general spaces, and practical computational methods of minimization in R^n. The primary emphasis is
on unconstrained problems, and while detailed computational algorithms or computer programs are not presented, references to
sources of computer programs are provided. The presentation is
rigorous with proofs included, and knowledge of the calculus of
variations and Banach spaces is necessary. Approximately one
exercise per page is included.

Daus, Paul H., and Whyburn, William M. ALGEBRA WITH APPLICATIONS
TO BUSINESS AND ECONOMICS. Addison-Wesley Series in Mathematics.
Reading, Mass.: Addison-Wesley, 1961. xi, 354 p. Appendix I: Equation
of a Plane, pp. 301-4. Appendix II: Division Process for the Square Root
of N, pp. 305-6. Appendix III: Tables, pp. 307-27. Answers to Problems,
pp. 329-47. Index, pp. 349-54.

A text for a one-semester course in college algebra for students
with minimal mathematics preparation. Topics include the real
number system; addition, subtraction, and multiplication; division
and fractions; linear and quadratic equations; laws of algebra;
matrix algebra and linear programming; exponents, radicals, and
logarithms; and mathematical induction, progressions, and the binomial theorem. Numerous worked examples are provided, and
the sections conclude with exercises.

Dean, Burton Victor; Sasieni, Maurice W.; and Gupta, Shiv. K. MATHE-
MATICS FOR MODERN MANAGEMENT. New York: Wiley, 1963. xiii,
442 p. Appendix, pp. 429-38. Index, pp. 439-42.

A text for a one-year course in mathematics for business and economics students. Topics include deterministic systems, including
differential calculus, optimization, Taylor's theorem, with appli-

cations to production, marketing, and equipment replacement problems; stochastic models, including integration, differential equations, and probability theory; linear models, including linear programming and the transportation and assignment problems; and the mathematics of finance. Examples are used freely to illustrate the concepts, and the sections and chapters conclude with exercises. The appendix is devoted to a further treatment of differential equations.

Dem'yanov, Vladimir F., and Malozemov, V.N. INTRODUCTION TO MINIMAX. Modern Analytic and Computational Methods in Science and Mathematics, vol. 32. New York: Wiley, 1974. vii, 307 p. Appendix I: Algebraic Interpretation, pp. 242–47. Appendix II: Convex Sets and Convex Functions, pp. 248–63. Appendix III: Continuous and Continuously Differentiable Functions, pp. 264–75. Appendix IV: Determination of the Point Nearest the Origin on a Polyhedron: Iterative Methods, pp. 276–95. Supplement: On Mandel'shtam's Problem, pp. 296–99. Notes, pp. 300–302. Bibliography, pp. 303–6. Subject Index, p. 307.

> Originally published in Russian in 1972 and translated by D. Louvish, this book treats several types of minimax problems. Chapters 1 and 2 are devoted to the problem of the best approximation of functions by algebraic polynomials based on Chebyshev interpolation. Chapter 3 discusses the linear minimax problem without constraints (discrete case), and chapters 4 and 5 treat the same problem with convex constraints. The transition from the discrete to the continuous case is made in chapter 6. The appendixes contain the requisite auxiliary material.

Diamond, Jay, and Pintel, Gerald. MATHEMATICS OF BUSINESS. Englewood Cliffs, N.J.: Prentice-Hall, 1970. viii, 375 p. Answers to Odd-numbered Problems, pp. 359–69. Index, pp. 371–75.

> A text for a one-semester course in business mathematics for students with minimal mathematics background. Topics include fractions and decimals, percentages, consumer economics, retailing and marketing mathematics, accounting, investment, data processing, and management. Detailed examples are provided, and the sections conclude with problems.

Dickson, Harold. VARIABLE, FUNCTION, DERIVATIVE: A SEMANTIC STUDY IN MATHEMATICS AND ECONOMICS. Atlantic Highlands, N.J.: Humanities Press, 1967. 176 p. References, pp. 167–68. Subject Index, pp. 169–74. Symbols, pp. 175–76. Author Index, p. 176.

> This book is concerned with semantic questions on the use of mathematics in economics, and it is useful to a varied audience, including economists, philosophers, and mathematicians. Topics include sets, variables, functions, derivatives, and differentials. Applications to economics are provided.

Dinwiddy, Caroline. ELEMENTARY MATHEMATICS FOR ECONOMISTS. Foreword by P.W. Bell. Series of Undergraduate Teaching Works in Economics, vol. 1. New York: Oxford University Press, 1967. xvi, 243 p. Answers to Exercises, pp. 218-32. Appendix: Tables, pp. 1-11.

A text, for an undergraduate course in mathematics for economists, which is divided into the following four parts: elementary algebra, introduction to calculus, arithmetic of growth, and some extensions of elementary mathematics, including matrix algebra and difference equations. Many elementary applications to economics are provided, and the chapters conclude with problems.

Dodes, Irving Allen. FINITE MATHEMATICS: A LIBERAL ARTS APPROACH. New York: McGraw-Hill, 1970. xi, 403 p. Appendix: Tables, pp. 373-80. Answers to Odd-numbered Problems, pp. 380-97. Index, pp. 397-403.

A text for a course in finite mathematics for liberal arts students who have a good background in high school algebra. A minimum of mathematics is presented, and emphasis is placed on applications in the areas of linear programming, games, Markov chains, and logarithms. Also included are one chapter each on the slide rule, the mathematics of management, and an introduction to computing.

Dorf, Richard C. MATRIX ALGEBRA: A PROGRAMMED INTRODUCTION. New York: Wiley, 1969. viii, 260 p. Answers to Exercises, pp. 236-57. Index, pp. 259-60.

A self-study designed to develop a skill in utilizing the algebra of matrices. Topics include matrix operations, linear equations and determinants, the rank, trace, and adjoint of a matrix, the inverse and charateristic equation of a matrix, and matrix transformations. The book consists mainly of worked problems, and the chapters conclude with summaries and exercises. It is useful to students in many fields, including economics. Assumes knowledge of high school algebra.

Dowsett, Wilfred Thomas. ELEMENTARY MATHEMATICS IN ECONOMICS. London: Sir Isaac Pitman, 1959. xxi, 250 p. Index, pp. 248-50.

A text for an undergraduate course in mathematical economics which is divided into the following three parts: a diagrammatic representation, a section mainly concerning the mathematics of equilibrium, and a section mainly concerning change through time. Topics include graphs and functions, demand and cost curves, competitive markets, matrices and determinants, simple functions and models of economic systems, difference equations, and differentiation (including partial differentiation). Numerous examples in economics are given, and the chapters conclude with references.

Draper, Jean E., and Klingham, Jane S. MATHEMATICAL ANALYSIS: BUSI-
NESS AND ECONOMIC APPLICATIONS. 2d ed. Revised by J.D. Weber.
New York: Harper and Row, 1972. x, 691 p. Algebra Review Problems,
pp. 669-74. Selected References, pp. 675-79. Index, pp. 681-91.

A text for a one-year course in quantitative techniques for students
in business and economics. The mathematical discussion of each
topic is followed by a discussion of its application in economics
and business, and a unique feature is the large number of examples
worked in detail. The eight chapter titles are: graphical representation; differential calculus: functions of more than one variable;
functions of one variable; integral calculus; differential equations;
difference equations; vectors and matrices; and applications of
matrix algebra. The sections conclude with problems.

Dyckman, Thomas R., and Thomas, L. Joseph. ALGEBRA AND CALCULUS
FOR BUSINESS. Englewood Cliffs, N.J.: Prentice-Hall, 1974. xii, 450 p.
Appendix, pp. 419-24. Answers, pp. 425-43. Index, pp. 445-50.

A text for a one-semester course in mathematics for students of
business administration. Topics include algebra and matrices with
applications to linear programming, differential and integral calculus, and set theory. This book is concerned with the communication of ideas and the reader's intuition is relied upon rather than
mathematical rigor. Problems with answers are given in the sections, and the chapters conclude with additional problems. The
appendix consists of tables.

Emerson, Lloyd S., and Paquette, Laurence R. FUNDAMENTAL MATHEMAT-
ICS FOR THE SOCIAL AND MANAGEMENT SCIENCES. Boston: Allyn and
Bacon, 1975. xii, 420 p. Appendix: Tables, pp. 357-87. Appendix: Answers
to Selected Problems, pp. 389-413. Index, pp. 415-20.

An introductory text for a one-year course in mathematics for students of the social and management sciences. The aim of this
text is to provide a good mathematical foundation as well as interesting examples of practical applications. Topics include elementary algebra and break-even analysis; vectors and matrices; determinants and solutions of systems of linear equations; linear programming; exponents, logarithms, and important nonlinear functions;
mathematics of finance; differential calculus with applications; max-
min theory and applications; integral calculus; introduction to probability; tree diagrams; and Markov analysis. The problems, which
follow the sections of the book, are set off in shaded rectangular
areas for ease of location.

Epstein, Bernard. LINEAR FUNCTIONAL ANALYSIS: INTRODUCTION TO
LEBESGUE INTEGRATION AND INFINITE-DIMENSIONAL PROBLEMS. Philadelphia: W.B. Saunders, 1970. x, 229 p. Appendixes, pp. 197-219. Some
Suggestions for Further Readings, pp. 221-22. Cited References, p. 223. Index, pp. 225-29.

An introduction to a very limited, and comparatively elementary, portion of functional analysis for students with limited mathematics training. Topics include metric spaces, Lebesgue measure and integration, normed linear spaces, linear functionals, operators on finite-dimensional spaces, and spectral theory in infinite-dimensional Hilbert spaces. The appendixes contain additional mathematical results such as Zorn's lemma, the Stieltjes integral, and the Weierstras approximation theorem.

Eves, Howard. ELEMENTARY MATRIX THEORY. Boston: Allyn and Bacon, 1966. xvi, 325 p. Bibliography, pp. 310-14. Index, pp. 315-25.

A text for a one-semester course in the theory of matrices for students in mathematics, social sciences, engineering, physics, business, and other fields. It is also useful as a text for summer institutes and in-service courses for teachers of mathematics. Assumes knowledge of only high school algebra, and problems follow the sections of the book.

Faddeev, Dmitrii K., and Faddeeva, Vera N. COMPUTATIONAL METHODS OF LINEAR ALGEBRA. Translated by R.C. Williams. San Francisco: Freeman, 1963. xi, 621 p. Supplement, pp. 556-57. Literature, pp. 558-611. Supplementary literature, pp. 612-16. Index, pp. 617-21.

Originally published in Russian, this book is an exposition of computational methods for solving a system of linear equations, the inversion of a matrix, and the complete and special eigenvalue problem. Useful mainly as a reference.

Fishburn, Peter C. MATHEMATICS OF DECISION THEORY. Methods and Models in the Social Sciences, vol. 3. Paris: Mouton, 1972. 104 p. References, pp. 95-99. Index, pp. 101-4.

A presentation of the mathematics used in axiomatic approaches in decision theory. Topics include set theory, binary relations, and choice functions; the mathematics of finite and infinite sets; topologies, algebras of sets, and mixture sets; and ordered groups and groupoids. The relationship between each mathematical topic and decision theory is pointed out. Knowledge of elementary set theory is helpful for comprehension.

Fisher, Robert C. AN INTRODUCTION TO LINEAR ALGEBRA. Belmont, Calif.: Dickerson, 1970. ix, 228 p. Appendixes, pp. 205-15. Answers to Selected Problems, pp. 217-24. Index, pp. 225-28.

An introduction to vector and inner product spaces, linear systems, determinants, and the spectral theorem for students with a calculus background. The appendixes are devoted to flow charts and theorems, and the sections conclude with problems.

Fowler, F. Parker, and Sandberg, E.W. BASIC MATHEMATICS FOR ADMIN-
ISTRATION. New York: Wiley, 1962. xvii, 339 p. Appendix I: Four-
place Logarithm Table, pp. 329-31. Index, pp. 333-39.

A text for a one-semester course in business mathematics with linear
programming as the main application. Topics include numbers, relations, and functions; vectors and matrices; linear systems and
linear programming; probability; exponents and logarithms; and
model building. The sections conclude with exercises.

Frank, Peter; Sprecher, David A.; and Yaqub, Adil. A BRIEF COURSE IN
CALCULUS WITH APPLICATIONS. New York: Harper and Row, 1971. x,
333 p. Appendixes, pp. 265-72. Tables, pp. 273-79. Answers to Selected
Exercises and Quizzes, pp. 280-326. Bibliography, p. 327. Symbol Index,
pp. 329-30. Index, pp. 331-33.

A text for an introductory course in mathematics for students who
are not majoring in mathematics but who need a working knowledge of calculus. The material is organized into the following
three categories: the basic theory of differential and integral
calculus and differential equations, applications, and statistics.
The sections of the book conclude with exercises, and the chapters
conclude with quizzes. The appendixes contain proofs of theorems
and formulas.

Franklin, Joel N. MATRIX THEORY. Prentice-Hall Series in Applied Mathematics. Englewood Cliffs, N.J.: Prentice-Hall, 1968. xii, 292 p. Index,
pp. 290-92.

An applications-oriented text for a one-year course in matrix algebra for graduates and undergraduates majoring in mathematics,
economics, science, or engineering who have knowledge of college algebra and calculus. The book is mathematically rigorous,
and simple numerical examples are used to illustrate the concepts.
Topics include determinants, the theory of linear equations, matrix
analysis of differential equations, unitary and normal matrices,
eigenvalues and eigenvectors, Jordan canonical form, variational
principles and perturbation theory, and numerical methods. The
sections conclude with problems consisting of numerical applications
and proofs of theorems.

Freund, John E. COLLEGE MATHEMATICS WITH BUSINESS APPLICATIONS.
2d ed. Englewood Cliffs, N.J.: Prentice-Hall, 1975. xi, 667 p. Appendix: Tables, pp. 611-37. Answers to Odd-numbered Exercises, pp. 638-56.
Index, pp. 657-67.

A text for a one-year course in college mathematics for students
of business and economics who have a high school algebra background. The concept of a mathematical model is stressed throughout, and an adequate foundation is provided for more advanced
work in statistics, operations research, inventory theory, mathemat-

ical programming, and other courses. Topics include systems of linear equations and inequalities, matrix algebra, linear programming, quadratic functions, exponential and logarithmic functions, periodic functions, sequences and limits, differential and integral calculus, probability and Bayes' rule, probability distributions, game theory, linear programming and the solution of a game, and simulation. Exercises are given at the ends of the sections, and specially marked problems introduce special topics and applications, and extra detail. For example, nonlinear depreciation is included in exercises on exponential functions, and the method of least squares is explained in exercises on partial differention. Most of the traditional material from the mathematics of finance is covered within the context of the mathematical ideas which are involved. Many concepts are illustrated with several examples.

Frisch, Ragnar. MAXIMA AND MINIMA: THEORY AND ECONOMIC APPLICATIONS. Chicago: Rand McNally, 1966. xii, 176 p. Appendix, pp. 171–73. Bibliography, p. 174. Index, pp. 175–76.

First published in French in 1960 and translated by Express Translation Service, this book consists of the development of the first and second order conditions for the extrema of functions of one variable and two or more variables, with and without constraints. The Lagrange multiplier technique is used for the cases with constraints. Applications to economics, including Pareto optimality, are included. This book concludes with a chapter on the theory of matrices and the calculation of determinants. Maxima and minima problems of the type that occur in linear and nonlinear and convex programming are not treated.

Fuller, Leonard E. BASIC MATRIX THEORY. Prentice-Hall Mathematics Series. Englewood Cliffs, N.J.: Prentice-Hall, 1962. ix, 245 p. Bibliography, pp. 225–27. Answers to Problems, pp. 228–36. Index, pp. 237–45.

A text for a course in matrix theory for graduates and undergraduates in a variety of disciplines, including social science. The first half of the book deals with matrix theory, and the remaining half with numerical computation techniques, including characteristic roots, inversion of matrices, and homogeneous forms. The concepts are illustrated in great detail, and the chapters conclude with problems.

Gandolfo, Giancarlo. MATHEMATICAL METHODS AND MODELS IN ECONOMIC DYNAMICS. New York: American Elsevier, 1971. x, 511 p. Appendixes, pp. 341–488. Bibliography, pp. 489–98. Author Index, pp. 499–502. Subject Index, pp. 503–11.

A comprehensive development of the mathematical methods of economic dynamics which is divided into the following two parts: difference equation models (linear with constant coefficients), and differential equation models. In both parts, consideration is given to first-, second-, and higher-order systems, and simultaneous equation systems. Among the applications are: Samuelson's multiplier-accelerator model, Metzler's inventory cycle model, and Leontief's dynamic model. The mathematical results are stated without proofs, and the appendixes are devoted to advanced mathematics, including the correspondence principle, stability of equilibrium and Liapunov's second method, and linear mixed differential-difference equations. Assumes knowledge of calculus. The chapters conclude with references.

Gelbaum, Bernard R., and March, James G. MATHEMATICS FOR THE SOCIAL AND BEHAVIORAL SCIENCES: PROBABILITY, CALCULUS, AND STATISTICS. Philadelphia: W.B. Saunders, 1969. xii, 337 p. Appendix, pp. 308-31. Bibliographical Appendix, pp. 333-34. Index, pp. 335-37.

A text for a one-year course in college mathematics for students in the social and behavioral sciences who have knowledge of intermediate algebra. Topics include sample spaces, probability spaces, counting, conditional probability, random variables on finite sample spaces, differentiation and integration, continuous random variables, estimation, hypothesis testing, and decision theory. The sections of the book conclude with exercises.

Gilligan, Lawrence G., and Nenno, Robert B. FINITE MATHEMATICS: AN ELEMENTARY APPROACH. Goodyear Series in Mathematics. Pacific Palisades, Calif.: Goodyear, 1975. ix, 463 p. Appendixes, pp. 409-35. Answers to Odd-numbered Exercises, pp. 436-57. Index, pp. 459-63.

A text for a course in finite mathematics for students in the social sciences, business, economics, or the liberal arts who have only a high school algebra background. Topics include logic, sets, counting, probability and statistics, vectors and matrices, linear programming and games, and an introduction to mathematical model construction. It contains numerous worked examples, and the sections contain problems for assignment. Each chapter concludes with a summary, vocabulary, and list of symbols. The appendixes contain a discussion of mathematical induction, summation notation, statistical tables, and tables of squares and square roots.

Gillis, Floyd E. MANAGERIAL ECONOMICS: DECISION MAKING UNDER CERTAINTY FOR BUSINESS AND ENGINEERING. Reading, Mass.: Addison-Wesley, 1969. vii, 296 p. Tables, pp. 268-93. Index, pp. 295-96.

A text for a course in business mathematics with applications to interest rate problems, break-even analysis, inventories, replacement, demand, firm costs, and pricing under perfect and imperfect competition. The chapters conclude with problems.

Girsanou, Igor Vladimirovich. ̄ LECTURES ON MATHEMATICAL THEORY OF EXTREMUM PROBLEMS. Lecture Notes in Economics and Mathematical Systems, vol. 67. Translated by B.T. Poljak. New York: Springer-Verlag, 1972. 136 p. Suggestions for Further Reading, pp. 124-28. References, pp. 129-36.

> Originally published in Russian, this book is an exposition of the functional-analytic approach to extremal problems. This method is used to derive extremum conditions for problems ranging from Pontryagin's ̄maximum principle in optimal control theory to duality theorems in linear programming.

Glaister, Stephen. MATHEMATICAL METHODS FOR ECONOMISTS. Lectures in Economics, 4. London: Gray-Mills, 1972. viii, 206 p. Suggested References, p. 194. Bibliography, p. 195. Answers to Odd-numbered Problems, pp. 196-204. Index, pp. 205-6.

> A text for an undergraduate course in mathematics for economists. It lies midway between a set of lecture notes and a full textbook. Liberal use is made of examples and illustrations in economics. Topics include vectors and matrices, determinants, the inverse of a matrix, systems of equations, eigenvalues, quadratic forms, sets and functions, differentiation and integration, functions of several variables, constrained and unconstrained optimization, and differential equations. The problems at the end of each chapter form an integral part of the text and are suitable for seminar classes.

Goldberg, Samuel. INTRODUCTION TO DIFFERENCE EQUATIONS WITH ILLUSTRATIVE EXAMPLES FROM ECONOMICS, PSYCHOLOGY, AND SOCIOLOGY. New York: Wiley, 1958. xii, 260 p. Selected References, pp. 246-47. Answers to Problems, pp. 248-54. Index, pp. 255-60.

> A reference for social scientists who have some facility with standard algebraic techniques and an acquaintance with trigonometry. Only a small part of the book requires knowledge of calculus for comprehension. Wherever possible, the mathematical topics are related to examples in the social sciences. Exercises follow the sections of the book.

Good, Richard A. INTRODUCTION TO MATHEMATICS. New York: Harcourt, Brace and World, 1966. xiv, 545 p. Appendix: A Table of Functional Values, pp. 508-12. Answers to Selected Exercises, pp. 513-39. Index, pp. 541-45.

> A first-year college text for a terminal introductory mathematics course for students in the biological and social sciences. It is divided into three parts as follows: linear mathematics, including systems of linear equations, linear transformations, and linear inequalities; the axiomatic approach to probability, including sets, logic, counting, powers and sequences, and random variables; elementary functions, including transcendental functions. A concept is introduced by means of examples, followed by proofs of theorems. Assumes knowledge of only high school algebra. Exercises follow the sections of the book.

Goodman, Adolph Winkler, and Ratti, J.S. FINITE MATHEMATICS WITH AP-
PLICATIONS. 2d ed., rev. New York: Macmillan, 1975. xiv, 541 p.
Appendixes, pp. 425-61. Tables, pp. 462-68. Answers to Exercises, pp. 469-527.
Index of Selected Symobols, pp. 528-30. Index, pp. 531-41.

    A text for a one-semester course in finite mathematics for students
who have a good high school algebra background. Topics include
logic, sets, combinatorial analysis, probability, stochastic processes
and Markov chains, vectors and matrices, linear programming, game
theory, graph theory, and input-output analysis. The appendixes
are devoted to functions, inequalities, mathematical induction, and
the gambler's ruin problem. There are seventy-three sets of exercises, each set with about one dozen problems. The appendix also
contains answers to exercises.

Gossage, Loyce C. BUSINESS MATHEMATICS--A COLLEGE COURSE. Cincinnati: South-Western, 1972. vi, 522 p. Appendixes, pp. 469-88. Answers to
Odd-numbered Problems, pp. 489-514. Index, pp. 515-22.

    A text for a two-semester course in business mathematics covering
the following topics: Numbers; fractions, decimals, percentage;
simple and compound interest; analysis of financial statements; payrolls; investments in real estate, securities, and insurance; and
descriptive statistics. The appendixes consist of an explanation of
the metric system, abbreviations and symbols used in business,
tables, and important equations. The sections of the book contain
problems, and the chapters conclude with review exercises.

Graham, Malcolm. MATHEMATICS: A LIBERAL ARTS APPROACH. New York:
Harcourt Brace Jovanovich, 1973. x, 278 p. Answers to Selected Exercises,
pp. 253-73. Index, pp. 274-78.

    A text for a course in college algebra for liberal arts students who
have a background in arithmetic. The objective of the book is to
develop an appreciation for mathematics as a creative science and
to provide an insight into the methods of mathematical reasoning.
Topics include logic and the nature of a proof; sets and numbers;
finite mathematical systems and the congruence relation; probability; topology; and mathematical machines. There are fifty-three
sets of exercises and nine review sets at the ends of the chapters.

_____. MODERN BUSINESS MATHEMATICS. 2d ed., rev. New York:
Harcourt Brace Jovanovich, 1975. xi, 449 p. Appendix: Logic and Mathe-
matical Reasoning, pp. 361-84. Answers to Selected Exercises, pp. 385-440.
Index, pp. 441-49.

    A text for a one- or two-semester course for teachers of grades K
through 8. It presents, from a mature viewpoint, all the mathematics included or closely related to the elementary school curriculum. Topics include set theory, relations and functions, number

systems, metric and nonmetric geometry, and statistics and probability. The sections contain exercises, and the chapters conclude
with review exercises.

Grawoig, Dennis E. DECISION MATHEMATICS. McGraw-Hill Accounting
Series. New York: McGraw-Hill Book Co., 1967. 370 p. Appendixes,
pp. 341-63. Index, pp. 365-70.

A text for a one-year course in mathematics for business economics students. Topics include matrix algebra and applications,
such as linear programming; differential and integral calculus; sets
and probability theory; and differential and difference equations.
No proofs are provided, and the subject matter is limited to those
areas currently used in decision models. The chapters conclude
with exercises, and the appendixes contain tables.

Gray, A. William, and Ulm, Otis M. APPLICATIONS OF COLLEGE MATH-
EMATICS. Beverly Hills, Calif.: Glencoe Press, 1970. viii, 359 p. Symbols, pp. 272-73. Glossary, pp. 274-77. Appendix A: Topics from Algebra,
pp. 278-80. Appendix B: The Cartesian Plane or Coordinate System, pp. 281-85.
Appendix C: Tables, pp. 286-325. Answers to Exercises, pp. 326-53. Index,
pp. 354-59.

A text for a course in mathematics for students of business, finance,
education, and the social sciences who have a minimum of three
years of high school mathematics. The topics include sets, relations, and functions; probability and statistics; sequences and series;
matrices and linear programming; applications of trigonometry; and
an introduction to differential and integral calculus. One chapter
on FORTRAN programs is included. The sections of the book conclude with exercises.

Graybill, Franklin A. INTRODUCTION TO MATRICES WITH APPLICATIONS
IN STATISTICS. Belmont, Calif.: Wadsworth, 1969. viii, 372 p. Index,
pp. 369-72.

Presents the following matrix algebra topics that are useful in the
theory of linear statistical models: linear transformations and characteristic roots; generalized inverse and conditional inverses; special matrices including patterned matrices; trace of a matrix; integration and differentiation; computing techniques; and nonnegative matrices and projections. Over 450 problems appear at the
ends of the twelve chapters, and more than eighty worked examples
are used to illustrate the mathematical results. This book can be
used as a text for a one-semester course in the theory of linear
statistical models, and as a supplementary text in courses in regression theory, analysis of variance, econometrics, and multivariate analysis. Assumes ability in matric algebra or linear algebra.

Grossman, Stanley I., and Turner, James E. MATHEMATICS FOR THE BIO-
LOGICAL SCIENCES. New York: Macmillan, 1974. xi, 512 p. Appendixes,
pp. 399-467. Tables, pp. 469-75. Answers to Selected Problems, pp. 477-506.
Index, pp. 507-12.

A text for an introductory mathematics course for majors in biology
and medicine. Topics include probability, vectors and matrices,
linear programming, Markov chains, game theory, difference equations, and differential equations. Only two chapters require familiarity with calculus. The appendixes could serve as a short course
in calculus. The sections of the book conclude with problems.

Hackert, Adelbert F. FINITE MATHEMATICS: FROM SETS TO GAME THEORY.
Lexington, Mass.: D.C. Heath, 1974. viii, 390 p. Appendix A: A Review
of Some Algebraic Concepts, pp. 331-42. Appendix B: Tables of Square Roots
and Areas of Standard Normal Distribution, pp. 343-44. Answers to Selected
Problems, pp. 347-87. Index, pp. 388-90.

A text for a one-semester course in finite mathematics for liberal
arts students who have little mathematics preparation. Topics include logic, abstract mathematical systems, number theory, matrix
algebra, probability, statistical analysis, game theory, and linear
programming. The sections of the book conclude with exercises,
and the chapters conclude with review exercises.

Hadley, George F. ELEMENTARY BUSINESS MATHEMATICS. Irwin Series in
Quantitative Analysis for Business. Homewood, Ill.: Richard D. Irwin, 1971.
xii, 667 p. Appendix I: English and Metric Units, pp. 641-51. Appendix II:
Roman Numeral System, pp. 652-53. Answers to Odd-numbered Problems,
pp. 654-67. Index, pp. 669-82.

A modern treatment of traditional mathematics courses taught in
junior colleges, teachers' colleges, and undergraduate schools of
business. It is designed for students from underprivileged classes
and emphasizes ways that business problems lead naturally to the
development of arithmetic.

_____. ELEMENTARY CALCULUS. San Francisco: Holden-Day, 1968. ix,
421 p. Tables, pp. 413-18. Index, pp. 419-21.

A text for a one-semester introductory calculus course for those
who do not wish to take a longer course typically offered to engineers and scientists. It avoids emphasis on applications to physics and engineering and is especially suited for social scientists.
The chapters conclude with references and problems.

Hadley, George F., and Kemp, Murray C. FINITE MATHEMATICS IN BUSI-
NESS AND ECONOMICS. New York: American Elsevier, 1972. ix, 521 p.
Index, pp. 518-21.

A text for a first course in college mathematics for students of business and economics who have a high school algebra and geometry background. The approach is set theoretic and the development is fairly rigorous with a tendency toward abstract reasoning in places. Many of the concepts are illustrated with business examples. Topics include set theory; vectors and matrices; linear equations and input-output analysis; linear programming and the simplex method; probability theory, including conditional probability models; statistical decision theory, including Bayes' law; and Markov chains. Chapters conclude with problems, arranged by the sections of the chapter, and references.

Haeussler, Ernest F., Jr., and Paul, Richard S. INTRODUCTORY MATHEMATICAL ANALYSIS: FOR STUDENTS OF BUSINESS AND ECONOMICS. Reston, Va.: Reston Publishing Co., 1973. xxxix, 600 p. Appendix A: Table of Powers-roots-reciprocals, pp. 537-39. Appendix B: Table of e^x and e^{-x}, pp. 540-41. Appendix C: Table of Selected Integrals, pp. 542-45. Answers to Odd-numbered Problems, pp. 547-90. Index, pp. 591-600.

A text designed to provide a mathematical foundation for students in business and economics who have a high school algebra background. Topics include equations and inequalities; functions and systems of linear equations; linear programming; exponential and logarithmic functions; limits and continuity; differential and integral calculus; multivariable calculus; indeterminate forms; sequences and series; and matrix algebra. Included are more than 650 examples which are explained in step-by-step detail, and more than 2,700 exercises which are structured to relate to the examples and material in the text.

Halmos, Paul R. MEASURE THEORY. The University Series in Higher Mathematics. New York: D. Van Nostrand, 1950. xi, 304 p. References, pp. 291-92. Bibliography, pp. 293-96. List of Frequently Used Symbols pp. 297-98. Index, pp. 299-304.

A unified treatment of measure theory applicable to mathematical analysis. The first half of the book assumes knowledge of elementary algebra and mathematical analysis, while point set topology and topological group theory are needed for the last half. The exercises at the end of the sections constitute an integral part of the book since many are extensions of the subject matter.

Hancock, Harris. THEORY OF MAXIMA AND MINIMA. 1917. Reprint. New York: Dover Publications, 1960. xiv, 193 p. Index, pp. 191-93.

Deals with maxima and minima problems of functions of one variable and of several variables with and without subsidiary conditions. In addition, these problems are examined for various types of functions, such as functions which have derivatives only on definite positions, homogeneous functions, and quadratic forms.

Hanes, Bernard. MATHEMATICS FOR MANAGEMENT SCIENCE. Mathematics and Quantitative Methods Series. Columbus, Ohio: Charles E. Merrill, 1962. xii, 226 p. Index, pp. 224-26.

A text for a one-semester course for undergraduates and graduates who have a good high school algebra background, and a reference for managers. Topics include differentiation and integration, ordinary differential equations, difference equations, series and probability distributions, matrix algebra, linear programming and the simplex method, functions of several variables, set theory, and an introduction to statistics, including the central limit theorem and Chebyshev's theorem. Theorems are not proved, and the mathematical results are illustrated with examples. The chapters conclude with problems.

Hanna, Samuel C., and Saber, John C. SETS AND LOGIC. Irwin Series in Quantitative Analysis for Business. Homewood, Ill.: Richard D. Irwin, 1971. 285 p.

A supplementary text for graduate and undergraduate courses in logic for liberal arts colleges and business schools. There are no mathematical prerequisites, and emphasis is placed on the development of the reader's ability to reason and think analytically. It includes new approaches in constructing Venn diagrams, the interpretation of mathematical theorems, the analogy between switching theory and the systems of sets and logic, and the analogy between decision trees and decision tables. Over 100 examples from mathematics, business, economics, law, medicine, sociology, psychology, and other fields are included, and over 250 exercises are presented.

Harary, Frank; Norman, Robert Z.; and Cartwright, Dorwin. STRUCTURAL MODELS: AN INTRODUCTION TO THE THEORY OF DIRECTED GRAPHS. New York: Wiley, 1965. ix, 415 p. Bibliography, pp. 392-95. Reference List of Principal Theorems, pp. 396-403. Glossary, pp. 404-10. Index, pp. 411-15.

An introduction to the theory of graphs in order to reveal as clearly as possible the usefulness of digraphs to the empirical investigator. It is useful to individuals working in the fields of computing, programming, information retrieval, automata, linguistics, cryptology, and electrical engineering. A good high school course in modern algebra and the facility in abstract thinking are the only prerequisites for comprehension. The chapters conclude with exercises.

Hart, William L. CALCULUS WITH APPLICATIONS TO SOCIAL AND LIFE SCIENCES. Cupertino, Calif.: Freel and Associates, 1973. xiv, 452 p. Appendixes: Tables, Answers, Index, pp. i-xlvi.

A text for a course in basic mathematics for freshman and sophomores in business, economics, and the other fields of the social sciences. It presents a balanced review of differential and integral calculus, including partial differentiation and double integrals, with applications mainly outside the fields of pure mathematics and the physical sciences. Other topics include infinite series and least squares. The sections of the book contain exercises, and the chapters conclude with review problems.

Hayes, Patrick. MATHEMATICAL METHODS IN THE SOCIAL AND MANAGERIAL SCIENCES. New York: Wiley-Interscience, 1975. xvi, 460 p. Appendix: Topics in Calculus, pp. 413-45. Selected References, pp. 447-48. Index, pp. 449-60.

A text for a one-year course in applied mathematics for students in management science who have knowledge of differential and integral calculus, and matrix algebra. It is divided into the following three parts: differential equations in the social and managerial sciences, convex analysis, and convex and geometric programming in management science. The appendix contains remedial material from the calculus of functions of several variables. Topics include sales and advertising; the spread of a rumor; price speculation; Phillips multiplier-accelerator effect; voting distribution; linear production models; economic lot size; equipment replacement; queueing; personnel selection; inventory control; and production functions with budget constraints. The presentation is fairly rigorous with many theorems proved, but graphs are used freely for illustration. Each of the ten chapters concludes with up to thirty exercises with answers provided for about one-third of the problems.

Henley, Ernest J., and Williams, Richard A. GRAPH THEORY IN MODERN ENGINEERING: COMPUTER AIDED DESIGN, CONTROL, OPTIMIZATION, RELIABILITY ANALYSIS. New York: Academic Press, 1973. xvi, 303 p. Appendixes, pp. 245-94. Bibliography, pp. 295-96. Index, pp. 297-303.

A fairly balanced coverage of the theory and applications of matrix- and graph-manipulation methods in engineering design and analysis. Applications are to control systems, linear and nonlinear optimization, sensitivity analysis, formulation of design algorithms, structural analysis and synthesis, and transportation problems. The computer programs developed for solving the problems are available from the authors. The chapters conclude with references.

Hermes, Henry, and LaSalle, Joseph P. FUNCTIONAL ANALYSIS AND TIME OPTIMAL CONTROL. Mathematics in Science and Engineering, vol. 56. New York: Academic Press, 1969. viii, 136 p. References, pp. 132-33. Subject Index, pp. 135-36.

A monograph on the main features of the linear time optimal problem as an application of mathematics. It is divided into the following three parts: a self-contained presentation of those aspects of functional analysis needed for the control problems, application to the linear time optimal control problem, and the nonlinear time optimal control problem. The presentation is rigorous, consisting of theorems with proofs, and illustrative examples. Some sections conclude with problems, and knowledge of differential and integral calculus, differential equations, and mathematical analysis is assumed.

Hilborn, C.E. MATHEMATICS FOR USE IN BUSINESS. Boston: Houghton
Mifflin, 1949. vii, 472 p. Tables, pp. 365–454. Answers to Odd-numbered
Problems, pp. 455–67. Index, pp. 469–72.

A text for a course in business mathematics dealing with elementary problems such as simple and compound interest, annuities, depreciation, and capitalized assets and costs. Numerous illustrations are used, and a comprehensive set of exercises are included
at the conclusions of the sections and chapters.

Hoel, Paul Gerhard. FINITE MATHEMATICS AND CALCULUS WITH APPLICA-
TIONS TO BUSINESS. New York: Wiley, 1974. ix, 446 p. Appendix,
pp. 398–99. Answers to Odd-numbered Exercises, pp. 400–441. Index,
pp. 443–46.

A text for a one-year course in mathematics for business, social
science, and biological science. The exposition relies heavily on
the student's intuition and his willingness to accept mathematical
results, and only proofs that are both simple and instructive are
included. The first half of the book is on finite mathematics and
includes topics such as functions and linear equations, matrices,
linear programming, probability, and game theory. The second
half takes up differential and integral equations, including partial
derivatives with applications to maxima and minima problems. Each
of the twelve chapters concludes with exercises which are grouped
according to the sections of the chapter. The appendix contains
tables of logarithms and the exponential functions.

Hoffmann, Laurence D. PRACTICAL CALCULUS FOR THE SOCIAL AND MANA-
GERIAL SCIENCES. New York: McGraw-Hill Book Co., 1975. ix, 406 p.

A text for a one-semester course in calculus for students in economics and management science who have knowledge of college
algebra. Topics include functions and graphs, differentiation,
exponential and logarithmic functions, integration, and functions
of several variables. A large number of examples reinforce the
techniques needed in solving practical problems. Proofs are intuitive and geometric whenever possible, and notational formalism
is kept to a minimum. The approach to the integral is from the
point of view of antidifferentiation.

Hogg, Robert V.; Schaeffer, Anthony J.; Randles, Ronald H.; and Hickman,
James C. FINITE MATHEMATICS AND CALCULUS WITH APPLICATIONS TO
BUSINESS AND THE SOCIAL SCIENCES. Menlo Park, Calif.: Cummings,
1974. xi, 402 p. Table 1: Random Numbers, pp. 382–85. Table 2: Growth
Function $(1+r)^n$, pp. 386–87. Table 3: Exponential Function, p. 388. Answers
to Selected Exercises, pp. 389–97. Index, pp. 399–402.

This book consists of a previous book by the same authors, FINITE
MATHEMATICS WITH APPLICATIONS TO BUSINESS AND THE
SOCIAL SCIENCES (see below) with two additional chapters, one
on differential calculus and one on integral calculus and continuous

random variables. It is suitable for a one-semester or two-semester course in basic mathematics for students in business and social
science. Assumes knowledge of only high school algebra. Exercises follow the sections of the book.

_____. FINITE MATHEMATICS WITH APPLICATIONS TO BUSINESS AND THE
SOCIAL SCIENCES. Menlo Park, Calif.: Cummings, 1974. ix, 308 p.
Table 1: Random Numbers, pp. 292-95. Table 2: Growth Function $(1+r)^n$,
pp. 296-97. Table 3: Exponential Function, p. 298. Answers to Selected
Exercises, pp. 299-305. Index, pp. 307-8.

A text for a one-semester course in basic mathematics for students
in business and social science. Each chapter begins with a description of a realistic problem that can be solved with the mathematical tools developed in the chapter. Visual displays, such as
pictures, graphs, tree diagrams, scatter diagrams, flow charts, and
network diagrams are used extensively. Topics include linear equations and inequalities, matrix algebra and linear programming, descriptive statistics, probability, network models, discrete random
variables, game theory, and the mathematics of finance. Exercises
follow the sections of the book.

Hohn, Franz E. ELEMENTARY MATRIX ALGEBRA. 2d ed. New York:
Macmillan, 1964. xiv, 395 p. Appendix I: The Notation Sigma and Product, pp. 355-65. Appendix II: The Algebra of Complex Numbers, pp. 366-72.
Appendix III: The General Concept of Isomorphism, pp. 373-75. Bibliography,
pp. 376-83. Index, pp. 385-95

A text for a course in matrix algebra for students in the social
sciences, engineering, physics, psychology, and other related
fields. Topics include determinants, inverses, vector spaces and
transformations, characteristic equation, bilinear quadratic, and
Hermitian forms. Assumes knowledge of only high school algebra.
The sections conclude with exercises.

Holtzman, Jack M. NONLINEAR SYSTEM THEORY: A FUNCTIONAL ANAL-
YSIS APPROACH. Englewood Cliffs, N.J.: Prentice-Hall, 1970. ix, 213 p.
References, pp. 203-8. Name Index, pp. 209-10. Subject Index, pp. 211-13.

An exposition of topics in nonlinear system theory based on functional analysis with emphasis on mathematical theorems. Assumes
knowledge of advanced calculus, differential equations, and linear
systems. This book is suitable for a graduate course.

Horst, Paul. MATRIX ALGEBRA FOR SOCIAL SCIENTISTS. New York: Holt,
Rinehart and Winston, 1963. xxi, 517 p. Index, pp. 505-17.

A text in matrix algebra for students, teachers, and research
workers who have only a high school algebra background. Mathematical rigor is sacrificed for the sake of simplicity and ease of
comprehension, and the author's own algebra of rectangular ar-

rays is emphasized. It is oriented toward the use of matrices in
multiple regression analysis, factor analysis, and analysis of variance. The chapters conclude with a summary section, list of
exercises, and answers to exercises.

Householder, Alston S. THE THEORY OF MATRICES IN NUMERICAL ANALY-
SIS. A Blaisdell Book in the Pure and Applied Sciences: Introduction to
Higher Mathematics. New York: Blaisdell, 1964. xi, 257 p. Bibliography,
pp. 205-48. Index, pp. 251-57.

A presentation of certain aspects of the theory of matrices that are
most useful in developing and appraising computational methods for
solving systems of linear equations and for finding characteristic
roots. Assumes familiarity with matrix algebra, the Cayley-Hamilton
theorem, and the related notions of a vector space, linear dependence, rank, and so forth. The chapters conclude with references,
problems, and exercises.

Howell, James E., and Teichroew, Daniel. MATHEMATICAL ANALYSIS FOR
BUSINESS DECISIONS. Irwin Series in Quantitative Analysis for Business.
Homewood, III.: Richard D. Irwin, 1971. xiv, 424 p. Appendix 1: Dif-
ferentiation Formulas, pp. 381-82. Appendix 2: Integration Formulas, p. 383.
Appendix 3: Notes on Probability, pp. 384-85. Appendix 4: Properties of
Exponents and Logarithms, pp. 386-87. Appendix 5: Tables, pp. 389-96.
Appendix 6: Answers to Odd-numbered Exercises, pp. 397-420. Index,
pp. 421-24.

A text in modern mathematics for students and business practitioners.
Topics include mathematics through integral calculus, probabilistic
models, financial investment models, and linear programming.

Huang, David S. INTRODUCTION TO THE USE OF MATHEMATICS IN ECO-
NOMIC ANALYSIS. New York: Wiley, 1964. xvi, 381 p. Answers to
Odd-numbered Exercises, pp. 263-75. Index of Mathematical Concepts,
pp. 277-79. Index of Economic Applications, pp. 280-81.

A text for a one-semester course in mathematics for economists
with a prerequisite of college algebra. The order of exposition
is calculus, followed by difference and differential equations, and
then by linear algebra with the main application being input-output
analysis. The presentation is not based on set theory. Exercises
follow the sections of the book, and the chapters conclude with
references.

Huffman, Harry. PROGRAMMED BUSINESS MATH. 3 bks. 3d ed. New York:
McGraw-Hill Book Co., 1975. Book 1, 224 p. Book 2, 224 p. Book 3,
304 p.

A presentation of post-high school programmed business mathematics.
Book 1 covers basic operations, problem solving, percentages, com-
missions, and reconciling statements. Book 2 covers interest, mar-

keting, payroll, and taxes. Book 3 covers ownership, depreciation, compound interest, securities, insurance, and statistics. Tests
are available for each book, and a teacher's manual is also available.

Hummel, Paul M., and Seebeck, Charles L., Jr. MATHEMATICS OF FINANCE.
3d ed., rev. New York: McGraw-Hill Book Co., 1971. x, 384 p. Appendix A: Bobtailed Multiplication, pp. 211-12. Appendix B: Common Logarithms, pp. 213-26. Appendix C: Progressions, pp. 227-36. Review Exercises,
pp. 237-44. Answers to Odd-numbered Problems, pp. 245-49. Tables,
pp. 251-61. Index, pp. 363-68.

A text for students in schools of commerce and business administration covering the mathematics of finance and investment. Topics
include simple and compound interest, annuities, perpetuities,
bonds, depreciation, and insurance. Assumes knowledge of college
algebra. The sections conclude with exercises.

Hyatt, Herman R., and Hardesty, James N. MODERN COLLEGE ALGEBRA.
Glenview, Ill.: Scott, Foresman, 1970. vi, 374 p. Appendix, pp. 340-44.
Answers to Selected Exercises, pp. 347-70. Index, pp. 371-74.

A text for a course in college algebra for students with varied
backgrounds and different plans for their future study. It begins
with a diagnostic text designed to assist students to pinpoint weaknesses in preparation, and each chapter concludes with a self-test
for which all answers are given. Topics include equations and
inequalities, relations and functions, analytic geometry, polynomial
functions, systems of equations and inequalities, matrices and determinants, exponential and logarithmic functions, sequences and
series, and probability. Numerous examples are used to illustrate
the concepts and problem-solving techniques. The sections conclude with exercises, and most chapters conclude with supplements
which extend the chapter content. The appendix consists of several useful tables, including common and natural logarithms.

Jaeger, John Conrad, and Starfield, A.M. AN INTRODUCTION TO APPLIED
MATHEMATICS. 2d ed., rev. New York: Oxford University Press, 1974.
xii, 505 p. Index, pp. 501-5.

A text for a course in basic mathematics needed for physics and
engineering students who have knowledge of statics, dynamics,
and calculus. It includes applications to biological and economic
problems, as well as to physics and engineering. Topics include
difference equations, ordinary differential equations with constant
coefficients, first order differential equations, partial differential
equations with numerical solutions, dynamical problems leading to
ordinary differential equations, vectors, particle dynamics, rigid
dynamics, Lagrange's equations, boundary value problems, Fourier
series and integrals, matrices, and the numerical solution of differential equations. The emphasis is on mathematical formulation

of problems that can lead to algorithms for solution by computer programming. The chapters conclude with examples.

James, David Edward, and Throsby, C.D. INTRODUCTION TO QUANTITATIVE METHODS IN ECONOMICS. New York: Wiley, 1973. xv, 335 p. Appendix A: Revision of Some Basic Algebra, pp. 295-99. Appendix B: Extensions of the Calculus, pp. 300-302. Appendix C: Extensions of Matrix Algebra, pp. 303-11. Appendix D: The Linear Regression Model in Matrix Terms, pp. 312-15. Answers to Exercises, pp. 316-25. Further Readings, pp. 326-28. Index, pp. 329-35.

An intermediate level text for a course in mathematics for economists. It is divided into the following five parts: simple functions and economic relationships, elementary calculus in economics, functions of several variables in economic analysis, linear economic models and matrix algebra, and elementary econometric models. It treats such topics as production and cost theory, supply and demand analysis, consumer behavior, the theory of the firm under different market structures, equilibrium analysis, the import multiplier, input-output analysis, linear programming and linear regression models. Assumes knowledge of college algebra. Each of the twenty-one chapters concludes with exercises.

James, Glenn, and James, Robert C. MATHEMATICS DICTIONARY. MULTILINGUAL EDITION. 3d ed., rev. New York: D. Van Nostrand, 1968. v, 517 p. Appendixes, pp. 395-446. French-English Index, pp. 447-65. German-English Index, pp. 466-81. Russian-English Index, pp. 482-502. Spanish-English Index, pp. 503-17.

A listing of 7,500 terms covering the entire range of mathematics from high school mathematics and geometry to the advanced university work in analysis and topology. The third edition includes topics such as category theory in which mathematicians are becoming more widely interested.

Jeffrey, Alan. MATHEMATICS FOR ENGINEERS AND SCIENTISTS. Applications of Mathematics Series. New York: Barnes and Noble, 1969. ix, 728 p. Answers to Selected Problems, pp. 700-720. Index, pp. 721-28.

An introductory mathematics text for engineering and science students covering standard topics through differential equations. The concepts are illustrated by numerical examples and graphs, and each chapter concludes with several dozen exercises. Assumes knowledge of college algebra.

Johnston, John B.; Price, G. Baley; and Van Vleck, Fred S. LINEAR EQUATIONS AND MATRICES. Addision-Wesley Series in Behavioral Sciences: Quantitative Methods. Reading, Mass.: Addison-Wesley, 1966. vii, 308 p. Answers to Selected Problems, pp. 285-300. Index, pp. 303-8.

A treatment of linear equations, algorithms, flow charts for computer programs, and their applications in the management and social sciences. One chapter is devoted to mathematical models in the social sciences and another chapter takes up models in business and economics. Assumes knowledge of high school mathematics. The sections conclude with exercises.

Jury, Eliahu Ibrahim. THEORY AND APPLICATION OF THE Z-TRANSFORM METHOD. New York: Wiley, 1965. xiii, 330 p. Appendix, pp. 278-300. Problems, pp. 301-320. Index, pp. 321-30.

A description of the Z-transform method of generating functions with applications to discrete system theory. Topics include the stability of linear and nonlinear discrete systems, periodic modes of oscillation in nonlinear discrete systems, Z-transform methods in approximation techniques, and applications to feedback systems, information and filtering theory, economics, linear sequential circuits, and Markov processes. The appendixes contain Z-transform pairs, total square integrals, and other special functions. The chapters conclude with references. Assumes knowledge of calculus and complex function theory.

Kaliski, Burton S. BUSINESS MATHEMATICS. New York: Harcourt Brace Jovanovich, 1972. xvii, 466 p. Appendix: Terms You Should Know, p. 429. Review of Problems for Unit VII, pp. 429-30. Glossary, pp. 431-39. Answers to Test Your Ability, pp. 441-55. Index, pp. 457-66.

A text for a one- or two-semester course in business arithmetic. It is divided into the following seven units: unit 1 reviews fundamental arithmetic procedures; units 2 through 7 deal with operating income, operating expenses, profits, other income and expenses, and accounting; and unit 7 introduces additional accounting and statistical analysis and the use of other number systems. The principles and concepts are developed step-by-step, and numerous sample problems are provided. Problems entitled "Test Your Ability" appear at intervals throughout each chapter, and the chapters conclude with review exercises.

Kattsoff, Louis O., and Simone, Albert J. FINITE MATHEMATICS WITH APPLICATIONS IN THE SOCIAL AND MANAGEMENT SCIENCES. New York: McGraw-Hill Book Co., 1965. xiv, 407 p. Appendix, pp. 349-61. Answers to Problems, pp. 363-97. Index of Applied Terms, pp. 399-402. Index of Mathematical Terms, pp. 403-7.

A text for a course in finite mathematics for students in liberal arts and business administration, and a supplementary text for graduate and undergraduate courses in linear programming and related quantitative analysis. Most of the applications are in business and economics, but a few are taken from sociology, psychology, education, and political science. Topics include sets, relations and functions, probability, matrices and vector spaces, linear relations and inequalities, nonlinear relations, sequences, linear program-

ming, and the simplex method. Problems are included in the
sections, and the appendix consists of tables of logarithms.

_____. FOUNDATIONS OF CONTEMPORARY MATHEMATICS WITH APPLI-
CATIONS IN THE SOCIAL AND MANAGEMENT SCIENCES. New York:
McGraw-Hill Book Co., 1967. xiv, 553 p. Appendix, pp. 491-503. Answers
to Problems, pp. 505-44. Index of Applied Terms, pp. 545-48. Index of
Mathematical Terms, pp. 549-53.

> A text for a one-year course in mathematics for social and manage-
> ment sciences with emphasis on applications. It is divided into the
> following five parts: finite mathematics, precalculus, calculus,
> linear algebra, and applied mathematics. The applied part deals
> with linear programming and the simplex method, and statistical
> decision theory including Bayes' decision rules. Examples are
> taken from economics, management, sociology, psychology, edu-
> cation, and political science. Some theorems are proved, while
> others are stated without proof. The sections of the book conclude
> with problems.

Keilson, Julian. GREEN'S FUNCTION METHODS IN PROBABILITY THEORY.
Griffin's Series of Statistical Monographs, no. 17. New York: Hafner, 1965.
viii, 220 p. Bibliography, pp. 199-204. Notes, pp. 205-17. Index,
pp. 218-20.

> A presentation of homogeneous Markov processes, both for discrete
> time (sums of independent, identically distributed random variables),
> and for continuous time (Wiener-Levy homogeneous diffusion pro-
> cesses and compound Poisson processes); and with inhomogeneous
> processes arising from them by the introduction of one or two ab-
> sorbing or impenetrable boundaries. The presentation is based on
> Green's function, and includes applications to practical problems
> (queues, inventories, and dams).

Kemeny, John G.; Mirkil, Hazleton; Snell, J. Laurie; and Thompson, Gerald
L. FINITE MATHEMATICAL STRUCTURES. Englewood Cliffs, N.J.: Prentice-
Hall, 1959. xi, 487 p. Index, pp. 481-87.

> A text for a one-year course in mathematics for students in physi-
> cal and engineering science who have one year of calculus. It
> can also be used for a one-semester course in linear algebra or a
> one-semester course in probability. Every section contains numer-
> ous examples and exercises, and wherever possible the main point
> of a section is illustrated by numerical examples. Topics include
> sets and functions, probability theory, linear algebra, convex
> sets, finite Markov chains, and continuous probability theory.

Kemeny, John G.; Schleifer, Arthur, Jr.; Snell, J. Laurie; and Thompson,
Gerald L. FINITE MATHEMATICS WITH BUSINESS APPLICATIONS. Prentice-
Hall Quantitative Methods Series. Englewood Cliffs, N.J.: Prentice-Hall,

1962. xii, 482 p. Appendix Tables, pp. 466–75. Index, pp. 477–82.

A text in finite mathematics in the context of business and industrial administration with a prerequisite of mathematical maturity
obtained from two and a half years of high school mathematics.
Topics treated include sets, probability theory, partitions and
counting, vectors and matrices, mathematics of finance and accounting, linear programming, and game theory. Each chapter
includes major business applications, exercises, and references.
The applications are in computer circuits, critical path method,
flow diagrams for computing and accounting procedures, Monte
Carlo simulation of decision processes, reliability, decision theory,
waiting line theory, and the simplex method for solving linear
programming problems. The sections conclude with exercises, and
the chapters conclude with suggested readings.

Kemeny, John G., and Snell, J. Laurie. MATHEMATICAL MODELS IN THE
SOCIAL SCIENCES. Introduction to Higher Mathematics. New York: Ginn
and Co., 1962. vi, 145 p. Appendixes, pp. 123–39. Index, pp. 143–45.

Presents eight mathematical models illustrating theories from economics, sociology, ecology, queueing, and dynamic programming.
The problem is first stated from a branch of the social sciences,
then a mathematical model (theory) is formed, and finally the consequences of the theory are deduced and the results interpreted.
This book is suitable for a one-semester mathematics course with
a prerequisite of one year of calculus and a course in finite mathematics. The chapters conclude with exercises, projects, and references. The appendixes include a fixed point theorem, utility
functions, finite Markov chains, generating functions, a combinatorial lemma, and functions of two variables.

Kemeny, John G.; Snell, J. Laurie; and Thompson, Gerald L. INTRODUC-
TION TO FINITE MATHEMATICS. 3d ed., rev. Englewood Cliffs, N.J.:
Prentice-Hall, 1974. xi, 484 p. Index, pp. 481–84.

A text for an introductory course in finite mathematics for students
in the behavioral and social sciences. The core material consists
of the first four chapters and includes the following main topics:
logic of statements, theory of sets and counting techniques, finite
probability, vectors and matrices, and the solution of simultaneous
equations. Other topics include computer programming with emphasis on the language BASIC, the central limit theorem, statistical inference, linear programming, theory of games and applications
in genetics, communication networks, marriage rules, economics
and finance, optimal harvesting of deer, and sequencing problems.
Proofs of theorems are included, and detailed examples with solutions are used to illustrate the concepts. The sections conclude
with exercises, and the chapters conclude with selected readings.

Kim, Thomas K. INTRODUCTORY MATHEMATICS FOR ECONOMIC ANALY-
SIS. Glenview, Ill.: Scott, Foresman, 1971. 502 p. Index, pp. 499-502.

A text for an undergraduate course in mathematical economics and
a supplement for micro- and macro-economic courses. It includes
elementary algebra, calculus, difference and differential equations,
and economic applications. Each chapter contains about thirty
problems for assignment.

Kingman, John Frank Charles, and Taylor, Samuel James. INTRODUCTION TO
MEASURE AND PROBABILITY. New York: Cambridge University Press, 1966.
x, 401 p. Index of Notation, pp. 395-96. General Index, pp. 397-401.

A reference for advanced topics in probability requiring a prerequisite of advanced calculus or real analysis. It is divided into the
following two parts: measure and integration, and probability
theory.

Klein, Erwin. MATHEMATICAL METHODS IN THEORETICAL ECONOMICS:
TOPOLOGICAL AND VECTOR SPACE FOUNDATIONS OF EQUILIBRIUM ANAL-
YSIS. New York: Academic Press, 1973. xix, 388 p. References, pp. 370-76.
Index, pp. 377-88.

A systematic presentation of topological and vector space methods
for students of theoretical economics as well as for research workers. It is divided into the following two parts: set theory and
point set topology, including the properties of spaces and mappings
and existence theorems of fixed points and applications to economic equilibrium of a fixed number of agents and commodities; and
vector spaces and vector space homomorphisms, and matrices, with
applications to the theory of multisectoral linear models. The
book concludes with a chapter on von Neumann's model of an expanding economy. The presentation is formal, consisting of definitions, theorems, remarks, and examples. Assumes knowledge of
one year of calculus and one semester of linear algebra, and some
familiarity with modern algebra and differential equations. Each
chapter concludes with exercises, problems, and notes.

Kline, Morris. MATHEMATICS: A CULTURAL APPROACH. Reading, Mass.:
Addison-Wesley, 1962. xv, 701 p. Table of Trigonometric Ratios, p. 679.
Answers to Selected Exercises, pp. 683-89. Index, pp. 693-701.

A text for a one-year terminal course for liberal arts students and
teachers of elementary mathematics for the secondary schools. The
emphasis is on ideas, rather than on formal mathematics, and the
use of symbols is kept to a minimum.

Kolman, Bernard, and Trench, William F. ELEMENTARY MULTIVARIABLE
CALCULUS. New York: Academic Press, 1971. vii, 505 p. Answers to
Selected Problems, pp. 475-500. Index, pp. 501-5.

An elementary treatment of the subject for a sophomore course in engineering, science, and mathematics. Topics include vectors and analytic geometry, differential calculus of real-valued and vector-valued functions, and integration. Difficult theorems are illustrated, rather than proved, and the exercises which follow the sections are both routine, for illustrating the material covered, and theoretical, for filling in the gaps in the presentation.

Kooros, A. ELEMENTS OF MATHEMATICAL ECONOMICS. New York: Houghton Mifflin, 1965. xii, 413 p. Answers to Selected Exercises, pp. 397-402. Selected Bibliography, pp. 403-6. Index, pp. 407-13.

A text for graduate and undergraduate courses in mathematical economics. It covers elementary algebra, calculus, differential and difference equations, matrix algebra, linear programming, and game theory. Each chapter includes many applications, and a set of from ten to twenty-five exercises.

Kreider, Donald L.; Kuller, Robert G.; Ostberg, Donald R.; and Perkins, Fred W. AN INTRODUCTION TO LINEAR ANALYSIS. Addison-Wesley Series in Mathematics. Reading, Mass.: Addison-Wesley, 1966. xvii, 773 p. Appendix I: Infinite Series, pp. 637-77. Appendix II: Lerch's Theorem, pp. 678-79. Appendix III: Determinants, pp. 680-99. Appendix IV: Uniqueness Theorems, pp. 700-722. Recommendations for Further Reading, pp. 723-24. Answers to Odd-numbered Exercises, pp. 725-66. Index, pp. 767-73.

A text for a one-year course in engineering mathematics for students who have knowledge of elementary calculus and analytic geometry. Topics include real vector spaces and linear transformations, linear differential equations, the Laplace transform, Euclidean spaces, Fourier series and their convergence, orthogonal series of polynomials, boundary-value problems for ordinary and partial differential equations and for Laplace's equation, and Bessel functions. Many mathematical results are stated without proof, and examples are used freely to illustrate the concepts. The sections of the book conclude with exercises.

Kuska, Edward A. MAXIMA, MINIMA, AND COMPARATIVE STATICS: A TEXTBOOK FOR ECONOMISTS. Foreword by B. Henry. London School of Economic Analysis. London: Weidenfeld and Nicolson, 1973. xvi, 263 p. Index.

A text for a course in mathematics for economists who have knowledge of elementary differential calculus. Topics include single and composite functions, differentiation, determinants, Jacobians, quadratic forms, constrained and unconstrained maxima and minima, and comparative statics. The chapters conclude with problems, and the solutions appear at the end of the book.

Lancaster, Kelvin. MATHEMATICAL ECONOMICS. Macmillan Series in Economics. New York: Macmillan, 1968. xiii, 411 p. References, pp. 385-92. Index, pp. 395-411.

A text for a course in mathematical economics and a reference for professional economists who have knowledge of calculus and matrix algebra. The first half of the book is devoted to mathematical analysis and considers optimization, static economic models, and dynamic economic models. The remaining half is devoted to mathematical reviews of set theory, various topics in linear algebra, difference and differential equations, and the calculus of variations. Each chapter contains a set of three to four exercises for assignment.

Leithold, Louis. THE CALCULUS WITH ANALYTIC GEOMETRY. 2d ed., rev. New York: Harper and Row, 1972. xvi, 1,064 p. Appendix, pp. A-2 to A-16. Answers to Odd-numbered Exercises, pp. A-17 to A-40. Index, pp. A-41 to A-50.

A comprehensive treatment of the subject in the following two parts: differential and integral calculus of a single variable and plane geometry suitable for a one-year course of nine or ten semester hours; and infinite series, the calculus of several variables, vectors in two and three dimensions, and a vector approach to solid analytic geometry, suitable for a course of five or six semester hours. It is designed for mathematics majors, as well as for students in engineering, the physical and social sciences, or nontechnical skills. Assumes knowledge of only high school algebra. The sections contain numerous exercises, and the chapters conclude with additional review exercises. The appendixes contain six tables of logarithms and functions, and also the Greek alphabet.

Levensen, Morris E. MAXIMA AND MINIMA. New York: Macmillan, 1967. xiv, 146 p. Bibliography, pp. 137-38. Index, pp. 139-46.

An elementary introduction to maxima and minima for students who have a background in elementary algebra and geometry. The presentation is simple and intuitive and is designed to bridge the gap between high school mathematics and college calculus. Exercises follow the sections of the book.

Lewis, J. Parry. AN INTRODUCTION TO MATHEMATICS FOR STUDENTS OF ECONOMICS. 2d ed., rev. New York: St. Martin's Press, 1969. xiii, 590 p. Appendixes, pp. 521-60. Bibliography, pp. 561-62. Answers to Exercises, pp. 563-72. Tables, pp. 574-84. Index, pp. 585-90.

A text for a one- or two-semester course in mathematics for economists. It is divided into the following six sections: algebra; trigonometry and geometry; calculus: functions of one variable; calculus: functions of many variables; mathematics of fluctuations and growth; and chiefly linear algebra. The appendixes deal with roots or equations of degree n, Newton's method for numerical solutions, determinants, Beta and Gamma functions, the cubic equation, factors of $Ax^2 + Bx + C$, the equation of degree n, and trigonometric results.

Lewis, Laurel J.; Reynolds, Donald K.; Bergseth, F. Robert; and Alexandra, Frank J., Jr. LINEAR SYSTEMS ANALYSIS. Continuing Education for Engineers. University of Washington Series. New York: McGraw-Hill Book Co., 1969. xvii, 489 p. Solutions to Selected Problems, pp. 430-79. Index, pp. 481-89.

A text and self-study for practicing engineers on a review of mathematical ideas and procedures that until recently has not been included in undergraduate education. The theories are developed from a background of specific examples. Topics include linear algebra and matrices, applications of matrices to differential equations, the Laplace transform and Fourier series, analogs and analog computers, feedback control systems, and engineering synthesis of linear systems. The chapters conclude with problems.

Lial, Margaret L., and Miller, Charles D. MATHEMATICS: WITH APPLICATIONS IN THE MANAGEMENT, NATURAL, AND SOCIAL SCIENCES. Glenview, Ill.: Scott, Foresman, 1974. viii, 564 p. Tables, pp. 528-41. Answers to Selected Exercises, pp. 542-59. List of Symbols, pp. 560-61. Index, pp. 562-64.

A text for a one-year course in finite mathematics with calculus for students of management, social science, and biology who have an intermediate high school algebra background. It is divided into the following parts: relations, sets, and functions; matrix theory and linear programming; probability, statistics, and decision theory; and the elements of calculus. A unique feature of this book is the use of many cases and examples to illustrate the mathematical ideas that are presented. The biological examples are from the areas of genetics, cancer research, ecology, and pollution control, as well as other fields. The sections of the book contain exercises, and some chapters have appendixes dealing with additional mathematical results. Each of the four parts concludes with references.

Locke, Flora M. COLLEGE MATHEMATICS FOR BUSINESS. 2d ed. New York: Wiley, 1974. xi, 365 p. Answers to Odd-numbered Problems, pp. 343-59. Index, pp. 361-65.

A text oriented toward community or personal needs or other professional courses in the business curriculum. Topics include percentage, trade and cash discounts, depreciation, payroll, federal income taxes, insurance, interest, sales and property taxes, and computer number systems. Sets of exercises are included in each chapter, and the chapters also conclude with summary exercises.

Lorch, Edgar R. PRECALCULUS: FUNDAMENTALS OF MATHEMATICAL ANAL-
YSIS. New York: Norton, 1973. xi, 380 p. Tables, pp. 305-21. Answers
and Solutions, pp. 323-76. Index, pp. 377-80.

> A text on precalculus mathematics for high school seniors or first-
> year college students, as well as for more advanced readers in en-
> gineering, biology, economics, sociology, and related fields. It
> is devoted to topics based on the structure of the real number sys-
> tem. The emphasis is on detailed treatments of the following
> functions: real, polynomial, algebraic, exponential, logarithmic,
> and circular. Other topics include trigonometry, mathematical
> induction, and approximation theory. Over 1,000 exercises are
> contained in the sections.

McAdams, Alan K. MATHEMATICAL ANALYSIS FOR MANAGEMENT DECI-
SIONS: INTRODUCTION TO CALCULUS AND LINEAR ALGEBRA. New York:
Macmillan, 1970. x, 354 p. Index, pp. 347-54.

> A text for graduate students of business administration who have
> little mathematics background. Topics include set theory, limits
> and continuity, maxima and minima, applications of the derivative,
> partial differentiation, exponential and logarithmic functions, in-
> tegration, linear programming, vectors, and matrices. It empha-
> sizes both elementary mathematics and applications, and some
> chapters have appendixes devoted mainly to applications. The
> chapters conclude with summaries, key concepts, and exercises.

McCready, Richard R. BUSINESS MATHEMATICS. 2d ed., rev. Belmont,
Calif.: Wadsworth, 1973. iii, 281 p. Appendix 1: Federal Service Exami-
nation Problems, pp. 261-64. Appendix 2: Compound Interest and Annuity
Tables, pp. 265-73. Answer Key, pp. 275-81.

> A text in business mathematics with emphasis on the businessman's
> point of view. Each chapter contains a brief discussion of the
> principles underlying a business activity and numerous problems
> with detailed solutions. Each major section contains problems
> which progress from the simple to the more complex, and review
> problems are also added frequently. The types of problems in-
> clude employee compensation, financial statement analysis, buying
> and selling goods, simple and compound interest, annuities, sink-
> ing fund and amortization, consumer credit, insurance, stocks and
> bonds, computer numbering systems, and metric measurement.

MacDonald, Peter. MATHEMATICS AND STATISTICS FOR SCIENTISTS AND
ENGINEERS. New York: D. Van Nostrand, 1966. xii, 299 p. Notation,
pp. xi-xii. Tables, pp. 292-95. Index, pp. 297-99.

> A text for a course in mathematics and statistics for students in
> engineering and related disciplines who have knowledge of elemen-
> tary differentiation and integration. Topics include rates of change,
> integration techniques, differentiation of functions of more than one

variable, application of differentiation to optimization and numerical solutions of algebraic equations, differential equations with
applications, applications of second order linear differential equations, simultaneous linear equations, types of probability distributions, statistical inference and estimation, curve fitting, and quality control. The presentation first develops a need for mathematics by means of descriptive examples, and then provides the
techniques for solving problems. The sections conclude with
exercises, and the chapters conclude with solutions to exercises
and references.

McDonald, T. Marll. MATHEMATICAL MODELS FOR SOCIAL AND MANAGE-
MENT SCIENTISTS. Boston: Houghton Mifflin, 1974. xii, 553 p. Answers
to Selected Problems, pp. 445-504. Appendix, pp. 505-50. Index, pp. 551-53.

A text for a two-semester course in finite mathematics and calculus
for students in the social and managerial sciences who have two
years of high school algebra. The presentation uses the intuitive
approach, with proofs omitted and concepts illustrated by examples.
Topics include sets and functions, mathematics of finance, probability, basic statistics, including simple least squares and the analysis of variance, systems of linear equations, linear programming,
and differential and integral calculus. The sections conclude with
problems. A unique feature is the inclusion of several computer-
calculated results of problems and simulations, and the instructor's
guide includes the BASIC language programs used to produce these
results.

McGinnis, Robert. MATHEMATICAL FOUNDATIONS FOR SOCIAL ANALY-
SIS. Foreword by E.F. Borgatta. New York: Bobbs-Merrill, 1965. xvi,
408 p. Subject Index, pp. 399-406. Name Index, pp. 407-8.

An introductory discussion of the following topics widely used in
the social sciences: sets, functions, relations, numbers, matrices,
scales, convergence, and differential and integral calculus. The
sections conclude with exercises, and the chapters conclude with
references.

MacLane, Saunders, and Birkhoff, Garrett. ALGEBRA. New York: Macmillan, 1967. xix, 598 p. Bibliography, pp. 575-76. Index, pp. 577-98.

A text for an undergraduate course in algebra based on abstract
and axiomatic methods. Topics include groups, rings, fields,
modules, vector spaces, tensor products, quadratic forms, Abelian
groups, lattices, and multilinear algebra. This book is not applications oriented, and many theorems are proved. The sections
conclude with exercises.

Maher, Laurence P., Jr. FINITE SETS: THEORY, COUNTING, AND AP-
PLICATIONS. Merrill Mathematics and Quantitative Methods Series. Colum-
bus, Ohio: Charles E. Merrill, 1968. xii, 110 p. Bibliography, pp. 104-6.
Index, pp. 107-10.

> A reference and supplementary text on set theory and related topics
> for those individuals not having any previous exposure to the sub-
> ject. The presentation is informal, and it emphasizes applications
> in a wide variety of examples and problems. Topics include pro-
> perties of sets and set operations, probability space and Boolean
> algebra, permutations and combinations, the counting of sets, and
> theorems of combinatorial analysis. Each of the five chapters
> concludes with a large number of questions and simple problems.

Maki, Daniel P., and Thompson, Maynard. MATHEMATICAL MODELS AND
APPLICATIONS: WITH EMPHASIS ON THE SOCIAL, LIFE, AND MANAGE-
MENT SCIENCES. Englewood Cliffs, N. J. Prentice-Hall, 1973. xv,
492 p. Appendix A: Topics in Calculus, pp. 453-58. Appendix B: Topics
in Differential Equations, pp. 459-70. Appendix C: Real Linear Spaces,
pp. 471-83. Index, pp. 485-92.

> An introduction to the theory and practice of building models in
> the social, life, and management sciences. It is useful as a text
> for undergraduate mathematics courses for these sciences. Topics
> include selected case studies in model construction, Markov chain
> models, models of linear optimization and computational aspects
> of such models, networks and flows, models of growth processes,
> growth models for epidemics, rumors, and queues. Assumes knowl-
> edge of finite mathematics and intuitive calculus. The sections
> conclude with exercises, and the chapters conclude with references.

Marcus, Marvin, A SURVEY OF FINITE MATHEMATICS. New York: Houghton Mifflin, 1969. x, 486 p. Answers to Quizzes and Selected Exercises,
pp. 397-482. Index, pp. 483-86.

> A text for a course in finite mathematics for students in the social
> and biological sciences, business administration, and liberal arts.
> The three chapter titles are: fundamentals, linear algebra, and
> convexity. Topics include sets and functions, linear equations,
> combinatorial matrix theory, linear programming, game theory,
> and Markov chains. There are about 150 worked examples in the
> text, and each of the nineteen sections concludes with a true-false
> quiz and a set of exercises. The book contains about 1,200 ex-
> ercises for assignment.

Margulis, B.E. SYSTEMS OF LINEAR EQUATIONS. Edited and translated by
J. Kristian and D.A. Levine. Popular Lectures in Mathematics, vol. 14. New
York: Macmillan, 1964. ix, 88 p. Answers to Exercises, pp. 86-88. No
Index.

Originally published in Russian in 1960, this book is an introduction to graphic and numerical solutions of systems of equations with a small number of variables, although some of the methods can be easily extended to arbitrary systems, including infinite systems. The meaning of inconsistent systems and their approximate solutions is also considered. This book is accessible to anyone with a good high school algebra background and is useful to those individuals desiring an introduction to this subject. Some of the chapters conclude with exercises.

Marriott, Francis Henry Charles. BASIC MATHEMATICS FOR THE BIOLOGICAL AND SOCIAL SCIENCES. Elmsford, N.Y.: Pergamon Press, 1970. xii, 229 p. Appendix A: A Note on Definitions, pp. 209-10. Appendix B: Infinite Series and Convergence, pp. 211-12. Appendix C: Tables of the Exponential and Natural Logarithmic Functions, pp. 213-16. Answers to Examples, pp. 217-20. Suggestions for Further Reading, pp. 221-22. References, pp. 223-24. Index, pp. 225-29.

A text for a college course in mathematics for students in the biological and social sciences. Topics include differential and integral calculus, scalars and vectors, complex numbers, and the simplest types of differential equations. Examples are taken from biology, economics, and related subjects, and from probability theory and physics. Statistical methods are not discussed. The chapters conclude with exercises.

Martin, Edley Wainright, Jr. MATHEMATICS FOR DECISION MAKING: A PROGRAMMED BASIC TEXT. 2 vols. Homewood, Ill.: Richard D. Irwin, 1969. Vol. I, 666 p.; vol. II, 659 p.

A text for a course in business mathematics for students with limited mathematics background. Volume 1, consisting of five main parts, is devoted mainly to linear relationships such as matrices and linear programming; and volume 2, consisting of four parts, is primarily on differential and integral calculus.

Martin, Hedley G. MATHEMATICS FOR ENGINEERING, TECHNOLOGY AND COMPUTING SCIENCE. Elmsford, N.Y.: Pergamon Press, 1970. ix, 361 p. Answers to Problems, pp. 337-54. Index, pp. 355-61.

A text for a course in mathematics for engineers, technologists, and others who have a background in differential and integral calculus. Topics include linear algebra, ordinary differential equations, vector analysis, and line and multiple integrals. The sections conclude with problems.

Mathews, Jerold C., and Langenhop, Carl E. DISCRETE AND CONTINUOUS METHODS IN APPLIED MATHEMATICS. New York: Wiley, 1966. xiii, 525 p. Index, pp. 521-25.

A text for a one-year course in applied mathematics for graduates and undergraduates who have at least one year of calculus. This book includes not only traditional topics in applied mathematics (ordinary differential equations, systems of ordinary differential equations, Fourier series, and vector calculus), but also discrete and finite models used in economics, genetics, psychology, management science, and industrial engineering. Among the applications are finite Markov chains, linear programming, and transportation networks. The presentation is fairly rigorous, with proofs of theorems provided. The sections conclude with problems, and the chapters conclude with references.

Maxwell, Edwin Arthur. A GATEWAY TO ABSTRACT MATHEMATICS. New York: Cambridge University Press, 1965. 139 p. Answers, p. 137. Index, p. 139.

An introduction to abstract mathematics with emphasis on digital arithmetic, groups, and modular arithmetic. It can be used as a self-study for anyone with a good high school algebra background, and it can serve as an introduction to the study of abstract algebra.

May, Albert E. MATHEMATICS OF FINANCE. New York: American Book Company, 1951. viii, 264 p. Appendix, pp. 143-52. Answers to Odd-numbered Problems, pp. 153-56. Tables, pp. 157-62. Index, pp. 163-64.

A text which attempts to build the fundamentals of investment theory on just a few simple important formulas, instead of the fifty or one hundred found in most standard texts. Students are taught the use of a few formulas for many types of problems. For example, each general annuity is reduced to an equivalent simple annuity by an appropriate manipulation. The appendix reviews logarithms, exponential equations, geometric progressions, and expansions of binomials.

May, Francis B. DEVELOPMENTS IN MATHEMATICS AND STATISTICS APPLICABLE TO BUSINESS PROBLEMS. Foreword by P[aul]. H. Rigby. University of Houston Studies in Business and Economics, no. 6. Houston: Center for Research in Business and Economics, University of Houston, 1960. v, 105 p. No Index.

A collection of four papers by F.B. May presented in 1959 in Houston at the U.S. Steel Symposium on mathematics and statistics for solving business problems. The titles of the papers are: "Probability and Its Uses in Operations Research"; "Linear Algebra"; "Calculus and Its Application to Economics"; and "Linear Programming." Each paper concludes with two bibliographies: elementary, and advanced. Assumes knowledge of basic statistics and elementary calculus.

May, Kenneth O. ELEMENTS OF MODERN MATHEMATICS. Addison-Wesley
Series in Mathematics. Reading, Mass.: Addison-Wesley, 1959. xvi, 607 p.
Appendix, pp. 586-97. Index, pp. 599-607.

A text for a course in modern mathematics for students familiar
with high school algebra and geometry and who are just beginning
a serious study of mathematics. The symbols of logic and set
theory are presented early and used throughout the book, and applications are made to a variety of fields, including the humanities, arts, biology, social sciences, physical sciences, and engineering. The mathematical rigor varies, the greatest formality
being applied to the simplest topics. Topics include an introduction to differential and integral calculus, probability, statistical
inference, and abstract algebra. The sections contain exercises
and problems with answers.

Meier, Robert C., and Archer, Stephen H. AN INTRODUCTION TO MATHE-
MATICS FOR BUSINESS ANALYSIS. New York: McGraw-Hill Book Co.,
1960. x, 283 p. Answer Key, pp. 267-78. Index, pp. 279-83.

A text and self-study on statistics and mathematics for businessmen
and others who have only a high school algebra background. Topics
include partial differentiation, integration, and probability. The
chapters conclude with exercises.

Merriman, Gaylord M., and Sterrett, Andrew. MATRICES AND LINEAR SYS-
TEMS: A PROGRAMMED INTRODUCTION. Menlo Park, Calif.: W.A.
Benjamin, 1973. xi, 436 p. Bibliography, pp. 425-26. Answers to Review
Questions, pp. 427-32. Greek Alphabet, p. 433. Index, pp. 434-36.

A text for either secondary school or college on the foundations of
linear algebra. It is based on lectures presented by the author at
a National Science Foundation Summer Institute for secondary
teachers at Miami University (Oxford, Ohio). The programmed
format encourages active participation by requiring an immediate
response by the student after the formal presentation of a concept.
Topics include special matrices, types of vector spaces, row space
and row rank of a matrix, and existence theorems for solutions of
linear systems. It is useful as a self-study for students in diverse
fields such as economics, sociology, biology, and engineering.
The chapters conclude with review exercises.

Miller, Ronald E. MODERN MATHEMATICAL METHODS FOR ECONOMICS
AND BUSINESS. International Series in Decision Processes. New York: Holt,
Rinehart and Winston, 1972. xii, 488 p. Answers to Odd-numbered Problems,
pp. 475-79. Index, pp. 481-88.

A text for a course in mathematics for business and social science
students. Topics include optimization with functions of one and
two variables, optimization in inventory and queueing models, and
linear and nonlinear programming. The chapters conclude with

summaries, problems, and references. Most chapters contain mathematical appendixes dealing with proofs and derivations.

Mills, Gordon. INTRODUCTION TO LINEAR ALGEBRA: A PRIMER FOR SOCIAL SCIENTISTS. Chicago: Aldine, 1970. xii, 226 p. Appendix A: Determinants, pp. 196–205. Appendix B: Solutions and Hints for Some of the Exercises, pp. 206–20. Appendix C: Suggestions for Further Reading, pp. 221–22. Index, pp. 223–26.

An introduction to finite-dimensional linear algebra for economists and other social scientists who have a background in high school algebra. The concepts are introduced by way of example, and most theorems are proved. Topics include set theory, systems of equations, algebra of finite-dimensional linear spaces, linear dependence and independence, matrices, linear programming, quadratic forms, and integral variables. The exercises in the sections are used to develop points not included in the text, or to provide practice in the algebraic formulation of applied problems.

Minrath, William R. HANDBOOK OF BUSINESS MATHEMATICS. New York: D. Van Nostrand, 1959. xii, 658 p. Index, pp. 647–58.

A reference and study devoted to methods for application of mathematics in solving practical business problems. The first section presents basic mathematical concepts, including calculus; the second section deals with computations that enter into everyday operations of business, including commercial discount, bank discount, simple and compound interest, annuities, depreciation, and amortization; and the third section is devoted to applications in real estate, stocks and bonds, insurance and installment loans, interpretation of financial statements, evaluation of securities, the use of computers, topics in operations research, and game theory.

Mizrahi, Abe, and Sullivan, Michael. FINITE MATHEMATICS WITH APPLICATIONS FOR BUSINESS AND SOCIAL SCIENCES. New York: Wiley, 1973. xi, 531 p., A–60, 1–4. Appendix: Tables, pp. 425–58; A Short Review of Number Systems and Their Properties, pp. 459–67. Answers, pp. A–1 to A–60. Index, pp. I–1 to I–4.

A text for a course in finite mathematics for students in varied disciplines who have minimal mathematics background. Real-life situations are presented first, followed by the mathematical models required to describe and solve them. Topics include sets, probability, linear equations and inequalities, linear programming, matrix algebra with applications to cryptography, demography, accounting, directed graphs, Markov chains, theory of games, basic statistics, and the mathematics of finance. The sections conclude with exercises, and the chapters conclude with exercises and references.

Moore, David S., and Yackel, James W. APPLICABLE FINITE MATHEMATICS.
Boston: Houghton Mifflin, 1974. xi, 398 p. Supplementary Readings,
pp. 371-72. Appendix: Table A, Binomial Probabilities, pp. 373-75. Answers
to Odd-numbered Exercises, pp. 377-95. Index, pp. 397-98.

> A text for a two-semester course in finite mathematics for students
> in the social, managerial, and behavioral sciences who have a
> prerequisite of high school algebra. It presents a unified body of
> material for applicability to model building with emphasis on for-
> mulation of problems and conceptual understanding of solutions,
> rather than on detailed solution methods which in practice are im-
> plemented by computer routines. Topics include probability, Markov
> chains, linear programming, game theory, and decision theory.
> Each chapter contains several projects designed for students who
> have access to a digital computer and who have knowledge of a
> language such as BASIC or FORTRAN. Each of the six chapters
> contains a theory section devoted to proofs of some theorems and
> other derivations and may be omitted without loss of continuity.
> In addition, some sections contain starred material not essential to
> the remainder of the text.

Munroe, M. Evans. MODERN MULTIDIMENSIONAL CALCULUS. Addison-
Wesley Series in Mathematics. Reading, Mass.: Addison-Wesley, 1963.
viii, 392 p. Answers, pp. 363-90. Index, pp. 391-92.

> A second-year calculus text devoted primarily to topics in multi-
> dimensional analysis. The emphasis is on concepts and methods,
> and exercises follow the sections of the book.

Murray, Francis J., and Miller, Kenneth S. EXISTENCE THEOREMS FOR OR-
DINARY DIFFERENTIAL EQUATIONS. New York: New York University Press,
1954. x, 154 p. Index, p. 154.

> A reference on existence theorems for ordinary differential equa-
> tions. The presentation is based on real function theory, and,
> for certain specialized results only, of elementary functions of a
> complex variable. The mathematical results provide a basis for
> the use of electronic differential analyzers for finding solutions of
> differential equations. The properties of solutions are also dis-
> cussed.

Nahikian, Howard M. TOPICS IN MODERN MATHEMATICS. New York:
Macmillan, 1966. viii, 262 p. Answers to Selected Problems, pp. 237-49.
List of Symbols, pp. 251-55. Index, pp. 257-62.

> A text for a one-semester course in modern mathematics covering
> topics of special interest to students of the life sciences, textiles,
> economics, and forestry. Topics include sets, Boolean algebra,
> probability, vectors and matrices, and functions with vector argu-
> ments. Exercises follow the sections, and the chapters conclude

with supplementary readings. Assumes knowledge of high school or college algebra and trigonometry.

Negus, Robert W. FUNDAMENTALS OF FINITE MATHEMATICS. New York: Wiley, 1974. xiii, 402 p. Answers to Selected Exercises, pp. 375–98. Index, pp. 399–402.

A text for a course in finite mathematics for students in business and social science. Topics include set operations, symbolic logic, relations and functions, matrices and vectors, linear inequalities and linear programming, probability, and Markov chains. The presentation is in terms of definition, explanation, and example, and no theorems are proved. The sections conclude with exercises.

Nikaido, Hukukane. CONVEX STRUCTURES AND ECONOMIC THEORY. Mathematics in Science and Engineering, vol. 51. New York: Academic Press, 1968. xii, 405 p. List of Symbols, pp. xi–xii. Bibliography, pp. 393–98. Author Index, pp. 399–400. Subject Index, pp. 401–5.

Mainly a reference for advanced students of mathematical economics and related fields on the theory of convex sets with applications to existence, optimality, stability, and uniqueness of solutions to linear and nonlinear systems, including the models of Leontief, von Neumann, and Walras. Topics include fixed point theorems, Frobenius-Perron's theorem, the Solow-Samuelson stability theorems, turnpike theorem, and the Jacobian matrix. The presentation is rigorous, with many proofs provided. Assumes knowledge of calculus, matrix algebra, and elementary topology.

_____. INTRODUCTION TO SETS AND MAPPINGS IN MODERN ECONOMICS. Translated by K. Sato. New York: American Elsevier, 1972. xiii, 343 p. Bibliography and References, pp. 335–38. Index, pp. 339–43.

Originally published in Japanese in 1960, this book is a graduate level text which introduces elementary concepts and theorems of topological sets and mappings and illustrates their most important applications to modern economics. The presentation relies heavily on Frobenius, separation, and fixed point theorems. The first part, modern static analysis, discusses interindustry analysis, linear programming, modern analysis of maximum and minimum problems, activity analysis of production, game theory and saddle point problems, the von Neumann model, and the analysis of consumer behavior. In the second part, the author analyzes the existence problem of an equilibrium solution in a Walrasian model. Assumes knowledge of differential and integral calculus and matrix algebra. Knowledge of elementary topology is also desirable, although this subject is taken up as needed.

Noble, Ben. APPLIED LINEAR ALGEBRA. Englewood Cliffs, N.J.: Prentice-Hall, 1969. xvi, 523 p. Hints and Answers to Selected Exercises, pp. 494–505.

Notes and References, pp. 506-10. Bibliography, pp. 511-14. Notation, pp. 515-16. Index, pp. 517-23.

> Presents the algebra of matrices, the theory of finite-dimensional vector spaces, and the basic results concerning eigenvectors and eigenvalues. It emphasizes topics that arise in both applications and numerical analysis, and it keeps the theory separate from the applications. This book is suitable as a text in linear algebra for students who have a calculus background. Exercises follow the sections of the book, and the chapters conclude with miscellaneous exercises.

Olive, Gloria. MATHEMATICS FOR LIBERAL ARTS STUDENTS. New York: Macmillan, 1973. xiii, 274 p. Appendix: The Real Number System, pp. 241-52. Suggestions for Further Reading, pp. 253-54. Hints and Answers to Selected Problems, pp. 255-70. Index, pp. 271-74.

> A text for a course in college mathematics for a mixed group of students, including those who have never had a course in high school mathematics. One half of the book deals with logic, sets, and geometry; and the other half takes up statistics, linear algebra, game theory, calculus, and computers. While the book is written for the novice, some problems may challenge students with a good high school algebra background.

O'Nan, Michael, with Setzer, C. LINEAR ALGEBRA. New York: Harcourt Brace Jovanovich, 1971. xiii, 385 p. Answers to Selected Exercises, pp. 375-82. Index, pp. 383-85.

> A text for a one-semester course in linear algebra for students who have the mathematical maturity gained in a one-year calculus course. It is designed for students in the physical sciences, mathematics, and engineering, and it takes a geometric approach to the study of real vector spaces and linear transformations in two- and three-dimensional space. The topics include linear equations, column vectors, matrices, determinants, vector spaces, linear transformations, inner product, and eigenvalues. The sections contain exercises.

Owen, Guillermo. MATHEMATICS FOR THE SOCIAL AND MANAGEMENT SCIENCES: FINITE MATHEMATICS. Philadelphia: W.B. Saunders, 1970. xi, 424 p. Appendix, pp. 399-417. Index, pp. 419-24.

> A text for an undergraduate course in finite mathematics for the social and management sciences. Topics include vectors and matrices, linear programming, set theory, probability, game theory, dynamic programming, and graphs and networks. The appendix is devoted to the solution of equations, the principle of induction, and exponents and logarithms. The sections conclude with problems. Numerous examples are worked out in detail, and proofs of theorems are also provided.

Owen, Guillermo, and Munroe, M. Evans. FINITE MATHEMATICS AND CAL-
CULUS: MATHEMATICS FOR THE SOCIAL AND MANAGEMENT SCIENCES.
Philadelphia: W.B. Saunders, 1971. x, 598 p. Appendix, pp. 535-59.
Answers to Selected Problems, pp. 561-92. Index, pp. 593-98.

 A text for a course in finite mathematics for students in the social
 and management sciences who have a good high school algebra
 background. Topics include sets, functions, and logic, with special reference to counting problems and the design of a computer;
 analytic geometry, with reference to linear equations and inequalities, break-even analysis, and nonlinear curves, with applications
 to supply and demand curves, product transformation curves, and
 compound interest; probability, including Bayes' formula, Poisson
 and normal distributions, and finite Markov chains; differential
 and integral calculus; vectors and matrices, including applications
 in input-output analysis, linear programming, and transportation
 problems. The emphasis is on applications as a means of motivat-
 ing the student. The sections conclude with exercises, and some
 chapters conclude with problems of higher dimensions. The appendix contains discussions of the solutions of equations, the principle
 of induction, and logarithm and other tables.

Parker, William Vann, and Eaves, James Clifton, MATRICES. New York:
Ronald Press, 1960. viii, 195 p. Bibliography, pp. 189-91. Index,
pp. 193-95.

 A text for an introductory course in the theory of matrices for students of mathematics and applied fields, including economics, engineering, statistics, psychology, operations analysis, and other
 fields. The necessary mathematics is developed as needed, and
 the classical approach to the theory by means of determinants is
 avoided. The sections conclude with problems.

Parthasarathy, T. SELECTION THEOREMS AND THEIR APPLICATIONS. Lecture Notes in Mathematics, vol. 263. New York: Springer-Verlag, 1972.
101 p. References, pp. 100-101. No index.

 A presentation of several selection theorems and their applications.
 Topics include continuous selections, measurable selections with an
 application to stochastic games, general theorems on selectors, von
 Neumann's measurable choice theorem, and the uniformization of
 sets in topological spaces. Assumes knowledge of measure theory
 and topology. The sections conclude with references.

Pearl, Martin. MATRIX THEORY AND FINITE MATHEMATICS. International
Series in Pure and Applied Mathematics. New York: McGraw-Hill Book Co.,
1973. xii, 450 p. Answers to Selected Exercises, pp. 423-42. Subject Index, pp. 443-48. Index of Symbols, pp. 449-50.

 A unified treatment of matrix algebra coupled with an introduction
 to three of the most important modern applications of matrices:

the theory of games, finite Markov chains, and the theory of
graphs. The first chapter on matrices contains most of the material encountered in a one-semester undergraduate course in vectors
and matrices. About one-half of the applications portion of this
book can be covered in one semester. Assumes knowledge of college algebra and elementary probability theory. The sections conclude with exercises, and the chapters conclude with references.

Pease, Marshall C. III. METHODS OF MATRIX ALGEBRA. Mathematics in
Science and Engineering, vol. 16. New York: Academic Press, 1965. xviii,
406 p. References and Recommended Texts, pp. 396-99. Subject Index,
pp. 401-6.

A text for students in the physical sciences and engineering who
are familiar with complex number theory. Topics include the
Jordan canonical form and other types of matrices, functions of a
matrix, singular and rectangular operators, the direct product and
the Kronecker sum, Sturm-Lioville systems, stability, and Markoff
matrices and probability theory. Assumes knowledge of calculus.
The chapters conclude with exercises.

Peck, Lyman C. BASIC MATHEMATICS FOR MANAGEMENT AND ECONOM-
ICS. Glenview, Ill.: Scott, Foresman, 1970. 323 p. Tables, pp. 276-87.
Answers to Selected Exercises, pp. 289-318. Index, pp. 319-23.

A survey of mathematics for students of business administration who
have a high school algebra background. Topics include linear
equations, matrices and inequalities; linear programming; mathematics of finance; probability; and differential and integral calculus. The sections conclude with exercises.

Peterson, John M. FINITE MATHEMATICS. New York: Holt, Rinehart and
Winston, 1974. viii, 307 p. Appendix A: Graphs of Inequalities, pp. 241-44.
Appendix B: Mathematical Induction, pp. 245-50. Appendix C: Tables,
pp. 251-66. Appendix D: Solutions to Selected Exercises, pp. 267-302.
Index, pp. 303-7.

A text for a one-semester course in finite mathematics with a prerequisite of collega algebra. Topics include logic, sets, combinatorial algebra, probability and statistics, vectors and matrices,
and linear programming and the theory of games. The sections
conclude with exercises, and the chapters conclude with review
exercises.

Pettofrezzo, Anthony J. MATRICES AND TRANSFORMATIONS. Teachers'
Mathematics Reference Series. Englewood Cliffs, N.J.: Prentice-Hall, 1966.
x, 133 p. Bibliography, pp. 112-13. Answers to Odd-numbered Exercises,
pp. 114-28. Index, pp. 129-33.

Presents the fundamental concepts of matrix algebra through eigenvalues and eigenvectors, first in an intuitive style, followed by a

more formal development. This book is useful for individual study, summer institutes for teachers in mathematics, and a one-semester college course in matrix algebra. The sections conclude with exercises.

Pfouts, Ralph W. ELEMENTARY ECONOMICS: A MATHEMATICAL APPROACH. New York: Wiley, 1972. Index, pp. 329-35.

A text for an economics principles course covering the usual topics in micro- and macro-theory, international economics, and public finance. It differs from most other principles texts in that the mathematical approach accompanies the traditional verbal and graphical approaches, enabling the instructor to go further in many of the subjects than is traditional in an elementary course. The level of mathematics used is high school algebra and elementary differential calculus. At only two points is mathematics beyond the elementary level used: the constrained maximization problems introduced in the discussion of utility theory, and second-order difference equations introduced by means of the multiplier-accelerator model. The chapters conclude with questions.

Pintel, Gerald, and Diamond, Jay. BASIC BUSINESS MATHEMATICS. Englewood Cliffs, N.J.: Prentice-Hall, 1972. v, 362 p.

A text for a college course in business mathematics covering the following topics: fractions and decimals, interest, consumer loans, retail and marketing, office procedures, mathematics of accounting, and the mathematics of data processing.

Piper, Edwin B.; Fairbank, Roswell E.; and Gruber, Joseph. APPLIED BUSINESS MATHEMATICS FOR CONSUMER AND BUSINESS USE. 9th ed. Cincinnati, Ohio: South-Western, 1970. 582 p. Drills for Accuracy and Speed, pp. 515-32. Supplementary Problems, pp. 533-56. Civil Service Problems, pp. 557-62. Interest Tables, p. 563. Glossary, pp. 564-68. Business Abbreviations and Symbols, pp. 569-70. Index, pp. 571-82.

A text for a high school course in elementary business mathematics, including personal finance, tax problems, retailing, probability, and computer mathematics. The material is divided into eighteen units that include eighty sections, and most sections are divided into two or more parts of convenient length for assignment.

Pipes, Louis A., and Hovanessian, Shahen A. MATRIX-COMPUTER METHODS IN ENGINEERING. New York: Wiley, 1969. xi, 333 p. Index, pp. 329-33.

A text for an undergraduate course in the application of mathematics to engineering problems to supplement the portion of the course on matrices. Assumes knowledge of differential and integral calculus, but knowledge of computer methods is not necessary. The

computer programs are written in FORTRAN or BASIC. The applications are in electrical, mechanical, and structural systems. The
chapters conclude with references and problems.

Pollard, Harry. APPLIED MATHEMATICS: AN INTRODUCTION. Addison-
Wesley Series in Mathematics. Reading, Mass.: Addison-Wesley, 1972. x,
99 p. Bibliography, p. 98. Index, p. 99.

A presentation of a small selection of problems of major importance, from a variety of fields, which are discussed in detail with
the aid of exercises and questions. Among the problems treated
are the two-body problem in astronomy, optimization, Hamilton's
and Lagrange's equations, Laplace's equation, the wave equation,
and the heat equation. This book is neither a text nor a book on
techniques, and knowledge of advanced calculus, including vector analysis through divergence theorems, is essential for understanding. The author states that the book is experimental.

Polya, George. MATHEMATICAL DISCOVERY: ON UNDERSTANDING,
LEARNING, AND TEACHING PROBLEM SOLVING. VOLUME 1. New York:
Wiley, 1962. xv, 216 p. Solutions, pp. 154-208. Hints to Teachers, and
to Teachers of Teachers, pp. 209-12. Bibliography, pp. 213-14. Index,
pp. 215-16.

A presentation of the study and means of solving elementary mathematical problems with the aim of improving the preparation of high
school mathematics teachers. Part 1 is on pattern, and part 2
deals with a general method of solution. Each chapter concludes
with about 100 examples and comments.

Puckett, Richard H. INTRODUCTION TO MATHEMATICAL ECONOMICS:
MATRIX ALGEBRA AND LINEAR MODELS. Lexington, Mass.: D.C. Heath,
1971. viii, 276 p. Index, pp. 273-76.

A text for a one-semester course in mathematical economics with
emphasis on linear models. The first five of the nine chapters
take up the necessary mathematics, and the remaining four chapters are devoted to input-output analysis, linear programming and
the computational aspects of the simplex method, duality and the
theory of games, characteristic roots and quadratic forms, activity
analysis, and the von Neumann model. Most of the theorems are
stated without proof. The sections include a few problems with
answers, and each chapter concludes with about one dozen exercises.

Rao, Calyampudi Radhakrishna, and Mitra, Sujit Kumar. GENERALIZED IN-
VERSE OF MATRICES AND ITS APPLICATIONS. Wiley Series in Probability
and Mathematical Statistics. New York: Wiley, 1971. xiv, 240 p. Bibliography on Generalized Inverses and Applications, pp. 219-33. Author Index,
pp. 235-36. Subject Index, pp. 237-40.

A text for a one-semester course on g-inverse of matrices and a supplementary text for a variety of courses such as matrix algebra, network theory, mathematical statistics, optimization problems, and numerical analysis. It attempts to bring together all the available results on "invertibility of singular matrices" under a unified theory and to discuss their applications. After describing the various types of g-inverse matrices and their properties, attention is turned to such topics as transformations, simultaneous reduction of a pair of Hermitian forms, estimation of parameters in linear models, conditions for optimality and validity of least-squares theory, and distribution of quadratic forms. The applications of g-inverse are in network theory, mathematical programming, and certain problems in mathematical statistics (discriminant functions when the dispersion matrix is singular and maximum likelihood extimation when the information matrix is singular). The presentation is rigorous with proofs of theorems provided. Assumes knowledge of linear algebra and mathematical statistics. The chapters conclude with complements.

Rassweiler, Merrill, and Rassweiler, Irene. FUNDAMENTAL PROCEDURES OF FINANCIAL MATHEMATICS. New York: Macmillan, 1952. vii, 260 p. Appendix I: Short Method Multiplication, pp. 237-40. Appendix II: Commutation Columns, pp. 241-50. Answers to Odd-numbered Problems, pp. 251-56 Index, pp. 257-60.

A text for a two-semester course in the mathematics of finance. It considers simple and compound interest, discounts, pricing, annuities, and life insurance. The chapters conclude with problems.

Read, Ronald C. A MATHEMATICAL BACKGROUND FOR ECONOMISTS AND SOCIAL SCIENTISTS. Prentice-Hall Series in Mathematical Economics. Englewood Cliffs, N.J.: Prentice-Hall, 1972. xvi, 1,024 p. Index, pp. 1009-24.

A comprehensive and illustrative text on elementary mathematics, including calculus, for students of economics and the social sciences. It differs from most textbooks in this category by its omission of statistics. Topics include geometry in n-dimensions, convex sets, combinations and probability, differential and difference equations, real and complex numbers, linear algebra, Markov chains, linear programming, theory of games, and graph theory. Throughout the book there are illustrative examples of economic significance. The sections conclude with exercises, and each chapter concludes with review exercises and references.

Rektorys, Karel, ed. SURVEY OF APPLICABLE MATHEMATICS. Translated and edited by the staff of the department of mathematics, University of Surrey. Cambridge, Mass.: MIT Press, 1969. 1,369 p. Comprehensive index.

A comprehensive survey of most, if not all, branches of applied mathematics undertaken by a team of seventeen Czech mathematicians. A useful reference for economists.

Rice, Louis A.; Mayne, F. Blair; and Deitz, James E. BUSINESS MATHEMA-
TICS FOR COLLEGES. 6th ed. Cincinnati: South-Western, 1973. iv, 236 p.
No index.

> A combination textbook and workbook in business mathematics.
> Topics include decimals and fractions, percentage in business,
> financial charges, payrolls and taxes, statistics and computers,
> financial statements, insurance, bonds, stocks, and annuities.
> Contains assignment sheets for solutions of most problems.

Richardson, Moses, and Richardson, Leonard F. FOUNDATIONS OF MATHE-
MATICS. 4th ed., rev. New York: Macmillan, 1973. xxvi, 582 p. List
of Symbols, pp. xxv–xxvi. Selected Bibliography, pp. 547–56. Table I:
Common Logarithms, pp. 557–58. Table II: Trigonometric Functions, p. 559.
Answers to Odd-numbered Problems, pp. 561–76. Index, pp. 577–82.

> A text for a survey course in mathematics for liberal arts students.
> Topics include logic, the number system, impossibilities and un-
> solved problems, analytic geometry, limits and calculus, trigonom-
> etry, probability and statistics, mathematical induction, vectors
> and matrices, and Euclidean and non-Euclidean geometry. The
> main applications are in games, linear programming, and Markov
> chains. Numerous exercises are provided, and the chapters con-
> clude with exercises.

Richardson, William H. FINITE MATHEMATICS. New York: Harper and Row,
1968. x, 191 p. Appendix A: What Is a Proof?, pp. 163–65. Appendix B:
Mathematical Induction, pp. 166–71. Appendix C: Tree Diagrams, pp. 172–76.
Appendix D: Answers to Selected Odd-numbered Exercises, pp. 177–85. Index,
pp. 189–91.

> A text for a one-semester course for students in liberal arts and
> business who have at least one year of high school algebra. The
> four chapter topics are mathematical logic, set theory, counting and
> the binomial theorem, and probability. The sections contain ex-
> ercises, and the chapters conclude with reviews and supplementary
> exercises.

Richman, Fred; Walker, Carol; and Wisner, Robert J. MATHEMATICS FOR
THE LIBERAL ARTS STUDENT. Contemporary Undergraduate Mathematics Series.
Belmont, Calif.: Brooks/Cole, 1967. viii, 190 p. Appendix on Notation and
Arithmetic Facts, pp. 159–82. Answers, pp. 183–87. Index, pp. 189–90.

> A text for a one- or two-semester terminal course in mathematics
> for liberal arts students who have no mathematics background.
> Topics include counting procedures, probability, numbers, exponen-
> tial growth, Fermat's theorem, logarithms, the use of computers,
> and descriptive statistics. Problems follow the sections of the book.

Ritchie, Robert W., ed. NEW DIRECTIONS IN MATHEMATICS. Englewood

Cliffs, N.J.: Prentice-Hall, 1963. 124 p. Appendix: Guests at the New Directions in Mathematics Conference, pp. 121-24.

A collection of four panel discussions of a conference entitled "New Directions in Mathematics," held at Dartmouth College on November 3-4, 1961, during the dedication of a new mathematics building. Of interest to economists is panel 3, "New Directions in Applied Mathematics," by P. Lax.

Roberts, Blaine, and Schulze, David L. MODERN MATHEMATICS AND ECO-NOMIC ANALYSIS. New York: Norton, 1973. x, 550 p. Index, pp. 535-50.

A text for graduate and undergraduate courses in mathematics for economists with an unusually broad coverage of topics. Included are treatments of differential and integral calculus, differential equations, existence and stability of solutions to static models, calculus of variations, mathematical programming (including dynamic and parametric), game theory (including n-person games), and control theory. The sections conclude with problems, and the chapters conclude with references. The presentation is fairly rigorous with proofs of simple theorems included.

Rockafellar, R. Tyrrell. CONVEX ANALYSIS. Princeton Mathematics Series, no. 28. Princeton, N.J.: Princeton University Press, 1970. xviii, 451 p. Comments and References, pp. 425-32. Bibliography, pp. 433-46. Index, pp. 447-51.

An exposition of the theory of convex sets and functions with emphasis on applications to extremum problems. Topics include systems of inequalities, the minimum or maximum of a convex function over a convex set, Lagrange multipliers, and minimax theorems. Assumes knowledge of linear algebra and real analysis. The presentation is fairly intuitive.

Rosenberg, Reuben Robert. COLLEGE MATHEMATICS FOR ACCOUNTING AND BUSINESS ADMINISTRATION. 2d ed. New York: McGraw-Hill Book Co., 1972. 230 p.

A text-workbook designed for a college course in business mathematics. The presentation is correlated with leading accounting textbooks and is particularly useful to students studying mathematics and accounting concurrently. Topics include fractions and percents, simple and compound interest, pricing merchandise, depreciation, payrolls, credit, annuities, insurance, accounting, and payroll records.

Rosenberg, Reuben Robert, and Poe, Roy W. COLLEGE BUSINESS MATHEMATICS: A PRACTICAL COURSE IN MATHEMATICS FOR COLLEGE STUDENTS IN BUSINESS. 5th ed. New York: McGraw-Hill Book Co., 1973. xv, 365 p.

A text for a course in business mathematics with emphasis on applications in accounting, data processing, and finance. Topics include fractions and decimals, purchase discounts, insurance, depreciation, financial statements, profit and dividends, stocks and bonds, descriptive statistics, debt repayment, real estate, consumer credit, income taxes, and social security.

Rosenstiehl, P., and Mothes, J. MATHEMATICS IN MANAGEMENT: THE LANGUAGE OF SETS, STATISTICS AND VARIABLES. Translated by A. Silvey. New York: American Elsevier, 1969. xiv, 392 p. Tables, pp. 387-89. Subject Index, pp. 390-92.

A text for a course in mathematics for business schools covering the following topics: set theory, events and probability, random variables, and common probabilistic models. Assumes no previous college mathematics. The chapters conclude with exercises and applications.

Rothman, Maurice. AN INTRODUCTION TO INDUSTRIAL MATHEMATICS. New York: Van Nostrand Reinhold, 1970. 370 p. Table of Standard Integrals, pp. 362-64. Index, pp. 365-70.

An introduction to industrial mathematics for students of mathematics who wish to become acquainted with the requirements of industry. It considers topics which are useful in finding solutions to industrial problems. These topics include series and convergence; partial fractions; determinants and matrices; solutions to systems of equations; Fourier series; differentiation; Euler's theorem; maxima and minima; line and surface integrals; coordinate geometry; solutions to differential equations; the Laplace transform; statistical inference; regression analysis; numerical integration and other methods in numerical analysis; operational research, including linear programming, network analysis, forecasting, and stock control; vector analysis; Green's theorem; mechanics; approximations; and binary arithmetic. While the presentation deals mainly with formulas useful for computations, there are some numerical examples especially in the chapter on numerical analysis. Chapters conclude with references for further reading.

Roueche, Nelda W. BUSINESS MATHEMATICS: A COLLEGIATE APPROACH. 2d ed., rev. Englewood Cliffs, N.J.: Prentice-Hall, 1973. xii, 591 p. Appendix A: Arithmetic, pp. 453-501. Appendix B: Tables, pp. 503-48. Answers, pp. 549-80. Index, pp. 581-91.

A text for a one-term business mathematics course in business programs in junior colleges and community colleges. It is divided into the following four parts: review, accounting mathematics, retail mathematics, and mathematics of finance. It contains numerous worked examples and over 1,000 problems for assignment. Topics include wages and payrolls, financial statement analysis, markup, discounts, simple and compound interest, annuities, sinking funds, and amortization.

Rutledge, William A., and Cairns, Thomas W. MATHEMATICS FOR BUSINESS ANALYSIS. New York: Holt, Rinehart and Winston, 1963. xii, 428 p. Tables, pp. 383–410. Answers to Odd-numbered Exercises, pp. 411-20. Index of Special Symbols, pp. 421-22. Index, pp. 423-28.

A text designed for one- or two-semester undergraduate courses in business mathematics covering the following topics: sets, numbers, calculus, probability, and matrix algebra with linear programming as the main application.

Saaty, Thomas L. MODERN NONLINEAR EQUATIONS. New York: McGraw-Hill Book Co., 1967. xv, 473 p. Index, pp. 459-73.

This book and NONLINEAR MATHEMATICS, by T.L. Saaty and J. Bram (see below), are suitable for a one-year course in nonlinear mathematics. From the theoretical viewpoint, it provides a perspective on the field of equations, and from the applied standpoint, it provides insights in formulating problems and in solving them. Topics include functional analysis, nonlinear difference equations, delay-differential equations, integral equations, and introdifferential and stochastic differential equations.

_____,ed. LECTURES ON MODERN MATHEMATICS. VOL. 1. New York: Wiley, 1963. ix, 175 p.

A collection of six expository lectures out of eighteen lectures presented at George Washington University, jointly sponsored by the university and the Office of Naval Research. Each lecture delineates a substantial research area and describes it broadly and comprehensively for an audience of mathematicians. The titles of the lectures are: "A Glimpse into Hilbert Space," by R.R. Halmos; "Some Applications of the Theory of Distribution," by L. Schwartz; "Numerical Analysis," by A.S. Householder; "Algebraic Topology," by S. Eilenberg; "Lie Algebras," by I. Kaplansky; and "Representation of Finite Groups," by R. Brauer.

_____. LECTURES ON MODERN MATHEMATICS. VOL. 2. New York: Wiley, 1964. ix, 183 p.

The second six lectures presented at George Washington University. The titles of the lectures are: "Partial Differential Equations with Applications in Geometry," by L. Nirenberg; "Generators and Relations in Groups--The Burnside Problem," by M. Hall, Jr.; "Some Aspects of the Topology of 3-Manifolds Related to the Poincare Conjecture,: by R.H. Bing; "Partial Differential Equations: Problems and Uniformization in Cauchy's Problem," by L. Garding; "Quasiconformal Mappings and Their Applications," by L. Ahlfors; and "Differential Topology," by J. Milnor.

_____. LECTURES ON MODERN MATHEMATICS. VOL. 3. New York: Wiley, 1965. ix, 321 p.

> The third and final volume containing six lectures presented at George Washington University. The lecture titles are: "Topics in Classical Analysis," by E. Hille; "Geometry," by H.S.M. Coxeter; "Mathematical Logic," by G. Kreisel; "Some Recent Advances and Current Problems in Number Theory," by P. Erdos; "On Stochastic Processes," by M. Loeve; "Random Integrals of Differential Equations," by J. Kampe de Feriet.

Saaty, Thomas L., and Bram, Joseph. NONLINEAR MATHEMATICS. International Series in Pure and Applied Mathematics. New York: McGraw-Hill Book Co., 1964. xv, 381 p. Appendix, pp. 365-72. Index, pp. 373-81.

> An attempt to unify the theory of nonlinear mathematics. The presentation is centered around existence and uniqueness theorems, characterizations (i.e., stability and asymptotic behavior), construction of solutions, convergence, approximation, and errors. The six chapter topics are linear and nonlinear transformations; nonlinear algebraic and transcendental equations; nonlinear optimization; nonlinear programming and systems of inequalities; nonlinear ordinary differential equations; introduction to automatic control and the Pontryagin principle; and linear and nonlinear prediction theory. It is written for mathematicians, engineers, operational analysts, and others who are familiar with functional analysis and the calculus of variations. The sections conclude with exercises, and the chapters conclude with references. The appendix contains a review of set theory and the Schouder-Tychonoff theorem.

Samelson, Hans. AN INTRODUCTION TO LINEAR ALGEBRA. Series in Pure and Applied Mathematics. New York: Wiley, 1974. xi, 265 p. Solutions, pp. 248-58. Appendix: Special Symbols, p. 259. Index, pp. 261-65.

> An intermediate level text for a one-semester course in linear algebra for a varied audience including mathematicians, physicists, statisticians, operation research majors, economists, engineers, and others. It develops two main aspects of linear algebra: geometry and linear equations, including quadratic forms. Topics include vectors, linear functionals and linear equations, and matrices; eigenvectors and eigenvalues; the Cayley-Hamilton theorem and the Jordan form; quadratic forms, inner product, and the spectral theorem. The style of presentation consists of theorems and propositions, followed by proofs, and assumes knowledge of one year of calculus, and matrix algebra.

Samuelson, Paul A. FOUNDATIONS OF ECONOMIC ANALYSIS. Cambridge, Mass.: Harvard University Press, 1947. Reprint. New York: Atheneum, 1965. xx, 447 p. Mathematical appendixes, pp. 357-439. Index, pp. 441-47.

An advanced mathematical treatment of classical microeconomic theory, including general equilibrium and dynamic analysis. Other topics include the theory of maximizing behavior, theory of cost and production, pure theory of consumer's behavior, transformations and composite commodities, and stability of equilibrium (linear and nonlinear systems). Many mathematical theorems which form the basis of economic theory are presented and illustrated, but not rigorously proved. Assumes knowledge of differential and integral calculus. This book is useful as a reference in courses in economic theory and mathematical economics.

Sasaki, Kyohei. INTRODUCTION TO FINITE MATHEMATICS AND LINEAR PROGRAMMING. Belmont, Calif.: Wadsworth, 1970. ix, 239 p. Appendix A: Determinants, pp. 214-18. Appendix B: Probability, pp. 219-27. Answers to Selected Problems, pp. 228-31. Author Index, p. 232. Subject Index, pp. 233-39.

A text for an introductory course in quantitative methods or mathematical programming for upper undergraduate of first-year graduate students in business, economics, and related social science courses. The mathematical concepts of a function, linear independence and dependence, and linear transformation are developed in the first three chapters; linear programming, the transportation and transshipment problems, parametric and integer programming, and game theory are treated in the next six chapters. Unique features of this book are the illustration of theoretical concepts with concrete examples and the intuitive reasoning behind the proofs of theorems. Assumes knowledge of high school algebra. Many sections contain exercises, and the chapters conclude with additional exercises and references.

Schkade, Lawrence L. VECTORS AND MATRICES. Columbus, Ohio: Charles E. Merrill, 1967. xii, 126 p. Index, pp. 123-26.

An introduction to topics from the theory of vectors and matrices that are most useful in formulating and solving problems related to management. Concepts are presented first in words, then in symbolic form, followed by proofs of theorems. Topics include vectors with applications to linear equations, determinants and techniques for finding the numerical value of an array of numbers, matrix algebra with applications to systems of linear equations, and applications of vectors and matrices to management (least squares, Markov models, networks, assignment models, and linear programming). Each of the eight chapters concludes with exercises.

Schmidt, Joseph William. MATHEMATICAL FOUNDATIONS FOR MANAGEMENT SCIENCE AND SYSTEMS ANALYSIS. Operations Research and Industrial Engineering. New York: Academic Press, 1974. xiii, 581 p. Appendix, pp. 567-75. Index, pp. 577-81.

A text for a course in the mathematics of management science and systems analysis for students who have an introductory knowledge of calculus. The topics include probability, real analysis, matrix analysis, classical optimization, the calculus of finite differences, and complex variables and transform methods. Applications to operations research are not stressed. Many theorems are stated and proved, and the chapters conclude with a large number of problems and references.

Schorling, Raleigh; Clark, John R.; and Lankford, Francis G., Jr. MATHE-MATICS FOR THE CONSUMER. New York: World Book, 1947. x, 438 p. Answers to Tests, pp. 422-32. Tables for Reference, pp. 433-34. Index, pp. 435-38.

A text for a course in consumer finance which is devoted to the simple computations necessary for everyday living. Topics include interest computation, consumer credit, investments, and taxation.

Schwartz, Jacob T. LECTURES ON THE MATHEMATICAL METHOD IN ANA-LYTICAL ECONOMICS. Mathematics and Its Applications, vol. 1. New York: Gordon and Breach, 1961. xi, 282 p. Index, pp. 279-82.

A presentation of some important applications of mathematics in economics. The book begins with the Leontief input-output model and then takes up business cycle theory following the ideas pioneered by L. Metzler and Keynes. Finally, general equilibrium is considered where the existence of Walrasian equilibrium is proved. It concludes with an equilibrium model combining neoclassical and Keynesian features. The presentation makes use of theorems and proofs, lemmas, corollaries, and definitions, but many mathematical applications in economics are presented without proof. The economic implications of the mathematical results are stressed. Assumes knowledge of the theories of matrices and linear transformations, calculus, and certain concepts from functional analysis, such as fixed point theorems.

Scott, Melvin R. INVARIANT IMBEDDING AND ITS APPLICATIONS TO OR-DINARY DIFFERENTIAL EQUATIONS: AN INTRODUCTION. Applied Mathematics and Computation Series. Reading, Mass.: Addison-Wesley, 1973. xviii, 215 p. Author Index, pp. 212-13. Subject Index, pp. 214-15.

An introduction to the theory of invariant imbedding and its application in solving linear and nonlinear boundary-value problems, linear initial-value problems, and systems of equations. The presentation is not rigorous mathematically, but requires knowledge of advanced calculus and ordinary differential equations. Extensive use is made of examples to illustrate the concepts. This book would be of interest to engineers, physicists, and applied mathematicians, and can be used as a supplementary text in courses in numerical analysis and ordinary differential equations. The chapters conclude with exercises and bibliographical discussions.

Searle, Shayle R. MATRIX ALGEBRA FOR THE BIOLOGICAL SCIENCES (INCLUDING APPLICATIONS IN STATISTICS). Series in Quantitative Methods for Biologists and Medical Scientists. New York: Wiley, 1966. xii, 296 p. Book List, p. 290. Index, pp. 291-96.

A text for a course in matrix algebra with the main applications being in regression and linear models. Few applications are presented specifically in the biological sciences, making this book suitable for a mixed group of readers, including students of economics and business. Topics include elementary matrix operations, determinants, inverse matrices, rank and linear independence, solution of simultaneous linear equations (especially systems not of full rank) in which the concept of a generalized inverse is introduced, latent roots and vectors, Jacobians, matrix functions, orthogonal and other types of matrices, and direct sums and products. Most of the theorems are proved and illustrated with examples. The chapters conclude with references, and all but two chapters conclude with exercises. Assumes knowledge of high school algebra.

Searle, Shayle R., and Hausman, Warren H.; et al. MATRIX FOR BUSINESS AND ECONOMICS. New York: Wiley-Interscience, 1970. xii, 362 p. Index of Applications, p. 355. General Index, pp. 357-62.

A text for a course in matrix algebra for business and economics students with no more than high school algebra preparation. Topics and applications include characteristic roots and vectors, normality and orthogonality, Markov chains, linear programming, regression analysis, and linear models. Statements and proofs of theorems are contained in the chapter appendixes. The chapters conclude with exercises and problems.

Seneta, Eugene. NON-NEGATIVE MATRICES: AN INTRODUCTION TO THEORY AND APPLICATIONS. New York: Wiley, 1973. x, 214 p. Appendixes: A, Some Elementary Number Theory; B, Some General Matrix Lemmas; C, Unitary Transformations; D, Some Real-variable Theory, pp. 182-95. Bibliography, pp. 196-206. Glossary of Notation and Symbols, p. 207. Author Index, pp. 209-10. Subject Index, pp. 211-14.

An attempt to bring together several aspects of the theory of nonnegative matrices for the benefit of both mathematicians and workers in the applied fields, such as economics, demography, and operations research. The first four chapters are concerned with nonnegative matrices, while the following two develop an analogous theory for infinite matrices. The mathematical prerequisites are: some knowledge of real variable theory, matrix theory, and complex numbers. The appendixes provide reviews of some needed mathematical concepts. Each section concludes with a bibliography, discussion, and exercises.

Shao, Stephen P. MATHEMATICS FOR MANAGEMENT AND FINANCE. 3d ed., rev. Cincinnati: South-Western, 1974. xi, 768 p. Appendix: Tables, pp. 563-720. Answers to Odd-numbered Problems, pp. 721-60. Index, pp. 761-68.

> A text for a one-year introductory course in the mathematics of business and finance for students who have had little or no high school algebra. It is divided into the following four parts: basic and modern mathematics, mathematics in business management, mathematics in investment-basic topics, and mathematics in investment-applications. Topics include decimals and fractions, systems of equations, progressions, logarithms, sets, probability, vectors and matrices, descriptive statistics, ratios and proportions, simple and compound interest, annuities, investment in bonds, depreciation and depletion, capitalization, life annuities, and life insurance. Each concept is illustrated with several examples. A college course based on this book can be used as a substitute for college algebra. The sections conclude with exercises.

Shapiro, Jesse M., and Whitney, D. Ransom. ELEMENTS OF CALCULUS. Merrill Mathematics Series. Columbus, Ohio: Charles E. Merrill, 1970. x, 275 p. Tables, pp. 243-54. Answers to Selected Problems, pp. 257-67. Index, pp. 271-75.

> A text on the introduction to calculus for students in the social, administrative, and biological sciences. Topics include sequences, functions, and limits; the derivative and its application; expansion of functions in series; integration; and functions of several variables. The presentation is fairly rigorous, and many theorems are proved. Assumes knowledge of college algebra. The sections of the book conclude with problems for assignment.

Siders, Ellis L. MATHEMATICS FOR MODERN BUSINESS AND INDUSTRY. New York: Holt, Rinehart and Winston, 1964. xiii, 262 p. Tables. Index, pp. 255-62.

> Applies elementary algebra to everyday business problems such as interest calculations, payrolls, and insurance. One year of high school algebra is helpful, but not essential for understanding. The chapters conclude with problems.

Simpson, Thomas M.; Pirenian, Zareh M.; Crenshaw, Bolling H.; and Riner, John. MATHEMATICS OF FINANCE. 4th ed. Englewood Cliffs, N.J.: Prentice-Hall, 1969. xvii, 572 p.

> A text for a course in business mathematics. Topics include ratios, proportions, and percents; exponents and radicals; quadratic equations; binomial theorem; elementary functions; logarithms; basic statistics; simple and compound interest and discount; annuities and life insurance; amortization and sinking funds; valuation of bonds; and depreciation. Numerous problems and exercises are included, and answers to odd-numbered problems are given.

Smart, D.R. FIXED POINT THEOREMS. Cambridge Tracts in Mathematics, no. 66. New York: Cambridge University Press, 1974. vi, 93 p. Bibliography, pp. 87-92. Index, p. 93.

An introduction to fixed point theorems and their applications in analysis. Since Banach space versions of most of the theorems are given, the reader is assumed to have knowledge of functional analysis. The presentation is rigorous, with theorems proved. This book is useful to anyone desiring information about fixed point theorems and their applications.

Smiley, Malcolm F. ALGEBRA OF MATRICES. Boston: Allyn and Bacon, 1965. xii, 258 p. Appendix: Introduction to Set Theory, pp. 229-48. Bibliography, p. 249. Index of Symbols, pp. 251-52. Index, pp. 253-58.

A text for a course in matrix algebra with topics of special interest for applications designed for students with a good calculus background. Topics include matrix polynomials and functions of matrices, vector spaces, finite games, Euclidean and unitary spaces, and quadratic forms. Each chapter concludes with over two dozen problems.

Smith, Karl J. FINITE MATHEMATICS: A DISCRETE APPROACH. Glenview, Ill.: Scott, Foresman, 1975. vi, 278 p. Appendix I: Review of Elementary Algebra, pp. 332-41; Appendix II: Mathematical Induction, pp. 342-47; Appendix III: Command Summary for BASIC, pp. 348-52; Appendix IV: Probability Distribution of a Binomial Random Variable, pp. 353-58; Selected Answers to Odd-numbered Problems, pp. 359-74. Index, pp. 374-78.

A nonrigorous text in finite mathematics with applications to business, sociology, and the behavioral sciences. Unique features are its emphasis on patterns, as revealed in examples and cases, and its inclusion of an optional chapter on computer programming using BASIC. Computer applications are integrated throughout in problems. Other topics include symbolic logic, set theory, probability, vectors and matrices, systems of equations and inequalities, Markov chains, and game theory. The sections of the book conclude with exercises, and the chapters conclude with summary outlines. Assumes knowledge of high school algebra.

Snyder, Llewellyn R., and Jackson, William F. ESSENTIAL BUSINESS MATHEMATICS. 6th ed. New York: McGraw-Hill Book Co., 1972. xiii, 578 p. Appendixes, pp. 487-546. Answers to Odd-numbered Problems, pp. 547-65. Index, pp. 567-78.

A text for a course in business mathematics for community colleges and business schools. It is based almost entirely on arithmetic and requires no knowledge of algebra. The aim of this book is to increase the student's knowledge and skill in the computation of practical financial problems of a business, civic, and personal nature, and to provide a sound basis for courses in accounting, investments, business finance, money and banking, insurance, real estate, statistics, retailing, and related subjects. A large number of problems are included in the sections. The appendixes include several tables, a number of business graphs, shortcuts in performing arithmetic, and useful formulas.

Springer, Clifford H.; Herlihy, Robert E.; and Beggs, Robert I. ADVANCED METHODS AND MODELS. Mathematics for Management Series, vol. 2. Foreword by G.L. Phillipe. Homewood, Ill.: Richard D. Irwin, 1965. xii, 273 p. General Index, pp. 271-73.

This volume, consisting of nine chapters, is devoted to the mathematics of compound interest, calculus of finite differences, sequences, differential and integral calculus, classical optimization techniques, simulation, mathematical programming, and multiple regression. Assumes knowledge of only high school algebra.

_____. BASIC MATHEMATICS. Mathematics for Management Series, vol. 1. Homewood, Ill.: Richard D. Irwin, 1965. xii, 225 p. General Index, pp. 221-25.

Volume 1 of this four-volume series is devoted to algebra (with aspects of analytic geometry), descriptive statistics, calculus, matrices, and set theory. Assumes knowledge of only high school algebra. Some chapters conclude with problems.

Springer, Clifford H.; Herlihy, Robert E.; and Mall, R.T. PROBABILISTIC MODELS. Mathematics for Management Series, vol. 4. Homewood, Ill.: Richard D. Irwin, 1966. xi, 301 p. Appendix A: Computer Programs, pp. 287-96. Appendix B: Standard Normal Distribution, pp. 297-98. Index, pp. 299-301.

An introduction to probabilistic models with applications and illustrations in Markov processes, queueing, decision theory, and simulation.

_____. STATISTICAL INFERENCE. Mathematics for Management Series, vol. 3. Homewood, Ill.: Richard D. Irwin, 1966. x, 352 p. Appendix A: A Description of the BASIC Computer Language, pp. 307-15. Appendix B: BASIC Computer Programs, pp. 316-48. General Index, pp. 349-52.

A review of statistics through hypothesis testing and interval estimation. Emphasis is placed on computer programs for solving problems.

Squire, William. INTEGRATION FOR ENGINEERS AND SCIENTISTS. Modern Analytic and Computational Methods in Science and Mathematics. New York: American Elsevier, 1970. 302 p. Appendixes, pp. 267-91. Author Index, pp. 293-97. Subject Index, pp. 299-302.

A reference for students and practicing scientists or engineers faced with the problem of evaluating integrals beyond those treated in elementary textbooks. The last three chapters could be used as a basis for a specialized second course in numerical methods. Assumes knowledge of undergraduate differential equations, and omits some important methods such as contour integration and the method of steepest descents. The appendixes consist of a list of doctoral dissertations on integration and integral equations, integration functions, and subroutines, and FORTRAN programs to aid in numerical evaluation of integrals. The sections conclude with problems, and the chapters conclude with bibliographic notes and comments.

Stelson, Hugh E. MATHEMATICS OF FINANCE. New York: D. Van Nostrand, 1957. xii, 327 p. Appendixes A–1 to A–9, pp. 195–206. Appendix Tables, pp. 207–315. Answers to Odd–numbered Exercises, pp. 316–24. Index, pp. 325–27.

A text for a course in mathematics for business students. It covers interest, discount, annuities, stocks and bonds, life insurance, and probability. Numerous examples are provided, and the sections, as well as the chapters, conclude with exercises. The appendixes contain derivations of various interest formulas, a formula for determining the number of years for money to double itself, and the accuracy of linear interpolations in tables of finance.

Stephenson, Geoffrey. MATHEMATICAL METHODS FOR SCIENCE STUDENTS. 2d ed. London: Longman, 1973. ix, 528 p. Answers to Problems, pp. 509–21. Index, pp. 523–28.

A text for a course in undergraduate mathematics for students in the physical sciences who are already familiar with analytical geometry, differential and integral calculus, and the use of sine, cosine, exponential, and logarithmic functions. Topics include real numbers and functions of a real variable, inequalities, limits and continuity, convergence of infinite series, Taylor and Maclauren series, hyperbolic functions, partial differentiation, line and multiple integrals, numerical integration, Fourier series, matrices, groups, vectors, solutions of algebraic and transcendental equations, ordinary differential equations, the Laplace transform, partial differential equations, the calculus of variations, and integral equations. This book is useful to students of the social sciences because of its broad coverage of topics.

_____. MATRICES, SETS AND GROUPS: AN INTRODUCTION FOR STUDENTS OF SCIENCE AND ENGINEERING. New York: American Elsevier, 1966. xi, 164 p. Further Reading, pp. 155–56. Answers to Problems, pp. 157–61. Index, pp. 163–64.

An elementary introduction to matrices, sets, and groups for undergraduates in science and engineering. Topics include matrix algebra, the inverse and related matrices, systems of linear equations, eigenvalues and eigenvectors, diagonalization of matrices, functions of matrices, cyclic and isomorphic groups, Cayley's theorem, and subgroups and cosets. Each concept is illustrated by one or more examples. This book is self-contained and requires only a good high school algebra background.

Stern, Mark E. MATHEMATICS FOR MANAGEMENT. Prentice-Hall Quantitative Methods Series. Englewood Cliffs, N.J.: Prentice-Hall, 1963 ix, 454 p. Index, pp. 451-54.

A text for a one-semester course for first-year graduate students in business who have a college algebra background. Topics include differential and integral calculus, differential and difference equations, matrix algebra, linear programming, utility, and the theory of games. A major application of the theory, such as production management or portfolio selection, is given at the conclusion of each of the ten chapters.

Stoer, Josef, and Witzgall, Christoph. CONVEXITY AND OPTIMIZATION IN FINITE DIMENSIONS I. New York: Springer-Verlag, 1970. ix, 293 p. Bibliography, pp. 269-85. Author and Subject Index, pp. 286-93.

A synopsis of the algebra of linear inequalities, the geometry of polyhedrons, the topology of convex sets, and the analysis of convex functions in an effort to provide a theoretical background for the mathematics of convex optimization. The emphasis is on linear and convex duality theory. The Farkas lemma on linear inequalities, Motzkin's description of polyhedrons, and Minkowski's supporting plane theorem are indispensable elementary tools which are contained in the first three chapters. The last two chapters (chapters 5 and 6) are devoted to two characteristic aspects of duality theory: conjugate functions of polarity on the one hand, and saddle points on the other hand. Assumes familiarity with basic topological concepts and the theory of real functions.

Stoll, Robert R. LINEAR ALGEBRA AND MATRIX THEORY. New York: Dover, 1969. xv, 272 p. References, p. 263. Index of Special Symbols, pp. 264-65. Index, pp. 267-72.

A text for advanced undergraduates and graduates in a course on the theory of matrices viewed against the background of modern algebra. This book is useful in a variety of fields, including economics, psychology, statistics, engineering, physics, and mathematics. The sections conclude with problems.

Stoll, Robert R., and Wong, Edward Tak-wah. LINEAR ALGEBRA. New York: Academic Press, 1968. x, 326 p. Symbols, p. x. Appendix: Notions of Set Theory, pp. 316-21. Index, pp. 322-26.

A text for a one-semester course in linear algebra for students in applied mathematics. Topics include inner product spaces, bilinear and quadratic forms, decomposition theorems for normal transformations, and applications to economics, physics, and chemistry. Assumes knowledge of calculus and linear differential equations. The sections conclude with exercises.

Stone, Harold S. DISCRETE MATHEMATICAL STRUCTURES AND THEIR APPLICATIONS. The SRA Computer Science Series. Chicago: Science Research Associates, 1973. xii, 403 p. Bibliography, pp. 388-92. Index, pp. 393-99. Index to Notation, pp. 400-402.

A text for a one- or two-semester course in discrete mathematics for computer-science and computer-engineering students. It consists of a condensed survey of algebraic structures, with applications to computers, and constitutes a significant part of a modern algebra text. Assumes knowledge of programming computers, in both high-level compiler language and in assembly language, for those portions dealing with computer arithmetic, sequential machines, compilers, and similar topics. The sections conclude with exercises.

Suprunenko, Dmitrii Alekseevich, and Tyshkevich, Regina Iosifouva. COMMUTATIVE MATRICES. Translated by Scripta Technica. New York: Academic Press, 1968. viii, 158 p. Bibliography, pp. 152-55. Subject Index, pp. 157-58.

Originally published in Russian, this book deals with the elementary properties of a system of commutative matrices, with general properties of commutative matrix algebras over an arbitrary field, and with certain classification questions pertaining to the theory of the full matrix of the field of complex numbers. The authors also present several unsolved problems in commutative algebra. This book is useful for scientists interested in the calculus of matrices.

Swartz, Clifford E.; with Swartz, P.A.; and Swartz, B.K. USED MATH FOR THE FIRST TWO YEARS OF COLLEGE SCIENCE. Englewood Cliffs, N.J.: Prentice-Hall, 1973. xii, 270 p. Appendixes, pp. 227-64. Index, pp. 265-70.

This book is not a text, but, rather, it is a reference and reminder on how to use math. It is designed for students who already possess a formal introduction to the topics. The chapter titles are: "Reporting and Analyzing Uncertainty," "Units and Dimensions," "Graphs," "The Simple Functions in Applied Math," "Statistics," "Quadratic and Higher Power Equations," "Simultaneous Equations," "Determinants," "Geometry," "Vectors," "Complex Numbers," "Calculus-Differentiation," "Integration," "Series and Approximations," "Some Common Differential Equations," and "Differential Operators." The appendixes consist of tables and formulas, primarily for engineering. The applications are mainly in engineering, but many topics are applicable in the social sciences.

Takayama, Akira. MATHEMATICAL ECONOMICS. Hinsdale, Ill.: Dryden Press, 1974. xxii, 744 p. Some Frequently Used Notations, pp. xii–xiii. Name Index, pp. 721–26. Subject Index, pp. 727–44.

A text for a graduate course in economic theory and mathematical economics and a reference for professional economists. It emphasizes the analytical and mathematical aspects of economic theory, and it begins at an elementary level and takes the student to the frontier of current research. The eight chapter titles are: "Development of Nonlinear Programming," "The Theory of Competitive Markets," "The Stability of Competitive Equilibrium," "Frobenius' Theorems," "Dominant Diagonal Matrices and Applications," "The Calculus of Variations and the Optimal Growth of an Aggregate Economy," "Multisector Optimal Growth Models," and "Developments of Optimal Control Theory and Its Applications." Many sections have appendixes which are devoted to mathematical reviews. The sections conclude with notes and references. Assumes knowledge of calculus and matrix algebra.

Taylor, Joan Gary; Applebaugh, Gwendolyn Neul; and Anderson, Dan. FINITE MATHEMATICS. San Francisco: Canfield Press, 1973. vi, 442 p. Appendixes, pp. 407–19. Answers to Selected Problems, pp. 420–38. Index, pp. 439–42.

A text in finite mathematics for students in the social and management sciences. Topics include sets, logic, permutations and combinations, probability, matrices, analytical geometry, statistics, and linear programming. Most sections contain exercises, and the chapters conclude with review exercises. The appendixes are devoted to properties of summation, squares and square roots, a method for calculating the median, and the standard normal table.

Telser, Lester G., and Graves, Robert L. FUNCTIONAL ANALYSIS IN MATHEMATICAL ECONOMICS: OPTIMIZATION OVER INFINITE HORIZONS. Chicago: University of Chicago Press, 1972. vii, 152 p. References, pp. 144–48. Index, pp. 149–52.

A collection of four self-contained chapters--the first and third previously published and the second and fourth previously reported. Each of the models involves the determination of the extremum of the present value or the expected present value of a series of revenues, costs, or utilities. In chapters 1 and 3 monopoly problems are considered as extremum problems over an infinite horizon in both discrete and continuous time, with and without constraints. Consumer demand for durable goods is studied as a dynamic economic optimization problem in chapter 2; and an abstract extremum problem is studied in chapter 4. Assumes knowledge of functional analysis and Banach spaces.

Tetra, B.C. BASIC LINEAR ALGEBRA. New York: Harper and Row, 1971. viii, 136 p. Answers to Some Exercises, pp. 127-34. Index, pp. 135-36.

A text on the introduction to vectors and matrices for college freshmen who possess knowledge of only high school algebra. Topics include linear equations, matrices and vectors, linear spaces, and determinants. It deals with abstract concepts such as linear dependence and independence, rank of a matrix, row-echelon form of a matrix, and the basis of a linear space. Each of the seventeen chapters contains about one dozen problems.

Theocharis, Reghinos D. EARLY DEVELOPMENTS IN MATHEMATICAL ECO-NOMICS. Foreword by Lord Robbins. New York: St. Martin's Press, 1961. x, 142 p. Bibliography, pp. 130-37. Index, pp. 139-42.

An assessment of the development of mathematical economics prior to Cournot, that is, prior to 1838. It is devided into periods as follows: the logicians; the probabilists; D. Bernoulli and utility theory; the Milanese school; the French contributions; Francesco Fuoco--the eclectic; the German contribution; and the British authors. The bibliography is divided into the following two parts: a bibliography of mathematical economics before Cournot, and other bibliography.

Theodore, Chris A. APPLIED MATHEMATICS: AN INTRODUCTION. MATH-EMATICAL ANALYSIS FOR MANAGEMENT. 3d ed., rev. Irwin Series in Quantitative Analysis for Business. Homewood, Ill.: Richard D. Irwin, 1975. xv, 611 p. Selected bibliography, pp. 578-80. Answers to Problems Marked with an Asterisk, pp. 581-601. Index of Subjects, pp. 605-11.

A text for a course in mathematics for business schools with em-phasis on quantitative analysis under conditions of uncertainty. It is divided into the following five parts: Boolean algebra with applications; algebra, analytic geometry, and functions; linear programming; calculus; and probability and Bayesian statistics. Concepts are illustrated with numerical examples. The sections contain over one dozen problems grouped according to levels of difficulty. Assumes knowledge of only high school algebra.

Thomas, James W., and Thomas, Ann M. FINITE MATHEMATICS. Boston: Allyn and Bacon, 1973. x, 342 p. Appendixes, pp. 307-12. Bibliography, pp. 313-15. Answers to Exercises, pp. 316-35. Notation Index, pp. 337-38. Index, pp. 339-42.

A text for a one-semester course in finite mathematics for students in business, biology, and the social sciences who have knowledge of high school algebra. Emphasis is on explicit explanation of the concepts, appealing to intuition wherever possible, followed

by examples, to provide insight and motivation, and finally by
exercises, to enable the student to learn by doing. Topics include logic, set theory, combinations and permutations, rules of
probability, Bayes' theorem, probability models, matrices, Markov
chains, linear programming, input-output analysis, and insurance
premium rates. Sections conclude with exercises. The appendixes
consist of tables of binomial probabilities and the axioms and definitions concerning the real numbers.

Thompson, Howard E. APPLICATIONS OF CALCULUS IN BUSINESS AND
ECONOMICS. Menlo Park, Calif.: W.A. Benjamin, 1973. xiii, 492 p.
Index, pp. 488-92.

A text for students in business and economics for use in conjunction with a course in calculus. It deals primarily with applications of mathematical concepts, and each application is accom-
panied by a discussion of the reasons for its interest and other
information necessary for the appreciation of the application.
The applications include the following areas: price theory, macroeconomics, corporate finance, inventory theory, marketing, and
other areas. The problems are designed to guide the student's
study of the chapter and to clear up difficulties that may be encountered.

Thorn, Richard S., and Kwitoski, Stephen A. BUSINESS MATHEMATICS. New
York: Random House, 1971. ix, 355 p. Final Examination, pp. 331-33.
Appendix: Federal Civil Service Entrance Examination, pp. 337-42. Answer
Key for A and B Problems, pp. 343-51. Index, pp. 353-55.

A text for a course in business mathematics in community colleges.
It is divided into the following four parts: useful fundamentals,
percentages in business, simple and compound interest, financial
statements and analysis, and special topics (logarithms, the slide
rule, and calculators and computers). The presentation is arranged
around a number of mathematical concepts which are first explained clearly, then symbolically, and then numerically. Each
chapter concludes with problems organized as a graded text. Assumes knowledge of high school arithmetic.

Thurston, Hugh. CALCULUS FOR STUDENTS OF ENGINEERING AND THE
EXACT SCIENCES. VOL. 1. Englewood Cliffs, N.J.: Prentice-Hall, 1962.
viii, 193 p. Solutions, pp. 165-90. Index, pp. 191-93.

This volume contains chapters A through E on differential and integral calculus, including one chapter on the applications of differentiation. The emphasis is on the techniques of calculus,
rather than on rigorous formal mathematics.

. CALCULUS FOR STUDENTS OF ENGINEERING AND THE EXACT
SCIENCES. VOL. 2. Englewood Cliffs, N.J.: Prentice-Hall, 1963. 208 p.
Solutions, pp. 187-206. Index, pp. 207-8.

Volume 2 contains chapters F through L on the calculus of several variables, differential equations, double limits, and differentials. Emphasis is on the techniques of calculus, rather than on formal mathematics. The sections conclude with exercises.

Tintner, Gerhard, and Millham, Charles B. MATHEMATICS AND STATISTICS FOR ECONOMISTS. 2d ed., rev. New York: Holt, Rinehart and Winston, 1970. xx, 485 p. Appendix A: Suggestions for Further Study, pp. 417-18. Appendix B: Answers to Selected Problems, pp. 419-53. Appendix C: Sources of Numerical Examples, pp. 454-55. Appendix D: Tables, pp. 456-70. Index of Names, pp. 471-72. Index of Mathematical and Statistical Terms, pp. 473-80. Index of Economic Terms, pp. 481-85.

A text for a course in mathematics and statistics for economics students who have knowledge of only high school algebra, but who are familiar with elementary economics. Intuitive proofs and demonstrations are often substituted for mathematical rigor. This book is divided into the following three parts: some applications of elementary mathematics to economics, calculus, and probability and statistics. Topics include systems of linear equations, logarithms, progressions, differentiation with economic applications, homogeneity, elements of integration, the calculus of variations, probability and random variables, sampling, tests of hypotheses, regression and correlation, and index numbers. Exercises follow the sections.

Traub, Joe Fred. ITERATIVE METHODS FOR THE SOLUTION OF EQUATIONS. Englewood Cliffs, N.J.: Prentice-Hall, 1964. xviii, 310 p. Glossary of Symbols, pp. xvi-xviii. Bibliography, pp. 291-304. Index, pp. 305-10.

A presentation of the general theory of iteration algorithms for the numerical solution of equations and systems of equations. Iteration functions are divided into four classes depending on whether or not they use new information at one or at several points or reuse old information. The known iteration functions are systematized and new classes of computationally effective iteration functions are introduced. Assumes knowledge of mathematical analysis and numerical methods.

Vainberg, Mordukhai M. VARIATIONAL METHODS FOR THE STUDY OF NONLINEAR OPERATORS. With a chapter on Newton's method, by L.V. Kantorovich and G.P. Akilov. Translated by A. Feinstein. San Francisco: Holden-Day, 1964. x, 323 p. Supplement, pp. 299-309. Bibliography, pp. 311-19. Index, pp. 321-23.

Originally published in Russian, this book is useful to mathematicians and to workers in applied sciences who are interested in nonlinear systems and their associated nonlinear equations. The emphasis is on proving existence theorems, largely concerning nonlinear integral operators, rather than on developing techniques for the actual solution of nonlinear equations. The supplement deals with the theory of Banach spaces, which is used freely in the presentation.

Valentine, Frederick A. CONVEX SETS. New York: McGraw-Hill Book Co., 1964. ix, 238 p. Appendix, pp. 195-208. Bibliography, pp. 209-32. Index, pp. 233-38.

A text for a one-semester graduate course in convex sets or related courses.

Varga, Richard C. MATRIX ITERATIVE ANALYSIS. Prentice-Hall Series in Automatic Computation. Englewood Cliffs, N.J.: Prentice-Hall, 1962. xiii, 322 p. Appendixes, pp. 298-304. Bibliography, pp. 305-17. Index, pp. 319-22.

A text for graduate students in mathematics, and a reference for research workers in applied fields, on solutions of matrix problems arising from discrete approximations to elliptic partial differential equations based on the mathematical theory of cyclic iterative methods. Topics include comparison theorems, successive over-relaxation iterative methods, derivation and solution of elliptic difference equations, alternating-direction implicit iterative methods, and matrix methods for parabolic partial differential equations. The sections conclude with exercises, and the appendixes contain examples of numerical solutions.

Vazsonyi, Andrew, and Brunell, Richard. BUSINESS MATHEMATICS FOR COLLEGES. Homewood, Ill.: Richard D. Irwin, 1974. ix, 404 p. Appendix: Alternate Assignments, pp. 307-401. Index, pp. 403-4.

A text which attempts to develop the mathematical skills to solve practical business problems. Solutions to examples are spelled out in detail, and the rules, formulas, and examples are set in frames so that the student can quickly find the techniques. Topics include basic arithmetic operations, decimals and fractions, percent, payrolls and taxes, retailing and depreciation, and financial management. Assignment sheets with space for solutions follow the sections of the book.

Vilenkin, Naum Ya. SUCCESSIVE APPROXIMATION. Translated by M.B.P. Slater and J.W. Teller. Popular Lectures in Mathematics, vol. 15. New York: Macmillan, 1964. ix, 70 p. Exercises, p. 69. Answers, p. 70. No index.

First published in Russian in 1961, this book is based on lectures given to ninth- and tenth-grade students at Moscow State University. It presents a number of methods for the approximate solution of equations. These methods, most of which are based on the concept of the derivative, include the methods of chords, the method of iteration, and successive approximation. It is accessible to anyone with a high school algebra background, and it is useful to anyone desiring an introduction to approximate methods of solution.

Washington, Allyn J. AN INTRODUCTION TO CALCULUS WITH APPLICA-TIONS. Menlo Park, Calif.: Cummings, 1972. ix, 355 p. Appendix A: Study Aids, pp. 309-11. Appendix B: Formulas from Geometry, pp. 312-13. Appendix C: Tables, pp. 314-23. Answers to Odd-numbered Exercises, pp. 325-49. Index, pp. 351-55.

A text for a one- or two-semester course in calculus for liberal arts students. Topics include analytic geometry, differentiation and integration of algebraic and elementary transcendental functions, partial derivatives, double integrals, basic statistics, and expansion of functions in series. The topics have been selected on the basis of their applications in business and social science and biology, as well as in the physical sciences and mathematics. This book contains over 350 worked examples and nearly 2,000 exercises.

Wayland, Harold. COMPLEX VARIABLES APPLIED IN SCIENCE AND ENGI-NEERING. New York: Van Nostrand Reinhold, 1970. xi, 350 p. Bibliography, pp. 335-36. Answers to Selected Exercises, pp. 337-43. Index, pp. 345-50.

A text for graduate and undergraduate courses in complex variables for students in engineering and applied science. It stresses the application of the theory of functions of a complex variable to solutions of the common second-order linear differential equations of mathematical physics and the special functions associated with them. Proofs of theorems are not rigorous, and many results are stated without proof. This book is useful to anyone desiring an introduction to the theory of functions of a complex variable from an applied viewpoint. The sections conclude with exercises.

Westlake, Joan R. A HANDBOOK OF NUMERICAL MATRIX INVERSION AND SOLUTIONS OF LINEAR EQUATIONS. Huntington, N.Y.: Kreiger, 1968. viii, 171 p. Appendix A: Glossary of Matrix Terminology, pp. 117-25. Appendix B: Theorems on Matrix Algebra, pp. 126-35. Appendix C: Test Matrices, pp. 136-57. References, pp. 159-64. Symbol Table, pp. 165-66. Index, pp. 167-71.

An account of a great many numerical methods for the inversion of matrices and solutions of systems of linear equations with emphasis on computer solutions.

Wheeler, Ruric E., and Peeples, W.D., Jr. MODERN MATHEMATICS FOR BUSINESS STUDENTS. Belmont, Calif.: Wadsworth, 1969. xii, 589 p. Appendix: Tables, pp. 480-527. Answers, pp. 529-82. Index, pp. 585-89.

A text for a one- or two-semester course in business mathematics.
Topics include linear equations, vectors and matrices, mathematics
of finance, random variables and probability distributions, linear
programming, theory of games and Markov analysis, differential
and integral calculus, and basic statistics. The intuitive approach
is used in place of the mathematical approach whenever possible.
Exercises are included in the sections, and the chapters conclude
with review exercises.

Widder, David Vernon. THE LAPLACE TRANSFORM. Princeton Mathematical
Series, vol. 6. Princeton, N.J.: Princeton University Press, 1946. x, 406 p.
Bibliography, pp. 392-97. Index, pp. 399-406.

Based on lectures given by the author on Dirichlet series and Laplace integrals. It is designed for students who have knowledge
of analysis comparable to that contained in the text of E.C.
Titchmarsh on the theory of functions. Topics include the Stieltjes
integral, moment problem, Tauberian theorems, bilateral Laplace
transform, and Stieltjes transform. The treatment is rigorous, and
useful as a reference on the theory, properties, and applications
of the Laplace transform.

Wilf, Herbert S. CALCULUS AND LINEAR ALGEBRA. New York: Harcourt,
Brace and World, 1966. xiv, 408 p. Appendix: Analytic Geometry,
pp. 381-92. Answers to Selected Exercises, pp. 393-404. Index, pp. 405-8.

A text for a one-year course in calculus and linear algebra for
scientists, including economists. Two-thirds of the book is devoted
to the calculus of functions of one variable while the remaining
one-third is devoted to topics in linear algebra. The final chapter takes up the differential calculus of several variables. The
sections conclude with exercises.

Yaari, Menahem E. LINEAR ALGEBRA FOR SOCIAL SCIENCES. Prentice-
Hall Series in Mathematical Economics. Englewood Cliffs, N.J.: Prentice-
Hall, 1971. xiv, 174 p. Index, pp. 168-74.

A text for a course in linear algebra for undergraduates and graduates in the social sciences. Topics include sets, linear and
finite-dimensional spaces, linear transformations, matrices, inequalities, permutations, characteristic vectors, characteristic
roots, and quadratic forms. The chapters conclude with problems.

Yamane, Taro. MATHEMATICS FOR ECONOMISTS: AN ELEMENTARY SUR-
VEY. 2d ed., rev. Englewood Cliffs, N.J.: Prentice-Hall, 1968. xvii,
714 p. Index, pp. 705-14.

A collection of topics from various branches of mathematics that
are frequently used in economics. It is divided into the following
four surveys: calculus, differential and difference equations,

matrix algebra, and statistics. The topics are discussed in a heuristic manner; the selection of topics has been based on the need
of economists; the emphasis is on surveying the mathematics, and
not on application; and the problems have been kept simple to
avoid complicated algebraic manipulation. Knowledge of the material in this book should assist one in reading the mathematically
oriented articles in such journals as the AMERICAN ECONOMIC
REVIEW and the REVIEW OF ECONOMIC STATISTICS. Assumes
knowledge of only high school algebra. The sections conclude
with problems.

Youse, Bevan K., and Stalnaker, Ashford W. CALCULUS FOR STUDENTS OF
BUSINESS AND MANAGEMENT. Scranton, Pa.: International Textbook,
1967. viii, 271 p. Tables, pp. 245-51. Answers to Odd-numbered Exercises,
pp. 255-67. Index, pp. 269-71.

A text for a one- or two-semester course in calculus through differential equations for undergraduates and graduates in business
administration and management. Numerous examples are used to
illustrate the concepts, and proofs of theorems are omitted if they
do not contribute to an understanding of the techniques of calculus. Exercises follow the sections of the book.

Part II

ECONOMICS

Chapter 2
MICROECONOMICS

Abraham, Claude, and Thomas, André. MICRO-ECONOMICS: OPTIMAL DECISION-MAKING BY PRIVATE FIRMS AND PUBLIC AUTHORITIES. Translated by D.V. Jones. Boston: Reidel, 1973. xix, 507 p. No index.

Originally published in French in 1970, this book is a text in microeconomics for students who have knowledge of differential and integral calculus. It is divided into two parts: general economic theory, including production, consumption, equilibrium, the social economic optimum, and discounting; and the theory of production, including marginal costs and investment choice, public investment, replacement of plant and machines and their amortization, the relationships between economic theory and accounting, and an introduction to decision theory. The models are formulated in mathematical terms. Chapters conclude with notes and references.

Allingham, Michael. EQUILIBRIUM AND DISEQUILIBRIUM: A QUANTITATIVE ANALYSIS OF ECONOMIC INTERACTION. Cambridge, Mass.: Ballinger Publishing Co., 1973. xi, 161 p. Appendixes, pp. 143-54. References, pp. 155-57. Index, pp. 159-61. About the Author, p. 161.

A presentation of a model of an economy in which the formulation of equilibrium is completely explained by the independent optimizing behavior of individual agents. It is divided into the following three parts: concepts, model, and analysis. The model is completely specified and numerical estimates of the parameters are obtained to explore the dynamic properties of the system. The appendix contains the notation, parameters, and solutions for the study. Knowledge of mathematical analysis is helpful.

Arrow, Kenneth J. ASPECTS OF THE THEORY OF RISK BEARING. Helsinki: Academic Book Store, 1965. 61 p. Historical and Bibliographical note, pp. 57-61. No index.

Three lectures delivered by the author in December 1963 in Helsinki under the sponsorship of the Yrjo Jahnsson Foundation. The first lecture is a proof of the expected utility theorem, and the

second is on the theory of risk aversion in which an attempt is
made to explain the holding of money by the fact that people
are basically risk averse. The third lecture contains the author's
views on insurance, risk, and resource allocation. The conclusion is that the price system fails to account adequately for the
risk bearing.

_____. ESSAYS IN THE THEORY OF RISK BEARING. Chicago: Markham,
1971. viii, 278 p.

Twelve essays dealing with economic behavior in the presence of
uncertainty built around the series of Yrjo Jahnsson Foundation
lectures delivered on December 16-18, 1963, in Helsinki, Finland.
Applications of risk to insurance, resource allocation, welfare,
medical care, production, and demand are considered. Knowledge
of probability theory and elementary differential and integral calculus is assumed.

_____. SOCIAL CHOICE AND INDIVIDUAL VALUES. 2d ed. Cowles
Foundation Monograph, 12. New Haven, Conn.: Yale University Press,
1963. 124 p.

A pioneering attempt to construct a formal procedure for passing
from a set of individual tastes and preferences to a pattern of
social decision making. Topics include: the social welfare function,
the compensation principle, and the general possibility theorem
for social welfare functions. The second edition consists of the
unrevised first edition (published in 1951) with an additional chapter, "Notes on the Theory of Social Choice, 1963," which contains the author's reflections, omissions from the first edition, and
comments on the literature published from 1951 to 1963.

Arrow, Kenneth J., and Hahn, Frank Horace. GENERAL COMPETITIVE ANALYSIS.
Mathematical Economic Texts, no. 6. San Francisco: Holden-Day, 1971.
xii, 452 p. Mathematical Appendixes: A, Positive Matrices; B, Convex and
Related Sets; C, Fixed Point Theorem and Related Combinatorial Algorithms,
pp. 370-427. Bibliography, pp. 428-37. Author Index, pp. 438-40. Subject Index, pp. 440-52.

A presentation in mathematical form of the basic theory of general
competitive equilibrium. Much of the book is devoted to the
proof of the existence and uniqueness of competitive equilibrium
under broad assumptions. The mathematical techniques used are
derived from the theory of functions of real variables and from
fixed point theorems. Stress is placed on the economic significance of the theorems. Each chapter concludes with notes on
the historical development of the subject matter treated.

Barrett, Nancy Smith. THE THEORY OF MICROECONOMIC POLICY. Lexington, Mass.: D.C. Heath, 1974. xiv, 308 p. Index, pp. 303-8.

A text for a microeconomic theory course beyond the introductory level for students who have a high school algebra background. It is divided into the following three parts: the theory of efficient resource allocation, tools of analysis, and beyond the static competitive model. In addition to the standard neoclassical topics a number of contemporary topics are considered, including behavior under uncertainty, linear programming and duality, input-output analysis, behavioral theories of the firm, and stability analysis. An entire chapter is devoted to capital theory and resource allocation over time. The use of mathematical notation is only slightly higher than that found in traditional works. Chapters conclude with questions and additional reading.

Baumol, William J. ECONOMIC DYNAMICS: AN INTRODUCTION. 3d ed., rev. New York: Macmillan, 1970. xix, 472 p. A Short Reading List, pp. 449-56. Answers, pp. 457-61. Index, pp. 465-72.

An introduction to changes in economic magnitudes over time which is divided into the following six parts: the magnificent dynamics, statics involving time, process analysis, single equation models, simultaneous equation models, and newer growth models. Although the mathematics is taken up as needed, prior knowledge of calculus and differential equations is helpful. Some chapters conclude with appendixes, and many sections contain exercises.

_____. ECONOMIC THEORY AND OPERATIONS ANALYSIS. 3d ed. Prentice-Hall International Series in Management. Englewood Cliffs, N.J.: Prentice-Hall, 1972. xiii, 626 p. Answers to Problems, pp. 611-15. Index, pp. 616-26.

A text for graduate and undergraduate courses in microeconomic theory with some applications to operations research. Part 1 is on analytic tools of optimization with an introductory example from inventory analysis; part 2 takes up microeconomic analysis; part 3 considers recent developments in mathematical economics, including input-output analysis and game theory; and part 4 is a short postscript on computers. Chapters contain exercises and references.

Becker, Gary S., with Grossman, M., and Michael, R.T. ECONOMIC THEORY. New York: Knopf, 1971. xii, 222 p. General Problems, pp. 210-13. Bibliography, pp. 214-18. Index, pp. 219-22.

A text in price theory for beginning graduate students who have knowledge of calculus. It is divided into the following four parts: demand analysis, supply of products, production and the demand for factors, and the supply of factors of production. Extended mathematical discussions are placed in footnotes and chapter appendixes, and all statements proved mathematically are also proved geometrically or verbally so that students with minimal mathematics training can follow. Sections of the book conclude with problems.

Black, Duncan. THE THEORY OF COMMITTEES AND ELECTIONS. New York: Cambridge University Press, 1958. xiii, 242 p. Index, pp. 239-42.

A summary of the pioneering work on the analytical theory of committees and elections by the author, including the theorem on single-peaked preferences. It is divided into two parts: the theory of committees and elections, and history of the mathematical theory of committees and elections (excluding proportional representation). There are no mathematical prerequisites.

Borch, Karl Henrik. THE ECONOMICS OF UNCERTAINTY. Princeton Studies in Mathematical Economics, vol. 2. Princeton, N.J.: Princeton University Press, 1968. vii, 227 p. Index, pp. 225-27.

An analysis of decision making under uncertainty with applications to economic problems. Topics include the Bernoulli principle, portfolio selection, the two-person zero-sum game and the general two-person game, and group decisions. Prerequisites include matrix algebra and elementary calculus. Chapters conclude with references.

Bowles, Samuel, and Kendrick, David [A.], et al. NOTES AND PROBLEMS IN MICROECONOMIC THEORY. Markham Economics Series. Chicago: Markham, 1970. xi, 200 p. No index.

A collection of annotated reading lists and sets of problems complete with detailed answers on microeconomic theory for use as a supplement to graduate theory courses and for preparation for the general examination for the Ph.D. degree in economics. Assumes knowledge of calculus. The problems cover consumer theory, production theory, markets, and income distribution. The problems are typical of those encountered in courses in mathematics for economists courses.

Bródy, András. PROPORTIONS, PRICES AND PLANNING: A MATHEMATICAL RESTATEMENT OF THE LABOR THEORY OF VALUE. New York: American Elsevier, 1970. 194 p. Appendix I: Definitions and Theorems, pp. 171-73. Appendix II: The Resolvent, pp. 175-78. Appendix III: Turnover Time and Life Span, pp. 179-85. References, pp. 187-89. Index, pp. 191-94.

A restatement of Marx's original theories of value in mathematical terms, and an attempt to relate these theories to modern quantitative economic concepts such as game theory, open and closed static and dynamic Leontief systems, linear programming, the mathematical theory of optimal processes, and other general equilibrium systems. The methodology draws heavily on the eigenvalue-eigenvector resolution of matrices. Rigorous mathematical theorems and proofs are located in the appendixes.

Buchanan, James M., and Tullock, Gordon. THE CALCULUS OF CONSENT: LOGICAL FOUNDATIONS OF CONSTITUTIONAL DEMOCRACY. Ann Arbor:

University of Michigan Press, 1962. x, 361 p. Appendix 1: Marginal Notes
on Reading Political Philosophy, by J.M. Buchanan, pp. 302-22. Appendix
2: Theoretical Forerunners, by G. Tullock, pp. 323-40. Notes, pp. 341-61.
No index.

> An attempt to explain the decision process of governmental allo-
> cation of resources by constitutional choice in four parts: the con-
> ceptual framework, the realm of social choice, analysis of decision-
> making rules, and the economics and ethics of democracy. One
> model attempts to determine if a given type of activity is to be
> collectivized or left private, and another model is developed to
> explain decision making under majority rule. This book makes a
> highly original contribution in the relation between economics
> and politics, and it provides valuable insights into the theory of
> government decision making.

Burstein, Meyer Louis. ECONOMIC THEORY: EQUILIBRIUM AND CHANGE.
New York: Wiley, 1968. xiv, 335 p. Author Index, pp. 327-29. Subject
Index, pp. 330-35.

> A text for a one-semester course in microeconomic theory for un-
> dergraduates who are familiar with elementary differential and in-
> tegral calculus and matrix algebra. It is divided into the follow-
> ing four parts: introduction, microeconomic theory, general equi-
> librium, and macrodynamics. Special features are: the theory of the
> firm from the viewpoint of activity analysis, integration of the theory
> of finance with the theory of the firm, monopoly theory as a special
> case of linear programming, integration of imperfect competition theory
> with information cost control and with Keynesian economics, and the
> use of the Pontryagin principle in standard optimization procedures.
> Economic laws and models are presented in mathematical form, and
> no proofs are provided. Chapters conclude with references.

Carlson, Sune. A STUDY ON THE PURE THEORY OF PRODUCTION. 1939.
Reprint. New York: Kelley and Millman, 1956. vii, 128 p. Selected Bibliography, pp. 127-28. No index.

> A presentation of the classical theory of production of a single
> firm, which includes joint production and joint costs, and an in-
> troduction to poly-periodic production in which inputs and costs
> of one period are connected with future outputs as well as pre-
> sent outputs. Most of the relationships are expressed mathemati-
> cally, and knowledge of calculus is assumed.

Champernowne, David Gawen. UNCERTAINTY AND ESTIMATION IN ECONOM-
ICS. VOL. 1. Mathematical Economics Texts, no. 2. San Francisco: Holden-Day,
1969. viii, 280 p. Selected Bibliography, pp. 274-75. Index, pp. 277-80.

> Contains chapters 1 to 9 of a three-volume series on uncertainty
> and estimation in economics. Topics include imperfect knowledge
> in economic theory, preference orderings and probability orderings,

probability and irrelevance, random variables, Bayes' procedure
for interpreting statistical evidence, and description of measurements. The mathematical results are usually stated without proof
and illustrated by example. Most chapters contain appendixes devoted to additional mathematical reviews, and the chapters conclude with up to two dozen problems each. Assumes knowledge
of calculus and matrix algebra.

_____. UNCERTAINTY AND ESTIMATION IN ECONOMICS. VOL. 2. Mathe-
matical Economics Texts, no. 3. San Francisco: Holden-Day, 1969. vi,
426 p. Selected Bibliography, pp. 421-22. Index, pp. 423-26.

Contains chapters 10 to 18 on topics in mathematical statistics
from an economic point of view, including testing hypothesis,
time series, and sequential sampling. One chapter is devoted to
an example of estimation from a hypothetical bivariate population
of 2,000 elements to illustrate cluster sampling, ratio estimates,
stratified sampling, sampling with paired clusters, and systematic
sampling. Assumes knowledge of partial differentiation, integral
calculus, and matrix algebra. Most chapters have appendixes devoted to additional mathematical reviews, and all chapters conclude with over one dozen problems each.

_____. UNCERTAINTY AND ESTIMATION IN ECONOMICS. VOL. 3. Mathe-
matical Economics Texts, no. 4. San Francisco: Holden-Day, 1969. v,
108 p. Selected Bibliography, p. 105. Index, pp. 107-8.

Contains chapters 19 to 22 of a three-volume series. Chapter
titles are: "Economic Effects of Uncertainty and Ignorance about
the Future," "Choice in Situations of Risk," "Planning in the
Face of Risk," and "Uncertainty, Finance, and the Management
of the Economy." Topics and applications include Bernoulli's
procedure applied to utility, portfolio selection, inventory storage,
project evaluation, and minimax procedures. Prerequisites are
differential and integral calculus and the calculus of probability.
Chapters conclude with problems.

Coddington, Alan. THEORIES OF THE BARGAINING PROCESS. Foreword by
G[eorge]. L[ennox]. S[harman]. Shackle. Chicago: Aldine, 1968. xx, 106 p.
Bibliography, pp. 99-101. Index, pp. 103-6.

An exposition of recent developments in the pure theory of bargaining. Topics include models of Edgeworth, von Neumann and
Morgenstern, Zeuthen, Nash, Pen, Foldes, and Bishop; a detailed
explanation of the CROSS model; the consistency of various bargaining models; and the use of game theory for solving bargaining
problems. Assumes knowledge of differential and integral calculus.

Cohen, Kalman J., and Cyert, Richard M. THEORY OF THE FIRM: RESOURCE
ALLOCATION IN A MARKET ECONOMY. Prentice-Hall International Series

in Management. Englewood Cliffs, N.J.: Prentice-Hall, 1965. xx, 406 p. Index, pp. 399-406.

A text for a one-semester course for undergraduates who have knowledge of calculus. It is divided into three parts: business firms, decision making and economic models; market structures and the theory of the firm; and new approaches to the theory of the firm. Stress is placed on the role of firms in the resource allocation process, explaining in detail the decision-making processes of firms in a variety of market structures. Each of the eighteen chapters concludes with references.

Coleman, James. THE MATHEMATICS OF COLLECTIVE ACTION. Methodological Perspectives Series. Chicago: Aldine, 1973. ix, 191 p. References, pp. 161-65. Computer Program and Output, pp. 167-85. Index, pp. 187-91.

Presents a mathematical basis for a theory of collective decisions. It first develops a concept of man's action in terms of causal processes and determinants of behavior, as illustrated in empirical studies in sociology. A second concept is developed in which man's action is based on preference, as illustrated in rational behavior of economic theory. The latter concept utilizes classical microeconomic theory, statistical decision theory, and the theory of games. Examples of the voting behavior of committees are included, as well as computer programs which solve the equilibrium equations and give the final results of the voting. Assumes knowledge of calculus and probability theory.

Cournot, Augustin. RESEARCHES INTO THE MATHEMATICAL PRINCIPLES OF THE THEORY OF WEALTH. Translated by N.T. Bacon. 1897. Reprint. Reprints of Economic Classics. New York: Augustus M. Kelley, 1960. xxiv, 213 p. Bibliography of Mathematical Economics from Ceva to Cournot (1711 to 1837), pp. 173-209. Index of Writers, pp. 211-13.

This book was first published in French in 1838. It includes an essay on Cournot and mathematical economics and a bibliography of mathematical economics by Irving Fisher. It is one of the first, if not the first, major work on the application of mathematics to economic analysis.

Cross, John G. THE ECONOMICS OF BARGAINING. Foreword by M[artin]. Shubik. New York: Basic Books, 1969. xvi, 247 p. Appendix A: Payoff Deterioration, pp. 223-26. Appendix B: The Case of Discontinuities in the Outcome Set, pp. 227-30. Appendix C: A utility Dependent Model, pp. 231-32. Appendix D: Pareto Optimality, pp. 232-33. Appendix E: An N-person Game, pp. 236-43. Index, pp. 245-47.

An attempt to treat the bargaining process in terms of economic theory in order to arrive at a determinate and useful theory. Topics include a review of models of bargaining; arbitration; the use of force, threats, and promises; bluffing; dynamic theory of

preagreement costs; and the control of disagreement. The application of theory is in oligopoly. Chapters conclude with notes
and references. Assumes knowledge of calculus.

Cyert, Richard M., and March, James G. A BEHAVIORAL THEORY OF THE
FIRM. Prentice-Hall Behavioral Sciences in Business Series. Englewood Cliffs,
N.J.: Prentice-Hall, 1963. ix, 332 p. Appendix A: Assumption, Predic-
tion, and Explanation in Economics, by R.M. Cyert and E. Grunberg, pp. 293-311.
Appendix B: Computer Models in Dynamic Economics, by K.J. Cohen and
R.M. Cyert, pp. 312-25. Index, pp. 327-32.

A presentation of the decision-making process of the business firm.
Topics include organizational goals, choice, and expectations;
price and output models; a model of rational managerial behavior;
a model of trust investment behavior; and prediction and explanation in economics. The presentation is nonmathematical, although
calculus is used in portions of the book.

Daniel, Coldwell III. MATHEMATICAL MODELS IN MICROECONOMICS.
Boston: Allyn and Bacon, 1970. x, 229 p. Index, pp. 215-28.

A presentation of the usual topics in contemporary microeconomic
theory of the firm, including general equilibrium, by means of
mathematical models. It is useful as a text in microeconomic
theory at the undergraduate or beginning graduate levels. Assumes knowledge of set theory, calculus, matrix algebra, and
probability theory at the level contained in books such as T. Yamane's
MATHEMATICS FOR ECONOMISTS (Prentice-Hall, 1968). Chapters
conclude with references.

David, Martin H.; Gates, William A.; and Miller, Roger F. LINKAGE AND
RETRIEVAL OF MICROECONOMIC DATA: A STRATEGY FOR DATA DEVELOP-
MENT AND USE. A Report of the Wisconsin Assets and Incomes Archives.
Lexington, Mass.: D.C. Heath, 1974. xv, 297 p. Appendix A: WAIS
Bibliography, pp. 163-64. Appendix B: Assignment of Identifiers to Taxpayers
and Beneficiaries, pp. 165-69. Appendix C: Beneficiary Record Structure,
pp. 171-72. Appendix D: Documentation on Program UPDATEAL, pp. 173-75.
Appendix E: Database Document for Merged Master File 1946-64, pp. 177-266.
Appendix F: Detecting Inconsistencies in the Master File, pp. 267-79. Notes,
pp. 281-86. Bibliography, pp. 287-90. General Index, pp. 291-93. Data
Descripter Index, pp. 295-96. About the Authors, p. 297.

A research report on the development of a large scale data collection system (data archive) using microeconomic data on individuals, families, and businesses from many sources, such as the
Social Security Administration and the Wisconsin Department of
Revenue. It develops procedures for retrieving data from disparate
document systems and for linking these data into a larger matrix
for wide variety of analytical purposes. Topics include a brief
history of the Wisconsin Assets and Incomes Archive Studies (WAIS);

the nature and nurture of linked information; methodology of data development; error detection and error correction; documents and procedures for the control of data processing and for the establishment of permanent records; and the use of pointers to link records in an archive. While the major portion of the book deals with procedures in constructing an archive, chapter 3 undertakes a statistical analysis of the data in the WAIS archive to show the range of problems that can be solved. Knowledge of basic statistics and computer programming, such as COBOL, is helpful.

Debreu, Gerard. THEORY OF VALUE: AN AXIOMATIC ANALYSIS OF ECONOMIC EQUILIBRIUM. Cowles Foundation Monograph, no. 17. New York: Wiley, 1959. xii, 114 p. References, pp. 103-7. Index, pp. 109-14.

A monograph which presents a rigorous, axiomatic development of two problems in the theory of value: (1) the price-making process in a free enterprise economy; and (2) the role of prices in an optimal state of an economy. The presentation is organized around the concept of the value function defined on a commodity space, and assumes knowledge of real analysis, topology, and set theory. The chapter topics are mathematics, commodities and prices, producers, consumers, equilibrium, optimum, and uncertainty. Chapters conclude with notes.

De Jong, Fritz J. DIMENSIONAL ANALYSIS FOR ECONOMISTS. Contributions to Economic Analysis, vol. 50. Amsterdam: North-Holland, 1967. xiii, 220 p. Mathematical Appendix: The Algebraic Structure of Dimensional Analysis, by W. Quade, pp. 143-99. Index of Names, pp. 201-6. Index of Subjects, pp. 207-20.

Describes and illustrates, by examples, dimensional analysis in economics. Except for the mathematical appendix, it is not a rigorous treatment and requires knowledge of only calculus. Nearly all examples are widely known and include the equation of exchange, supply of labor, Cobb-Douglas and CES production functions, the monetary model of the Netherlands Bank, simple and compound interest, and applications to econometrics. The exposition begins with simple concepts and examples, and proceeds to more advanced grades of rigor; the most advanced topic is reached in the mathematical appendix where the algebraic structure of dimensional analysis is developed. Dimensional analysis is an aid to the formulation of economic models.

Dierker, Egbert. TOPOLOGICAL METHODS IN WALRASIAN ECONOMICS. Lecture Notes in Economics and Mathematical Systems, vol. 92. New York: Springer-Verlag, 1974. iv, 130 p. Some Standard Notation, p. 126. References, pp. 127-30.

A presentation of general equilibrium theory in terms of topological concepts, primarily differentiable manifolds and mappings.

Topics include regular equilibria, Scarf's example, Debreu's theorem on the finiteness of the number of equilibria of an economy,
continuity of the Walras correspondence for demand functions, density of trasversal intersection, and regular economies. Assumes
knowledge of topology.

Dmitriev, V.K. ECONOMIC ESSAYS ON VALUE, COMPETITION AND UTIL-
ITY. Translated by D. Fry. Edited with an introduction by D.M. Nuti.
New York: Cambridge University Press, 1974. 231 p. Bibliography, pp. 221-24.
Index, pp. 225-31.

Originally published in Russian in 1904, this book consists of three
essays by the author who is considered the founder of Russian mathematical economics. In the first essay, "The Theory of Value of
David Ricardo," the author anticipates much of the Leontief and
Sraffa systems. The second essay, "The Theory of Competition of
Augustine Cournot," contains many ideas which form the basis of
present-day developments of imperfect or monopolistic competition
and oligopoly. The final essay, "The Theory of Marginal Utility,"
deals with the marginal utility background of demand.

Ehrenberg, A.S.C. REPEAT-BUYING: THEORY AND APPLICATIONS. New
York: American Elsevier, 1972. xviii, 322 p. Appendix A: A Worked Example, pp. 263-89. Appendix B: Some Useful Tables, pp. 291-310. Bibliography, pp. 311-15. Author Index, pp. 317-18. Subject Index, pp. 319-22.

A presentation of empirical and theoretical results concerning consumers' repeat-buying behavior and frequently bought brand-name
goods. It is based on the author's research and emphasis is placed on
the application of the negative binomial and logarithmic probability
distributions. The book is divided into the following parts: introduction, repeat-buying, practical applications, mathematical theory, buying more than one brand, and research and development.
Assumes knowledge of calculus and probability theory.

Elliott, Jan Walter. ECONOMIC ANALYSIS FOR MANAGEMENT DECI-
SIONS. The Erwin Series in Economics. Homewood, Ill.: Richard D. Irwin,
1973. xi, 378 p. Appendix, pp. 353-61. Index, pp. 365-78.

A text for a course in managerial economics which attempts to
integrate the tools of econometrics with microeconomic theory.
It is divided into three parts: tools for measuring economic relationships, including a survey of the available methods such as
statistical inference and regression; analysis of demand, cost, and
market prices with emphasis on applications of the tools; analysis
of economic behavior of firms, including linear programming and
simulation. Assumes knowledge of calculus. Chapters conclude
with questions. The appendix consists of statistical tables.

Evans, Griffin C. MATHEMATICAL INTRODUCTION TO ECONOMICS. New
York: McGraw-Hill Book Co., 1930. xi, 177 p. Appendix I: Bibliography

of Collateral Reading in English, pp. 165-66. Appendix II: Economics and
the Calculus of Variations, pp. 167-73. Index, pp. 175-77.

> A presentation of the following economic topics by means of simple
> mathematical methods: monopoly, competition, demand and cost,
> change of units, variable price, rates of change, equation of ex-
> change, utility, theory of production, economic dynamics, appli-
> cation of integration, and an introduction to the calculus of varia-
> tions. One of the early texts, this book is now mainly of his-
> torical interest.

Fabrycky, Wolter J., and Thuesen, G.J. ECONOMIC DECISION ANALYSIS.
Englewood Cliffs, N.J.: Prentice-Hall, 1974. x, 390 p. Appendixes,
pp. 347-84. Index, pp. 385-90.

> A text for a variety of college courses, such as industrial eco-
> nomics, business economics, agriculture and forest economics, and
> engineering technology economics. It presents the technique for
> optimizing decisions in four parts: introduction, in which basic
> concepts such as supply and demand are discussed; evaluating
> economic equivalence, where interest calculations, depreciation
> and taxes, and benefit-cost analysis are considered; estimates,
> risk, and uncertainty, including decision theory; and economic
> decision models, including break-even analysis, inventory and
> queueing problems, and linear programming. Assumes knowledge
> of college algebra. The appendixes consist of an illustration of
> the simplex method, methods of forming mutually exclusive alter-
> natives, interest tables, and selected references.

Fair, Ray C. THE SHORT-RUN DEMAND FOR WORKERS AND HOURS. Contributions to Economic Analysis, vol. 59. New York: American Elsevier,
1969. xii, 225 p. Data Appendix, pp. 209-21. Bibliography, pp. 222-23.
Index, pp. 224-25.

> A model of the short-run demand for worker and for hours paid per
> worker is developed and tested in an effort to measure increasing
> returns to labor services and of increasing short-run returns to scale.
> The model was estimated using seasonally unadjusted monthly data
> for seventeen, three-digit U.S. manufacturing industries. Assumes
> knowledge of econometric techniques.

Ferguson, Charles E. THE NEOCLASSICAL THEORY OF PRODUCTION AND
DISTRIBUTION. New York: Cambridge University Press, 1969. xviii, 384 p.
References, pp. 365-77. Author Index, pp. 379-81. Subject Index, pp. 382-84.

> A text for a course in microeconomics with a prerequisite of dif-
> ferential calculus. It is divided into two parts: the macroeco-
> nomic theories of distribution, and technical progress. A systematic
> and thorough statement of the neoclassical theory of production and
> distribution is provided.

Fishburn, Peter C. DECISION AND VALUE THEORY. Publications in Operations Research, no. 10. New York: Wiley, 1964. xvi, 451 p. Author Index, pp. 439-41. Subject Index, pp. 443-51.

A formulation of the theory of decision based on expected relative values or expected utilities of strategies with the criterion of choice being the maximization of the decision maker's expected relative value. The concept of independence or additivity in value theory is considered in detail. The book is based primarily on the author's research experience in the field. Assumes knowledge of intermediate probability theory. Chapters conclude with references.

_____. THE THEORY OF SOCIAL CHOICE. Princeton, N.J.: Princeton University Press, 1973. xii, 264 p. References, pp. 255-59. Index, pp. 261-64.

Presents several types of social choice functions based on preferences of individuals in a given society. It is divided into three parts: social choice between two alternatives, which includes a variety of majority-like functions; simple majority social choice, which focuses on social choice among many alternatives when two-element feasible subset choices are based on simple majority; and a general study of aspects and types of social choice functions for many alternatives. Assumes knowledge of elementary set theory and linear algebra.

_____. UTILITY THEORY FOR DECISION MAKING. Publications in Operations Research, no. 18. New York: Wiley, 1970. xiv, 234 p. Answers to Exercises, pp. 215-21. References, pp. 223-27. Author Index, pp. 229-30. Subject Index, pp. 231-34.

A self-contained book on utility theory which includes practically all the necessary mathematics. It is based on binary relations of set theory, but concepts from algebra, group theory, topology, probability theory, and the theory of mathematical expectations are also used. The book is divided into the following three parts: utilities without probabilities, expected-utility theory, and states of the world. Topics include: additive utilities on finite and infinite sets, preferences on homogeneous product sets, expected utility for strict partial orders, expected utility for probability measures, additive expected utility, and Savage's expected-utility theory. Chapters conclude with exercises, which form an integral part of the book.

Fisher, Franklin M., and Shell, Karl. THE ECONOMIC THEORY OF PRICE INDICES: TWO ESSAYS ON THE EFFECTS OF TASTE, QUALITY, AND TECHNICAL CHANGE. Economic Theory and Mathematical Economics Series. New York: Academic Press, 1972. xv, 117 p. Index, pp. 115-17.

Essay 1 is on the pure theory of the true cost-of-living index which is based on the theory of consumer demand. Under given assumptions about consumer tastes, the Paasche formula will be a more reliable

guide to the true cost-of-living index than the Laspeyres, and
under other assumptions the Laspeyres will be a more reliable
guide. The introduction of new products and quality change are
also considered. Essay 2 attempts to build a theory of national-
output deflators based on the theory of production, and treats
problems such as changing production possibilities, brought about
by technical change and a change in factor supplies, new goods,
and quality change. Proofs of theorems and lemmas are included,
but these are not rigorous from a mathematical viewpoint. Essay
2 makes use of the general envelope theorem. Each essay concludes with footnotes and references, and knowledge of calculus
and matrix algebra is assumed.

Fisher, Irving. MATHEMATICAL INVESTIGATIONS IN THE THEORY OF
VALUE AND PRICE (1892). APPRECIATION AND INTEREST (1896). 1925.
Reprint. Reprints of Economic Classics. New York: Augustus M. Kelley, 1961.
iv, 126 p. and vii, 98 p.

The first part consists of a reprint from the TRANSACTIONS OF
THE CONNECTICUT ACADEMY, vol. 9, July 1892, and includes
four appendixes (pp. 90-124), and five reviews (pp. 125-26) taken
from different sources. The second part is a reprint of a publication by the American Economic Association in 1896. It contains an appendix of statistical tables (pp. 93-98). This book is
one of the first publications on the use of differential calculus in
economics.

Fouraker, Lawrence E., and Siegel, Sidney. BARGAINING BEHAVIOR. New
York: McGraw-Hill Book Co., 1963. ix, 309 p. Bibliography, pp. 211-14.
Appendixes, pp. 215-306. Index, pp. 307-9.

A report on an experiment in economic conflict of the unequal-
strength, bilateral monopoly type and the oligopoly type. For
each type, the theoretical formulation is presented, the experimental procedures are outlined, and the experimental tests are presented and analyzed. Experiments are conducted on several models
in each type. For example, oligopoly experiments are performed
with duopoly and triopoly models under both complete and incomplete information. The appendixes consist of a tabulation of the
experimental results. This study sheds light on the conditions
which lead to Pareto optimal conflict resolutions, the effect of
the amount of information on equilibrium solutions of price and
quantity, and the relationship beween the number of participants
and equilibrium solutions.

Frank, Charles R., Jr. PRODUCTION THEORY AND INDIVISIBLE COMMOD-
ITIES. Princeton Studies in Mathematical Economics, no. 3. Princeton,
N.J.: Princeton University Press, 1969. xii, 141 p. Appendix, pp. 122-37.
Index, pp. 139-41.

An attempt to apply modern mathematical techniques to production theory with emphasis on indivisibilities and increasing returns to scale. One chapter is devoted to integer activity analysis of production and allocation. The appendix contains all the relevant mathematical definitions, lemmas, and theorems and proofs, making this book fairly self-contained for anyone possessing knowledge of set theory, elementary calculus, and matrix algebra. References are cited in footnotes throughout the text.

Frisch, Ragnar. THEORY OF PRODUCTION. Translated by R.I. Christophersen. Chicago: Rand McNally, 1965. xiv, 370 p. Analytical Summary, pp. 346-66. Index, pp. 367-70.

This book presents the historical development of the theory of production up to the late 1930s and omits later developments such as input-output analysis, constant elasticity of substitution production functions, and empirically estimated production functions. Assumes knowledge of differential and integral calculus, and matrix algebra. The analytical summary consists of definitions of terms arranged according to chapter.

Gale, David. THE THEORY OF LINEAR ECONOMIC MODELS. New York: McGraw-Hill Book Co., for the Rand Corporation, 1960. xxi, 330 p. List of Notations, pp. xv-xvii. Bibliography, pp. 323-26. Index, pp. 327-30.

A text for undergraduates and graduates in a variety of fields, including mathematics, economics, and engineering, and also a reference dealing with linear models. Topics include linear and integral linear programming; two-person games; and linear models of exchange and production, including the Leontief model and von Neumann's model of an expanding economy. The models are introduced by describing the economic problem of interest, followed by assumptions and simplifications, and abstract formulations. Assumes knowledge of calculus and matrix algebra. Each chapter concludes with bibliographical notes and about two dozen exercises.

Goodwin, Richard Murphy. ELEMENTARY ECONOMICS FROM A HIGHER STANDPOINT. New York: Cambridge University Press, 1970. x, 199 p. Appendix: Mathematical Formulation of the Model, pp. 191-98. Index, p. 199.

A text for second year college students from a variety of disciplines, including the social sciences and mathematics. It deals with fairly advanced topics in economic theory using primarily geometric techniques. Topics include the determination of relative prices and outputs, the turnpike theorem, optimal consumption and growth, labor theory of value, capital accumulation, technical progress, input-output analysis, and dynamic programming. Knowledge of vector analysis and linear algebra helpful.

Graaf, J. De V. THEORETICAL WELFARE ECONOMICS. New York: Cambridge University Press, 1967. xii, 178 p. Index of Authors, pp. 123-74. Index of Subjects, pp. 175-78.

> A fairly complete account of the formal theory of welfare economics with the chapter appendixes containing the most rigorous mathematical arguments.

Griliches, Zvi, and Ringstad, Vidar. ECONOMIES OF SCALE AND THE FORM OF THE PRODUCTION FUNCTION: AN ECONOMETRIC STUDY OF NORWEGIAN MANUFACTURING. New York: American Elsevier, 1971. ix, 204 p. Appendix A: Comparative Results Industry by Industry, pp. 128-73. Appendix B: Selected Results from Experimental Runs, pp. 174-93. Appendix C: Various Bias Formulae, pp. 194-98. Appendix D: Error-in-the-variables Bias in Nonlinear Contexts, pp. 199-202. References, pp. 203-4. No index.

> The estimation of production functions of the Cobb-Douglas and CES forms for the Norwegian economy using cross-section studies and data from the 1963 Norwegian Census of Manufacturing Establishments. The appendixes contain analyses of individual industry results, reports on the results of alternative definitions of variables, alternative criteria of sample selection, and proofs of certain statistical formulas. Assumes knowledge of mathematical economics and econometrics.

Hadar, Josef. ELEMENTARY THEORY OF ECONOMIC BEHAVIOR. Addison-Wesley Series in Management Science and Economics. Reading, Mass.: Addison-Wesley, 1966. ix, 332 p.

> A text for an undergraduate course in microeconomic theory with a prerequisite of high school algebra. The graphic illustrations are supplemented by mathematical derivations requiring no calculus. For example, incremental changes in quantities, prices, and so forth, are denoted by the delta notation rather than the derivative. It covers the usual topics in intermediate microeconomic theory, including Walrasian general equilibrium, but does not include extensions to linear programming, input-output analysis, and decision theory. Chapters conclude with summaries, references, and exercises.

_____. MATHEMATICAL THEORY OF ECONOMIC BEHAVIOR. Reading, Mass.: Addison-Wesley, 1971. vii, 375 p. Author Index, pp. 367-69. Subject Index, pp. 370-75.

> A text for a graduate course in microeconomic theory covering the usual topics with extensions to linear programming. Chapters conclude with bibliographic notes. One chapter is devoted to a review of matrix algebra and calculus.

Hansen, Bent. A SURVEY OF GENERAL EQUILIBRIUM SYSTEMS. Economics Handbook Series. New York: McGraw-Hill Book Co., 1970. xiii, 238 p. Index, pp. 229-38.

> A survey of the most important general equilibrium systems encountered in contemporary economic literature, with emphasis on structure and specifications, economic interpretation, and origin of the systems associated with the names of Walras, Cassel, Lindahl, Keynes, Hicks, Samuelson, Patinkin, Leontief, von Neumann, and Solow, as well as other writers. All models are formulated in terms of systems of equations, and knowledge of elementary calculus is assumed. Chapters conclude with references.

Hempenius, Anton Leendert. MONOPOLY WITH RANDOM DEMAND. Rotterdam: Rotterdam University Press, 1970. viii, 87 p. Appendixes, pp. 74-83. References, p. 85. Index, p. 87.

> A study of monopoly in which there is one random component: demand. Various types of random demands are employed: homoskedastic demand with linear and nonlinear costs; heteroskedastic demand; and demand based on lognormal distribution. The seven appendixes are devoted to mathematical derivations and proofs of theorems. Assumes knowledge of elementary differential and integral calculus and probability theory.

Henderson, James M., and Quandt, Richard E. MICROECONOMIC THEORY: A MATHEMATICAL APPROACH. 2d ed., rev. Economic Handbook Series. New York: McGraw-Hill Book Co., 1971. xvi, 431 p. Appendix: Mathematical Review, pp. 383-420. Index, pp. 423-31.

> A text for a course in microeconomic theory for graduates and undergraduates who possess knowledge of differential and integral calculus. It consists of a mathematical treatment of classical microeconomic theory and some of the modern approaches based on linear models, including linear programming, game theory, and input-output analysis. Emphasis is on methods and applications rather than on detailed proofs, although some theorems, such as the existence of competitive equilibrium, are proved. Chapters conclude with exercises and references.

Hibdon, James E. PRICE AND WELFARE THEORY. New York: McGraw-Hill Book Co., 1969. xv, 492 p. Mathematical Appendix, pp. 459-80. Index, pp. 481-92.

> A text for a course in microeconomics for undergraduates and graduates who possess knowledge of the rudiments of calculus and college algebra. It covers the traditional topics, including linear programming and input-output analysis. The style is primarily nonmathematical, although some economic laws are presented in terms of equations. Chapters conclude with references.

Hildenbrand, Werner. CORE AND EQUILIBRIA OF A LARGE ECONOMY. Princeton Studies in Mathematical Economics, no. 4. Princeton, N.J.: Princeton University Press, 1974. viii, 251 p. Summary of Notation, pp. 234-35. Bibliography, pp. 236-46. Name Index, pp. 247-48. Subject Index, pp. 249-51.

A presentation of the relationship between two concepts of equilibrium for an economy: the core, which is a cooperative equilibrium concept, and Walras equilibrium, which is a noncooperative concept. It is divided into two parts: mathematics, and
economics. The mathematics part is devoted to set theory, metric
spaces, Euclidean spaces, and measure theory. The economics
part consists of four chapters covering the topics demand; exchange,
which presents core and Walras equilibria; limit theorems on the
core; and economies with production. The presentation is rigorous,
with proofs of most theorems provided. Sections conclude with
problems in which theorems must be proved, and mathematical
derivations required. Chapters conclude with notes. Assumes
knowledge of advanced calculus or real analysis.

Horowitz, Ira. DECISION MAKING AND THE THEORY OF THE FIRM. New York: Holt, Rinehart and Winston, 1970. xii, 468 p. Index, pp. 455-68.

A text for a course in microeconomic theory which attempts to
bridge the gap between neoclassical price theory and managerial
decision making. Both mathematics and graphs are used, and the
extensions include linear, integer, and quadratic programming;
alternatives to profit maximization; decision theory under risk and
uncertainty; and Markov processes and queueing theory. Assumes
knowledge of differential and integral calculus, matrix algebra,
and elementary probability theory. Chapters conclude with from
one to two dozen exercises each.

Johansen, Leif. PRODUCTION FUNCTIONS: AN INTEGRATION OF MICRO AND MACRO SHORT RUN AND LONG RUN ASPECTS. Contributions to Economic Analysis, vol. 75. Amsterdam: North-Holland, 1972. ix, 274 p. Bibliography, pp. 265-69. Index, pp. 270-74.

A critical examination and reformulation of production theory using
data from the operation of the Norwegian tanker fleet. It makes
use of sectors of production consisting of many micro units producing a homogeneous product; for each sector various concepts
of production functions are considered and the relationships between the micro/macro and short-run/long-run aspects are examined. Assumes knowledge of differential and integral calculus.

Johnson, Harry G. THE TWO-SECTOR MODEL OF GENERAL EQUILIBRIUM. Yrjo Jahnsson Lectures. Chicago: Aldine-Atherton, 1971. 118 p. Appendixes, pp. 87-118. No index.

A geometrical development of a two-sector general equilibrium
model based on the author's 1970 Yrjo Jahnsson lectures delivered

in Helsinki. The model is applied to the problems of income distribution and economic growth. The three appendixes are devoted to the mathematics of one-sector models of income distribution and growth, general equilibrium analysis of excise taxes, and general equilibrium with public goods.

Kalecki, Michal. STUDIES IN ECONOMIC DYNAMICS. London: Allen and Unwin, 1943. 92 p. No index.

A collection of the following five essays by the author: costs and prices, the short-term and long-term rate of interest, a theory of profits, the "pure" business cycle, and the trend. Assumes knowledge of elementary calculus.

Kassouf, Sheen. NORMATIVE DECISION MAKING. Prentice-Hall Foundations of Administration Series. Englewood Cliffs, N.J.: Prentice-Hall, 1970. vii, 88 p.

A brief introduction to modern decision making by the manager. Topics include decision making under uncertainty, objective and subjective probabilities with an application to portfolio selection and in the absence of probabilities. Chapters conclude with problems. Knowledge of probability theory is helpful.

Katzner, Donald. STATIC DEMAND THEORY. Macmillan Series in Economics. New York: Macmillan, 1970. x, 242 p. Appendix, pp. 179-220. References, p. 221. Answers to Exercises, pp. 223-29.

A self-contained text on demand analysis for individuals possessing knowledge of advanced calculus and matrix algebra. In static demand analysis the consumer is assumed to make decisions and purchase commodities once per period and the demand functions describe his choices under all market contingencies at that time. The formal theorem-proof style of exposition is employed. Topics include foundations of utility, utility maximization, integrability, special kinds of utility functions, Ockham's razor, portfolio selection, and stochastic demand theory. The appendix presents many mathematical results and theorems which are not readily available in the literature. The exercises at the ends of chapters 2 through 7 contain many examples and proofs omitted from the text, and represent a wide range of difficulty with hints supplied for the hardest exercises. Useful as a supplementary text for microeconomic theory courses. Chapters conclude with references.

Kemp, Murray C. A CONTRIBUTION TO THE GENERAL EQUILIBRIUM THEORY OF PREFERENTIAL TRADING. Foreword by I. Svennilson. Contributions to Economic Analysis, vol. 61. Amsterdam: North-Holland, 1969. 149 p. Appendix: Optimal Tariffs and Optimal Consumption Taxes, pp. 141-47. References, p. 148. Index, p. 149.

An attempt to show that the positive and normative problems posed
by preferential trading and investing arrangements can be approached
by standard general-equilibrium methods under the assumptions of:
(1) a stationary world, and (2) trade and investment are conducted
in barter terms. Although many graphs are used to illustrate concepts, knowledge of calculus and mathematical economics is necessary for portions of the book. Chapters conclude with notes on
additional mathematical concepts.

Knudsen, Neiels Chr. PRODUCTION AND COST MODELS OF A MULTI-PRODUCT
FIRM: A MATHEMATICAL PROGRAMMING APPROACH. Adense University
Studies in History and Social Sciences, vol. 13. Denmark: Odense University Press, 1973. ix, 300 p. Summary in Danish, pp. 290-96. References,
pp. 297-300. No Index.

A text for a course in production and cost theory for a mixed
group of students, including mathematicians, statisticians, economists, and operations researchers. Topics include convex programming, Kuhn-Tucker theory, Fenchel-Rockafellar theory, production and cost models of a multiproduct firm, activity analysis models of production, supply and demand functions, models
for decentralized planning, and decomposition methods. Assumes
knowledge of matrix algebra, calculus, and mathematical analysis.

Kogiku, Küchlrö Chris. MICROECONOMIC MODELS. New York: Harper and
Row, 1971. x, 309 p. Appendix: Mathematical Reviews, pp. 221-89. Bibliography, pp. 292-300. Index of Symbols, pp. 303-4. Subject Index, pp. 305-9.

A presentation of contemporary microeconomic models in three
parts: marginalist microeconomic models, requiring knowledge of
calculus and determinants; linear models, including von Neumann's
dynamic general equilibrium model, requiring a background in
linear programming; and set-theoretical models. Most of the mathematical results are presented without proof. The appendix is devoted to mathematical reviews arranged by parts, making this book
largely self-contained for all those with a college algebra background.

Koyck, Leendert Marinus. DISTRIBUTED LAGS AND INVESTMENT ANALYSIS.
Contributions to Economic Analysis, vol. 4. Amsterdam: North-Holland, 1954.
vii, 111 p. Appendix: Sources of Data, p. 111. No index.

An introduction to distributed lags in investment analysis. It attempts to measure the time-shape of lags of economic relations in
American industry--railway freight traffic, electric light and power,
cement, steel, and petroleum. Assumes knowledge of elementary
mathematical statistics and college algebra. This book is available
from University Microfilms, a Xerox company, Ann Arbor, Michigan.

Kuenne, Robert E. THE THEORY OF GENERAL ECONOMIC EQUILIBRIUM. Princeton, N.J.: Princeton University Press, 1963. xv, 590 p. Bibliography, pp. 569-80. Index, pp. 581-90.

A presentation of the theory of general microeconomic systems for the practicing economist. It is divided into three parts: introduction to general economic analysis, historical development of neoclassical general equilibrium analysis, and extensions of neoclassical analysis into the development of operational frameworks, the spatial and temporal dimensions, and the proofs of existence and uniqueness of solutions. The extensions include input-output analysis (open, closed, regional, and interregional), dynamic input-output models, and linear programming. The theorems are derived in a more space-consuming fashion than those found in mathematics books, and only those theorems of special importance to the presentation are included. Assumes knowledge of calculus, matrix algebra, differential equations, and set theory.

Lesourne, Jacques. ECONOMIC ANALYSIS AND INDUSTRIAL MANAGEMENT. Prentice-Hall International Series in Management. Englewood Cliffs, N.J.: Prentice-Hall, 1963. xxii, 631 p. Author Index, pp. 619-21. Subject Index, pp. 622-31.

An English translation of TECHNIQUE ECONOMIQUE ET GESTION INDUSTRIELLE, second edition, Paris. It considers economic theory as applied to management in three parts: econometrics in business, cost structure, and problems of synthesis. The tools applied are: amortization, marginal analysis, linear programming, forecasting, time-series analysis, and the concept of economic optimum. Assumes knowledge of elementary calculus; other mathematical topics are contained in chapter appendixes and introduced as needed. Chapters conclude with references. Useful as a supplement in courses in operations research and applied economics.

Lund, Philip J. INVESTMENT: THE STUDY OF AN ECONOMIC AGGRE-GATE. Mathematical Economics Texts, no. 7. San Francisco: Holden-Day, 1971. vii, 167 p. References, pp. 153-60. Author Index, pp. 161-63. Subject Index, pp. 165-67.

A text for a course in investment theory for undergraduates and graduates who possess knowledge of calculus. The chapter titles are: "Introduction," "Theories of Aggregate Investment," "Statistical and Questionnaire Studies," "Estimation of Log Distributions," and "Econometric Studies."

Lutz, Friedrick, and Lutz, Vera. THE THEORY OF INVESTMENT OF THE FIRM. Princeton, N.J.: Princeton University Press, 1951. xi, 253 p. Index, pp. 250-53.

An attempt to integrate the theory of production with the theory of capital at the individual firm level. It is argued that the

entrepreneur should maximize the present value of expected profits. Topics are analyzed in the order in which they appear on
the balance sheet--investments in goods in process, inventories and
fixed equipment, and the optimum method of finance. It also
considers topics such as capital intensity, sources of funds, investment demand schedule, risk and uncertainty, the cash balance,
assets valuation, and the distinction between capital and income.
Assumes knowledge of calculus.

Mack, Ruth P. PLANNING ON UNCERTAINTY: DECISION MAKING IN
BUSINESS AND GOVERNMENT ADMINISTRATION. New York: Wiley-
Interscience, 1971. xi, 233 p. Index, pp. 223-33.

A largely nonmathematical exposition of decision making in three
parts: elements of statistical decision theory, including conditional
probability and decision trees; decision agents, value judgments
and rational choice, the time dimension, and the cost of uncertainty; examples of decisions, including the IBM decision to introduce the 360 computer line, methods of reducing the cost of
uncertainty, and the role of learning models. There are no mathematical prerequisites.

Magnússon, Guðmunder. PRODUCTION UNDER RISK: A THEORETICAL STUDY.
Uppsala, Sweden: Almqvist and Wiksells Boktryckeri, 1969. 286 p. Appen-
dix I: Derivation of Certainty Equivalents under Expected Utility, pp. 245-48.
Appendix II: A Note on the Production Function, The Possibility Set of the
Firm, and The Distinction between Stochastic and Decision Variables, pp. 249-52.
Appendix III: Numerical Examples on the Model in Chapter 9, pp. 253-56. List
of Symbols, pp. 257-60. List of Figures, p. 261. List of Tables, p. 262. Refer-
ences, pp. 263-78. Name Index, pp. 279-82. Subject Index, pp. 283-86.

A modification of the riskless theory of the firm to include risk
and uncertainty. It develops the optimum conditions of a firm
producing a single good (one decision variable) and two goods
(two decision variables) under conditions of uncertainty. Topics
include uncertainty about the effectiveness of market factors, two-
stage processes under uncertainty about production quantities, product diversification, and the effect of government measures under
uncertainty. Assumes knowledge of matrix algebra and calculus.
Chapters conclude with summaries and notes.

Malinvaud, Edmond. LECTURES ON MICROECONOMIC THEORY. Translated by
A. Silvey. Advanced Textbooks in Economics, vol. 2. New York: American
Elsevier, 1972. ix, 318 p. Appendix: The Extrema of Functions of Several
Variables with or without Constraint on the Variables, by J.C. Milleron,
pp. 299-313. Index, pp. 315-18.

Originally published in French under the title of LECONS DE THEORIE
MICROECONOMIQUE in 1969, this book is a text for microeconomic
theory courses for students who have knowledge of calculus and

finite mathematics. Topics include classical economic theory and
the modern developments such as game theory, uncertainty, economies with many agents, and public goods. It covers almost completely the different viewpoints of general equilibrium.

Marschak, Jacob, and Radner, Roy. ECONOMIC THEORY OF TEAMS. Cowles
Foundation for Research in Economics at Yale University, monograph no. 22.
New Haven, Conn.: Yale University Press, 1972. x, 345 p. References,
pp. 335-38. Index, pp. 339-45.

A presentation of efficient ways of providing information and of
allocating it among decision makers who constitute a team. It is
divided into three parts: single-person decision problems, team-
organization problems, and optimality and viability in a general
model of organization. Useful to anyone interested in decision
making in groups of people. Assumes knowledge of calculus,
probability, and statistical inference.

Marschak, Thomas, and Selten, Reinhard. GENERAL EQUILIBRIUM WITH PRICE-
MAKING FIRMS. Lecture Notes in Economics and Mathematical Systems,
vol. 91. New York: Springer-Verlag, 1974. xi, 246 p. References, pp. 245-46.

A development of static general equilibrium in an economy whose
agents are consumers and noncolluding firms. Most of the effort
consists of formulating alternative equilibrium concepts and studying the conditions under which each type of equilibrium exists.
The book does not deal with dynamics or the welfare aspects of
the equilibriums studied. Requires knowledge of point-set topology.

Modigliani, Franco, and Cohen, Kalman L. THE ROLE OF ANTICIPATION
AND PLANS IN ECONOMIC BEHAVIOR AND THEIR USE IN ECONOMIC
ANALYSIS AND FORECASTING. Studies in Business Expectations and Planning, no. 4. University of Illinois Bulletin, vol. 58, no. 38. Urbana: University of Illinois, 1961. 166 p. Glossary of Symbols, pp. 153-55. Glossary of Terms, pp. 156-57. Bibliography, pp. 158-66. No index.

An exploration of the uses of statistical data bearing on anticipations and plans of firms regarding decision making. It is divided
into two parts: the role of anticipations and plans in entrepre-
neurial decision making and the nature of the "relevant" horizon;
and role of anticipatory data in economic analysis and forecasting.
Included is a model for utilizing ex ante data in economic analysis and forecasting. Assumes knowledge of matrix algebra and
statistical decision theory.

Morishima, Michio. EQUILIBRIUM, STABILITY, AND GROWTH: A MULTI-
SECTORAL ANALYSIS. New York: Oxford University Press, 1964. xii,
227 p. Appendix: Generalization of the Perron-Frobenius Theorems for Nonnegative Square Matrices, pp. 195-215. References, pp. 216-24. Index,
pp. 225-27.

A collection of the author's revised papers previously published in journals on Leontief's input-output system. Chapter titles are: "Comparative Static Analysis of the Simplest Input-output System," "Stability Analysis of the Walras-Leontief System," "A Dynamic Leontief System with Neoclassical Production Functions," "An Alternative Dynamic System with a Spectrum of Techniques," "Workability of Generalized von Neumann models of Balanced Growth," and "Balanced Growth and Efficient Program of Very-long-run Growth." Assumes knowledge of differential and integral calculus and linear algebra.

Morishima, Michio; Murata, Y.; Nosse, T.; and Saito, M. THE WORKING OF ECONOMETRIC MODELS. New York: Cambridge University Press, 1972. ix, 339 p. Bibliography, pp. 330-35. Index, pp. 337-39.

An empirical study of the application of general equilibrium in four parts: a nonlinearized microeconomic model of the Keynesian type constructed from time series data for the United States for 1902-52; a mixed Keynes-Leontief, input-output model for estimating government expenditure multipliers; a general equilibrium analysis of prices and outputs in Japan, 1953-65, based on L. Klein's nonsubstitution theorem; and an input-output analysis of disguised unemployment in Japan in 1951-56, based on the Keynes-Leontief model. Assumes knowledge of econometric techniques.

Mukerji, Anil. THE LOGIC OF REVEALED PREFERENCE. New York: Paragon Book Gallery, 1965. ix, 167 p. References, pp. 159-61. Index, pp. 163-67.

A monograph on the theory of revealed preference which sets forth the necessary and sufficient axioms to ensure that the consumer has a complete preference ordering. Empirical applications are omitted. Many theorems are proved, and knowledge of set theory and differential calculus is assumed.

Murphy, Roy E., Jr. ADAPTIVE PROCESSES IN ECONOMIC SYSTEMS. Mathematics in Science and Engineering, vol. 20. New York: Academic Press, 1965. xvi, 209 p. Bibliography, pp. 197-203. Subject Index, pp. 205-9.

An attempt to formulate an economic theory when decision makers do not possess full information about the parameters of an economic system. Use is made of stochastic processes, dynamic programming, communications theory, information theory, and thermodynamics. The presentation is both descriptive and highly notational from a mathematical viewpoint. The applications are to adaptive economic processes, investment model, multiactivity capital allocation processes, and the concept of an economic state space. The book concludes with a summary of the postulates for individual adaptive behavior and collective adaptive behavior. Familiarity with probability theory and mathematical analysis is helpful. Chapters conclude with references.

Naylor, Thomas H., and Vernon, John M. MICROECONOMICS AND DECI-
SION MODELS OF THE FIRM. The Harbrace Series in Business and Econom-
ics. New York: Harcourt, Brace and World, 1969. xiii, 482 p. Author
Index, pp. 471-76. Subject Index, pp. 477-82.

A decision-oriented graduate and undergraduate text for microeconomic theory and managerial economics courses, and a reference for workers in management science, operations research,
and business administration. It is divided into seven parts as
follows: introduction; marginal analysis models of the firm; mathematical programming models of the firm (including nonlinear and
integer programming); dynamic and probabilistic models, including
an introduction to the calculus of variations and Pontryagin's
maximum principle; computer simulation; investment models; and
other models, including games and behavioral models. Two chapters on risk and uncertainty were contributed by E.T. Byrne, Jr.
Some of the chapters contain mathematical appendixes, and all
chapters conclude with references. Assumes knowledge of calculus and probability theory.

Nerlove, Marc. ESTIMATION AND IDENTIFICATION OF COBB-DOUGLAS
PRODUCTION FUNCTIONS. Chicago: Rand McNally, 1965. vi, 193 p.
Author Index, pp. 191-93.

Presents the problems of estimation of parameters of Cobb-Douglas
type production functions under different assumptions. Topics include partial identification: the Marschak-Andrews approach;
estimation for perfectly competitive and regulated industries: Klein's
approach; reduced form estimation; and identification by means of
dynamic models of producer behavior. Assumes knowledge of calculus, the calculus of variations, and matrix algebra. Chapters
conclude with notes, and the last chapter also concludes with an
appendix on estimation of dynamic economic relationships from
time series of cross-sections.

Oakford, Robert V. CAPITAL BUDGETING: A QUANTITATIVE EVALUATION
OF INVESTMENT ALTERNATIVES. New York: Ronald Press, 1970. vii,
276 p. Bibliography, pp. 235-36. Appendixes, pp. 237-69. List of Symbols,
pp. 270-71. Index, pp. 272-76.

A reference for students, practicing engineers, governmental offi-
cials, and corporate officers who are involved in capital budgeting procedures. Topics include decisions based on complete and
incomplete information, borrowing and investment decisions, sensitivity analysis, and probability treatment of uncertainty. Familiarity with linear algebra and intermediate probability theory is necessary. The appendixes consist of mathematical reviews and tables.
Chapters conclude with problems.

Oxenfeldt, Alfred Richard; Miller, David; and Shuchman, Abraham. INSIGHTS INTO PRICING FROM OPERATIONS RESEARCH AND BEHAVIORAL SCIENCE. Belmont, Calif.: Wadsworth, 1961. 124 p. Index, pp. 122–24.

A nonmathematical introduction of operations research and behavioral science concepts into the field of business. These concepts are: decision theory, game theory, pricing in competitive bidding, perception and pricing, attitudes and pricing, and group membership and pricing.

Page, Talbot. ECONOMICS OF INVOLUNTARY TRANSFERS: A UNIFIED APPROACH TO POLLUTION AND CONGESTION EXTERNALITIES. Lecture Notes in Economics and Mathematical Systems, vol. 85. New York: Springer-Verlag, 1973. xi, 159 p. List of Symbols, pp. 114–15. Appendix, pp. 116–54. Selected Bibliography, pp. 155–59.

Pollution and congestion are analyzed as special cases of involuntary transfers—a concept which originated in Pigou's "uncompensated disservices." The three chapter titles are: "A Selected Review of the Literature," "A Theory of Involuntary Transfers Which Incorporates the Pigovian Position," and "Air Pollution in London, an Exemplary Involuntary Transfer." The models use linear and nonlinear distributed lag equations in a regression model. The appendix presents the numerical estimates of the parameters of the model.

Pattanaik, Prasanta K. VOTING AND COLLECTIVE CHOICE. SOME ASPECTS OF THE THEORY OF GROUP DECISION-MAKING. New York: Cambridge University Press, 1971. viii, 184 p. Bibliography, pp. 172–78. Glossary of Logical and Set-theoretic Symbols, p. 179. Erratum, p. 180. Index, pp. 181–84.

An examination of recent developments in the theory of social choice with emphasis on the problems raised by K.J. Arrow's impossibility theorem. Three different methods for solving these problems are presented: a social ordering of preferences based on restricted individual preference, the substitution of a best alternative social ordering for the complete social ordering requirement, and the inclusion of preference intensities as a factor underlying social choice. Assumes knowledge of set theory.

Phlips, Louis. APPLIED CONSUMPTION ANALYSIS. Advanced Textbooks in Economics, vol. 5. New York: American Elsevier, 1974. xi, 279 p. References, pp. 265–74. Index, pp. 275–79.

An attempt to reformulate demand theory to make it suitable for econometric applications. It is divided into the following two parts: (1) static demand, including utility, additive and separable utility functions, the structure of preferences, restrictions and demand functions, empirical implementation, and cost of living indexes; and (2) dynamics, including single-equation models, dynamic demand systems, substitution, complementarity, dynamic cost of living indexes, and an intertemporal approach. Assumes knowledge of calculus and, in the case of starred sections, matrix algebra. Exercises are included.

Powell, Alan A. EMPIRICAL ANALYTICS OF DEMAND SYSTEMS. Lexington, Mass.: D.C. Heath, 1974. xiv, 149 p. Bibliography, pp. 137-43. Index, pp. 145-49.

A survey of recent developments in demand theory with emphasis on the estimation of demand parameters where detailed accounts are given only for linear equations. After an introduction to demand theory and separable utility functions, the author considers estimation of Stone's system, the Rotterdam school models of Barten and Theil, and models which utilize the Brown and Heien's S-Branch utility tree. The final chapter introduces intertemporal trade-offs. It is suitable for a supplementary text in a macroeconomic theory course. Assumes knowledge of linear algebra, calculus, and statistical estimation theory.

Quirk, James P., and Saposnik, Rubin. INTRODUCTION TO GENERAL EQUILIBRIUM THEORY AND WELFARE ECONOMICS. Economic Handbook Series. New York: McGraw-Hill Book Co., 1968. 221 p. Name Index, pp. 217-18. Subject Index, pp. 219-21.

Presents the relatively recent developments in general equilibrium analysis without utilizing advanced mathematics. Topics include welfare economics, and the existence, uniqueness, and stability of competitive equilibrium with proofs of theorems included. It is designed as a text for a graduate course in general equilibrium or as a supplementary text for graduate courses in mathematical economics or microeconomic theory.

Rader, Trout. THEORY OF GENERAL ECONOMIC EQUILIBRIUM. New York: Academic Press, 1972. xx, 362 p. List of Notational Conventions and Basic Symbols, pp. xix-xx. References, pp. 343-52. Author Index, pp. 353-55. Subject Index, pp. 357-62.

A text for a two-semester course in microeconomic theory for graduates and a supplementary text for mathematical economics. It is organized into four parts: introduction, production, consumption, and trade, with the techniques of game theory, input-output analysis, and linear programming included. The presentation is rigorous with proofs of theorems provided, requiring prerequisites of calculus and linear algebra. The sections contain problems designed to strengthen one's grasp of economics. The exercises are mainly on the mathematics of the book.

Raiffa, Howard. DECISION ANALYSIS: INTRODUCTORY LECTURES ON CHOICES UNDER UNCERTAINTY. Addison-Wesley Series in Behavioral Science: Quantitative Methods. Reading, Mass.: Addison-Wesley, 1968. xxiii, 309 p. Index, pp. 301-9.

An elementary presentation of Bayesian decision theory assuming knowledge of only high school algebra. Covers practically the same ground as APPLIED STATISTICAL DECISION THEORY (Boston: Graduate School of Business Administration, Harvard University, 1961) by H. Raiffa and R. Schlaifer, but in nonmathematical style. Emphasis is placed on applications of decision theory to real life situations. Some chapters conclude with projects and appendixes.

Rothenberg, Jerome. THE MEASUREMENT OF SOCIAL WELFARE. Englewood Cliffs, N.J.: Prentice-Hall, 1961. xii, 357 p. Bibliography, pp. 337-52. Index, pp. 353-57.

An examination of the possibilities for a useful welfare economics rather than merely a formally correct one. It is divided into five parts: the problem of social choice, including Arrow's general impossibility theorem; ordinal utility analysis in which the compensation principle is stressed; cardinal utility analysis where topics such as preference intensity and expected utility are discussed; a mixed model where it is argued that single-peakedness of preference ordering is unreasonable; and the relevance criteria as applied to the values prevailing in a community. A reasonably complete survey of the welfare literature during the period 1935-60 is provided. Assumes knowledge of college algebra.

Scarf, Herbert E., with Hansen, T. THE COMPUTATION OF ECONOMIC EQUI-LIBRIA. Cowles Foundation for Research in Economics at Yale University Monograph 24. New Haven, Conn.: Yale University Press, 1973. ix, 249 p. Appendix 1: A FORTRAN Program for the Replacement Step of section 6.3., pp. 233-37. Appendix 2: An Estimate of the Number of Iterations Required by the Algorithm, pp. 238-40. References, pp. 241-45. Index, pp. 247-49.

A monograph which provides a general method for the explicit numerical solution of the neoclassical (Walrasian) model of general equilibrium. Other topics include numerical applications of Brower's theorem and degeneracy. Useful to anyone interested in the techniques of economic planning and whose work requires the solution of nonlinear systems of equations. Assumes knowledge of fixed point theorems in mathematical analysis.

Schmalensee, Richard. APPLIED MICROECONOMICS: PROBLEMS IN ESTIMATION, FORECASTING, AND DECISION-MAKING. San Francisco: Holden-Day, 1973. ix, 118 p. No index.

A description of a sequence of computer-based exercises in applied microeconomics to supplement courses in economic theory and managerial economics. Topics include statistical estimation and pricing in competition, monopoly, and oligopoly markets. Assumes knowledge of calculus, and access to a computer program for multiple regression analysis is necessary. All other programs employed are presented in the instructor's manual. Chapters conclude with exercises.

Sen, Amartya Kumar. COLLECTIVE CHOICE AND SOCIAL WELFARE. Mathematical Economics Texts, no. 5. San Francisco: Holden-Day, 1970. xii, 225 p. Bibliography, pp. 201-18. Name Index, pp. 221-22. Subject Index, pp. 223-25.

> A presentation of the problem of aggregating the preferences of individuals into a collective or social preference function. It synthesizes widely scattered articles and other works for use in welfare economics, theory of economic policy and planning, political theory of the state, and moral philosphy. Topics include collective choice and rationality, social welfare functions, value and choice, conflicts and dilemmas, aggregation problems, and majority choice and related systems. Many theorems are proved.

Sengupta, S. Sankar. OPERATIONS RESEARCH IN SELLERS' COMPETITION: A STOCHASTIC MICROTHEORY. Publications in Operations Research, no. 12. New York: Wiley, 1967. xvi, 228 p. Appendixes, pp. 213-22. References, pp. 223-25. Index, pp. 227-28.

> A development of the structure of optimal decision rules with respect to pricing, selling costs, management of production and inventories, and capital investment. The approach is analytical, rather than empirical, and emphasis is on integrating probability theory with the classical theory of the firm, with extensions to renewal and Markov models. Useful as a supplement in courses in microeconomic theory and management. Assumes knowledge of integral calculus and probability theory. The appendixes contain the derivation of sales probabilities in differentiated oligopoly and mathematical results on the concavity of average profit functions.

Shephard, Ronald William. COST AND PRODUCTION FUNCTIONS. Princeton, N.J.: Princeton University Press, 1953. vii, 104 p. No index.

> A presentation in mathematical terms of several types of production and cost functions. Topics include dual determination of production functions from cost functions, constraints on the factors of production, homothetic production functions, Cobb-Douglas production functions, the problem of aggregation, and dynamics of monopoly. Assumes knowledge of calculus.

_____. THEORY OF COST AND PRODUCTION FUNCTIONS. Princeton Studies in Mathematical Economics, no. 4. Princeton, N.J.: Princeton University Press, 1970. xi, 308 p. References, pp. 292-93. Appendix 1: Mathematical Concepts and the Theorems for Semi-continuity and Quasi-concavity (convexity), pp. 295-97. Appendix 2: Mathematical Concepts and Propositions for Correspondences, pp. 298-300. Appendix 3: Utility Functions, pp. 301-5. Index, pp. 306-8.

> A presentation of the mathematical economic theory of production and costs with the main topic being the analysis of substitution between the factors to achieve a given output. Topics include

the distance function of a production structure, the factor and price minimal cost functions, production correspondences, the aggregation problem for cost and production functions, duality of cost and production structures, and dualities for production correspondences. Assumes knowledge of advanced calculus.

Shubik, Martin. STRATEGY AND MARKET STRUCTURE: COMPETITION, OLIGOPOLY, AND THE THEORY OF GAMES. Foreword by O. Morgenstern. New York: Wiley, 1959. xviii, 387 p. Appendix A; Utility Theory, pp. 339-42. Appendix B: Welfare "Fair Division" and Game Theory, pp. 343-57. Appendix C: Notes for the Text, pp. 359-78. Index of Names, pp. 379-80. Index, pp. 381-87.

A unified approach to the various theories of competition and markets using primarily game theory. Although it is intended mainly for economists, parts of the book would be of interest in management science, operations research, and law. It is divided into two parts: classical approaches to game theory and oligopoly, and oligopoly in a game theory context with applications to the market structure of several American industries. Assumes knowledge of calculus. Chapters conclude with notes.

Simmons, Peter J. CHOICE AND DEMAND. New York: Wiley, Halsted Press, 1974. vi, 120 p. References, pp. 115-18. Index, pp. 119-20.

An axiomatic development of demand theory with emphasis on the preference structure of the individual, revealed preference and integrability, and the structure of preference orderings. Assumes knowledge of set theory and differential calculus.

Somermeyer, W.H., and Bannink, R. A CONSUMPTION-SAVINGS MODEL AND ITS APPLICATIONS. Contributions to Economic Analysis, vol. 79. New York: American Elsevier, 1973. xvi, 431 p. References, pp. 410-16. Author Index, pp. 417-18. Subject Index, pp. 419-31.

An application of the Modigliani-Brumberg theory of saving to data collected in a 1960 savings survey of the Netherlands. Topics include the theory of optimal allocation of consumption and savings over time based on an additive-separable utility function, a microsavings model based on an addi-log utility function, a macrosavings model, and the effects of bioligical demographic factors on savings in the Netherlands, and in Mexico for the years 1949-66. Assumes knowledge of calculus and matrix algebra.

Takayama, Takashi, and Judge, George G. SPATIAL AND TEMPORAL PRICE AND ALLOCATION MODELS. Contributions to Economic Analysis, vol. 73. New York: American Elsevier, 1971. xx, 528 p. Appendixes, pp. 433-521. Author Index, pp. 522-24. Subject Index, pp. 525-28.

Deals with the specification of models that may be employed in analyzing pricing and allocation problems over space and time. It is divided into the following six parts: mathematical program-

ming and competitive equilibrium, linear programming and allocation models, single product spatial price equilibrium models,
multiproduct spatial price equilibrium models, general competitive spaceless and spatial equilibrium models, and pricing and
allocation over time. The twelve appendixes are devoted to
mathematical reviews, proofs of theorems, algorithms for solving
programming problems, and extensions of some of the results to
nonlinear programming problems. Assumes knowledge of calculus
and linear algebra.

Theil, Henri, et al. OPTIMAL DECISION RULES FOR GOVERNMENT AND
INDUSTRY. Studies in Mathematical and Managerial Economics, vol. 1.
Chicago: Rand McNally, 1964. xvii, 364 p. Bibliography, pp. 357-60.
Index, pp. 361-64.

Deals with the theory of linear decision rules and their applications at the microeconomic and the macroeconomic levels. Topics
include static theory of quadratic preferences and linear constraints
with applications in antirecession for the United States in the
1930s and in production and employment scheduling; dynamic theory of linear decision rules with applications in production and
employment scheduling; application of dynamic theory of linear
decision rules to macroeconomic decision problems--mainly the
development of a four-year optimal strategy for the United States
and three-year strategies for the Netherlands; and the development
of strategies for multiperson problems. Assumes knowledge of
matrix algebra, elementary probability theory, and calculus.

Vandermeulen, Daniel C. LINEAR ECONOMIC THEORY. Englewood Cliffs,
N.J.: Prentice-Hall, 1971. xii, 543 p. Index, pp. 533-41. Frequently
Used Symbols: Explanation and References, pp. 542-43.

An intermediate level text for undergraduates in microeconomic
theory. It uses linear programming to derive neoclassical microeconomic models of the firm, consumer, and market, and also to
develop the essential core of general equilibrium and welfare
theory. By the use of computational techniques of linear programming, economic analysis is carried out in a sequence of
piecewise linear solutions that are extended to their smooth economic theory counterparts. The relationship between the primal
and dual linear programs provides the unifying framework for the
entire presentation. Assumes knowledge of college algebra. Each
of the twenty chapters concludes with exercises designed to test
the student's mastery of the material and his ability to derive important results.

Von Weizsacker, C.C. STEADY STATE CAPITAL THEORY. Lecture Notes in
Operations Research and Mathematical Systems, vol. 54. New York: Springer-
Verlag, 1971. iii, 102 p. References to the Literature, pp. 97-102.

A text for a short course in capital theory based on steady states
of stationary and exponentially growing economies. It is divided

into the following four parts: capital theory without capital, circulatory capital, fixed capital, and general equilibrium of steady states. Assumes knowledge of calculus.

Wallace, James P., and Sherret, A. ESTIMATION OF PRODUCT ATTRI-BUTES AND THEIR IMPORTANCE. Lecture Notes in Economics and Mathematical Systems, vol. 89. New York: Springer-Verlag, 1973. vi, 94 p. No index.

Presents a new methodology for using product attributes as a means of evaluating consumer choice. Part 1 discusses the methodology for estimating the relative importance of product attributes by means of statistical inference, and part 2 provides an application of the methodology to mode choice process in transportation planning.

Walras, Léon. ELEMENTS OF PURE ECONOMICS, OR THE THEORY OF SO-CIAL CHOICE. Translated by William Jaffé. 1954. Reprint, Reprints of the Classics. New York: Augustus M. Kelley, 1969. 620 p. Translator's Notes, pp. 497-558. Table of Corresponding Sections, Lessons, and Parts, pp. 559-63. Collation of Editions, pp. 565-610. Subject Index, pp. 612-18. Index of Names, pp. 619-20.

A reissue of Walras's classic statement of the general theory of value as presented in the 1874, 1889, 1896, and 1900 editions, and the EDITION DEFINITIVE of 1926.

White, Douglas John. DECISION THEORY. Chicago: Aldine, 1969. ix, 185 p. References, pp. 179-82. Index, pp. 183-85.

A presentation of the main results of a research project designed to ascertain the content of decision theory. Topics include theories of choice, value, and uncertainty; decidability; practical considerations in decision analysis; information for decision; pragmatic aspects of decision theory; and mathematical models and decisions. The presentation is largely descriptive and knowledge of calculus is assumed. Chapters conclude with summaries.

Wu, Shih-Yen, and Pontney, Jack A. AN INTRODUCTION TO MODERN DEMAND THEORY. New York: Random House, 1967. ix, 270 p. Mathematical Appendix, pp. 247-63. Index, pp. 265-70.

This book considers, from a mathematical viewpoint, the more familiar topics of demand theory under static certainty: cardinal and ordinal utility, the Marshallian demand curve, price and income displacement analysis in 2- and n-good cases, and elasticity. It also examines the newer developments, including the axiomatic derivation of utility functions, revealed preferences, Kuhn-Tucker maximization, von Neumann-Morgenstern utility indexes, and Friedman-Savage utility-of-income analysis. Assumes knowledge of differential calculus and matrix algebra. The appendix is devoted to special mathematical techniques. Chapters conclude with notes and references.

Chapter 3

MACROECONOMICS

Adelman, Irma. THEORIES OF ECONOMIC GROWTH AND DEVELOPMENT. Stanford, Calif.: Stanford University Press, 1961. vii, 164 p. Notes, pp. 149-55. Index, pp. 159-64.

An examination of the growth theories of Smith, Ricardo, Marx, and Schumpeter, and a presentation of the author's own neo-Keynesian growth model. Knowledge of differential equations is helpful.

Agarwala, Ramgopal. AN ECONOMETRIC MODEL OF INDIA 1948-1961. Foreword by R.J. Ball. London: Frank Cass, 1970. xvi, 188 p. Appendix I: The Compilation of the Data, pp. 163-74. Bibliography, pp. 175-85. Index, pp. 185-88.

A presentation of macroeconomics in the context of the Indian economy. It also sets forth a macro model of the economy consisting of twenty-four equations and applies it through simulation. Assumes knowledge of econometrics.

Allen, R.G.D. MACRO-ECONOMIC THEORY: A MATHEMATICAL TREATMENT. New York: St. Martin's Press, 1967. xii, 420 p. Exercises: Solutions, pp. 411-15. Index, pp. 416-20.

A text for graduate and undergraduate courses in macroeconomics with extensive applications to growth and international trade. It is divided into four parts: the tools; short-period equilibrium: Keynes and the classics; long-period equilibrium: growth models; and medium-period disequilibrium: cycle models. The treatment is confined to deterministic models, and it omits problems based on the Turnpike theorem. Assumes knowledge of differential and difference equations. References are contained in the footnotes throughout the book, and the chapters conclude with eight or nine problems each.

Arrow, Kenneth J., and Kurz, Mordecai. PUBLIC INVESTMENT, THE RATE OF RETURN, AND OPTIMAL FISCAL POLICY. Foreword by J.L. Fisher.

Macroeconomics
Macroeconomics

Baltimore: Johns Hopkins Press, for Resources for the Future, 1970. xxviii, 218 p. Bibliography, pp. 209-13. Index, pp. 215-18.

An application of the Pontryagin maximum principle to an infinite horizon neoclassical growth model with a single input which can be consumed or used as government or private investment. The optimal growth path is shown to be obtained under perfect capital markets and the private utility maximization of a single representative consumer by a government policy of a lump-sum taxation and government expenditure. It includes an illustration of optimization using dynamic programming. This book can be used as supplementary reading in mathematical economics courses, and requires knowledge of the calculus of variations.

Bailey, Martin J. NATIONAL INCOME AND THE PRICE LEVEL: A STUDY IN MACROTHEORY. New York: McGraw-Hill Book Co., 1962. ix, 304 p. Appendix: The Concept of Income, pp. 269-99. Index, pp. 301-4.

A presentation of the theory of national income determination within a framework of equilibrium analysis. Topics include: general equilibrium with imperfect markets, policies for economic stabilization, the effects of the rate of interest and real cash balances on consumption, and expectations and adjustments to change: a key link in cumulative movements of the economy. The appendixes to the chapters are devoted to second-order autoregressive systems, and a discussion of equipment durability. Assumes knowledge of differential calculus. Chapters conclude with references.

Batra, Raveendra N. STUDIES IN THE PURE THEORY OF INTERNATIONAL TRADE. New York: St. Martin's Press, 1973. xiii, 355 p. Author Index, pp. 351-52. Subject Index, pp. 353-55.

A text for a graduate or advanced undergraduate course in trade theory with a prerequisite of differential calculus. After a review of classical international trade, the author takes up the new theories that have emerged during the previous decade, including the implications of factor market imperfections, the theory of imperfect product markets, the concept of effective protection, the repercussions of the introduction of intermediate goods, and non-traded goods models. Some of the chapters have appendixes devoted to formal mathematical treatments of the theories. Chapters conclude with references and supplementary readings.

Bowers, David A., and Baird, Robert N. ELEMENTARY MATHEMATICAL MACROECONOMICS. Prentice-Hall Series in Mathematical Economics. Englewood Cliffs, N.J.: Prentice-Hall, 1971. xv, 304 p. Answers to Exercises, pp. 292-99. Index, pp. 300-304.

A text for a course in macroeconomics for graduates and undergraduates who have knowledge of partial differentiation and the total differential. Chapters 1 through 7 restate the classical theory of demand in elementary mathematical terms, while chapters 8 through 10 present the theory of aggregate supply and its relationship to aggregate demand in a manner which is different from that of most other textbooks. Chapter 11 presents the older Harrod–Domar analysis and the newer neoclassical growth models; and chapter 12 deals with the requirements for equilibrium in an "open" economy. Chapters conclude with references and exercises.

Branson, William H. MACROECONOMIC THEORY AND POLICY. New York: Harper and Row, 1972. xiv, 460 p. Index, pp. 447-60.

A text for a graduate and undergraduate course in macroeconomics which integrates the verbal, graphical, and mathematical approaches, and takes the equilibrium view of the interrelationships which exist in the economy. It is divided into the following four parts: introduction to macroeconomics; national income determination; the static equilibrium model; sectoral demand functions and extensions of the basic model; and growth with full employment: aggregate growth models. Assumes knowledge of elementary differential and integral calculus. Chapters conclude with references.

Brems, Hans. LABOR, CAPITAL, AND GROWTH. Lexington, Mass.: D.C. Heath, 1973. xvii, 188 p. Epilogue, pp. 173-74. Notes, p. 175. Index, pp. 179-88.

Surveys classical, Marxian, linear, and neoclassical models of economic growth. Assumes knowledge of calculus. Chapters conclude with notes and references, and some chapters have appendixes with empirical measurements.

_____. OUTPUT, EMPLOYMENT, CAPITAL, AND GROWTH. New York: Harper and Row, 1959. xiii, 349 p. Epilogue, p. 339. Index, pp. 343-49.

One of the early texts for a course in economic theory which attempts to restate and occasionally extend neoclassical models of output, employment, capital, and growth with the use of linear programming and input-output analysis. It is divided into the following four parts: the Keynesian model, the demand for input, disaggregation of the Keynesian model, and dynamization of the Keynesian model. Assumes knowledge of elementary calculus and differential equations.

_____. QUANTITATIVE ECONOMIC POLICY: A SYNTHETIC APPROACH. 1968. New York: Wiley, 1968. xii, 514 p. Index, pp. 505-14.

A selection of mathematical models from the fields of microeconomic theory, macroeconomic theory, capital theory, and international trade. The emphasis is on solvability and synthesis, and the following models are included: Cobb-Douglas, duopoly, labor-management bargaining, Austrian model of general equilibrium, static open and closed Leontief systems, growth model of von Neumann, and vintage growth models. The mathematical results are illustrated both numerically and graphically. Assumes knowledge of calculus and differential equations. Chapters conclude with references.

Burmeister, Edwin, and Dobell, A. Rodney. MATHEMATICAL THEORIES OF ECONOMIC GROWTH. Foreword by R[obert]. M. Solow and a contribution by S.J. Turnovsky. Macmillan Series in Economics. New York: Macmillan, 1970. xix, 444 p. Appendix: Mathematical Properties of Negative Squares Matrices, pp. 437-40. Index, pp. 441-44.

A text for a graduate course in economic growth and a supplementary reading in graduate economic theory and capital theory courses. It brings together widely scattered research and textual material in a unified treatment, and includes such topics as one- and two-sector growth models, multisector growth models, Leontief models with alternative techniques, neoclassical multisector models without joint production, Turnpike theorems, and optimal economic growth using control theory. Assumes knowledge of differential equations, calculus, and matrix algebra. Chapters conclude with exercises, answers, and references.

Burrows, Paul, and Hitiris, Theodore. MACROECONOMIC THEORY: A MATHEMATICAL INTRODUCTION. New York: Wiley, 1974. xiii, 210 p. Table of symbols, pp. 195-98. Answers to exercises, p. 199. Bibliography, pp. 200-202. Author Index, pp. 203-4. Subject Index, pp. 205-10.

A text for a macroeconomic theory course for graduates and undergraduates who have knowledge of partial differentiation, difference equations, and matrix algebra. The objective of this book is to bridge the gap between the verbal/diagrammatic exposition of macroeconomic theory and the mathematical treatment commonly found in journal articles. It begins with a step-by-step development of the expenditure, monetary, and the production and employment sectors, and then takes up aggregate supply and demand, followed by the external sector. Other topics include dynamic models of income and price determination, business cycles, and growth theory. Chapters conclude with exercises.

Chacholiades, Miliades. THE PURE THEORY OF INTERNATIONAL TRADE. Chicago: Aldine, 1973. xvii, 451 p. Name Index, pp. 445-46. Subject Index, pp. 447-51.

A text for a course in international trade for students who are familiar with calculus, although the geometric approach is extensively used. Some chapters contain appendixes devoted to proofs of essential theorems, and all chapters conclude with references.

Chakravarty, Sukhamoy. CAPITAL AND DEVELOPMENT PLANNING. Foreword by P[aul]. Samuelson. Cambridge, Mass.: MIT Press, 1969. xx, 344 p.

Appendix A: The Classification of Variational Problems in Economics, pp. 267-305. Appendix B: Basic Ideas in the Theory of Optimal Control, pp. 306-18. Appendix C: Nonnegative Indecomposable Square Matrices, pp. 319-24. Appendix D: Separation Theorems on Convex Sets, pp. 325-27. Bibliography, pp. 329-40. Author Index, p. 341. Subject Index, pp. 343-44.

Presents theories of economic growth based on the original research of F. Ramsey. Topics include optimal growth paths for finite and infinite horizons, optimal programs of capital accumulation in a two-sector model, multisector planning models with linear technologies and nonlinear utility functions, and the Leontief dynamic growth model. It is designed for the economic theorist and planner, rather than the mathematician, and is accessible to anyone possessing a working knowledge of calculus.

Chalmers, James A., and Leonard, Fred H. ECONOMIC PRINCIPLES; MACROECONOMIC THEORY AND POLICY. Foreword by L[awrence]. R. Klein. New York: Macmillan, 1971. xix, 460 p. Data Appendix, pp. 431-35. Answers to Questions, pp. 436-53. Index, pp. 455-60.

A quantitative-oriented text for an introductory macroeconomic course. It uses algebra, rather than geometry, for model building of the U.S. economy and subsequent testing of the models using macroeconomic data. It covers the standard topics in introductory macroeconomics.

Dasgupta, Ajit K., and Hagger, A.J. THE OBJECTIVES OF MACRO-ECONOMIC POLICY. New York: Macmillan, 1971. List of Main Symbols, pp. 516-17. Index of Subjects, pp. 519-27. Index of Authors, pp. 528-29.

A fairly advanced text on macroeconomics for undergraduates and graduates who have knowledge of calculus. It has a strong British emphasis, with British institutions providing the framework, and with problems and empirical research drawn from British experience. Topics include internal balance and external balance in which topics such as the consumption function, fixed-capital and housing investment functions, inventory investment, stabilization policy and balance of payments policy are considered; problems of price stability and economic growth; and interrelationships among the following objectives: internal balance and price stability, internal balance and external balance, and economic growth and price stability.

Dernburg, Thomas Frederick, and Dernburg, Judith Dukler. MACROECONOMIC ANALYSIS: AN INTRODUCTION TO COMPARATIVE STATICS AND DYNAMICS. Reading, Mass.: Addison-Wesley, 1969. x, 292 p. Index, pp. 288-92.

A text for an undergraduate course in macroeconomics for students who have knowledge of differential and integral calculus. Part 1 covers comparative statics macroeconomic analysis; part 2 deals with single-equation dynamics; and part 3 is concerned with multi-sectoral static and dynamic models.

Domar, Evsey D. ESSAYS IN THE THEORY OF ECONOMIC GROWTH. New York: Oxford University Press, 1957. ix, 272 p. Index, pp. 263-72.

A collection of nine essays by the author written during the period 1944-56, each essay representing an application of the rate of growth as an analytical device to a specific economic problem. The papers represent important contributions to the theory of income and employment as originally put forth by Keynes. The concluding chapter presents a Soviet model of growth.

Eckaus, Richard S., and Rosenstein-Rodan, P.N., eds. ANALYSIS OF DEVELOPMENT PROBLEMS: STUDIES OF THE CHILEAN ECONOMY. Contributions to Economic Analysis, vol. 83. New York: American Elsevier, 1973. xii, 430 p. References, pp. 417-26. Subject Index, pp. 427-30.

A collection of fifteen papers on economic development of Chile based on research sponsored by the ODEOLAN of the government of Chile and the Center of International Studies, MIT, from 1968 to 1970. The papers are grouped under the following headings: essays on methodology, estimates of shadow prices, sectoral studies, and studies of the effects of inflation. A substantial part of the research is devoted to the development of multisector linear programming models. Among the authors are: E. Bacha, J.R. Behrman, S. Bitar, P.B. Clark, C. Diaz-Alejandro, A.M. Jul, F. Setan, L. Taylor, and H. Trivelli.

Fujino, Shozaburo. A NEO-KEYNESIAN THEORY OF INFLATION AND ECONOMIC GROWTH. Lecture Notes in Economics and Mathematical Systems, vol. 104. New York: Springer-Verlag, 1974. 96 p. No index.

An investigation of the processes of cyclical fluctuations, inflation, and economic growth, and a study of the relationship between short-run and long-run analyses of the economy. It presents a model on stagflation. All models are depicted mathematically, and knowledge of calculus is assumed.

Green, H.A. John. AGGREGATION IN ECONOMIC ANALYSIS: AN INTRODUCTORY SURVEY. Princeton, N.J.: Princeton University Press, 1964. ix, 129 p. List of References, pp. 121-26. Index, pp. 127-29.

An attempt to identify the basic propositions underlying the problem of aggregation in economics, and to survey a variety of contributions. It is divided into the following three parts: grouping of variables in a single utility or production function, aggregation of economic relations, and the measurement of capital and inconsistent aggregation. It is useful as a supplementary reading in courses such as macroeconomics, econometrics, and mathematical economics. Assumes knowledge of calculus and matrix algebra.

Gupta, Kanhaya Lal. AGGREGATION IN ECONOMICS: A THEORETICAL AND EMPIRICAL STUDY. Rotterdam: Rotterdam University Press, 1969. ix, 92 p. Appendix A: Mathematical Notes, pp. 74-78. Appendix B: Basic Data, pp. 79-85. Appendix C: Implicit Sampling Errors, pp. 86-87. Bibliography, pp. 88-90. Index, pp. 91-92.

> An analysis of the nature, size, and causes of the aggregation bias in macro parameters, and an evaluation of the relative predictive effectiveness of micro and macro models. It also sets forth rules for partitioning a given micro system.

Haavelmo, Trygve. A STUDY IN THE THEORY OF ECONOMIC EVOLUTION. Contributions to Economic Analysis, vol. 3. New York: American Elsevier, 1954. viii, 114 p. Index, pp. 112-14.

> A presentation of the basic ideas of economic development in the following six parts: introduction, simple models of economic growth, deterministic theories of evolutionary dissimilarities, the stochastic approach, theories of interregional relations, and some specifications upon a more flexible theoretical framework. Assumes knowledge of elementary calculus.

Hamberg, Daniel. MODELS OF ECONOMIC GROWTH. New York: Harper and Row, 1971. x, 246 p. Index of Names, pp. 239-40. Index of Subjects, pp. 241-46.

> A text for a graduate course in growth theory for students with one year of calculus and some understanding of differential equations. It is divided into the following three parts: the Harrod-Domar, neoclassical, and Cambridge growth models; saving and economic growth; and induced invention and economic growth. Nearly all topics are associated with one-sector growth models.

Hanson, James. GROWTH IN OPEN ECONOMIES. Lecture Notes in Operations Research and Mathematical Systems, vol. 59. New York: Springer-Verlag, 1971. iv, 127 p. Bibliography, pp. 125-27.

> After reviewing post-World War II literature on terms of trade, the author summarizes the results of recent work on two-sector growth models. A closed two-sector growth model is developed and applied to a two-country world. Sufficient conditions for long-run balanced growth are developed under a variety of assumptions about factor intensities, and the two-country model is extended to the case where the two economies are growing at different natural rates.

Heesterman, A.R.G. FORECASTING MODELS FOR NATIONAL ECONOMIC PLANNING. Boston: Reidel, 1970. ix, 132 p. List of Tables, p. 128. Bibliography, pp. 129-32.

> A presentation of the following four types of econometric models:

macroeconomic short-term models, macroeconomic long-term models, sectorized short-term models, and sectorized long-term models. The representative models include input-output, Harrod-Domar, Tinbergen's method of instruments and targets, and Theil's quadratic programming approach. Assumes knowledge of calculus and linear algebra. Chapters conclude with notes.

Inagaki, M. OPTIMAL ECONOMIC GROWTH: SHIFTING FINITE VERSUS INFINITE TIME HORIZON. Contributions to Economic Analysis, no. 63. New York: American Elsevier, 1970. xvi, 196 p. Mathematical Appendix, pp. 165-79. Bibliography, pp. 181-93. Subject Index, pp. 194-96.

Presents models that try to maximize welfare over time and offers a systematic appraisal of utility maximization of infinite time as originally put forth by F.P. Ramsey. Part 1 works out the logical structure of the problem of optimal economic growth. Part 2 is a collection of five self-contained papers on optimal economic growth which have hitherto appeared only in mimeographed form as publications of the Netherlands Economic Institute. The appendix contains proofs of theorems. Assumes knowledge of advanced calculus or real analysis. It is concluded that utility maximization over infinite time does not provide an acceptable definition of optimal economic growth.

Johansen, Leif. A MULTI-SECTORAL STUDY OF ECONOMIC GROWTH. Contributions to Economic Analysis, 21. Amsterdam: North-Holland, 1960. viii, 177 p. Name Index, p. 173. Sector Index, p. 174. Subject Index, pp. 175-77.

An exposition of disaggreative analysis of economic growth with a view to the implementation using existing statistical data. Topics include the formal structure of the multisector model, the input-output table, the production and demand structures, the solution matrix T, and the observed values for the exogenous variables. The statistical data for input-out analysis pertain to the 'Norwegian economy for 1950. Assumes knowledge of calculus and input-output analysis.

Kalecki, Michal. THEORY OF ECONOMIC DYNAMICS: AN ESSAY ON CYCLICAL AND LONG-RUN CHANGES IN CAPITALIST ECONOMY. 2d ed., rev. London: Allen and Unwin, 1965. 176 p. Statistical appendix pp. 163-73. Subject Index, pp. 175-76.

This book covers substantially the same ground of Kalecki's two previous books, ESSAYS IN THE THEORY OF ECONOMIC FLUCTUATIONS (New York: Farrar & Rinehart, 1939) and STUDIES IN ECONOMIC DYNAMICS (New York: Farrar & Rinehart, 1944). It is divided into the following six parts: degree of monopoly and distribution of income, determination of profits and national income, the rate of interest, determination of investment, the business cycle, and long-run economic development. Assumes knowledge of calculus.

Kemp, Murray C. THE PURE THEORY OF INTERNATIONAL TRADE. Englewood Cliffs, N.J.: Prentice-Hall, 1964. viii, 324 p. Appendix on Price
Elasticities, pp. 301-03. Solution to Problems, pp. 303-15. Glossary of
Symbols, pp. 316-18. Index of Authors, pp. 319-21. Index of Names,
pp. 322-24.

A text for a graduate course in international trade for students
who have knowledge of mathematical economics. It is divided
into the following six parts: introduction, barter trade between
fully employed economies, the gains from international trade and
investment, money and the market for foreign exchange, trade
between underemployed economies, and the econometrics of
international trade.

_____. THE PURE THEORY OF INTERNATIONAL TRADE AND INVESTMENT.
Englewood Cliffs, N.J.: Prentice-Hall, 1969. viii, 359 p. Appendix on
Price Elasticities, pp. 352-53. Author Index, pp. 355-56. Subject Index,
pp. 357-59.

A text for graduates and undergraduates on the barter theory of
international trade and investment. It is divided into the following four parts: an analysis of barter trade between fully employed
economies; growth theory; gains from trade and optimal trade and
investment; and the rate of exchange, the terms of trade, and the
balance of payments. Assumes knowledge of calculus. The chapters conclude with problems and references.

Kogiku, Küchlrö Chris. AN INTRODUCTION TO MACROECONOMIC MODELS.
New York: McGraw-Hill Book Co., 1968. viii, 235 p. Bibliography,
pp. 198-205. Appendix: Mathematical Tools, pp. 206-24. References,
pp. 224-25. Index, pp. 227-35.

A theoretical presentation of macroeconomic models in which each
model is evaluated with respect to the following points: assumptions regarding the goods, money, labor, and capital stock markets; which variables are determined within the model, and which
ones are assumed given; whether all the variables are determined
simultaneously by the entire system, or whether some variables are
determined within subsystems; the model's static and dynamic properties; and implications for economic policy. The book is divided into the following three parts: basic concepts in macromodel
analysis, models of income determination, and models of economic
growth. The presentation uses more mathematics than traditional
presentations, and one year of college mathematics provides sufficient background. The appendix reviews mathematical concepts.
The chapters conclude with exercises.

Kuh, Edwin, and Schmalensee, Richard L. AN INTRODUCTION TO APPLIED
MACROECONOMICS. New York: American Elsevier, 1973. xv, 229 p.
Appendix A: Variables Appearing in the Model, pp. 207-13. Appendix B:
The Equations of the Model, pp. 215-23. Index, pp. 225-29.

A text for a graduate course in macroeconomics centered around a medium-sized quarterly model of the U.S. economy employing eight-two endogenous and thirty-five exogenous variables, forty-two stochastic equations, and forty identities. After an introduction to distributed lag models, attention is given to detailed analysis of personal consumption expenditures, fixed investment spending, business inventories, international trade, income distribution, unemployment, wages and prices, and monetary and fiscal policy. The model can be simulated on a time sharing system.

Kurihara, Kenneth K. APPLIED DYNAMIC ECONOMICS. London: George Allen and Unwin, 1963. 122 p. Appendix I: Mr. Barna on British Economic Growth, p. 109. Appendix II: Professor Haavelmo on Economic Evolution, pp. 110-12. Appendix III: Professor Lewis on Economic Growth, pp. 116-18. Index, pp. 119-22.

A collection of twelve papers previously published by the author in journals grouped under the following headings: Problems of Developed Economies (four papers), Problems of Developing Economies (four papers), and International Prosperity and Progress (four papers). One paper presents a linear programming model of growth.

_____. MACROECONOMICS AND PROGRAMMING. London: Allen and Unwin, 1964. 100 p. Appendix: Greek Letters Used in Mathematical Economics, p. 93. Selected References, pp. 94-95. Index, pp. 97-100.

A supplementary reading for students of macroeconomic theory and mathematical programming. Part 1, macroanalysis and policy, contains a critical discussion of the theories of income--employment, trade cycles, and general prices. Part 2, macrolinear programming, applies mathematical programming (linear and nonlinear) techniques to optimization problems in employment, investment, and growth. Topics include multiplier analysis, a cyclical growth model, input-output analysis, linear programming, and parametric nonlinear programming. Assumes knowledge of calculus and matrix algebra.

Leamer, Edward E., and Stern, Robert M. QUANTITATIVE INTERNATIONAL ECONOMICS. Boston: Allyn and Bacon, 1970. xii, 209 p. Index of Names and Authors, pp. 201-4. Index of Subjects, pp. 205-9.

A guide and reference for economics graduate students, academicians, and practicing economists in private and governmental employment which presents the various methods for quantitative measurement of the most important relationships in the areas of international trade and finance. Topics include the theory and measurement of the elasticity of substitution in international trade, forecasting and policy analysis with econometric models, theory and measurement of trade dependence and interdependence, and

the constant-market-share analysis of export growth. Assumes
knowledge of international trade and econometrics. The chapters
conclude with references.

Merkies, A.H.Q.M. SELECTION OF MODELS BY FORECASTING INTER-
VALS. Translated by M. von Holten-de-Wolff. Boston: Reidel, 1973. vi,
136 p. Appendix A, pp. 130–32. Bibliography, pp. 133–36. No index.

Originally published in Dutch in 1972, this book compares macroeconomic models on the basis of their forecasting abilities. Topics
include forecasting as a selection problem, the criteria for selection, the set of admitted models and the way in which the selection is carried out, and examples. The main applications are the
forecasts of export surplus of the Netherlands and the gross national product for 1966 and 1967 based on 1948–65 data. Models
with lagged endogenous and lagged exogenous variables, and models
that can be derived from lagged models are among those considered.
Assumes knowledge of econometrics, calculus, and matrix algebra.
The chapters conclude with notes. The appendix contains a computer program in ALGOL.

Morishima, Michio. EQUILIBRIUM, STABILITY, AND GROWTH: A MULTI-
SECTORAL ANALYSIS. New York: Oxford University Press, 1964. xii,
227 p. Appendix: Generalizations of the Perron-Frobenius Theorems for Nonnegative Square Matrices, pp. 195–215. References, pp. 216–24. Index,
pp. 225–27.

A treatment of equilibrium and growth using various forms of Leontief's input-output system. The systems considered are: static
input-output, stability of the Walras-Leontief system, a dynamic
Leontief system with neoclassical production functions, a generalized von Neumann model of balanced growth, and balanced
growth and efficient program of very long-run growth. The treatment is rigorous with proofs of theorems provided. Assumes knowledge of calculus and matrix algebra.

Mosak, Jacob L. GENERAL-EQUILIBRIUM THEORY IN INTERNATIONAL
TRADE. 1944. Reprint. Cowles Foundation Monograph, 7. Ann Arbor:
University Microfilms, A Xerox Company, 1970. xiii, 187 p. Bibliography,
pp. 181–83. Index of Names, p. 185. General Index, pp. 186–87.

Attempts to extend the general equilibrium system for a closed exchange economy to the theory of exchange in the case of trade
between several countries in order to construct a market equilibrium in an international economy. Assumes knowledge of calculus and matrix algebra.

Mundall, Robert A. MONETARY THEORY: INFLATION, INTEREST, AND
GROWTH IN THE WORLD ECONOMY. Pacific Palisades, Calif.: Goodyear,
1971. ix, 189 p. Index, pp. 187–89.

An attempt to provide an alternative theory of inflation and growth to the Keynesian system of unemployment equilibrium. Although the presentation is largely nonmathematical, knowledge of calculus is necessary to understand some of the more rigorous portions.

Negishi, Takashi. GENERAL EQUILIBRIUM THEORY AND INTERNATIONAL TRADE. Studies in Mathematical and Managerial Economics, vol. 13. New York: American Elsevier, 1972. viii, 284 p. Subject Index, pp. 279-81. Author Index, pp. 282-84.

Many of the chapters appear to be revisions of articles published by the author in journals. The wide range of topics includes optimality, existence, and stability of competitive equilibrium; increasing returns; externalities; monopolistic competition; the second best theorem; and the dichotomy of price and monetary theories. Proofs of theorems are provided, and knowledge of advance calculus is assumed. The chapters conclude with references.

Neher, Philip A. ECONOMIC GROWTH AND DEVELOPMENT: A MATHEMATICAL INTRODUCTION. New York: Wiley, 1971. xiv, 322 p. Index, pp. 321-22.

A text (or supplementary text) for undergraduate courses in intermediate macro theory, mathematical economics, and growth and development. It is essentially an exploration of two single-sector models: a neoclassical model of an advanced economy, and a dynamic model of a primitive economy. Special topics treated include: proportional growth, constant returns to scale and technical change, balanced and unbalanced growth, the theory of optimal economic growth using the calculus of variations, and growth in an open economy. The author presents theories in nonmathematical form and then in mathematical form on the assumption that the student knows some calculus or is willing to learn it. The chapters conclude with exercises and references.

Ølgaard, Anders. GROWTH, PRODUCTIVITY AND RELATIVE PRICES. Research Study of Institute of Economics, University of Copenhagen, no. 10. Amsterdam: North-Holland, 1966. viii, 309 p. Name index, pp. 307-8. Subject Index, p. 309.

A survey of the macroeconomic theory of growth together with some original contributions by the author. Topics include a survey of one-sector models of growth with a distinction between calendar-year effect, the vintage effect, and the age effect upon efficiency; two-sector models in which one sector produces a consumer good and the other a good which may be used for consumption as well as for investment purposes; an empirical analysis of the Danish terms of trade in foreign trade during the last century; sector terms of trade and the gains resulting from changes in those terms within a static open Leontief model with exogenous prices;

empirical findings on sector terms of trade of Danish industries
and U.S. agriculture; and sector terms of trade and technical
progress. Assumes knowledge of graduate level economic theory
and econometric techniques.

Olsen, Erling. INTERNATIONAL TRADE THEORY AND REGIONAL INCOME
DIFFERENCES: UNITED STATES 1880-1950. Contributions to Economic Analysis, vol. 70. Amsterdam: North-Holland, 1971. xv, 221 p. List of
Symbols, pp. xiii-xv. Appendix A: Elasticities, pp. 185-87. Appendix B:
The Computer Programs, by B. Anderson, pp. 189-98. Appendix C: Sensitivity Analysis on a Hybrid Computer, by B. Anderson and E. Olsen,
pp. 199-218. Index, pp. 219-21.

A text for a course in international trade which is divided into
three parts. Part 1, international trade theory, attempts to explain why nations trade and how trade affects international differences in income. Part 2, regional income differences, presents
an economic model in time and space focusing on regional income
differences. Part 3 evaluates the model using data for the United
States from 1880 to 1950, and estimates parameters, such as substitution, distribution, and efficiency parameters. Appendix A shows
the sensitivity of the model to a change in the value of one of
the parameters. Assumes knowledge of mathematical economics
and econometrics, and some familiarity with computer programming.

Papandreou, Andreas G[eorge]. INTRODUCTION TO MACROECONOMIC
MODELS: AN ANALYTICAL APPROACH. Training Seminar Series, 5.
Athens, Greece: Center of Planning and Economic Research, 1965. 200 p.
Appendix: Special Topics in Linear Algebra, pp. 171-200. No index.

A monograph on deterministic model construction in macroeconomics. Topics include static and dynamic model construction; the
ordering of variables in a model; equilibrium and its displacement;
and the nexus between statics and dynamics. Problems of aggregation are ignored. Assumes knowledge of calculus and linear
algebra.

Pitchford, John D. POPULATION IN ECONOMIC GROWTH. Contributions
to Economic Analysis, vol. 85. New York: American Elsevier, 1974. viii,
280 p. Index, pp. 275-80.

A survey of that part of economic theory which is concerned with
population and economic growth, and an attempt to construct a
new theory on the relationship between population growth, resource
use, and capital accumulation. It is divided into the following
three parts: population theory, population growth models, and
the determination of net fertility; population control and optimum
population growth; and population control in an optimal growth
process. It is argued that a divergence between individual and
social values may warrant population control. The final chapter

incorporates age structure into population models. Assumes knowledge of calculus. The chapters conclude with references.

Qayum, Abdul. NUMERICAL MODELS OF ECONOMIC DEVELOPMENT. Rotterdam: Rotterdam University Press, 1966. ix, 106 p. No index.

A collection of ten unpublished and published articles by the author on simple models of economic development. All models are illustrated numerically, and knowledge of calculus is helpful for comprehension.

Reichardt, Helmut. OPTIMIZATION PROBLEMS IN PLANNING THEORY. Center of Planning and Economic Research Lecture Series, no. 24. Athens: Center of Planning and Economic Research, 1971. 79 p. References, pp. 77-79. No index.

A brief mathematical presentation of the following four models of optimal economic planning: a one-sector model with a linear production function, a one-sector model with neoclassical production function, a two-sector model which maximizes consumption, and a two-sector model with minimum time problems. Assumes knowledge of the calculus of variations.

Sen, Amartya Kumar. CHOICE OF TECHNIQUES: AN ASPECT OF THE THEORY OF PLANNED ECONOMIC DEVELOPMENT. 2d ed. New York: Oxford University Press, 1962. 127 p. Appendix A: Choice of Agriculture Techniques, pp. 90-97. Appendix B: Strumilin on Time Preference, pp. 98-101. Appendix C: Technique for the Cotton-weaving Industry in India, pp. 102-14. Appendix D: The Ambar-Charkha Is a Technique of Cotton-spinning, pp. 115-19. Appendix E: On the Application of Programming, pp. 120-23. Note to Chapter 8, pp. 123-24. Index, pp. 125-27.

A presentation of the problem of choice of production techniques in an underdeveloped economy planning for economic growth. It begins with a simple model involving only one type of consumer good and the production of capital goods by unassisted labor, and then introduces alternative sets of time series of income and employment and multiplicity of commodities. Later, technical change, and international trade are introduced into the model. Some chapters have mathematical appendixes, and knowledge of calculus is assumed.

Solow, Robert M. CAPITAL THEORY AND THE RATE OF RETURN. Introduction by J[an]. Tinbergen. New York: American Elsevier, 1963. 98 p. References, p. 98. No index.

A collection of the Professor F. De Vries lectures given by the author in 1963 in Rotterdam. The lectures are nonmathematical and deal with the rate of return to capital, technical progress, and the aggregate production function. Calculations of rates of return for U.S. and German economies are given.

_____. GROWTH THEORY: AN EXPOSITION. New York: Oxford University Press, 1970. iv, 109 p. Bibliography, pp. 107-9. No index.

A modified version of six lectures (the Radcliffe lectures) delivered at the University of Warwick in 1968 and 1969. The titles are: "Characteristics of Steady States," "A Variable Capital/Output Ratio," "A Model without Direct Substitution," "A Model with Two Assets," "Economic Policy in a Growth Model," and "Aspects of Economic Policy." The lectures represent an introduction to aggregate theory of growth with some ideas of the direction of future research in this area. Assumes knowledge of calculus.

Tinbergen, Jan, and Bos, Hendricus Cornelis. MATHEMATICAL MODELS OF ECONOMIC GROWTH. Economics Handbook Series. New York: McGraw-Hill Book Co., 1962. x, 131 p. Appendix: Symbols Used, pp. 119-26. Bibliography, p. 127. Index, pp. 129-31.

A presentation of the following mathematical models of economic growth: (1) one scarce factor: one sector; (2) several scarce factors: one sector, (3) several sectors: fixed prices: no substitution; (4) several sectors: fixed prices: substitution; (5) several sectors: variable prices; and (6) several sectors and several regions. While mathematics is used freely, emphasis is on the economic significance of the models, particularly on their use for policy decisions. Assumes knowledge of calculus and linear algebra.

Tobin, James. ESSAYS IN ECONOMICS: VOLUME I: MACROECONOMICS. Markham Economic Series. New York: American Elsevier, 1971. xvi, 526 p. Index, pp. 515-26.

A collection of twenty-four professional papers in macroeconomics written by the author over the past thirty years. The papers were originally published in books and journals, such as ECONOMETRICA and the REVIEW OF ECONOMICS AND STATISTICS, and are grouped under the following headings: macroeconomic theory (seven papers), economic growth (five papers), money and finance (twelve papers). A few papers require knowledge of calculus and matrix algebra for comprehension.

Vanek, Jaroslav. MAXIMAL ECONOMIC GROWTH: A GEOMETRIC APPROACH TO VON NEUMANN'S GROWTH THEORY AND THE TURNPIKE THEOREM. Ithaca, N.Y.: Cornell University Press, 1968. xi, 122 p. Index, pp. 121-22.

Translates the theory of efficient growth--specifically, von Neumann's growth model with its turnpike extension--into simple, nonmathematical language. This approach should serve to acquaint a wider audience with the nature of the von Neumann model.

Wan, Henry Y., Jr. ECONOMIC GROWTH. The Harbrace Series in Business and Economics. New York: Harcourt Brace Jovanovich, 1971. xiii, 428 p. Appendix: A Mathematical Review, pp. 381-424. Index, pp. 425-28.

A text for a graduate course on economic growth with a prerequisite of one year of calculus. The presentation synthesizes growth theories since 1958 with optimal growth theory, employing consistent mathematical terminology. Topics include post-Keynesian and several neoclassical models and their extensions. A review of the necessary mathematics is contained in the appendix. Many chapters also have mathematical appendixes, and all chapters conclude with references and problems.

Whewell, William. MATHEMATICAL EXPOSITION OF SOME DOCTRINES OF POLITICAL ECONOMY. Reprints of Economic Classics. New York: Augustus M. Kelley, 1971. 105 p. No index.

Contains the following three reprints of publications by William Whewell which first appeared in TRANSACTIONS OF THE CAMBRIDGE PHILOSOPHICAL SOCIETY: (1) "Mathematical Exposition of Some Doctrines of Political Economy," 1829. 32 p.; (2) "Mathematical Exposition of Some of the Leading Doctrines in Ricardo's PRINCIPLES OF POLITICAL ECONOMY AND TAXATION," 1831. 44 p.; and (3) "Mathematical Exposition of Some Doctrines of Political Economy--Second Memoirs," 1850. 22 p. The following article, not previously published, is also included: "Mathematical Exposition of Certain Doctrines of Political Economy--Third Memoirs," 1850. 7 p.

Yntema, Theodore Otte. A MATHEMATICAL REFORMULATION OF THE GENERAL THEORY OF INTERNATIONAL TRADE. Chicago: University of Chicago Press, 1932. xii, 120 p. Appendix I: Some Mathematical Theorems, pp. 96-102. Appendix II: A Theorem in Linear Equations, pp. 103-13. Bibliography, pp. 114-15. Glossary of Symbols, pp. 117-18. Index, pp. 119-20.

A presentation of equilibrium conditions in international trade in terms of simultaneous equations. Provisions are made in the equations to accommodate service items, credit transactions, duties and bounties, transportation costs, rival and complementary demand and supply, and monopoly. A technique is also presented for studying the effects of distrubances to equilibrium, and applied to the general case involving m countries and n commodities. Assumes knowledge of calculus and matrix algebra.

Zarembka, Paul. TOWARD A THEORY OF ECONOMIC DEVELOPMENT. Mathematical Economics Texts, no. 9. San Francisco: Holden-Day, under the auspices of the Institute of International Studies, University of California, Berkeley, 1972. xii, 249 p. Appendix A: The Asymptotic Capital/Output Ratio for the Secondary Sector, pp. 219-21. Appendix B: Exact Solution of the Model of Chapter 2, pp. 222-24. Appendix C: The Pontryagin Problem for Chapter 4, pp. 225-29. Appendix D: Solution for the Growth Rate of Real Wages (section 6.1), pp. 230-31. Appendix E: The Profit Function for the Three-factor CES Production Function (section 8.3), pp. 232-34. Bibliography, pp. 235-41. Index, pp. 243-49.

A contribution to the theory of economic development and a textbook for development courses. It consists of an analysis of partial and general equilibrium problems during economic development, with stress on the transformation from agriculture to industry. Part 1 examines the theory of surplus labor; part 2 studies the structure of a closed economy with the aid of a dual economic model; part 3 introduces the external sector and explores the relationship between trade and development; and part 4 examines the actual and empirical aspects of the agricultural sector. Assumes knowledge of modern mathematical economics.

Chapter 4

INPUT-OUTPUT ANALYSIS

Almon, Clopper, Jr.; Buckler, Margaret B.; Horwitz, Lawrence M., and Reimbold, Thomas C. 1985: INTERINDUSTRY FORECASTS OF THE AMERI-CAN ECONOMY. Lexington, Mass.: Heath, Lexington Books, 1974. xiii, 251 p. List of Tables, p. xi. Appendix A: End Tables, pp. 203-43. Index, pp. 245-50. About the Author, p. 251.

A projection of outputs of the sectors of the U.S. economy for each year between 1973 and 1985, using input-output tables of 185 sectors. The forecasts are those for the final demand components, business investment, foreign trade, and labor productivity. The main conclusions are that the labor force, labor productivity, and gross national product will increase at steadily declining rates during this period. Assumes knowledge of matrix algebra and regression.

_____. 1973-1985 IN FIGURES. SUPPLEMENT TO 1985: INTERINDUSTRY FORECASTS OF THE AMERICAN ECONOMY. Lexington, Mass.: Lexington Books, 1974. 190 p.

This report contains two additional sets of tables. The first presents plots of output, investment, employment, and labor productivity for each of the 90 investment industries. The second presents a matrix listing which shows, for each of the 185 industries, who buys its product and how each buyer's purchase will change from 1973 to 1985.

Arrow, Kenneth J., and Hoffenberg, Marvin, et al. A TIME SERIES ANALYSIS OF INTERINDUSTRY DEMANDS. Contributions to Economic Analysis, 17. Amsterdam: North-Holland, for the Rand Corporation, 1959. vii, 290 p. Appendixes, pp. 136-86. Bibliography, pp. 287-90.

Presents an input-output study of the U.S. economy based on studies undertaken at the Rand Corporation over the period 1951 to 1957. The model consists of sixty-six purchasing industries based on the 1947 Interindustry Relations Study (BLS Report no. 33, 1953), and the data used in the analysis consisted of time series for the years 1929 to 1950. The appendixes present the data used to implement the study.

Bacharach, Michael. BIPROPORTIONAL MATRICES AND INPUT-OUTPUT CHANGE. University of Cambridge Department of Applied Economics, monograph 16. New York: Cambridge University Press, 1970. xii, 170 p. Guide to Notation, pp. x-xii. References, pp. 167-70. No index.

An investigation of the behavior of input-output relations through time based on a model consisting of two parts: a descriptive part, and a pair of identities for estimating the parameters of the descriptive part. It also examines the implications of the model for input-output projections and Markovian programming. The final chapter presents an estimate of an input-output matrix for Britain for 1960 obtained by a biproportional transformation of a corresponding table for 1954 subject to given intermediate output and input constraints for 1960. Assumes knowledge of differential and integral calculus, matrix algebra, and statistical inference.

Bharadwaj, Ranganath. STRUCTURAL BASIS OF INDIA'S FOREIGN TRADE. A STUDY SUGGESTED BY THE INPUT-OUTPUT ANALYSIS. Series in Monetary and International Economics, no. 6. Bombay: University of Bombay, 1962. xi, 121 p. Selected Bibliography, pp. 113-18. Index, pp. 119-21.

An investigation of the structural basis of Indian foreign trade using input-output techniques with a twenty-sector table. It begins with a concise statement and rigorous proof of the factor-proportion theory of international trade, and it attempts to verify empirically a hypothesis regarding the structural basis of India's foreign trade derived from the Heckscher-Ohlin model.

Bródy, András, and Carter, Anne P., eds. INPUT-OUTPUT TECHNIQUES. New York: American Elsevier, 1972. viii, 600 p. Appendix: Summary Reports on Standardized Input-output Tables, pp. 583-600. No index.

A collection of thirty-two papers presented at the fifth international input-output conference held in Geneva in 1971. The papers are grouped under the following headings: environmental analysis (three papers), population and manpower (three papers), foreign trade in national models (three papers), multiregional and multi-national systems (three papers), prices and financial analysis (four papers), national planning (four papers), applied dynamic systems (three papers), and theoretical questions (three papers). Among the authors are: A. Bródy, B.R. Bergman, H.D. Evans, D. Ford, W. Leontief, P.N. Mathur, and H. Wessels.

Bruno, Michael. INTERDEPENDENCE, RESOURCE USE AND STRUCTURAL CHANGE IN ISRAEL. Jerusalem: Bank of Israel, 1962. 324 p.

An application of input-output analysis to the economy of Israel.

142

Input-output matrices at three levels of disaggregation (77x77, 42x42, and 20x20) are presented, together with the definitions, description of estimating procedures, and the structure of inputs of the primary factors. Direct and indirect import requirements per unit of final demand, as well as an estimate of the total rate of return to capital by branch of economic activity are also given. Structural changes in production from 1958 to 1960 are traced out and compared with actual developments. The final chapter presents a forecast for 1964.

Cameron, Burgess. INPUT-OUTPUT ANALYSIS AND RESOURCE ALLOCATION. New York: Cambridge University Press, 1968. vii, 109 p. Appendix: Determinants and Matrices, pp. 96-104. List of Symbols, pp. 105-6. Bibliography, p. 107. Index, p. 109.

This book is an attempt to combine Keynesian macroeconomic theory of employment with the open, input-output model with the aim of describing market equilibrium. It presents to economists with only a high school algebra background the fruits of input-output analysis and linear programming and also the strategy of development.

Carter, Anne P. STRUCTURAL CHANGE IN THE AMERICAN ECONOMY. Harvard Studies in Technology and Society. Cambridge, Mass.: Harvard University Press, 1970. xviii, 292 p. Figures, pp. xiii-xvi. Tables, pp. xvii-xviii. Appendix A: Sensitivity of Measures of Changing Industrial Specialization to the Assumed Structure of Final Demand, pp. 223-24. Appendix B: Classification Schemes, pp. 225-36. Appendix C: 38-order Data, pp. 237-49. Appendix D: 76-order Data, pp. 250-74. References, pp. 275-82. Index, pp. 283-92.

A summary of one and one-half decades of analytical work on structural changes (changes in input-output coefficients) in the American economy. It is divided into two parts: structural change and industrial specialization between 1939 and 1961 with emphasis on the period 1947-58, and with much of the descriptive material summarized in a set of double-log scatter diagrams; and structural change and economic efficiency as measured by capital/labor ratios. This report shows that while some industries have declined relative to others--aluminum, concrete, and plastics, for example, have been substituted for steel, cooper, and timber-- the input-output structure of the economy has changed slowly during a period characterized by rapid technological advances.

Carter, Anne P., and Bródy, A[ndrás]., eds. APPLICATIONS OF INPUT-OUTPUT ANALYSIS. VOLUME 2. New York: American Elsevier, 1972. x, 387 p. Statistical References by Counting, p. 387. No index.

Volume 2 contains eighteen papers presented at the fourth international conference on input-output techniques held in Geneva

January 8-12, 1968. The papers are grouped under the following headings: nation-wide analysis, regional and interregional analysis, stability of coefficients, forecasting coefficients, and accounting and data processing. This volume was published in honor of Wassily Leontief and includes the following contributors: A. Aidenoff, C. Almon, R. Beals, M. Börlin, M. Bruno, R. Buzunov, H.O. Carter, H.B. Chenery, C. Dougherty, A. Duval, E. Fontela, M. Fraenkel, A. Gabus, G. Gehrig, J.M. Henderson, D. Ireri, S. Lombardini, A.S. Manne, C.F. Menezes, A. Middelhoek, Wm. H. Miernyk, Y. Murakami, I. Ozaki, P. Sevaldson, S. Shishido, K. Tokoyama, J. Tsukui, B. Vaccara, C. Velay, T. Watanabe, and T. Weisskopf.

_____. CONTRIBUTIONS TO INPUT-OUTPUT ANALYSIS. VOLUME 1. New York: American Elsevier, 1972. x, 345 p. Statistical References by Counting, p. 345. No index.

Published in honor of Wassily Leontief, volume 1 contains sixteen papers presented at the fourth international conference on input-output techniques held in Geneva January 8-12, 1968. The papers are grouped under the following headings: dynamic analysis, structural change, interregional analysis, price analysis, basic framework, and demography and education. The contributors are: M. Augustinovics, J. Bénard, A. Bródy, A.P. Carter, A. Chakravarti, R.H. Day, A. Ghosh, T. Gigantes, Y. Guillaume, T. Heidhves, L. Hejl, V. Kossov, O. Kyn, W. Leontief, P.N. Mathur, K.R. Polenske, B. Sekerka, G. Simon, A.D. Smirnov, R. Stone, and J.W. Waelbroeck.

Catanese, Anthony J. SCIENTIFIC METHODS OF URBAN ANALYSIS. Urbana: University of Illinois Press, 1972. xiv, 336 p. Appendixes, pp. 311-24. Bibliography, pp. 325-30. Author Index, pp. 331-32. Subject Index, pp. 332-36.

A text for undergraduate and graduate courses in urban analysis with emphasis on applications. It is divided into the following five parts: scientific method; predictive and estimating models, including probabilistic models; optimizing models, including models based on classical calculus, linear, nonlinear, dynamic, and stochastic programming; simulation and gaming; and urban information systems. Elementary knowledge of matrix algebra and differential calculus is assumed, although much of the necessary mathematics is reviewed as needed. Chapters conclude with exercises and references. The appendixes consist of statistical tables. This book is useful for a mixed audience, including economists, sociologists, administrators, engineers, mathematicians, and operations research analysts.

Chakravarty, Sukhamoy. THE LOGIC OF INVESTMENT PLANNING. Contributions to Economic Analysis, vol. 18. New York: American Elsevier, 1959. xii, 170 p. References, pp. 169-70.

A presentation of the issues involved in multisector investment
planning based on the Harrod-Domar model, the Mahalanobis
models, the Leontief dynamic model, and von Neumann's model
of an expanding economy. Emphasis is given to input-output
relations, the role of gestation lags in the production process,
the intersectoral investment deliveries, and the questions of international trade. Assumes knowledge of mathematical economics.

Chenery, Hollis B., and Clark, Paul G. INTERINDUSTRY ECONOMICS.
New York: Wiley, 1959. xv, 345 p. Summary of Notation, pp. xiv-xv.
Index, pp. 337-45.

A unified presentation of interindustry techniques and their empirical applications. The first part consists of an introduction to
interindustry theory, starting with the simplest input-output model,
with more complicated extensions requiring four additional input-
output models and four linear programming models. The second
part takes up applications of interindustry analysis in the United
States, Italy, Denmark, Argentina, and Columbia. One chapter
is devoted to interregional input-output analysis. Assumes knowledge of matrix algebra. Chapters conclude with references.

Emerson, M. Jarvin, et al. THE INTERINDUSTRY STRUCTURE OF THE KAN-
SAS ECONOMY. Office of Economic Analysis and Kansas Department of
Economic Development Planning Division Report, no. 21. Manhattan: Depart-
ment of Economics, University of Kansas, 1969. vii, 221 p. No index.

This report discusses the construction and use of input-output models
and presents a table for the Kansas economy, using 1965 data,
displaying sixty-nine processing sectors, eight final demand sectors,
and six final payment sectors. Direct, indirect, and induced requirements' coefficients are computed.

Farag, Shawki M. INPUT-OUTPUT ANALYSIS: APPLICATIONS TO BUSINESS
ACCOUNTING. Foreword by V.K. Zimmerman. Monograph 5. Urbana: Center for International Education and Research in Accounting, University of Illinois,
1967. 130 p. Bibliography, pp. 121-30. No index.

An investigation of input-output analysis as a social accounting
system and an assessment of its applications and potentialities to
business accounting. It is divided into two parts. The first part,
consisting of three chapters, presents the tenets of input-output
theory and application; the second part, consisting of four chapters, considers possible applications of input-output data and methods
in business accounting. Assumes knowledge of matrix algebra.

Frisch, Ragnar. PLANNING FOR INDIA: SELECTED EXPLORATIONS IN
METHODOLOGY. Foreword by P.C. Mahalanobis. Indian Statistical Series,
no. 8. New York: Asia Publishing House, 1960. ii, 87 p. Appendix: A
Flow Chart, p. 87. No index.

This book contains selections and excerpts from a number of memoranda written by the author in 1954 and 1955 on various aspects of national planning at the Indian Statistical Institute. The main techniques that are developed and applied are input-output analysis and the author's logarithmic potential method of linear programming. A 26x26 input-output matrix for India for 1950-51 is illustrated.

Ghosh, A. EXPERIMENTS WITH INPUT-OUTPUT MODELS: AN APPLICATION TO THE ECONOMY OF THE UNITED KINGDOM, 1948-55. New York: Cambridge University Press, 1964. xv, 148 p. Technical appendix to chapter 2, pp. 132-41. References, pp. 142-43. Index, pp. 144-47.

An empirical analysis of the structure of production in postwar Britain based on the input-output table for 1948. The first three chapters present the open input-output model and a 47x47 table for 1948. In chapters 4 and 5, the author compares input-output techniques with other forecasting models. Chapter 6 is devoted to the aggregation problem; and chapter 7 considers the problem of minor transactions. The remaining seven chapters are devoted to a decomposition of the input-output model into two-submodels, the construction of a matrix of capital input coefficients from the augmented transaction matrix, a comparison of the Leontief system of homogeneous production functions with the linear nonhomogeneous case, price projections using input-output models, the problem of shortages in industries, and changes in input coefficients.

. PLANNING PROGRAMMING AND INPUT-OUTPUT MODELS: SELECTED PAPERS ON INDIAN PLANNING. University of Cambridge Department of Applied Economics Monograph 15. Cambridge: At the University Press, 1968. viii, 166 p. Index, pp. 165-66.

A collection of fourteen articles by the author, previously published in journals. Articles 1 and 2 survey the techniques used in Indian planning, mainly from an econometric viewpoint. Articles 3 to 8 are concerned with planning problems in the Indian economy, and articles 9 to 14 deal with empirical applications of some standard models as well as some models developed by the author in seeking numerical solutions to various problems. Among the techniques discussed are input-output analysis, mathematical programming, and multisector growth models. Assumes knowledge of econometric methods.

Ghosh, A., with Chakrabarti, A. PROGRAMMING AND INTERREGIONAL INPUT-OUTPUT ANALYSIS: AN APPLICATION TO THE PROBLEM OF INDUSTRIAL LOCATION IN INDIA. University of Cambridge Department of Applied Economics Monograph, no. 22. New York: Cambridge University Press, 1973. vii, 104 p. Appendixes to chapters, pp. 73-100. References, p. 101. Index, pp. 102-4.

An application of linear programming models to the study of the optimal pattern of regional production, exchange of commodities, and the location of industries in India. It is devoted to a review of fixed-coefficient models in interregional problems, an empirical application of these models using Indian data, a description of various linear programming models and their empirical applications, a discussion of dual solutions and their implications, and suggestions for further research.

Harmston, Floyd K., and Lund, Richard E. APPLICATION OF AN INPUT-OUTPUT FRAMEWORK TO A COMMUNITY ECONOMIC SYSTEM. University of Missouri Studies, vol. 42. Columbia: University of Missouri Press, 1967. xi, 124 p. Appendix A: Illustrative Questionnaire, pp. 105-7. Notes, pp. 108-9. Bibliography, pp. 110-16. Index, pp. 117-24.

A general treatment of the problems involved in constructing an input-output table for a community of as few as 15,000 inhabitants. A 15-sector input-output table for a hypothetical community of 20,000 inhabitants is used as an illustration. The multipliers developed represent dollar turnover or money velocity, and are not the usual production multipliers employed in most input-output tables. The framework for the hypothetical input-output table is based on the authors' experiences in small community systems in the state of Wyoming. Assumes knowledge of introductory input-output analysis.

Harris, Curtis C. THE URBAN ECONOMIES, 1985: A MULTIREGIONAL MULTI-INDUSTRY FORECASTING MODEL. Lexington, Mass.: D.C. Heath, 1973. xvi, 231 p. Appendix Table, pp. 201-23. Index, pp. 225-30. About the Author, p. 231.

This book presents an experimental multiregional, multi-industry forecasting model using ninety-nine industry sectors, four labor sectors, sixty-nine equipment purchasing sectors, twenty-eight construction sectors, eight government sectors, four age sectors, and two race sectors. Data for the model was collected for each county in the United States, although forecasts are made only for the counties classified in Standard Metropolitan Areas. The study presents the theoretical framework, equations of the model, data estimating procedures, estimates of the parameters, and forecasts of regional impacts of reduced defense expenditures. The appendix presents selected projections (industry, population, per capita income, and so forth, for standard metropolitan areas with a 1970 population of over one million. The projections are for 1970, 1975, 1980, and 1985.

Harris, Curtis C., and Hopkins, Frank E. LOCATIONAL ANALYSIS: AN INTERREGIONAL MODEL OF AGRICULTURE, MINING, MANUFACTURING, AND SERVICES. Lexington, Mass.: D.C. Heath, 1972. xiv, 303 p. Notes, pp. 223-32. Appendix A: Procedure for Estimating County Supply by Input-output Sector, pp. 233-49. Appendix B: Procedures for Estimating County

Demand, pp. 251-82. Appendix C: Procedures for Estimating the Coefficients of the Objective Function of the Transportation Problem by Input-output Sector, pp. 283-97. Index, pp. 299-303.

An interregional, multi-industry model designed to explain industrial location. Multivariable regression analysis is used to examine the importance of various explanatory variables on location as measured by the change in output and a linear programming transportation algorithm is used to obtain marginal transportation costs. The book is designed for advanced students and practitioners in the fields of regional economics, economic geography, regional science, industrial organization, management science, and social accounting. Assumes knowledge of statistics, linear programming, location theory, and social income accounting.

Hatanaka, Michio. THE WORKABILITY OF INPUT-OUTPUT ANALYSIS. Foreword by O. Morgenstern. Ludwigshafen am Rhein: Fachverlag Fur Wirtschaftstheorie und Okonometrie, 1960. 310 p. Bibliography, pp. 300-306. Index, pp. 307-10.

A monograph which presents statistical tests of the coefficients of the 1947 static U.S. input-output table, using time series of outputs and final demands. It is concerned with the analysis and testing of various kinds of errors in input-output models, such as prediction errors and tests concerning the size and structure of errors. Assumes knowledge of econometric techniques.

Isard, Walter. METHODS OF REGIONAL ANALYSIS: AN INTRODUCTION TO REGIONAL SCIENCE. New York: Wiley, 1960. xxix, 784 p. List of Tables, pp. xxiii-xxv. List of Figures, pp. xxvi-xvii. List of Maps, pp. xxviii-xxix. Author Index, pp. 759-64. Subject Index, pp. 765-84.

A presentation of several techniques of regional analysis, including interregional and regional input-output techniques and interregional linear programming. The book is mostly descriptive and requires knowledge of matrix algebra only in the chapters dealing with linear programming and input-output analysis. Topics include population analysis, interregional flow analysis, balance of payments, regional cycle and multiplier analysis, and gravity, potential, and spatial interaction models. The other contributors are: D.F. Bramhall, G.A.P. Carrothers, J.H. Cumberland, L.N. Moses, D.O. Price, and E.W. Schooler. Chapters conclude with references.

_____, et al. GENERAL THEORY: SOCIAL, POLITICAL, ECONOMIC, AND REGIONAL, WITH PARTICULAR REFERENCE TO DECISION-MAKING ANALYSIS. The Regional Science Study Series. Cambridge, Mass.: MIT Press, 1969. xliii, 1,040 p. Appendix, pp. 867-992. Name Index, pp. 993-96. Subject Index, pp. 997-1040.

A presentation of a general theory on the social, political, and economic structure and function of regions within the framework of decision-making analysis drawing heavily on game theory and general competitive equilibrium theory. It is useful to operations research analysts, economists, sociologists, political scientists, urban and regional planners, geographers, and psychologists. It is divided into two parts: part 1 deals with extensions of partial equilibrium theory and consists mainly of the applications of game theory to regional analysis. Part 2 presents the foundations for a general theory with the theory itself forming the content of chapters 13 (the verbal statement) and 14 (the mathematical statement). The appendix contains proofs of theorems, less interesting cases, and other material of an abstract nature. The mathematical equations and notation are confined to the appendix and chapter 14. Assumes mathematical maturity comparable to that obtained in linear algebra and calculus courses. Chapters conclude with references.

Isard, Walter, and Cumberland, John H., eds. REGIONAL ECONOMIC PLANNING: TECHNIQUES OF ANALYSIS FOR LESS DEVELOPED AREAS. Foreword by R. Gregoire. Paris: European Productivity Agency of the Organization for European Economic Co-operation, 1961. 450 p. Appendixes, pp. 431-42. Index of Authors, pp. 443-44. Subject Index, pp. 445-50.

A collection of eighteen papers presented at the first study conference on problems of economic development organized by the European Productivity Agency and held in Bellagio, Italy, from June 19 to July 1, 1960. The papers are grouped into five parts: introduction ("Regional and National Economic Planning and Analytical Techniques for Implementation," by W. Isard and T. Reiner); the regional, national, and institutional setting (one paper from each of the delegates from Greece, Italy, Spain, Turkey, Yugoslavia, and an additional paper by A. Mayne of the Puerto Rico Planning Board); basic problems of regions in the process of economic development (two papers); techniques of analysis (six papers); and synopsis and summary (two papers). Discussion papers are also included. The techniques discussed in the papers include linear programming, input-output analysis, and benefit-cost analysis. Among the authors are: O. Eckstein, L. Rodwin, P.N. Rosenstein-Rodan, M. Rossi-Doria, and R. Stone.

Kendrick, John W., with Carson, C.S. ECONOMIC ACCOUNTS AND THEIR USES. New York: McGraw-Hill Book Co., 1972. xi, 339 p. Appendixes, pp. 305-27. Name Index, pp. 329-31. Subject Index, pp. 333-39.

An up-to-date presentation of the theory and application of national economic accounts with references to the U.S. and UN, systems of accounts. The specific accounting systems discussed include national income and product, flow-of-funds, national wealth, and input-output analysis. The problems of international accounts and comparisons are examined, and an estimate of gross world product (GWP) is made for a recent year. The possibilities of disag-

gregation on a regional basis to permit analysis of subnational economies are considered. The two appendixes present the U.S. summary national income and product accounts for 1968, and a description of the UN system of accounts.

Lee, Tong Hun; Moore, John R.; and Lewis, David P. REGIONAL AND IN-
TERREGIONAL INTERSECTORAL FLOW ANALYSIS. THE METHOD AND AN
APPLICATION TO THE TENNESSEE ECONOMY. Knoxville: University of
Tennessee Press, 1972. vii, 164 p. Appendix A: Tables of Basic Data,
pp. 109-46. Appendix B: The Form of Interview Questionnaires Used for
Manufacturing Firms, pp. 147-57. Bibliography, pp. 159-61. Index, pp. 162-64.

A report on the construction of an input-output model for Tennessee
and a three-region intersectoral model of the state (interregional intersectoral model). The regional model is limited to the rows-only
approach. It deals with both the conceptual problems of the model
and the data-collection problems. The seven chapter titles are:
"Introduction," "Recent Developments in Interindustry Analysis,"
"Approach to Empirical Implementation," "The State-wide Model:
The Empirical Results," "The Interregional Model: The Empirical
Results," "Summary," and "Conclusions." Assumes knowledge of
matrix algebra and econometric techniques. This study finds that
the final demand multiplier for business investment exerts the
most important stimulus to employment and that expenditures of
state and local governments follow closely as next most important.

Leontief, Wassily W. INPUT-OUTPUT ECONOMICS. New York: Oxford
University Press, 1966. viii, 257 p. No index.

A collection of reprints of eleven essays on input-output economics (some originally published in SCIENTIFIC AMERICAN and the
HARVARD BUSINESS REVIEW) written by the author between 1947
and 1965. Four of the articles are nontechnical and deal with
the basic concepts of economics with applications to wages, profits,
prices, and taxes. The remaining seven articles, requiring some
knowledge of high school algebra and elementary matrix algebra,
deal with the construction and use of input-output tables for the
U.S. economy, the economic effects of disarmament, structure
of American trade, and multiregional input-output analysis. Useful as a supplementary text for courses in input-output economics
and macroeconomics.

_____. THE STRUCTURE OF THE AMERICAN ECONOMY, 1919-1939: AN
EMPIRICAL APPLICATION OF EQUILIBRIUM ANALYSIS. 2d ed., enl. New
York: Oxford University Press, 1951. xvii, 264 p. Appendix 1: Simplified
Example of a Closed Economic System, pp. 221-22. Appendix 2: Statistical
Sources and Methods of Computation, pp. 223-44. Appendix 3: Basic Tables,
pp. 245-59. Index, pp. 261-64.

The second edition consists of the first edition plus four additional

chapters--each chapter consisting of an article published in the
QUARTERLY JOURNAL OF ECONOMICS or the AMERICAN
ECONOMIC REVIEW. It is divided into the following four parts:
quantitative input and output relations in the economic system
of the United States in 1919 and 1929, the theoretical scheme,
data and variables in the American economic system 1919-29,
and applications of input-output technique to the American economic system in 1939. This book is a pioneer work in input-output
studies.

_____, et al. STUDIES IN THE STRUCTURE OF THE AMERICAN ECONOMY:
THEORETICAL AND EMPIRICAL EXPLORATIONS IN INPUT-OUTPUT ANALY-
SIS. New York: Oxford University Press, 1953. x, 561 p. Appendixes 1
to 3, pp. 486-541. Table Index, pp. 545-47. Chart Index, pp. 548-49.
Name Index, pp. 550-51. Subject Index, pp. 552-61.

A progress report consisting of twelve chapters and three appendixes
on the construction of an input-output table for the American economy, and on the analysis of the structure of the economy. Each
chapter is written by one of the research workers on the project
(some authors contributed more than one chapter), and classified
under the following parts: static and dynamic theory, consisting
of three chapters by W.W. Leontief; the extensions of input-
output techniques to interregional analysis, consisting of one chapter by W.W. Leontief and one by W. Isard; the capital structure
of the American economy, consisting of one chapter by R.N. Grosse
and one by P.G. Clark; one chapter by each of the following:
H.B. Chenery, M. Holzman, A.P. Grosse, and A.R. Kinstin;
and a concluding chapter by J.S. Duesenberry and H. Kistin. The
appendixes consist of tables showing the capital structure of American industries for 1939, input-output classification for the 1947
input-output study, and price elasticities from budget studies.
Four fold-in tables of the input-output analysis for 1939, capital
stock, and industry classifications are included.

Lukács, Ottó, et al., eds. INPUT-OUTPUT TABLES: THEIR COMPILATION
AND USE. Budapest, Hungary: Akademiai Kiado, 1962. 292 p. No index.

A collection of twenty-one papers presented at Branch A of the
Statistical Scientific Conference on June 1-5, 1961, in Budapest
at the Hungarian Academy of Sciences. The conference was organized by the Hungarian Statistical Office and the Hungarian
Economic Society, and was attended by representatives of statistical offices, universities, and scientific institutions of the USSR,
and Eastern and Western Europe. The papers fall into two groups:
those dealing with the establishment and processing of input-output
tables, and those dealing with their use. Tables for Hungary, the
USSR, and Poland are presented; applications of input-output analysis in planning, efficiency computations, price computations, and
the aluminum and building industries are included; and several
issues related to input-output analysis, such as international com-

parisons of tables and the relation of the tables to national income accounting, are examined. The authors include: M. Augusztinovics, A. Bródy, P. Havas, Z. Kenessey, J. Kornai, Z. Roman, J. Rudolph, A. Rácz, and B. Szybisz.

Mathur, Purushottam Narayan, and Bharadwaj, Ranaganath, eds. ECONOMIC ANALYSIS IN INPUT OUTPUT FRAMEWORK: WITH INDIAN EMPIRICAL EX-PLORATIONS. Vol. 1. Foreword by V.M. Dandekar. Poona, India: Input Output Research Association, Gokhale Institute of Politics and Economics, 1967. xv, 256 p. Appendix I: List of Participants, pp. 247-48. Appendix II: List of Papers Discussed in Seminars, pp. 248-49. Index, pp. 250-56.

A collection of seventeen papers from among those presented at the first input-output seminar sponsored by the Input-output Research Association and held in Poona in 1965. The papers are grouped under the following headings: flow and stock tables, dynamic analysis, regional and industry studies, and international trade. Also included is an introductory paper, "The Input-output Economics--a Resume," by R. Bharadwaj and P.N. Mathur. The papers are concerned with dynamic input-output systems for planning, the derivation and use of shadow prices, methodology for finding import substitution, regional analysis, and Indian applications, including a twenty-nine-sector flow table for 1959-60, a matching capital coefficient matrix, a labor input row divided into eleven occupational classes, and five eighteen-sector regional tables covering all India.

Mathur, Purushottam Narayan, and Venkatramaiah, P., eds. ECONOMIC ANALYSIS IN INPUT OUTPUT FRAMEWORK WITH INDIAN EMPIRICAL EX-PLORATIONS. Vol. 2. Poona, India: Input Output Research Association, Gokhale Institute of Politics and Economics, 1969. vii, 233 p. Index, pp. 223-33.

A collection of fourteen papers presented at the second seminar on input-output analysis held in Poona, India, in 1967. The papers are grouped under the following headings: flow tables, stock tables, regional and location studies, prognostication and planning, and international trade. In addition, there are two introductory papers: "Input-output Framework for Explorations in Theoretical and Empirical Research," by P.N. Mathur; and "Methodological Survey of the Applications of Input Output in Developing Countries," by R. Bharadwaj. While volume 1 dealt with a resume of the techniques of input output and its application to problem areas, volume 2 attempts to expand the horizon of the use of these techniques. The papers are concerned primarily with the structural coefficients of input output analysis and on the application of this method to problems of capital theory and growth

Mennes, L.B.M.; Tinbergen, Jan; and Waardenburg, J. George. THE ELE-MENT OF SPACE IN DEVELOPMENT PLANNING. Foreword by H.C. Bos.

New York: American Elsevier, xiii, 340 p. Appendix 1: The Characteriza-
tion of Spatial Units, pp. 269-73. Appendix 2: Data and Solutions of the
Mexican Model, pp. 274-79. Appendix 3: Mathematical Remarks on the Decomposition Procedure in Chapters 4 and 6, pp. 280-95. Appendix 4: Heavy
Products, pp. 296-313. Appendix 5: Classification of Sectors, pp. 314-17.
Appendix 6: Cost Coefficients, pp. 318-19. References, pp. 320-23. List
of Symbols, pp. 324-34. Author Index, p. 335. Subject Index, pp. 336-40.

> Presents methods for dealing with the element of space in medium-
> or long-term economic development planning ranging from world
> planning with a continental subdivision to regional planning within
> a country. A linear programming development model of a closed
> world economy is formulated as a transportation problem in which
> transportation costs are minimized. A second linear programming
> model is then used to find the optimal pattern of transportation
> flows. The model is generalized to an open national economy.
> A hierarchical model is developed for subdivisions of a nation or
> region in which input-output analysis is employed. Assumes knowl-
> edge of mathematical programming, input-output analysis, and
> econometric theory. This book is useful to all mathematically
> oriented regional and development economists.

Miernyk, William H. THE ELEMENTS OF INPUT-OUTPUT ANALYSIS. New
York: Random House, 1965. xi, 156 p. Index, pp. 153-56.

> A presentation of the essentials of input-output analysis entirely
> in nonmathematical terms, the last chapter covering the rudiments
> of input-output mathematics. Topics include applications of input-
> output analysis, international developments, regional and inter-
> regional input-output analysis, and frontiers of input-output anal-
> ysis. Chapters conclude with references.

_____. SIMULATING REGIONAL ECONOMIC DEVELOPMENT: AN INTER-
INDUSTRY ANALYSIS OF THE WEST VIRGINIA ECONOMY. Lexington, Mass.:
D.C. Heath, 1974. xxii, 337 p. Appendixes, pp. 208-323. Selected Bib-
liography, pp. 327-30. Notes, pp. 333-37.

> Presents a forty-eight-sector, 1965 state input-output table and a
> projection of the state's output and employment for 1975. The
> appendixes consist of the questionnaires used to obtain the data
> from the firms, input-output and aggregate income accounts, and
> matrix tables.

Netherlands Economic Institute, ed. INPUT-OUTPUT RELATIONS: PROCEED-
INGS OF A CONFERENCE ON INTER-INDUSTRIAL RELATIONS HELD AT
DRIEBERGEN, HOLLAND. Leiden, Netherlands: H.E. Stenfert Kroese, 1953.
ix, 234 p. No index.

> A collection of ten papers presented at a conference in Driebergen
> in September 1950 under the sponsorship of the Rockefeller Foun-
> dation. The papers are grouped under two headings: theory (four

papers), and applications (six papers). The papers on applications deal with input-output studies in Norway, the United Kingdom, the Netherlands, and the United States. The authors are: O. Aukrust, T. Barna, W.D. Evans, E.Glaser, R.M. Goodwin, T.C. Koopmans, W. Leontief, G.F. Lueh, J. Sandee, D.B.J. Schouter, R. Stone, and J.E.G. Utting.

Polenske, Karen R. STATE ESTIMATES OF TECHNOLOGY, 1963. Volume 4 of Multiregional Input-Output Analysis. Lexington, Mass.: D.C. Heath, 1974. xix, 521 p. Appendix A: Industrial Classification, pp. 129-49. Appendix B: Regional Classifications, pp. 151-56. Appendix C: 1963 State Estimates of Technology, pp. 157-467. Appendix D: 1947, 1958, 1963 State Estimates of Outputs, Employment, and Payrolls, pp. 469-509. Bibliography, pp. 511-19. About the Author, p. 521. No index.

A presentation of the 1963 estimates of state technologies for use in the Harvard multiregional input-output model. Technology data are assembled for eighty-seven industries and for fifty-one regions (fifty states and the District of Columbia). Sources of data and methodologies for agriculture, mining, and new construction technologies are also explained. This book was written with the assistance of C.W. Anderson, O. Dixon, R.M. Kubarych, M.M. Shirley, and J.W. Wells.

Quesnay, Francois. QUESNAY'S TABLEAU ECONOMIQUE. Clifton, N.J.: Augustus M. Kelley, for the Royal Economic Society and the American Economic Review, 1972. Irregularly paged. No index.

A facsimile reprint of the third edition of Quesnay's TABLEAU, and an English translation by M. Kuczynski and R.L. Meek of the original work published in 1768. It also contains facsimiles and and translations of the first and second editions.

Richardson, Harry W. INPUT-OUTPUT AND REGIONAL ECONOMICS. New York: Wiley, Halsted Press, 1972. 294 p. Bibliography, pp. 261-83. Name Index, pp. 184-86. Subject Index, pp. 287-94.

Describes regional input-output techniques and shows the usefulness and limitations of input-output as a practical tool for regional economic analysis. The first part is on the theory of input-output and multiplier analysis and interregional models. The second part deals with the collection and use of data; and the third part treats applications of input-output techniques to economic impact analysis, regional forecasting, linear programming, environmental problems, and urban models. It is written for the regional economist and it is relatively nontechnical.

Rodgers, John M. STATE ESTIMATES OF OUTPUTS, EMPLOYMENT, AND PAYROLLS, 1947, 1958, 1963. Preface by K.R. Polenske. Volume 2 of Multiregional Input-Output Analysis. Lexington, Mass.: D.C. Heath, 1972.

xvii, 253 p. List of Abbreviations, p. xvii. Appendix A: Classification Tables, pp. 113-22. Appendix B: Values of Manufacturing Sector Shipments and Output: 1947, 1958, and 1963, pp. 123-27. Appendix C: 1947, 1958, and 1963 State Estimates of Outputs, Employment, and Payrolls, pp. 129-241. Bibliography, pp. 243-52. About the Author, p. 253.

Presents the estimates of state outputs, employment, and payrolls for 1947, 1958, and 1963 for eighty-one input-output industries. It describes the estimation procedures and data sources used in preparing the state statistics. The appendixes contain the tabulations of output, employment, and payroll data by state and industry.

Rogers, Andrei. MATRIX ANALYSIS OF INTERREGIONAL POPULATION GROWTH AND DISTRIBUTION. Berkeley and Los Angeles: University of California Press, 1968. xiv, 119 p. Index, pp. 117-19.

A monograph on the analysis of interregional population growth and distribution by means of the algebra of matrices. It begins by presenting the demographer's well-known matrix model of population growth, and then applies the model to forecasting population, stability problems, intervention, and migration. Knowledge of linear algebra is useful. Chapters conclude with references.

_____. MATRIX METHODS IN URBAN AND REGIONAL ANALYSIS. San Francisco: Holden-Day, 1971. xiii, 508 p. Index, pp. 500-508.

A text for an undergraduate course in urban and regional analysis for students who have knowledge of only high school algebra. It emphasizes computations, rather than theorems and proofs, and it illustrates how matrix algebra is used to develop interregional or intraregional models of growth and development, input-output analysis, statistical analyses, and models of linear programming, games, and networks. Characteristic roots and vectors are used in the analysis of stable states in demographic and economic models, and factor analysis is presented as axis rotation in multidimensional space. The appendix to chapter 3 presents FORTRAN programs for simple matrix operations. Chapters conclude with references and exercises.

Rudra, Ashok. RELATIVE RATES OF GROWTH--AGRICULTURE AND INDUSTRY. Series in Economics, no. 14. Bombay: University of Bombay, 1961. 91 p.

The author uses a 30x30 input-output table of the Indian economy for 1960-61 to analyze the effects that would be generated if industry is made to grow at a relatively higher rate of growth than that attained by the agricultural sector. The techniques used include Leontief's interindustry analysis and Richard Stone's linear expenditure system.

Sadler, Peter; Archer, Brian; and Owen, Christine. REGIONAL INCOME MULTIPLIERS. Bangor Occasional Papers in Economics, no. 1. Bangor, Wales: University of Wales Press, 1973. xii, 109 p. Appendix: Sources and Methods, pp. 98-109. No index.

> A monograph with three aims: (1) to apply input-output models to small regions for planning purposes; (2) to contribute to theoretical knowledge by the empirical testing of a hypothesis; and (3) to study the impact of the aluminum smelter upon the economy of Anglesey. Regional multipliers are developed and used to estimate changes in income and demand. Assumes knowledge of elementary input-output analysis.

Schaffer, William A., et al. INTERINDUSTRY STUDY OF THE HAWAIIAN ECONOMY. Honolulu: Department of Planning and Economic Development, State of Hawaii, 1972. 104 p.

> Presents an interindustry model of the State of Hawaii based on fifty-four industry sectors during 1967. The study devotes a chapter each to an introduction, a review of input-output analysis, a summary of the Hawaiian economy, the input-output model, and conclusions. The four appendixes are devoted to the more technical aspects of the estimating procedure, the input-output model and associated analytical tools, the survey and adjustments, and the industry definitions and sources of data.

Scheppach, Raymond C., Jr. STATE PROJECTIONS OF THE GROSS NATIONAL PRODUCT, 1970, 1980. Volume 3 of Multiregional Input-Output Analysis. Lexington, Mass.: D.C. Heath, 1972. xvii, 271 p. Appendix A: Input-output Industry Numbers, Tables, and Related SIC Codes, pp. 99-104. Appendix B: 1970 and 1980 State-projections of Final Demands (MRIO data set 3), pp. 105-259. Bibliography, pp. 261-69. About the Author, p. 271.

> The components of gross national product that are estimated are: personal consumption expenditures, gross private capital formation, net inventory change, net foreign exports, state and local government net purchases of goods and services, and Federal government purchases. Assumes knowledge of matrix algebra and econometric techniques.

Steiss, Alan Walter. MODELS FOR THE ANALYSIS AND PLANNING OF URBAN SYSTEMS. Lexington, Mass.: Lexington Books, 1974. xi, 352 p. Notes, pp. 325-45. Author Index, pp. 347-48. Subject Index, pp. 349-52.

> An exploration of a number of models that might serve as a frame of reference for interdisciplinary studies of complex urban problems. Included are the methodologies of operations research, systems analysis, structural-functional analysis, urban activity systems, dynamic change models, cybernetics, and general systems theory. Some mathematical notation is used in only one chapter. The models

are reviewed from the following standpoints: historical development, verbal description, limitations, advantages, and applicability to urban problems.

Stone, Richard. INPUT-OUTPUT AND NATIONAL ACCOUNTS. Foreword by M. Gilbert. Paris: Organization for European Economic Co-operation, 1961. 202 p. Bibliography, pp. 191-202. No index.

An examination of the conceptual problems involved in subdividing the national accounts into industry sectors. The first part, chapters 1 to 5, is concerned with problems of definition, classification, and arrangement; the second part, chapters 6 to 13, deals with input-output models and their applications. One chapter is devoted to a review of matrix algebra. It illustrates input-output analysis in the following countries: Denmark (1947), Japan (1951), Spain (1954), New Zealand (1952-53), Australia (1953-54), France (1951), and the United Kingdom (1954).

Stone, Richard; Bates, John; and Bacharach, Michael. INPUT-OUTPUT RELATIONSHIPS 1954-66. Programme for Growth, no. 3. London: Chapman and Hall, 1963.

Published by the department of applied economics, University of Cambridge. This book is devoted to a projection of the input-output relationships for 1954 into those expected in 1966. A new method for the projection of the matrix of input coefficients is presented based on substitution among nonprimary inputs (effect of substitution) and on variation of the ratio between total primary inputs and the total nonprimary inputs (effect of fabrication). The study uses a two-way classification of interindustry flows by commodity and industry.

Stone, Richard, and Croft-Murray, Giovanna. SOCIAL ACCOUNTING AND ECONOMIC MODELS. London: Bowes and Bowes, 1959. 88 p. Appendix: A List of Works Cited, pp. 85-88. No index.

A sequel to Meade and Stone's NATIONAL INCOME AND EXPENDITURE (London: Oxford University Press, 1944). In four chapters, it presents a brief description of social accounting, including the input-output table, together with an introduction to economic model building. Assumes knowledge of matrix algebra and elementary calculus.

Tilanus, Christiaan Bernhard. INPUT-OUTPUT EXPERIMENTS. THE NETHERLANDS 1948-1960. Economic Series, vol. 5. Rotterdam: Rotterdam University Press, 1966. viii, 141 p. References, pp. 134-39. Samenvatting, p. 140. Index, p. 141.

An empirical presentation of input-output analysis for readers encountering the subject for the first time. It begins with a small numerical example to introduce the basic concepts, and then describes the thirteen input-output tables for the Netherlands for the years 1948 through 1960. The tables contain thirty-five industries, seven primary sectors, and six final demand sectors. Forecasts of intermediate demand are made.

Treml, Vladimir G.; Gallik, Dimitri M.; Kostinsky, Barry L.; and Kruger, Kurt W. THE STRUCTURE OF THE SOVIET ECONOMY: ANALYSIS AND RECONSTRUCTION OF THE 1966 INPUT-OUTPUT TABLE. Praeger Special Studies in International Economics and Development. New York: Praeger Publishers, 1972 xxiv, 686 p. Appendix A: Correlation of 1959 and 1966 Sector Classifications, pp. 555–62. Appendix B: Commodity-establishment Adjustments, pp. 563–76. Appendix C: Price Adjustments for Foreign Trade Conversion Coefficient, pp. 577–81. Appendix D: Labor Input Coefficients for Aggregated Sectors, pp. 583–602. Appendix E: Input Coefficients, pp. 603–7. Appendix F: Adjustment of Retail Trade Data, pp. 609–30. Appendix G: Test Results, pp. 631–39. Bibliography: Western Sources, pp. 641–45. Bibliography: Soviet Sources, pp. 645–55. Index of Names, pp. 657–60.

An exhaustive reconstruction and analysis of the Soviet 1966 110-sector input-output table made possible by support from the U.S. Arms Control and Disarmament Agency. The result is a 76-sector ex post table for the Soviet economy for 1966 in value terms, consisting of 6,156 separate elements (some of which are zeros) of which 827 elements are estimated. The first four chapters are devoted to descriptions of Soviet input-output studies, with emphasis on the description and assessment of the 1966 ex post table in value terms. The rest of the book is devoted to interpretations of available input-output data and descriptions of the various elements entering into the process of reconstructing the 76-sector table. The complete reconstructed 76-sector 3-quadrant flow table, including labor and fixed capital data, the matrix of direct material, labor and capital input coefficients, and the matrix of full (direct plus indirect) input coefficients is appended at the end of the book.

Van Duijn, Jacob J. AN INTERREGIONAL MODEL OF ECONOMIC FLUCTUATIONS. Lexington, Mass.: D.C. Heath, 1972. x, 187 p. List of References, pp. 179–81. Index, pp. 183–87.

A study of regional economic fluctuations in a closed economy. It considers the questions of how regional and national fluctuations of output and unemployment are affected by the economic characteristics of regions and how the central government can attain its goal of interregional equity and economic stability at the lowest level of unemployment. The mathematics prerequisites are calculus and matrix algebra. Several interregional simulation models are developed and applied to hypothetical data. There are no real life applications.

Van Wickeren, Alfred. INTERINDUSTRY RELATIONS: SOME ATTRACTION MODELS: A CONTRIBUTION TO REGIONAL ECONOMICS. Foreword by L.H. Klaassen and J.H.P. Paelinck. Rotterdam: Rotterdam University Press, 1973. xiii, 231 p. List of Symbols, pp. 214-17. Bibliography, pp. 219-23. List of Subjects, pp. 225-28. Summary, pp. 229-31.

A study of industrial location divided into two parts: theory, and application which involves the estimation of population parameters of the reduced form equations relating to selected Dutch industries. Chapter 7 is a reprint of an article, "An Attraction Analysis for the Austrian Economy," published in the journal REGIONAL AND URBAN ECONOMICS (vol. 2, no. 3, 1972). Assumes knowledge of matrix algebra and econometric techniques.

Westphal, Larry E. PLANNING INVESTMENTS WITH ECONOMIES OF SCALE. Contributions to Economic Analysis, vol. 69. New York: American Elsevier, 1971. xxiii, 380 p. Appendix A: Proof of Price Decomposition of the Non-convex Dual, pp. 291-92. Appendix B: An Illustrative Numerically Specified Non-convex Model, pp. 293-305. Appendix C: Formal Statement of the Capital Accumulation Path Models, pp. 306-19. Appendix D: The Dual Problem for CAPMA (Capital Accumulation Path Model A), pp. 320-27. Appendix E: Data Sources for the Capital Accumulation Path Model, pp. 328-49. Appendix F: The Rate of Return to Investment in CAPMA and a Comparison of Optimal Investment Policies for Decreasing Costs Sectors Using a One-sector Model, pp. 353-64. Selected Bibliography, pp. 365-72. Index, pp. 373-80.

The development of an economy-wide model for use in planning the scheduling and scale of major investment projects in less developed economies which would have effects throughout the entire economy. It illustrates the methodology by the planning of a petrochemical complex and an integrated steel mill in South Korea. The model is a dynamic input-output optimizing model having the mathematical structure of a mixed integer-continuous variable programming problem, the function of integer variables being to introduce economies of scale. The model includes production, import, export, and capacity expansion for each sector. Assumes knowledge of mathematical economics and econometric methods.

Yamada, Isamu. THEORY AND APPLICATION OF INTERINDUSTRY ANALYSIS. Economic Research Series, no. 4. Tokyo: Kinokuniya Bookstore, 1961. x, 254 p. Index, pp. 253-54.

A study of input-output analysis which is divided into two parts: part 1, theories of interindustry analysis; and part 2, applications of interindustry analysis, with special reference to the Japanese economy. The Topics included in part 1 are a synthesis between the physical quantity system and the pure system, the problem of aggregation of industries, interregional input-output analysis, and the problem of international trade. Part 2 develops economic structures for 1951, 1953, and 1955 for the Japanese economy using economic circulation graphs, and then proceeds to the analysis for 1951 and 1954 using the method of cost coefficients. It concludes with a chapter on disaggregation for purposes of economic planning. Assumes knowledge of matrix algebra and econometric techniques.

Yan, Chiou-Shuang. INTRODUCTION TO INPUT-OUTPUT ECONOMICS. Principles of Economics Series. New York: Holt, Rinehart and Winston, 1969. x, 134 p. Index, pp. 130-34.

An elementary treatment of input-output economics covering the theoretical framework, empirical applications and statistical problems, and variations in input-output models. Topics include problems and practices of preparing input-output tables, basic input-output techniques, and regional input-output models. Matrix algebra is used to supplement verbal discussion.

Chapter 5

MODELS

Ando, Albert; Fisher, Franklin M.; and Simon, Herbert A. ESSAYS ON THE STRUCTURE OF SOCIAL SCIENCE MODELS. Cambridge, Mass.: MIT Press, 1963. iv, 172 p. Index, pp. 169-72.

A collection of six reprints of articles published by the authors in the following sources: STUDIES IN ECONOMETRIC METHODS (Wiley, 1953), ECONOMETRICS, INTERNATIONAL ECONOMIC REVIEW, and POLITICAL SCIENCE REVIEW. One additional article by F.M. Fisher, "Properties of the von Neumann Ray in Decomposable and Nearly Decomposable Technologies," appears for the first time. The first two essays deal with the problem of causality; the second two with the problem posed by the aggregation of economic data; the fifth essay presents and proves two theorems on the aggregation into indexes and applies them to two examples from political science; and the last two essays extend the results to the von Neumann model of production. Assumes knowledge of econometric theory.

Beach, Earl Francis. ECONOMIC MODELS: AN EXPOSITION. New York: Wiley, 1957. xi, 227 p. Glossary, pp. 204-5. Answers to Exercises, pp. 206-18. Table A. Student's t Distribution, pp. 219-20. Table B. Chi-square Distribution, p. 221. Index of Names, pp. 223-24. Subject Index, pp. 225-27.

A development of both mathematical and econometric models. Examples of the former type include linear and nonlinear models, such as equilibrium of supply and demand and monopoly revenue and profit; continuous dynamic models, such as the Domar macro model; sequence models, such as the cobweb model. Examples of the latter include simple and multiple regression. Assumes knowledge of basic statistics and elementary calculus. Each of the ten chapters concludes with exercises and readings.

Bergstrom, Abram R. THE CONSTRUCTION AND USE OF ECONOMIC MODELS. Applied Mathematics Series. London: English Universities Press, 1967. x, 131 p. Index, pp. 129-31.

A brief survey of a wide selection of mathematical economic models which are suitable for statistical estimation and which can be used as a basis for the prediction and regulation of the behavior of actual economies. Assumes knowledge of calculus, differential equations, and statistical theory, but no familiarity with economics is necessary. Among the models considered are dynamic models for a single industry based upon Cobb-Douglas production functions and log-linear demand equations, a trade-cycle model of the Phillips type, the Solow single-sector and the Meade two-sector growth models, models of cyclical growth which combine various feed-back relations representing monetary and fiscal policies, the Solow putty-putty and the Jorgenson putty-clay models for vintage capital with Cobb-Douglas production functions, the Arrow model for the acquisition of technical knowledge, and multisector models of the Walras and Leontief types. The book concludes with a discussion of methods of estimating the parameters of the models, including simultaneous equation estimation methods.

Blalock, Hubert M., Jr. THEORY CONSTRUCTION: FROM VERBAL TO MATHEMATICAL FORMULATIONS. Prentice-Hall Methods of Social Science Series. Englewood Cliffs, N.J.: Prentice-Hall, 1969. xi, 180 p. Appendix A: Theory Building and the Statistical Concept of Interaction, pp. 155-65. Appendix B: Some Elementary Calculus, pp. 166-77. Index, pp. 178-80.

Presents elementary guidelines for the construction of scientific models with emphasis on theories that are deductive and causal. Among the topics discussed are distributed lags and feedback, recursive systems, simultaneous equations, and identification. Topics are based on case studies drawn from the work of J.C. Coleman, T.K. Hopkins, R.K. Merton, L.F. Richardson, H.L. Zetterberg, and others. It does not treat empirical or statistical procedures. Assumes knowledge of elementary differential and integral calculus, differential equations, and matrix algebra.

_____, ed. CAUSAL MODELS IN THE SOCIAL SCIENCES. METHODOLOGICAL PERSPECTIVES. Chicago: Aldine-Atherton, 1971. xi, 515 p. Index, pp. 509-15.

A collection of twenty-seven articles, many of which are reprints from journals and elsewhere, dedicated to P.F. Lazarsfeld. They are grouped under the following headings: simple recursive models (four articles), simultaneous-equation techniques (seven articles), the causal approach to measurement error (eight articles), and other complications (four articles). Many papers assume knowledge of matric algebra, least squares procedures, and familiarity with the simultaneous-equation approach. Among the contributors are: H.A. Simon, H.M. Blalock, Jr., A.S. Goldberg, O.D. Duncan, T.C. Koopmans, F.M. Fisher, A. Ando, H.L. Costner, and M.T. Hannan.

Duesenberry, James; Fromm, Gary; Klein, Lawrence R.; and Kuh, Edwin, eds.
THE BROOKINGS MODEL: SOME FURTHER RESULTS. New York: American Elsevier, 1969. xxi, 519 p. Author Index, pp. 512-14. Subject Index, pp. 515-19.

The third volume in a series describing the specification, estimation,
solution, and simulation of a large-scale structural model of the
U.S. economy. It contains thirteen papers by noted authors on
the applications of the Brookings model to problems in economies
grouped under the following headings: expenditure and output
(three essays), income distribution (two essays), fiscal sector (two
essays), monetary sector (two essays), complete system solutions
(two essays), and policy implications (two essays). An introductory paper consists of a progress report by E. Kuh. The authors
are: R.E. Bolton, F. DeLeeuw, M.K. Evans, P.J. Dhrymes, F.
Fromm, S.H. Goldfeld, A. Kisselgoff, L.R. Klein, D.T. Kresge,
E. Kuh, C. Lovell, M.D. McCarthy, A.L. Nagar, and P. Taubman.

———. THE BROOKINGS QUARTERLY ECONOMETRIC MODEL OF THE
UNITED STATES. Chicago: Rand McNally, 1965. xv, 776 p. Key to
Abbreviations, pp. 739-64. Author Index, pp. 765-69. Subject Index,
pp. 770-76.

A collection of eighteen essays which constitute a complete model
of the U.S. economy which explains the variations in GNP and
its major components, as well as major price movements, employment, and wage rates. The essays are grouped under the following six headings: producer investment decisions (four essays); consumer expenditure decisions (two essays); income distribution, price
and wage determination and labor force (three essays); subsectors:
foreign trade and agriculture (two essays); monetary and fiscal (two
essays); and estimation, simulation, aggregation and the complete
model (four essays). Knowledge of advanced econometric theory
is necessary for proper understanding of the details of the models
and of the equations. The authors are: E.W. Adams, A. Ando,
L. Baissonneault, E.C. Brown, P.G. Darling, F. DeLeeuw, J.
Duesenberry, M. Dutta, R. Eisner, K.A. Fox, C.C. Holt, E. Kuh,
D.W. Jorgenson, L.R. Klein, S. Lebergott, M.C. Lovell, S.J.
Maisel, R.R. Rhomberg, C.L. Schultze, Y. Shinkai, G.R. Sparks,
D.B. Suits, J.L. Tryon.

Fiar, Ray C. A SHORT-RUN FORECASTING MODEL OF THE UNITED STATES
ECONOMY. Lexington, Mass.: D.C. Heath, Lexington Books, 1971. xiii,
264 p. Appendix A, pp. 247-51. Appendix B, pp. 253-56. References,
pp. 257-59. About the Author, p. 261. Index, pp. 263-64.

A description of a short-run forecasting model of the U.S. economy
consisting of seven equations explaining expenditures components of
gross national product, two equations explaining the level of housing starts, one employment equation explaining the difference between the establishment-based employment data and the household-
survey employment data, two labor force participation equations,

one price equation, six identities, and one production function. There are nineteen endogenous variables and sixteen exogenous variables. The forecast results of the model are compared with those of the Wharton and Office of Business Economics models. The appendixes contain some of the data used in the analysis, and estimates of the seven expenditure equations.

Fromm, Gary, and Klein, Lawrence R. THE BROOKINGS MODEL: PERSPECTIVE AND RECENT DEVELOPMENTS. New York: American Elsevier, 1975. xii, 679 p. Author Index, pp. 667-71. Subject Index, pp. 673-79.

Fifteen papers presented at a conference held at the Brookings Institute in February 1972. The papers are grouped under the following five headings: perspective on econometric modeling (two papers), studies with the Brookings model (four papers), complete system (two papers), comparison of models (three papers), and new developments in econometric modeling (four papers). Topics include history of econometric model building, consumptions functions for short-run models, a disaggregated quarterly model of United States trade and capital flows, the theory of the firm equations in the Brookings and Wharton models, predictive abilities of a large model, criteria for evaluation of econometric models, the Maryland interindustry forecasting model, the input-output sector of the Wharton annual and industry forecasting model, and uses of dynamic input-output macroeconomic models. The contributors are: A. Ando, C. Almon, J.S. Duesenberry, V.G. Duggal, G. Fromm, B.G. Hickman, S.H. Hymans, L.R. Klein, S.Y. Kwack, M.D. McCarthy, S.M. Menshikov, F. Modigliani, R.S. Preston, P.A. Samuelson, G.R. Schink, P. Taubman, and J. Waelbroeck.

Fromm, Gary, and Taubman, Paul. POLICY SIMULATIONS WITH AN ECONOMETRIC MODEL. Foreword by K. Gordon. Washington, D.C.: The Brookings Institute, 1968. xiv, 179 p. Appendixes, pp. 125-76. Index, pp. 177-79.

The second in a series of volumes describing the structure, solutions, and simulation of the Brookings model. It contains detailed descriptions of the results of statistical testing and simulation of the model. The five chapter titles are: "Structure of the Model," "Complete System Solutions and Simulation," "Analysis of Excise Tax Changes," "Excise Tax Reduction: Simulated Results," and "Evaluation of Alternative Policies." The appendixes present the system equations and a list of variables and definitions, coding of Brooking industries, analysis of price behavior after excise tax changes, and adjustment of model equations for excise tax reductions.

Goldberger, Arthur S. IMPACT MULTIPLIERS AND DYNAMIC PROPERTIES OF THE KLEIN-GOLDBERGER MODEL. Contributions to Economic Analysis, vol. 19. New York: American Elsevier, 1959. 138 p. No index.

An attempt to reduce the Klein-Goldberger model into a form in which direct and indirect effects can be evaluated using reduced form equations. The model is used to evaluate determinants of economic change, and to trace out time paths of response of several exogenous stimuli.

Haitovsky, Yoel; Treyz, George; and Su, Vincent. FORECASTS WITH QUARTERLY MACROECONOMIC MODELS. National Bureau of Economic Research Studies in Business Cycles, no. 23. New York: Columbia University Press, for the National Bureau of Economic Research, 1974. xix, 353 p. Appendix (microfiche).

A monograph that examines macroeconomic forecasts and models to advance scientific inquiry and to improve evaluation techniques and further development in this area. It is based primarily on models and forecasts of the Office of Business Economics and the Wharton School of Finance and Commerce. In part 1 the authors examine econometric model forecasts based on observed rather than on projected values for the exogenous variables when no subjective judgment is used to adjust the equations of the model. In part 2 the authors examine the models and forecasts with all of the adjustments and values for the exogenous variables. A procedure is presented for decomposing the forecast error for each variable into several components. Assumes knowledge of econometric techniques of simultaneous equation models.

Holte, Fritz C. ECONOMIC SHOCK-MODELS: STUDIES IN THE THEORY OF THEIR CONSTRUCTION. Oslo: Norwegian Universities Press, 1962. 116 p. List of Symbols, pp. 110-16. No index.

A presentation of the following three studies on the construction and use of shock-models: causal relations and assumptions about latent variables; central problems in the choice of shock-models; and the use of an aggregated relation for prediction purposes. Both stochastic and nonstochastic models are considered and the variables are grouped into endogenous and exogenous classes. The models are abstract and no empirical applications are considered. Assumes knowledge of elementary calculus and matrix algebra.

Hyvärinen, Lassi P. MATHEMATICAL MODELING FOR INDUSTRIAL PROCESSES. Lecture Notes in Operations Research and Mathematical Systems, vol. 19. New York: Springer-Verlag, 1970. vi, 122 p. References, pp. 121-22. No index.

A survey of different approaches in developing models to describe the behavior of industrial processes in terms of controllable variables. The seven chapter titles are: "Basic Concepts," "Optimizing Models," "Methods of Optimum Search," "Design of Experiments," "Dynamic Covariance Analysis," "Principal Component Analysis," and "Regression Analysis." It does not include the theory of optimal control, stability of control systems, nor techniques in data acquistion. Assumes knowledge of elementary probability theory and mathematical statistics.

Kenkel, James L. DYNAMIC LINEAR ECONOMIC MODELS. New York: Gordon and Breach, 1974. xvii, 380 p. Bibliography, pp. 374-77. Index, pp. 378-80.

A text and reference on the theory of linear difference equations and their applications in the construction and analysis of dynamic economic models. It is written mainly for economists, and assumes knowledge of higher mathematics, including complex variables, linear algebra, and the theory of linear difference equations. It can also be of use to students of mathematics who are interested in the theory of linear difference equations. Topics include the solution of difference equations; equilibrium, stability, and solution of linear systems of difference equations; distributed lag models with emphasis on estimation and interpretation of the parameters; acceleration models; impact and delay multipliers; an introduction to optimal control; and forecasting with autoregressive models. Many of the most important theorems are proved, and numerous examples are provided throughout the text in order to show various applications in economics.

Klein, Lawrence R. ECONOMIC FLUCTUATIONS IN THE UNITED STATES, 1921-1941. Cowles Foundations Monograph, 11. New York: Wiley, 1950. ix, 174 p. Appendix, pp. 135-68. Index, 11. 169-74.

A presentation of several economic models including models based on single equations and systems of equations, profit maximizing theory of the firm, and the theory of the household; and the following three models: simple three equation system (Model I), reduced form model (Model II), and a large structural model (Model III). The latter three models are tested using data for the U.S. economy for 1920 and 1941.

Klein, Lawrence R., and Golberger, Arthur S. AN ECONOMETRIC MODEL OF THE UNITED STATES 1929-1952. Contributions to Economic Analysis, vol. 9. New York: American Elsevier, 1955. xv, 165 p. Explanation of Symbols, pp. xiv-xv. Appendix I: The Basic Time Series, pp. 115-41. Appendix II: The Tax-transfer Functions, pp. 142-55. Appendix III: Residuals from Extimated Equations, pp. 156-59. Index, pp. 161-65.

An econometric model of the United States which can be applied to practical economic problems such as business cycle forecasting. As a guide to future use, it includes a table of the difference between actual values of the main endogenous variables and those estimated by the structural equations when observed values of the explanatory variables are inserted for the years 1949-54. The model consists of twenty endogenous variables, including gross private domestic capital formation, national income, and an index of hourly wages, and eighteen exogenous variables, including population, number of government employees, and government expenditures. There are twenty equations in the model. Assumes knowledge of econometric theory.

Koerts, J., and Abrahamse, Adrian P.J. ON THE THEORY AND APPLICATION OF THE GENERAL LINEAR MODEL. Rotterdam: Rotterdam University Press, 1969. ix, 185 p. Tables of the Durbin-Watson Bounds Test, pp. 175-78. List of References, pp. 179-83. Index, pp. 184-85.

A theoretical examination of several questions concerning the general linear model for applied research workers. In particular, it deals with the nature of the probability distribution of the correlation coefficient R squared and discusses its usefulness for statistical inference. It also examines various tests for autocorrelation and considers the consequences of the normality assumption of these disturbances. The mathematical and statistical derivations are carefully prepared and several of the eight chapters have mathematical appendixes. Assumes knowledge of mathematical statistics.

Künstman, Albert. TRUNCATION OF LONG-TERM DECISION MODELS. Rotterdam: Rotterdam University Press, 1971. viii, 135 p. References, pp. 126-28. Samenvortting, pp. 129-32. Index, pp. 133-35.

A report on a procedure for truncating a decision model, that is, replacing a long-term decision model by a medium-term model such that the optimization in both cases leads to the same first-period decision. The model attempts to show how medium- and long-term decisions influence first-period decisions, and how these influences can be corrected. Since optimization over long periods gives more information than needed, it would be better to truncate the decision model in such a way that only a limited number of periods would be needed, while approximating the same first-period optimal policy. Topics include mathematical formulation of the problem, formulation of the model, an experiment with a truncated target function, and the influence of separate link variables. Simulation is the main tool used. Assumes knowledge of college algebra and matrix algebra. The advanced mathematics is confined in the chapter appendixes for the interested reader.

McCarthy, Michael D. THE WHARTON QUARTERLY ECONOMETRIC FORE-CASTING MODEL MARK III. Studies in Quantitative Economics, no. 6. Philadelphia: Economic Research Unit, Department of Economics, Wharton School of Finance and Commerce, University of Pennsylvania, 1972. 199 p. Appendix: Description of the Variables for the Wharton Quarterly Forecasting Model MARK III, pp. 190-97. References, pp. 198-99. No index.

A description of the third version of the Wharton Quarterly Econometric Forecasting Model (MARK III). Chapter 1 consists of a two-page introduction; chapter 2 presents a flow chart description of MARK III, followed by a description and single equation analysis of the individual equations in the model. Chapter 3, co-authored with V.G. Duggal, presents an analysis of the simulation properties of the model and is divided into two parts: a study of the multiplier properties, and an analysis of the error properties. Chapter 4 presents an analysis of various estimators for large econometric models, including the ordinary least squares and two-stage least squares. Chapter 5 describes the use of the model in an actual forecasting environment.

Packer, Arnold H. MODELS OF ECONOMIC SYSTEMS: A THEORY FOR THEIR DEVELOPMENT AND USE. Cambridge, Mass.: MIT Press, 1972. xiii, 236 p. Bibliography, pp. 220-31. Index, pp. 233-36.

Describes a general procedure for using computerized models of socioeconomic systems with examples drawn primarily from economic and fiscal policy. The purpose of this procedure is to enable the use of computer technology to bring decision analysis and econometrics to socioeconomic policy making. It is divided into three parts: adaptive modeling processes, applications to resource allocation problems using input-output analysis and linear programming, and illustration of applications to macroeconomic stabilization problems. The procedures are cast in game theory form with two or more decision makers. The following prerequisites are helpful: mathematical programming, econometrics, and computer programming.

Papandreou, Andreas G[eorge]. FUNDAMENTALS OF MODEL CONSTRUCTION IN MACRO-ECONOMICS. Training Seminar Series, 1. Athens, Greece: C. Serbinis Press, for the Center of Economic Research, 1962. x, 172 p. Reading Suggestions, pp. 171-72. No index.

Presents rules for model construction with examples in macro-economics. The relationship between explanatory and policy-oriented models as well as that between static and dynamic models is explored at length in terms of the ordering of the variables appearing in them. Some well-known models, such as the Harrod model, are presented. Assumes knowledge of the rudiments of difference equations.

Prachowney, Martin F.J. A STRUCTURAL MODEL OF THE U.S. BALANCE OF PAYMENTS. Contributions to Economic Analysis, no. 60. New York: American Elsevier, 1969. 176 p. Appendix 1: Data Sources, pp. 125-42. Appendix 2: Diagrams of Actual and Predicted Values of the Dependent Variables, pp. 143-57. Appendix 3: A Theoretical Framework of the Monetary-fiscal Policy Mix for the Balance of Payments, pp. 159-64. Appendix 4: Changes in the Endogenous Variables Caused by the Assumed Monetary-fiscal Policy Mix, pp. 165-66. Appendix 5: A Hypothetical Model of Capital Movements Based on Portfolio Adjustments, pp. 167-70. Bibliography, pp. 171-74. Subject Index, pp. 175-76.

The construction and estimation of a model of the foreign sector

of the U.S. economy for evaluating the determinants of all major
items in the U.S. balance of payments, including the merchandise
trade and capital flows. The foreign trade model is also incorporated into a small domestic sector of the U.S. economy in order
to examine the interrelationship between the two sectors. The
purpose of the model is to understand and solve the balance of
payments deficit. The four chapter titles are: "Introduction,"
"The Theoretical Framework," "Empirical Estimates," and "U.S.
Balance of Payments Policies." The determinants are: merchandise imports and exports; foreign travel expenditures; transportation,
private remittances, and other services; direct international investment; long-term portfolio capital movements; interest and dividends
earned abroad; short-term capital movements; and other long-term
movements. Assumes knowledge of econometric methods.

Preston, Ross S. THE WHARTON ANNUAL AND INDUSTRY FORECASTING
MODEL. Studies in Quantitative Economics, no. 7. Philadelphia: Depart-
ment of Economics, Wharton School of Finance and Commerce, University of
Pennsylvania, 1972. 321 p. References, pp. 318-21. No index.

A description of the Wharton Annual Forecasting Model, including
the structure of the model; simultaneous equations, errors, and
multiplier properties; and projections from 1971 to 1980. The
model consists of 155 stochastic equations, 191 definitional identities, 346 endogenous, and 90 exogenous variables.

Rigby, Paul H. MODELS IN BUSINESS ANALYSIS. Merrill's Mathematics
and Quantitative Methods Series. Editors' preface by V.E. Cangelosi and
M.J. Hinich. Columbus, Ohio: Charles E. Merrill, 1969. ix, 102 p.
Index, pp. 101-2.

A presentation of some conceptual problems of model building in
business. The six chapter titles are: "Concept of the Model,"
"Model Building in Scientific Thinking," "Models and Decision
Making," "Model Building," "Definitional Models," and "Behavioral
Models." Topics include types of models, purposes achieved by
model construction, the role of models in scientific inquiry, problems of model building (including identification and specification),
and the relationship among static, dynamic, deterministic, and
stochastic models. Examples are used, and knowledge of high
school algebra is assumed.

Rivett, Patrick. PRINCIPLES OF MODEL BUILDING: THE CONSTRUCTION
OF MODELS FOR DECISION ANALYSIS. New York: Wiley, 1972. x,
141 p. References, pp. 137-38. Index, pp. 139-41.

An exploration of the stages by which decision models may be constructed and a constructive criticism of the techniques of present-
day model building. Topics include: influences on the decision-
taker, classification of models, models of investment, sequential
decisions, decision and utility theory, competitive problems, fore-

casting and simulation procedures, and organizational objectives. The presentation is nonmathematical, but knowledge of the terminology and concepts of basic statistics and operations research is assumed.

Rutledge, John. A MONETARIST MODEL OF INFLATIONARY EXPECTATIONS. Lexington, Mass.: D.C. Heath, 1974. xv, 119 p. Appendix, pp. 101-5. Notes, pp. 107-15. Index, p. 119. About the Author, p. 121.

A study of investor forecasting which focuses on profit maximizing behavior in the choice of forecasting mechanism. It analyzes the ways that market traders form anticipations of the future state of inflation, and empirical evidence is presented to indicate that autoregressive models of forecasts are misspecified, resulting in bias and inconsistent estimates of the parameters in the models. A new model is tested for forecasting inflation which utilizes information such as expectations of investors regarding the behavior of the money supply. Assumes knowledge of elementary calculus, matrix algebra, and statistical estimation theory.

Searle, Shayle R. LINEAR MODELS. Wiley Series in Applied Probability and Statistics. New York: Wiley, 1971. xxi, 532 p. Literature Cited, pp. 515-22. Statistical Tables, pp. 523-27. Index, pp. 529-32.

A self-contained book on linear model techniques for analyzing unbalanced data, that is, data having unequal numbers of observations in subclasses. The numerous examples and exercises contained in chapters 3 to 8 utilize hypothetical data requiring only simple arithmetic for analysis. Topics include full rank and nonfull rank regression models, regression on dummy variables, and variance estimation from unbalanced data. Assumes knowledge of matrix algebra and mathematical statistics.

Steele, Joe L. THE USE OF ECONOMETRIC MODELS BY FEDERAL REGULATORY AGENCIES. Foreword by H.H. Liebhafsky. Lexington, Mass.: D.C. Heath, 1971. xvii, 104 p. Bibliography, pp. 93-96. Notes, pp. 97-102. Index, pp. 103-4.

An analysis and evaluation of the first econometric model to be employed in a rate-making proceeding before a federal agency. The model was used by the Federal Power Commission in the 1961 PERMIAN BASIN AREA RATE CASE which dealt with natural gas rates. The book also analyzes the potential usefulness of econometric models as tools of analysis in rate-making proceedings before other regulatory agencies. The model used by the Federal Power Commission is a linear multiple regression equation in the logarithms of the variables with six independent variables.

Suits, Daniel B. THE THEORY AND APPLICATION OF ECONOMETRIC MODELS. Training Seminar Series, 3. Athens, Greece: Center of Economic Research, 1963. xiii, 147 p. Bibliography, pp. 145-47. No index.

A presentation of the nature and use of econometric models in relatively nontechnical terms for the student of general economics. It is divided into three parts: part 1 (two chapters) is devoted to the relationship of economic models to economic theory, and to the ways that models are used for forecasting and policy analysis. Part 2 (three chapters) presents a thirty-two-equation model of the U.S. economy with an explanation of how forecasts are made with it. Part 3 (one chapter) consists of a brief treatment of some of the more technical matters of econometric analysis, including the problem of identification and a short survey of estimation techniques. Assumes knowledge of college algebra and basic statistics.

Theil, Henri. ECONOMIC FORECASTS AND POLICY. Contribution to Economic Analysis, vol. 15. Amsterdam: North-Holland, 1958. xxxi, 562 p. Index, pp. 557-62.

A report dealing with three basic issues in econometric models based on the Netherlands experience: (1) the predictive quality of models; (2) the relationship between decision making and the econometric model which is used to formulate predictive statements; and (3) the estimation of systems of simultaneous equations. The book concludes with two chapters on the use of forecasts for policy purposes. Most chapters conclude with appendixes devoted to applications and mathematical and statistical reviews. Assumes knowledge of advanced calculus, matrix algebra, and statistical decision theory.

Chapter 6

COLLECTIONS OF ARTICLES, PAPERS, ESSAYS,

AND OTHER CONTRIBUTIONS

Adelman, Irma, and Thorbecke, Erik, eds. THE THEORY AND DESIGN OF
ECONOMIC DEVELOPMENT. Baltimore: Johns Hopkins Press, 1966. x,
427 p. Index, pp. 419-27.

A collection of fourteen papers presented at a conference on economic development sponsored by the Center for Agricultural and
Economic Development, Iowa State University, in November 1965.
The papers are grouped under the following two headings: development theory and strategy, and development planning and programming. The applications discussed can serve as reference material
for planning agencies throughout the world, as well as for courses
in economic development. Assumes knowledge of probability theory
and mathematical economics.

Arrow, Kenneth J., ed. SELECTED READINGS IN ECONOMIC THEORY FROM
ECONOMETRICA. Cambridge, Mass.: MIT Press, 1971. viii, 448 p. Name
Index, pp. 441-42. Subject Index, pp. 443-48.

The second in a series of four volumes of selected readings from
ECONOMETRICS. It contains twenty-two articles, each by a different author, on original research and inquiry, all published prior
to 1966. Topics of the articles include income distribution, consumer demand, stability of equilibrium, the bargaining problem,
resource allocation, expectations, utility discrimination, social
welfare functions, separable utility, aggregation, risk-aversion,
and two-level planning. Among the authors are: J.H. Blau, G.
Debreu, T.C. Koopmans, J. Kornai, W. Leontief, R. Duncan
Luce, P.A. Samuelson, and A. Wald. Some of the articles assume knowledge of advanced calculus, mathematical analysis, set
theory, linear algebra, and probability theory.

Arrow, Kenneth J.; Karlin, Samuel; and Suppes, Patrick, eds. MATHEMATI-
CAL METHODS IN THE SOCIAL SCIENCES, 1959. Stanford Mathematical
Studies in the Social Sciences, 4. Stanford, Calif.: Stanford University Press,
1960. viii, 365 p. No index.

A collection of twenty-three scientific papers by well-known authors in economics, management science, and psychology presented at the first Stanford Symposium on Mathematical Methods in the Social Sciences at Stanford University, June 24, 1959. The nine papers in economics deal with the application of mathematical methods in such areas as resource allocation, stability, capital accumulation and technical change, and consumer behavior. In management science, two papers are on inventory problems, one on queueing theory, and one on an algorithm for a nonlinear programming problem. Of the ten papers in psychology, one deals with sto-chastic learning models, two papers are on the theory of measure-ment and testing, six papers represent contributions to mathemati-cal learning theory, and one paper is devoted to utility theory. Assumes knowledge of differential and integral calculus, linear algebra, and intermediate probability theory.

Ball, Richard J., ed. THE INTERNATIONAL LINKAGE OF NATIONAL ECO-NOMIC MODELS. Contributions to Economic Analysis, vol. 78. New York: American Elsevier, 1973. xii, 467 p. Index, pp. 465-67.

A collection of twelve hitherto unpublished studies by authors re-presenting countries involved in project LINK, dealing with the problem of integration of national econometric models. The papers are grouped under the following five parts: the theory of inter-national linkage (three papers), the general models of project LINK (three papers), models of trade, capital, and services (three papers), application of bilateral linkage (two papers), and the operation of total linkage (one paper). An introductory article by B.G. Hickman, L.R. Klein, and R.R. Rhomberg is also in-cluded. The countries participating in project LINK include the United States, France, Italy, Japan, Canada, Belgium, United Kingdom, West Germany, Finland, Austria, Australia, Sweden, the Netherlands, and the United Nations.

Banerji, Ranan B., and Mesarovic, Mihajlo B., eds. THEORETICAL APPROACHES TO NON-NUMERICAL PROBLEM SOLVING. Lecture Notes in Operations Re-search and Mathematical Systems, vol. 28. New York: Springer-Verlag, 1970. vi, 466 p. No index.

A collection of nineteen papers presented at the Fourth Systems Symposium of the Systems Research Center at Case Western Reserve University on November 19-20, 1968. The papers are grouped under the following headings: overviews, problems in foundations, current research, and new applications. Included among the topics 'treated are: heuristic search programs, dynamic programming, game theory, the use of computer programs in solving problems, and methodology of problem solving. Most of the papers require little formal mathematics training.

Barna, Tibor, ed. STRUCTURAL INTERDEPENDENCE AND ECONOMIC DE-
VELOPMENT. Preface by W[assily]. Leontief. New York: St. Martin's Press,
1963. x, 365 p. No index.

A collection of seventeen papers presented at the third international input-output conference held in September 1961 in Geneva.
The conference was sponsored jointly by the Secretariat of the
United Nations and the Harvard Economic Research Project and
was attended by over 200 economists and statisticians from forty-
one countries. The papers are grouped under the following headings: models of economic development, regional models, input-
output techniques and national planning, and problems of estimation and statistics.

_____. THE STRUCTURAL INTERDEPENDENCE OF THE ECONOMY. Fore-
word by R. Tremelloni and prefatory note W[assily]. Leontief. New York:
Wiley, 1956. viii, 429 p. No index.

A collection of twenty-one papers presented at the second international conference on input-output analysis in Varenna, Italy,
on June 27-July 10, 1954. The papers are grouped under the
following headings: methods of analysis, social accounting systems, national experiences, and special applications. The authors
are: O. Aukrust, L.S. Berman, L.P. Blanc, K. Byerke, V. Cao-
Pinna, H.B. Chenery, W.D. Evans, E.W. Gilboy, W. Leontief,
E. Malinvaud, G. Morton, H. Markowitz, P.N. Rasmussen, J.
Sandee, P. Sevaldson, and R. Stone.

Baumol, William J., and Goldfield, Stephen M., eds. PRECURSORS IN
MATHEMATICAL ECONOMICS: AN ANTHOLOGY. Series of Reprints of
Scarce Works on Political Economy, no. 19. London: London School of Eco-
nomics and Political Science, 1968. xiii, 389 p. Index of Names, pp. 387-89.

A collection of thirty-four papers on mathematical economics dating
from 1738 to 1936 which the editors believe to be classic writings.
The papers are grouped under the following five headings: theory
of games, theory of utility and welfare economics, pricing and the
theory of the firm, general equilibrium theory, and the theory of
distribution and taxation. The writings of Bernouilli, Antonelli,
Pareto, Ramsey, Edgeworth, Walras, Cournot, Wald, and von
Neumann are included.

Beckenbach, Edwin F., ed. APPLIED COMBINATORIAL MATHEMATICS. University of California Engineering and Physical Sciences Extension Series. New
York: Wiley, 1964. xxii, 608 p. Answers to Multiple-choice Review Problems, p. 583. Author Index, pp. 585-90. Subject Index, pp. 591-608.

A collection of eighteen contributions, each by a different author,
on applied combinatorial mathematics. The contributions are grouped
under the following headings: computation and evaluation, counting and enumeration, control and extremization, and construction

and existence. Each contribution concludes with references and a set of multiple choice review questions. Among the authors are: L. Breiman, R. Harary, R.E. Kalaba, D.H. Lehmer, E.W. Montroll, G. Polya, J. Riordan, A.W. Tucker, and J. Wolfowitz.

Bernd, Joseph L., ed. MATHEMATICAL APPLICATIONS IN POLITICAL SCIENCE, IV. Charlottesville: University Press of Virginia, 1969. 83 p.

A collection of four technical articles presented at the 1968 summer institute on Mathematical Applications in Political Science at Virginia Polytechnic Institute. Articles 1 and 2 are on methodology; article 3 is concerned with an application of spatial analysis; and article 4 presents a mathematical exploration of ecological regression.

Beshers, James M., ed. COMPUTER METHODS IN THE ANALYSIS OF LARGE-SCALE SOCIAL SYSTEMS. 2d ed., rev. Cambridge, Mass.: MIT Press, 1968. vii, 266 p. Index, pp. 263-66.

A collection of twenty-three papers presented at a conference held in October 1964 at the Massachusetts Institute of Technology. The revised edition contains the original papers with revisions and some extensions. Papers are on the design and parameter estimation of large-scale social systems, social data systems, and demographic applications.

Bhagwati, Jagdish N., and Eckaus, Richard S., eds. DEVELOPMENT AND PLANNING: ESSAYS IN HONOUR OF PAUL ROSENSTEIN RODAN. London: George Allen and Unwin, 1972. 343 p. P.N. Rosenstein-Rodan: Bibliography, pp. 336-39. List of Authors, p. 340. Index, pp. 341-43.

Eighteen essays, previously unpublished, in honor of P.N. Rosenstein-Rodan grouped under the following seven headings: growth and development economics (two essays), development and planning (five essays), income distribution and regional development (four essays), development and international trade (three essays), cost-benefit analysis (one essay), labour productivity: international differences and short-run behavior (two essays), and value theory (one essay). Knowledge of mathematical economics is helpful for some of the papers. Among the authors are J.H. Adler, J.N. Bhagwati, R.S. Eckaus, B. Hansen, E.E. Hagen, C.P. Kindleberger, L. Lefeber, I.M.D. Little, F. Modigliani, P.A. Samuelson, R.M. Solow, and J. Tinbergen.

Bicksler, James L., and Samuelson, Paul A., eds. INVESTMENT PORTFOLIO DECISION-MAKING. Foreword by P[aul].A. Samuelson. Lexington, Mass.: D.C. Heath, 1974. xiii, 369 p. About the Editors, p. 369. No index.

A collection of twenty-four reprints of articles previously published in journals and elsewhere, grouped under the following headings:

theory of choice, dynamic portfolio choice frameworks, the assessment problem, the efficient markets hypothesis, the time-state-preference valuation framework, the capital asset pricing model, the two-parameter capital asset pricing model, and ex post portfolio performance. In addition, an introductory essay, "Theory of Portfolio Choice and Capital Market Behavior: An Introductory Survey," by J.L. Bicksler, is included. These articles represent a well-planned excursion into the rigorous and scholarly literature of investments. Among the authors are: K.J. Arrow, K. Borch, E.F. Fauna, E.S. Phelps, P.A. Samuelson, K.V. Smith, and J. Tobin.

Borch, Karl Henriki, ed. THE MATHEMATICAL THEORY OF INSURANCE: AN ANNOTATED SELECTION OF PAPERS ON INSURANCE PUBLISHED 1960-1972. Foreword by K.J. Arrow. Lexington, Mass.: D.C. Heath, 1974. xvi, 372 p. Index of Names, pp. 369-70. Index of Subjects, pp. 371-72.

A collection of twenty-three articles published by the author in journals and elsewhere grouped under the following headings: the optimal form of reinsurance contracts (two papers), reciprocal reinsurance arrangements (six papers), the reinsurance market (six papers), the dynamic theory of insurance (six papers), and insurance capital markets (three papers). These papers are based on the application of decision theory and game theory to problems of insurance, and assume knowledge of differential and integral calculus and probability theory.

Borch, Karl Henriki, and Mossin, Jan, eds. RISK AND UNCERTAINTY. New York: St. Martin's Press, 1968. xv, 455 p. Index, pp. 451-55.

A collection of twenty-one papers representing the proceedings of a conference held by the International Economic Association in Bratislava, Czechoslovakia, in 1966. The papers are arranged under the following headings: economic decisions under uncertainty, general decision theory, group decisions and market mechanisms, uncertainty and national planning, and sequential decision problems. The prerequisites for some of the articles include calculus and probability theory.

Bos, Hendricus Cornelis, ed. TOWARDS BALANCED INTERNATIONAL GROWTH. New York: American Elsevier, 1969. x, 329 p. Index, pp. 327-29.

A collection of sixteen heretofore unpublished essays dedicated to Professor Jan Tinbergen on the occasion of his retirement as director of the Netherlands Economic Institute in 1968. The twenty-three contributors are colleagues and staff who have worked at the institute under Tinbergen's guidance on development problems since 1955. Among the topics treated are optimum investment in underdeveloped economies, interindustry relations, effectiveness of project versus plan aid, balanced and maximum growth in dualistic economies, optimum international distribution of production, and trade flows and geographical distance. Among the contributors are: H.C. Box, S. Chakravarty, M. Inagaki, L.H. Klaasen, H. Linnemann, J.P. Pronk, J.S. Ramaer, and J. Serck-Hanssen.

Bos, Hendricus Cornelis; Linnemann, H.; and De Wolff, P., eds. ECONOMIC STRUCTURE AND DEVELOPMENT: ESSAYS IN HONOUR OF JAN TINBERGEN. New York: American Elsevier, 1973. x, 283 p. No index.

A collection of fifteen heretofore unpublished essays in honor of Jan Tinbergen on his retirement from his Chair at the Netherlands School of Economics. The topics dealt with include targets and instruments under uncertainty (by L. Johnsen), a scalar measure of social income (by K.A. Fox and P. van Moeseke), consumer demand (by H. Theil), a linear macro production function (by J. Sandee), the comparative cost theory of international trade (by W. Leontief), development planning (by S. Chakravarty), Hungarian experience in long-term planning (by J. Kornai), and optimal economic growth and exhaustible resources (by T.C. Koopmans).

Boulding, Kenneth E., and Spivey, W. Allen, eds. LINEAR PROGRAMMING AND THE THEORY OF THE FIRM. New York: Macmillan, 1960. viii, 227 p. Bibliography, pp. 217-24. Index, pp. 225-27.

A collection of seven essays presented at a seminar for undergraduate college teachers of economics sponsored by the Ford Foundation and the University of Michigan and held in the summer of 1958 at the university. The purpose of the seminar was to examine the applications of linear programming and other operations research techniques to economics. In addition to the editors, the authors are: S. Cleland, C.W. Kwang, H.H. Jenny, C.M. White, and Y.L. Wu.

Brown, Murrary, ed. THE THEORY AND EMPIRICAL ANALYSIS OF PRODUCTION. National Bureau of Economic Research, Studies in Income and Wealth, no. 31. New York: Columbia University Press, 1967. x, 515 p. Author Index, pp. 501-3. Subject Index, pp. 504-15.

A collection of nine papers on production functions presented at a conference on production functions in the United States and Canada. Topics of the papers include Cobb-Douglas and CES production functions, vintage effects and the time path of investment in production relations, and production function and economic growth.

Buchler, Ira R., and Nutini, Hugo, eds. GAME THEORY IN THE BEHAVIORAL SCIENCES. Pittsburgh: University of Pittsburgh Press, 1969. xiii, 268 p. A Bibliography with Some Comments, by M[artin]. Shubik, pp. 253-61. Index, pp. 263-68.

A collection of twelve papers on game theory in the behavioral sciences, most of which were presented at the conference on applications of the theory of games in the behavioral sciences held at McGill University, Montreal, on August 15-17, 1966. The papers are grouped under the following headings: applications, experimental games, and applications of related approaches. Some of the applications are in anthropology and cultural processes.

Buckley, Walter, ed. MODERN SYSTEMS RESEARCH FOR THE BEHAVIORAL SCIENTIST: A SOURCEBOOK. Foreword by A. Rapoport. Chicago: Aldine, 1968. xxv, 525 p. Selected References, pp. 514-19. Index, pp. 521-25.

A collection of fifty-nine articles previously published in journals and elsewhere grouped under the following headings: general systems research: overview; parts, wholes, and levels of integration; systems, organization, and the logic of reflections; information, communication, and meaning; self-regulation, and self-direction in psychological systems; and self-regulation and self-direction in sociocultural systems. The authors include R.L. Ackoff, C.W. Churchman, K.W. Deutsch, J. von Neumann, A. Rapoport, and N. Wiener.

Bursk, Edward C., and Chapman, John F., eds. NEW DECISION-MAKING TOOLS FOR MANAGERS: MATHEMATICAL PROGRAMMING AS AN AID IN THE SOLVING OF BUSINESS PROBLEMS. Cambridge, Mass.: Harvard University Press, 1963.

A collection of seventeen reprints of articles on decision making grouped under the following headings: general (five articles), finance (two articles), marketing (four articles), product strategy (three articles), and production (three articles). Topics treated include mathematical programming, PERT, capital budgeting, simulation, inventory planning, product diversification, quality control, design of experiments, and production scheduling. Among the authors are: H.I. Ansoff, Wm. J. Baumol, G.A. Busch, J.F. Magee, R.W. Miller, F. Modigliani, R. Schlaifer, and H.A. Simon.

Byrne, R.F.; Charnes, A[braham].; Cooper, W[illiam].W[ager].; Davis, O.A.; and Gilford, Dorothy, eds. STUDIES IN BUDGETING: BUDGETING INTER-RELATED ACTIVITIES-2. Studies in Mathematical and Managerial Economics, vol. 11. New York: American Elsevier, 1971. xiii, 392 p. No index.

A collection of nine papers (four not previously published) arranged under two headings: capital budgeting, and budgeting by public agencies. Topics include capital budgeting under risk, a chance-constrained approach to capital budgeting, the payback period and capital budgeting decisions, defense budgeting, and an empirical study of congressional appropriations. Among the authors are: R.F. Byrne, A. Charnes, W.W. Cooper, O.A. Davis, F.S. Hillier, O. Kortanek, B. Näslund, T. Ruefli, and H.M. Weingartner.

Carter, Charles Frederick; Meredith, G.P.; and Shackle, George Lennox Sharman, eds. UNCERTAINTY AND BUSINESS DECISIONS: A SYMPOSIUM ON THE LOGIC, PHILOSOPHY AND PSYCHOLOGY OF BUSINESS DECISION-MAKING UNDER UNCERTAINTY. 2d ed., rev. and enl. Liverpool, England: Liverpool University Press, 1957. x, 158 p. Index, pp. 153-58.

A collection of fifteen contributions on business decision making divided into two parts. Part 1 consists of seven contributions delivered at the meeting of the British association at Liverpool in 1953 and that made up the first edition. Part 2 contains the remaining eight papers consisting of reprints of previously published articles, condensed accounts of books, or entirely original contributions. Topics include methodology and logic of decision making, the mathematical tools, expectations in economics and business, and the impact of uncertainty. Knowledge of probability is helpful. Among the authors are: C.F. Carter, D.J. O'Connor W.R. Dunlop, W.B. Gallie, G.P. Meredith, A.D. Roy, G.L.S. Shackle, and J.W.N. Watkins.

Charlesworth, James C., ed. MATHEMATICS AND THE SOCIAL SCIENCES: THE UTILITY AND INUTILITY OF MATHEMATICS IN THE STUDY OF ECONOMICS, POLITICAL SCIENCE, AND SOCIOLOGY. A SYMPOSIUM SPONSORED BY THE AMERICAN ACADEMY OF POLITICAL AND SOCIAL SCIENCES. Foreword by James C. Charlesworth. Philadelphia: American Academy of Political and Socience, 1963. v, 121 p.

Consists of the following six essays which were presented at the symposium: "Mathematics in Economics," by L. Hurwicz; "Limits to the Uses of Mathematics in Economics," by O. Morgenstern; "The Use of Mathematics in the Study of Political Science," by O. Benson; "Mathematics and Political Science," by A. Hacker; "Uses of Mathematics in Sociology," by H. White; and "Limits to the Uses of Mathematics in the Study of Sociology," by D. Martindale.

Chenery, Hollis B., et al., eds. STUDIES IN DEVELOPMENT PLANNING. Harvard Economic Studies, vol. 136. Cambridge, Mass.: Harvard University Press, 1971. xii, 422 p. References, pp. 405-17. Index, pp. 419-22.

A collection of seventeen essays by eighteen scholars, resulting from the Project for Quantitative Research in Economic Development of the Center for International Affairs at Harvard University, grouped under the following four headings: general planning models, international trade and external resources in the development process, problems of planning policy and allocation in agriculture and education, and the empirical foundations of planning and growth. Topics include optimization models in planning models, substitution and economies of scale in planning models, problems of interregional and intersectoral allocation, foreign aid, optimal allocation of investment and education, demand elasticities, and growth effects of changes in labor quality and quantity. The authors include M. Bruno, C. Gotsch, D. Kendrick, L. Landau, A. MacEwan, L. Taylor, and T.E. Weisskopf.

Chipman, John S.; Hurwicz, Leonid; Richter, Marcel K.; and Sonnenschein, Hugo F., eds. PREFERENCES, UTILITY AND DEMAND: A MINNESOTA SYMPOSIUM. Harbrace Series in Business and Economics. New York: Harcourt Brace Jovanovich, 1971. ix, 510 p. A Selected Bibliography, pp. 437-92. Author Index, pp. 495-98. Subject Index, pp. 501-510.

Part 1 consists of fifteen papers by nine authors (seven economists and two mathematicians) on the mathematical foundations of the classical theories of preference, utility, and demand written between the period 1956 and 1968. Part 2 contains annotated translations of classic papers by Antonelli, Volterra, Pareto, Frisch, and Alt. Part 3 presents a comprehensive bibliography of published works on theories of preferences, utility, and demand prepared by M. Aoki, J.S. Chipman, and P.C. Fishburn. Except for two essays in part 1, higher mathematics, such as mathematical analysis and topology, is necessary for comprehension.

Christ, Carl F., et al., eds. MEASUREMENT IN ECONOMICS: STUDIES IN MATHEMATICAL ECONOMICS AND ECONOMETRICS IN MEMORY OF YEHUDA GRUNFELD. Stanford, Calif.: Stanford University Press, 1963. xiv, 319 p. Yehuda Grunfeld: In Memoriam, pp. ix-xi. Bibliography of Yehuda Grunfeld, pp. xiii-xiv. Author Index, pp. 315-16. Subject Index, pp. 317-19.

A collection of twelve original essays grouped under the following headings: theory and measurement of consumption, theory and measurement of production, theory and measurement of monetary phenomena, and econometric methodology. The authors are C.F. Christ, M. Friedman, L.A. Goodman, Z. Griliches, A.C. Harberger, N. Liviatan, J. Mincer, Y. Mundlak, M. Nerlove, D. Patinkin, L.G. Telser, and H. Theil. The topics treated by the papers include tests of the permanent income hypothesis, estimation of products and behavioral functions, least-squares estimation of transition probabilities, and portfolio selection. Assumes knowledge of calculus.

Cochrane, James L., and Zeleny, Milan, eds. MULTIPLE CRITERIA DECISION MAKING. Columbia: University of South Carolina Press, 1973. xiv, 816 p. Bibliography, pp. 779-96. Index, pp. 797-816.

Fifty-eight papers on advanced techniques of resolving multiple goal conflicts presented at Capstone House, University of South Carolina, on October 26-27, 1972. The majority of the papers deal with mathematical models, and many are concerned with the relationship between formalized decision-making techniques (computer and mathematical analysis and human judgment). The contributors represent the fields of economics, management science, and mathematics.

Cootner, Paul H., ed. THE RANDOM CHARACTER OF STOCK MARKET PRICES. Cambridge, Mass.: MIT Press, 1964. ix, 510 p. References, pp. 507-10. No index.

A collection of twenty-one essays grouped under the following headings: origins and justifications of the random walk theory (two papers), refinement and empirical testing (six papers), the random walk hypotheses reexamined (eight papers), and the statistical analysis of option price (five papers). Among the authors are L. Bachelier, A. Cowles, E.F. Fama, C.W.J. Granger, M.G. Kendall, A.B. Moore, O. Morgenstern, M.F.M. Osborne, H.V. Roberts, C.M. Sprenkle, Wm. Steiger, and H. Working.

David, Paul A., and Reder, Melvin W., eds. NATIONS AND HOUSEHOLDS IN ECONOMIC GROWTH: ESSAYS IN HONOR OF MOSES ABRAMOVITZ. New York: Academic Press, 1974. xiii, 411 p. Author Index, pp. 407-11.

A collection of fifteen essays by noted authors in honor of Moses Abramovitz on his sixtieth birthday. The papers are grouped under the following two headings: microeconomic foundations (seven papers), and macroeconomic performance: growth and stability (eight papers). The topics treated include individual and social welfare significance of quantitative indexes of economic growth, economic-demographic interdependence, the maximum principle, the process of economic growth, monetary policy, economic fluctuations and economic indicators. Some of the papers require knowledge of differential and integral calculus. Among the authors are K.J. Arrow, P.A. David, M. Friedman, B.G. Hickman, G.H. More, and T. Scitovsky.

Day, Richard H., and Robinson, Stephen M., eds. MATHEMATICAL TOPICS IN ECONOMIC THEORY AND COMPUTATION. Philadelphia: Society for Industrial and Applied Mathematics, 1972. vi, 149 p. No index.

Seven papers presented at the symposium on mathematical economics sponsored by the Office of Naval Research at the 1971 fall meeting of the Society for Industrial and Applied Mathematics held at the University of Wisconsin, October 11-13. The papers are grouped under the following three headings: the 1971 John von Neumann lecture, consisting of the paper, "The Mathematics of Speculative Price," by P.A. Samuelson; mathematical models of economic growth (three papers); and computation of fixed points, with economic applications (three papers). Topics include a renewal model of economic growth, equilibrium growth of dynamic models, optimal growth under uncertainty, computation of a capital stock invariant under optimization, a search for the fixed points of a continous mapping, and techniques for computing fixed points of continuous mappings. Assumes knowledge of calculus and mathematical analysis. The contributors are: J.S. Chipman, D. Gale, T. Hansen, M.M. Jeppson, T.C. Koopmans, O.H. Merrill, and R. Radner.

Dean, Robert D; Leahy, William H.; and McKee, David L., eds. SPATIAL
ECONOMIC THEORY. New York: Free Press, 1970. ix, 365 p. Index,
pp. 357-65.

A collection of twenty articles on spatial economics, written
since 1970 and previously published in journals. The articles
are grouped under the following headings: least cost theory
(three papers), locational interdependence or spatial competition
(three papers), market area analysis (four papers), locational
equilibrium analysis (four papers), and general equilibrium (three
papers). In addition, there are two introductory articles and
one concluding article. Some of the articles assume knowledge
of differential and integral calculus and matrix algebra. Among
the authors are S. Enke, M.L. Greenhut, E.M. Hoover, H.
Hotelling, W. Isard, G.G. Judge, L.N. Moses, and T. Takayama.

Deane, R.S., ed. A NEW ZEALAND MODEL: STRUCTURE, POLICY USE
AND SOME SIMULATION RESULTS. Research Paper no. 8. Wellington:
Reserve Bank of New Zealand, 1972. 92 p. No index.

A collection of five papers summarizing the results of an effort to
construct a quarterly macroeconomic model of the New Zealand
economy using ninety-five equations. The main paper deals with
the simulation results, while other papers are devoted to uses of
macroeconometric model simulation, structurability of the model,
and recent developments in large model extimation. The authors
are R.S. Deane, D.E.A. Giles, C. Gillion, M.A. Lumsden, and
A.B. Sturm.

Dickinson, John P., ed. PORTFOLIO ANALYSIS: A BOOK OF READINGS.
Lexington, Mass.: D.C. Heath, 1974. ix, 236 p. Bibliography, pp. 227-36.
No index.

Eleven reprints of articles previously published in journals which
attempt to summarize the present state of portfolio theory. Also
included is one additional article heretofore unpublished, "Portfolio Theory: An Overview," by J.P. Dickinson. The articles
are grouped under the following four headings: statistical aspects
of portfolio analysis (two articles), the mean-variance criterion:
some modifications (three articles), portfolio selection and performance (three articles), and portfolio theory and capital budgeting
(three articles). Topics include distribution and independence of
successive rates of return from the British equity market; entropy,
market risk, and the selection of efficient portfolios; modular portfolio selection; measurement of porfolio performance under uncertainty; and portfolio adjustments and capital budgeting criteria.
Assumes knowledge of calculus, matrix algebra, and regression

analysis. Among the authors are R.A. Brealey, J.F. Brewster, C.B. Chapman, H. Levy, G.C. Philippatos, M. Sarnat, and C.J. Wilson.

Drèze, Jacques H., ed.; with Delbaen, F.; Gevers, L.; Guesnerie, R.; and Sondermann, D. ALLOCATION UNDER UNCERTAINTY: EQUILIBRIUM AND OPTIMALITY. New York: Wiley, 1974. xxiv, 256 p. No index.

Fourteen papers presented at a workshop in economic theory held in Bergen, Norway, during July and August 1971 and sponsored by the International Economic Association. The papers are grouped under the following five headings: individual decisions (three papers), general equilibrium (two papers), individual risks in large markets (three papers), optimum investment with asset markets (three papers), and short-run equilibrium with money (three papers). Topics include axiomatic theories of choice, decisions under uncertainty, optimum accumulation under uncertainty, stochastic preferences and general equilibrium, competitive equilibrium of the stock exchange and Pareto efficiency, discount rates for public investments under uncertainty, and continuity of the expected utility. Most of the papers assume knowledge of probability theory and calculus. The authors are Y. Caspi, F. Delbaen, J.H. Drèze, L. Gevers, J.M. Grandmont, R. Guesnerie, J.Y. Jaffray, E. Malinvaud, J.A. Mirrlees, T. de Montbrial, A. Sandmo, and D. Sondermann.

Eichhorn, Wolfgang; Henn, R.; Opitz, O.; and Shephard, Ronald William, eds. PRODUCTION THEORY: PROCEEDINGS OF AN INTERNATIONAL SEMINAR HELD AT THE UNIVERSITY OF KARLSRUHE, MAY–JULY, 1973. Lecture Notes in Economics and Mathematical Systems, vol. 99. New York: Springer-Verlag, 1974. ix, 386 p. Author Index, pp. 375–77. Subject Index, pp. 379–86.

Twenty papers presented at a seminar on production theory grouped under the following five headings: production functions; homotheticity, quasilinearity, and technical progress; linear multisectoral production models; nonlinear multisectoral production models; production correspondences; and topics related to production theory. Seven of the nineteen participants were from Canada, France, the Netherlands, Sweden, and the United States; the remaining participants were from West Germany. Among the authors are M.J. Beckmann, R. Fare, K. Hellwig, J. Marschak, and R.W. Shephard.

English, J. Morley, ed. ECONOMICS OF ENGINEERING AND SOCIAL SYSTEMS. New York: Wiley, 1972. ix, 321 p. Index, pp. 315–21.

A collection of twelve lectures presented during the spring of 1969 by university extension at the University of California under the statewide lecture series program. Topics covered include optimization, input-output analysis, economic forecasting, and cost control. Assumes knowledge of operations research techniques.

Fox, Karl A.; Sengupta, Jati K.; and Narasimham, G.V.L., eds. ECONOMIC MODELS, ESTIMATION AND RISK PROGRAMMING: ESSAYS IN HONOR OF GERHARD TINTNER. Lecture Notes in Operations Research and Mathematical Economics, vol. 15. New York: Springer-Verlag, 1969. viii, 461 p. Selected Bibliography of Gerhard Tintner, pp. 456-61.

A collection of nineteen essays by twenty-two authors in honor of Gerhard Tintner grouped under the following headings: introductory essays, economic models and applications, estimation of econometric models, and stochastic programming methods in economic models. The following two introductory essays are also included: "The Econometrics Work of G. Tintner," by J.K. Sengupta; and "The Invisible Revolution in Economics," by K. Fox. Some of the papers are reprints of articles previously published. The other contributors are S.N. Afriat, R.L. Anderson, A. Charnes, Wm. W. Cooper, K.A. Fox, T. Haavelmo, H.H. Hall, E.O. Heady, D.W. Jorgenson, D.W. Katzner, L.R. Klein, T.C. Koopmans, P.V. Moeseke, G.V.L. Narasimham, J.P. Quirk, C.J. Steilberg, H. Theil, J. Tinbergen, T. Wang, H. Wold, and A. Zellner.

Georgescu-Roegen, Nicholas. ANALYTICAL ECONOMICS: ISSUES AND PROBLEMS. Foreword by P.A. Samuelson. Cambridge, Mass.: Harvard University Press, 1966. xvi, 434 p. Analytical Index, pp. 417-29. Index of Names, pp. 431-34.

A collection of five hitherto unpublished papers and twelve previously published papers by the author grouped under the following headings: some orientation issues in economics (five unpublished papers); choice: utility and expectation (six papers); special topics of production (four papers); and economic development (two papers). Included among the papers is "Mathematical Proofs of the Break-down of Capitalism," originally published in ECONOMETRICS in 1960.

Goldberger, Arthur S., and Duncan, Otis D., eds. STRUCTURAL EQUATION MODELS IN THE SOCIAL SCIENCES. Quantitative Studies in Social Relations. New York: Seminar Press, 1973. xv, 358 p. References, pp. 349-58.

A collection of fifteen papers presented at a conference on structural equations cosponsored by the Social Science Research Council and the Social Systems Research Institute of the University of Wisconsin on November 12-16, 1970. The topics treated include the identification problem, procedures for estimating population parameters, multiple indicator models, factors in the process of occupa-

tional achievement, and senate voting scaling problems. The
authors are H.L. Costner, D.D. Duncan. D.L. Featherman, A.S.
Goldberger, Z. Griliches, R.M. Hauser, N.W. Henry, J.E.
Jackson, K.G. Joreskog, D.A. Kenny, K.C. Land, Wm. M.
Mason, M. Nerlove, R. Schuessler, T.P. Schultz, H. Theil,
and D.E. Wiley.

Goreux, Louis M., and Manne, Alan S., eds. MULTI-LEVEL PLANNING:
CASE STUDIES IN MEXICO. Foreword by H.B. Chenery. New York: American Elsevier, 1975. viii, 556 p. Subject Index, pp. 553-56.

Twenty-two contributions dealing with empirical planning models
for the Mexican economy arranged under the following five headings: overview (three contributions), multisector models (six contributions), the energy sector (three contributions), the agricultural sector (seven contributions), and decomposition algorithms and
multilevel planning (three contributions). This book is a collab-
orative effort among a number of institutions, such as the Banco
de Mexico and the International Bank for Reconstruction and Development, and several contributions are revisions of papers presented at meetings, such as the Second World Congress of the
Econometric Society. Among the techniques discussed are input-
output analysis, linear programming, integer programming, and
regression analysis. The contributors are L. Barraza, L.M.
Bassoco, J.H. Duloy, G.F. de la Garza, Y. Franchet, R.A.
Inman, D.B. Keesing, J. Kornai, G.P. Kutcher, R.D. Norton,
T. Rendon, L. Solis, S.T. Reyes, J.A. Valencia, and D.L.
Winkelmann.

Gossling, W.F., ed. INPUT-OUTPUT IN THE UNITED KINGDOM. London:
Frank Cass, 1970. xxi, 209 p. Index of Authors and Discussants, p. 200.
List of General and Recent References, pp. 201-5. Subject Index, pp. 206-9.

A collection of nine papers on the construction and uses of input-
output analysis presented at a conference under the auspices of
the econometrics department, University of Manchester, in 1968.

Griliches, Zvi, ed. PRICE INDEXES AND QUALITY CHANGE. Foreword by
J.C. Partee. Studies in New Methods of Measurement. Edited for the Price
Statistics Committee, Federal Reserve Board. Cambridge, Mass.: Harvard Uni-
versity Press, 1971. x, 287 p. Bibliography, pp. 275-8. Index, pp. 283-87.

A collection of eight papers (some previously published) on theoretical and empirical problems involved in the measurement of
price and changes in product quality arranged in three categories:
introduction of hedonic price indexes and tast and quality change
in the pure theory of the true-cost-of-living index, regression approach to price change measurement, and use of second-hand market prices for the measurement of quality change. The authors
are P. Cagan, P.J. Dhrymes, F.M. Fisher, Z. Griliches, R.E.

Hall, I.B. Kravis, R.E. Lipsey, K. Shell, and J.E. Triplett. The bibliography is divided into the following two parts: price indexes and quality change, and other words cited. Assumes knowledge of calculus and regression theory.

Hansen, W. Lee. ed. EDUCATION, INCOME, AND HUMAN CAPITAL. National Bureau of Economic Research, Studies in Income and Wealth. New York: Columbia University Press, 1970. x, 320 p. Notes on Contributors, pp. 307-9. Author Index, pp. 311-14. Subject Index, pp. 315-20.

A collection of most of the papers presented at the Conference on Education and income held at the University of Wisconsin on November 15-16, 1968, jointly sponsored by the university's department of economics and the Conference on Income and Wealth. The papers are grouped under the following three headings: education and production functions, education and the distribution of income, and education and human capital in international economics. Assumes knowledge of elementary calculus.

Hardt, John P., et al., eds. MATHEMATICS AND COMPUTERS IN SOVIET ECONOMIC PLANNING. Yale Russian and East European Studies, 5. New Haven: Yale University Press, 1967. xxii, 298 p. References Cited, pp. 267-91. Contributors, p. 291. Index, pp. 293-98.

A collection of four nontechnical papers on the uses of mathematics and computers in Soviet economic planning presented at a conference at the University of Rochester in May 1965. The topics treated include: information and control, input-output analysis, linear programming, and optimizing models for multiperiod programming.

Hart, P.E.; Mills, G.; and Whitaker, J.K., eds. ECONOMETRIC ANALYSIS FOR NATIONAL ECONOMIC PLANNING. Foreword by R.S. Clarke. Preface by R.C. Tress. London: Butterworths, 1964. xii, 320 p. No index.

A collection of eleven papers presented at the sixteenth symposium of the Colston Research Society held at the University of Bristol, April 6-9, 1964. The first four papers are British studies dealing with econometrics of inflation and aggregate growth models and industrial production functions, and the next three represent three different national planning models: a short-term forecasting model of the United States, a short-run model of the Netherlands, and a model of resource allocation devised as a contribution to the French Fifth Plan. The following three studies deal with technical rigidities on the optimal rates of investment and growth, integration of econometric planning decisions, and the relationships between family composition, prices, and expenditure patterns. The final paper is concerned with the stability of economic relationships over time. Each paper is followed by a discussion paper. The contributors include H. Diehl, L. Johansen, J. Johns-

ton, L.R Klein, G. Menges, A. Moustacchi, A.P. Parten,
J.J. Post, G. Pyott, W. Sadowski, J.D. Sargan, R. Stone,
and P.J. Verdoorn.

Hawkes, Nigel, ed. INTERNATIONAL SEMINAR ON TRENDS IN MATHE-
MATICAL MODELING. Lecture Notes in Economics and Mathematical Systems,
vol. 80. New York: Springer-Verlag, 1973. vi, 288 p. No index.

A collection of twenty contributions on predictive techniques presented to an international seminar held in Venice on December
13-19, 1971. Only a few of the papers require a background
in calculus. They are grouped under the following headings:
general topics for future research, the planning of systems, simulation modeling and game theory, studies of computer use and
methodology for modeling, and human interaction on modeling.
Topics discussed in the papers include central planning, simulation
to test environmental policy, global equilibrium, socioeconomic
modeling, and models of historical processes. The authors are
K.A. Bagrinovsky, J. Barraud, P. Costa, N.C. Dalkey, S. Enzer,
R. Faure, Y. Friedman, D.C. Gazis, O. Helmer, M.G. Kendall,
D.L. Meadows, N.N. Moiseev, Y.N. Parlovsky, V. Piasentin,
J. Randers, M. Shubik, N. Teodorescu, A. Toffler, and Y.I.
Zhuravlev.

Heady, Earl O., ed. ECONOMIC MODELS AND QUANTITATIVE METHODS
FOR DECISIONS AND PLANNING IN AGRICULTURE: PROCEEDINGS OF AN
EAST-WEST SEMINAR. Ames: Iowa State University Press, 1971. xiii, 518 p.
Participants, pp. 502-6. Index, pp. 507-18.

A collection of forty-seven papers by authors from countries from
both the East and the West presented at a conference held at
Keszthely, Hungary, during the summer of 1968. The papers are
grouped under the following headings: foundation and background
in planning models, problems and potentials at the micro level,
regional modes of planning and development, experiments and experiences with national planning models for agriculture, formulation
of national models, and gaps between possibilities for improvement
in performance. Assumes knowledge of finite mathematics and
elementary calculus.

Herndon, James F., and Bernd, Joseph L., eds. MATHEMATICAL APPLICA-
TIONS IN POLITICAL SCIENCE, VI. Charlottesville: University Press of Virginia, 1972. vii, 142 p. No index.

A collection of four papers presented at the 1970 summer institute
in mathematical applications in political science at Virginia Polytechnic Institute with the support of the National Science Foundation. In addition, an introductory paper by J.F. Herndon and
J.L. Bernd is included. The topics treated by these four papers
are linear and nonlinear functions as criteria for distinction be-

tween judicial decisions in different legal systems, calculus of voting, models of coalition formulation in voting bodies, and three-person coalitions and three-person games. The authors are J. Brams and Wm. H. Riker, F. Kort, R.D. McKelvey, and P.C. Ordeshook.

_____. MATHEMATICAL APPLICATIONS IN POLITICAL SCIENCE, VII. Charlottesville: University Press of Virginia, 1974. viii, 84 p. No index.

Four papers presented at the 1971 summer institute in mathematical applications in political science under the support of the National Science Foundation and sponsored by the Virginia Polytechnic Institute and State University. Topics treated include voting behavior, dimensional analysis in congressional voting, dimensionality and change in judicial behavior, and the uses of statistics in aggregate data analysis. The authors are A.R. Clausen, L.S. Mayer, C.L. Taylor, and S.S. Ulmer.

Hickman, Bert G., ed. ECONOMETRIC MODELS OF CYCLICAL BEHAVIOR. VOLUMES I AND II. Studies in Income and Wealth, no. 36. New York: Columbia University Press, for the National Bureau of Economic Research, 1972. 1,246 p. Vol. 1, xii, 598 p.; vol. 2, xi, pp. 601-1246. A Note on Scientific Method in Forecasting, by V.L. Bassie, pp. 1211-18. Author Index, pp. 1219-21. Subject Index, pp. 1223-46.

A collection of ten papers presented at the Conference on Econometric Models of Cyclical Behavior at Harvard University on November 14-15, 1969, under the joint sponsorship of the SSRC Committee on Economic Stability and the National Bureau of Economic Research Conference on Research in Income and Wealth. Volume 1, which comprises part 1 of the two-volume series, consists of three papers which describe and analyze stochastic and nonstochastic simulations of the OBE, Wharton, and Brookings models, and one paper which describes the OBE, Wharton, and FMP models by the NBER. Volume 2, comprising parts 2 and 3 of the series, contains papers on the dynamic properties of the Wharton model, the effects of aggregation over time on dynamic characteristics of an econometric model, and the predictive and forecasting performance of U.S. econometric models.

_____. QUANTITATIVE PLANNING OF ECONOMIC POLICY. Foreword by R.D. Calkins. Washington, D.C.: Brookings Institute, 1965. xiii, 279 p. References, pp. 267-69. Conference Participants, pp. 271-72. Index, pp. 273-79.

Eleven papers on quantitative economic policy presented at an international conference held at the Brookings Institute on August 19-24, 1963. The conference was sponsored by the Social Science Research Council (SSRC) Committee on Economic Stability, with financial support from the Ford Foundation. The purpose of the conference was to present recent developments in quantitative policy planning and to appraise experiences in this field in France, Japan, and the Netherlands. The contributors are B. Cazes, K.A. Fox, W. Hessel, B.G. Hickman, C.C. Holt, E.S. Kirschen, J.K. Sengupta, S. Shishido, H. Theil, E. Thorbecke, C.A. van den Beld, and T. Watanabe. The models discussed in the papers include single-equation, simultaneous-equation (Brookings–SSRC and Klein–Goldberger models), and input–output models.

Horwich, George, and Samuelson, Paul A., eds. TRADE, STABILITY, AND MACROECONOMICS: ESSAYS IN HONOR OF LLOYD A. METZLER. Economic Theory and Mathematical Economics. New York: Academic Press, 1974. xviii, 558 p. Author Index, pp. 554-58.

A collection of twenty-two contributions, all hitherto unpublished, arranged under the following headings: international trade (six papers), mathematical economics (six papers), inventory fluctuations (three papers), macromonetary theory (four papers), and growth (three papers). Among the contributors are E. Ames, K.J. Arrow, G. Bilkes, J.A. Carlson, G.L. Childs, J.S. Chipman, G. Horwich, H.G. Johnson, M.C. Kemp, M.C. Lovell, T. Negishi, W.F. Berg, J.P. Quirk, E.A. Thompson, H. Uzawa, H.Y. Wan, Jr., and Wm. E. Wehrs. Calculus is the main prerequisite for most of the papers.

International Computation Center, Rome. LONG RANGE PLANNING: INTERNATIONAL SYMPOSIUM. New York: Gordon and Breach, 1967. xix, 531 p.

A collection of forty-one papers presented at a symposium in Paris in September 1965. About one half of the papers appear in French and the remainder in English. The topics covered include criteria and organization for planning, input–output analysis, project evaluation, depreciation and the influence of taxation, networks, forecasting, linear programming and input–output applications in planning, and planning risks.

Intriligator, Michael D., ed. FRONTIERS OF QUANTITATIVE ECONOMICS. Vol. 1. Contributions to Economic Analysis, vol. 71. New York: American Elsevier, 1971. xi, 471 p. Subject Index, pp. 469-71.

A collection of thirteen hitherto unpublished papers presented at the Econometric Society meeting in New York in December 1969. The papers are grouped under the following headings: economic methodology and theory (three papers), econometric techniques (five papers), and quantitative approaches to traditional topics in economics (two papers on monetary policy, and one paper on each of the following topics: capital markets, industrial organization, and econometric studies in history). Among the topics treated are externalities, information systems, theory of voting, forecasting with a large econometric model, Bayesian statistics, and simulation. The contributors are K.J. Arrow, R.W. Clower, E.F. Fama, A.N. Halter, M.L. Hayenga, H.G. Johnson, L.R. Klein, J. Marschak, J.J. Manetsch, T.H. Naylor, C.R. Plott, T.J. Rothenberg, L. Weiss, G. Wright, and A. Zellner. The introduction is by Michael D. Intriligator.

Intriligator, Michael D., and Kendrick, David A., eds. FRONTIERS OF QUANTITATIVE ECONOMICS. Vol. 2. Contributions to Economic Analysis, vol. 87. New York: American Elsevier, 1974. xii, 580 p. Subject Index, pp. 571–80.

Nine papers presented at the winter meeting of the Econometric Society held in Toronto, Canada, in 1972. The papers are grouped under the following three headings: economic methodology and theory (three papers); econometric techniques (two papers); and quantitative approaches to traditional topics in economics (four papers). An introduction by the editors and from three to four comments on each paper are also included. Topics include market equilibrium and uncertainty, applications of duality theory, distributed lags, investment, health economics, multisector models for development planning, and labor market discrimination. Most of the papers are survey-type contributions with emphasis given to recent developments in various fields and topics, and a few papers assume knowledge of advanced econometric techniques. The contributors are R.L. Basmann, W.E. Diewert, M.S. Feldstein, R.B. Freeman, L. Hurwicz, M.D. Intriligator, D.W. Jorgenson, D.A. Kendrick, A.S. Manne, R. Radner, and C.A. Sims.

Judge, George G., and Takayama, Takashi, eds. STUDIES IN ECONOMIC PLANNING OVER SPACE AND TIME. Contributions to Economic Analysis, vol. 82. New York: American Elsevier, 1973. xii, 727 p. Subject Index, pp. 724–27.

A collection of thirty-five hitherto unpublished essays on economic planning grouped under the following four headings: national and international planning models, linear programming models over space and time, nonlinear models over space and time, and public and private planning models. The contributions represent both the conceptual framework and applications, such as macroeconometric models, linear and recursive programming, regional and dynamic input-output analysis in transportation, agriculture, and trade, and location of industries. Among the authors are M.A. Abe, H.E. Buchholz, H. Correa, S. Czamanski, R.H. Day, J.W.B. Guise, E.O. Heady, G.G. Judge, P.A. Samuelson, J. Tsukui, and J. Tinbergen.

Kalecki, Michal. SELECTED ESSAYS ON THE ECONOMIC GROWTH OF THE SOCIALIST AND THE MIXED ECONOMY. New York: Cambridge University Press, 1972. vii, 176 p. Index, pp. 171–76.

A collection of fifteen essays by the author, about half of which were originally published in journals. Part 1 (eleven essays) is on growth theory in a socialist economy; part 2 (two essays) is devoted to investment planning and project selection; and part 3 (two essays) deals with a mixed economy. Assumes knowledge of elementary calculus.

Kendall, Maurice G[eorge]., ed. MATHEMATICAL MODEL BUILDING IN ECONOMICS AND INDUSTRY. Series 2. London: Charles Griffin, 1970. 277 p. No index.

A collection of fourteen papers presented at a conference organized by Scientific Control Systems in London in June 1968 and June 1969, and in Montreal in October 1969. The papers dealt with applications in airlines, public utilities, finance, forecasting, marketing, and management. The authors are R.J. Ball, P.C. Briant, E.S.M. Chadwick, R.H.E. Duffett, C.G. Edge, J. Gratwick, P.A.B. Hughes, R.W. Linder, E.F. Mellen, G.H. Orcutt, F.G. Pyatt, and M.K. Wood. Knowledge of calculus and probability theory is necessary to understand some of the papers.

Klein, Lawrence R., and Ohkawa, Kazushi, eds. ECONOMIC GROWTH: THE JAPANESE EXPERIENCE SINCE THE MEIJI ERA. Homewood, Ill.: Richard D. Irwin, 1968. xv, 424 p. Program of the Conference, pp. 423-24.

A collection of thirty-eight papers (twenty-two by Japanese and sixteen by foreign scholars) presented at the International Conference on Economic Growth--A Case Study of the Japanese Experience on September 5-10, 1966, at the Japan Economic Research Center in Tokyo. The papers are grouped under the following headings: nineteenth-century and prewar economic growth, postwar economic growth, and summary remarks.

Koopmans, Tjalling C. SCIENTIFIC PAPERS OF TJALLING C. KOOPMANS. Edited by Martin Beckmann, Carl F. Christ, and Marc Nerlove. New York: Springer-Verlag, 1970. xii, 600 p. Publications of Tjalling Charles Koopmans, pp. 595-600. No index.

Twenty-eight selected scientific papers by Tjalling C. Koopmans orginally published in journals and elsewhere dealing with the following three general topics: statistical identification and estimation of parameters in econometric models, activity analysis, and intertemporal utility maximization and the optimum allocation of resources over time. Some of the papers are coauthored with other authors. Knowledge of calculus is assumed in several papers.

_____, ed. ACTIVITY ANALYSIS OF PRODUCTION AND ALLOCATION: PROCEEDINGS OF A CONFERENCE. Cowles Foundation Monograph, no. 13. New York: Wiley, 1951. xiv, 404 p. References, pp. 381-85. Index of Names, p. 387. Subject Index, pp. 389-404.

A collection of twenty-five papers by members of the Cowles Foundation arranged under the following headings: theory of programming and allocation, including linear programming and input-output models (ten papers); application of allocation models (six papers); mathematical properties of convex sets (four papers); and problems of computation (five papers). Most of the papers assume knowledge of calculus and matrix algebra. The contributors are K.J. Arrow, G.W. Brown, Y. Brozen, A. Coale, G.B. Dantzig, R. Dorfman, D. Gale, M.A. Geisler, N. Georgescu-Roegen, M. Gerstenhaber, C. Hildreth, T.C. Koopmans, H.W. Kuhn, O. Morgenstern, S. Reiter, P.A. Samuelson, H.A. Simon, H. Smith, A.W. Tucker, and M.K. Wood.

Krupp, Sherman Roy, ed. THE STRUCTURE OF ECONOMIC SCIENCE: ESSAYS ON METHODOLOGY. Englewood Cliffs, N.J.: Prentice-Hall, 1966. vii, 282 p. Index, pp. 277-82.

A collection of nontechnical essays grouped under the following headings: theory and dispute in economics, mathematics and observation in economics, the boundaries of economic theory, and value premises in economics. The authors are Wm. J. Baumol, K.E. Boulding, R.B. Brandt, M. Bronfenbrenner, J.M. Buchanan, C.W. Churchman, E. Grunberg, S.R. Krupp, K. Lancaster, F. Machlup, H. Margenau, L. Nahers, J. Rothenberg, and E. Rotwein.

Kuhn, H[arold].W[illiam]., and Tucker, A.W., eds. LINEAR INEQUALITIES AND RELATED SYSTEMS. Annals of Mathematics Studies, no. 38. Princeton, N.J.: Princeton University Press, 1956. xxi, 322 p. A Bibliography on Linear Inequalities and Related Subjects, pp. 305-22.

A collection of eighteen papers by noted authors on the mathematical theory of linear inequalities arranged under the following categories: mathematical results which form a basis for the models, mathematical questions that have arisen from economic theory, and the economic applications of the models. The authors are G.B. Dantzig, R.J. Duffin, K. Fan, L.R. Ford, Jr., D.R. Fulkerson, D. Gale, A.J. Goldman, I. Heller, A.J. Hoffman, J.B. Kruskal, H.W. Kuhn, H.D. Mills, G.L. Thompson, A.W. Tucker, and P. Wolfe.

Lamberton, Donald McLean, ed. ECONOMICS OF INFORMATION AND KNOWLEDGE: SELECTED READINGS. Baltimore: Penguin Books, 1971. 384 p. Further Readings, pp. 366-76. Acknowledgments, p. 377. Author Index, pp. 378-80. Subject Index, pp. 381-84.

A collection of eighteen reprints of articles arranged under the following headings: surveys (two articles), economic organization (four articles), information and efficiency (two articles), information policy (four articles), international aspects (two articles), business planning (three articles), and conclusion (one article). The papers focus on the microeconomic aspects of information and knowledge, and attempt to reveal the failures of traditional economics. Assumes knowledge of calculus for some of the papers. Among the authors are K.E. Boulding, H. Demsetz, K. Grossfield, R.A. Jenner, B.S. Loasby, J. Marschak, M.V. Posner, R. Rees, M. Shubik, and G.J. Stigler.

Lange, Oskar Richard. PAPERS IN ECONOMICS AND SOCIOLOGY 1930–1960. Foreword by M. Kalecki. Translated and edited by P.F. Knightsfield. Elmsford, N.Y.: Pergamon, 1970. ix, 600 p. Bibliography of Works: 1925–1963, pp. 587–600.

This translation from Polish is a collection of selected papers previously published by Oskar Lange over a period of thirty years. The papers are grouped under the following headings: Marxist and socialist theory; political economy and socialism; economic theory; economic–mathematical models, econometrics, and statistics; and economic science in the service of practice.

Lange, Oskar Richard; McIntyre, Francis; and Yntema, Theodore [Otte]., eds. STUDIES IN MATHEMATICAL ECONOMICS AND ECONOMETRICES. Chicago: University of Chicago Press, 1942. 292 p. No index.

A collection of twenty essays written in memory of Henry Schultz and grouped under the following four headings: Henry Schultz (four essays), economic theory (seven essays), statistical theory (three essays), and econometrics (six essays). Among the topics treated by these essays are the contributions of H. Schultz, capital and value theory, estimates of demand and cost functions, marginal utility, risk and uncertainty, regression analysis, and expenditure patterns of families of different types. The authors include J. Dean, A.G. Hart, M. Friedman, Wm. Jaffé, O.R. Lange, G.K.K. Link, P.A. Samuelson, J.H. Smith, G. Tintner, and W.A. Wallis. Assumes knowledge of calculus and regression analysis.

Layard, Richard, ed. COST-BENEFIT ANALYSIS: SELECTED READINGS. Penguin Modern Economics Readings. Baltimore: Penguin Books, 1972. 496 p. Exercises, pp. 473–77. Further Readings, p. 479. Acknowledgments, p. 481. Author Index, pp. 483–88. Subject Index, pp. 489–96.

A collection of eighteen reprints of articles grouped under the following headings: general surveys (two essays), measuring cost and benefits when they occur (six essays), the social time preference rate and the social opportunity cost of capital (five essays), the treatment of risk (two essays), the treatment of income distribution (one essay), and the case of the third London airport (two essays). An introductory paper by Richard Layard is included. Among the authors are K.J. Arrow, R.J. Dorfman, M.S. Feldstein, A.C.

Harberger, E.J. Mishan, R.A. Musgrave, A.R. Priest, A.K. Sen, and R. Turvey.

Lazarsfeld, Paul F., ed. MATHEMATICAL THINKING IN THE SOCIAL SCIENCES. New York: Russell and Russell, 1954. 444 p. Notes, pp. 416-38. Index, pp. 439-44.

A collection of eight essays, hitherto unpublished, on the use of mathematics in economics, sociology, and psychology. Topics include probability models for analyzing time changes in attitudes and for imitative behavior and distribution of status, social behavior models of Rashensky, factor analysis, scaling attitudes, latent structure and models of optimization and adaptive behavior. Some of the essays require knowledge of calculus and matrix algebra, and most of them assume familiarity with basic statistics. The authors are T.W. Anderson, J.S. Coleman, L. Guttman, P.F. Lazarsfeld, J. Marschak, N. Raskevsky, and H.A. Simon.

Lazarsfeld, Paul F., and Henry, Neil W., eds. READINGS IN MATHEMATICAL SOCIAL SCIENCE. Cambridge, Mass.: MIT Press, 1966. 371 p. Major Reference Books in Mathematical Social Science, p. 371. No index.

A collection of nineteen reprints of articles by mathematicians and sociologists which are grouped under the following five headings: introduction (one article), problems of measurement (four articles), the mathematical study of small homogeneous groups (five articles), the mathematical study of more complex groups (five articles), the models analyzing processes (four articles). The prerequisites for many of the articles consist of matrix algebra, probability theory, and finite mathematics. The authors include R.R. Bush, J.S. Coleman, H. Guetzkow, J.G. Kemeny, P.F. Lazarsfeld, J. Marschak, F. Mosteller, M. Shubik, H.A. Simon, and G. Rasch. This book is useful to anyone interested in applying operations research in the social sciences, and can be used as a supplement in operations research courses.

Los, Jerzy, and Los, Maria W., eds. MATHEMATICAL MODELS IN ECONOMICS. New York: American Elsevier, 1974. xvii, 483 p. No index.

A collection of thirty-five papers representing the proceedings of the symposium on mathematical models in economics held from February to July 1972, and the conference on von Neumann models held July 10-15, 1972, in Warsaw. The symposium and conference were organized by the Polish Academy of Sciences. The papers are grouped under the following headings: generalizations of the von Neumann models (twelve papers), dynamical problems of an expanding economy (ten papers), convex analysis and its application to the economic models (five papers), and different topics (eight papers). The contributors represented many countries from throughout the world and include mathematicians and economists.

Among the authors are R.A. Dana, I.V. Evstigneev, V. Klee, J. Los, O. Morgenstern, R.T. Rockafellar, J. Zabczyk, and A. Zauberman.

McGuire, C.B., and Radner, Roy, eds. DECISION AND ORGANIZATION: A VOLUME IN HONOR OF JACOB MARSCHAK. Studies in Mathematical and Managerial Economics, vol. 12. New York: American Elsevier, 1972. x, 361 p. Publications of Jacob Marschak, pp. 337-41. References, pp. 343-49. Name Index, pp. 351-52. Subject Index, pp. 353-61.

A collection of fourteen chapters by various authors (some authors contributed two chapters) on the theory of rational choice; designed to bridge the gap between graduate price theory courses and the theory of decision and organization. The topics of the papers include the structure of alternatives, characteristics of preference orderings, complications associated with time and uncertainty, communication conflict, the theory of teams, and decentralized systems. Assumes knowledge of calculus, Matrix algebra, and elementary probability. The authors are K.J. Arrow, M.J. Beckman, G. Debreu, L. Hurwicz, T.C. Koopmans, C.B. McGuire, T.A. Marschak, R. Radner, H. Scarf, and H.A. Simon.

Machol, Robert E., and Gray, Paul, eds. RECENT DEVELOPMENTS IN INFORMATION AND DECISION PROCESSES. New York: Macmillan, 1962. x, 197 p.

Twelve papers presented at a symposium on information and decision processes held at Purdue University in April 1961. Topics include the mathematics of self-organizing systems, dynamic programming, deferred and Bayesian decision theories, and the estimation of reliability. The authors are R. Bellman, K.L. Chung, P.A. Diamond, B. Dunham, R. Fridshal, H. Goode, T.C. Koopmans, S. Moriguti, J.H. North, H. Raiffa, H. Robbins, E. Samuel, L. Savage, M. Tribus, N. Wiener, and R. Williamson.

Majumdar, Tapas, ed. GROWTH AND CHOICE: ESSAYS IN HONOR OF U.N. GHOSAL. Foreword by B. Datta. New York: Oxford University Press, 1969. xiv, 104 p. No index.

Eight essays by former students of U.N. Ghosal on the occasion of his retirement from the professorship of political science at Presidency College, Calcutta. All the essays are linked together by the realization that growth problems involve a variety of choice problems, and that the theory of social choice is necessary to define the content of growth, including the optimal growth path for finite planning horizons. Two other essays deal with problems, such as inflation of a growing economy. The final three essays deal with the theory of economic policy, including the role of majority decisions in choice. Assumes knowledge of elementary differential and integral calculus.

Malinvaud, Edmond, and Bacharach, M.O.L., eds. ACTIVITY ANALYSIS IN
THE THEORY OF ECONOMIC GROWTH. New York: St. Martin's Press,
1967. xv, 334 p. References, p. xv. Index, pp. 331-34.

> Eleven papers, plus a summary report, presented at a conference
> held by the international economic association in Cambridge,
> England, in July 1963. The papers are grouped under the fol-
> lowing headings: studies on growth theory, planning theory, and
> planning experiences. The authors are M. Allais, M.O.L.
> Bacharach, S. Chakravarty, E.O. Heady, L. Hurwicz, J. Kornai,
> L.W. McKenzie, E. Malinvaud, R. Radner, S. Rafael, N.S.
> Randhawa, and J.R.N. Stone. Assumes knowledge of calculus
> and mathematical analysis.

Manne, Alan S., ed. INVESTMENTS FOR CAPACITY EXPANSION: SIZE,
LOCATION, AND TIME-PHASING. Studies in the Economic Development
of India, no. 5. London: Allen and Unwin, 1967. 239 p. Name Index,
pp. 237-38. Subject Index, pp. 238-39.

> Fourteen papers by A.S. Manne and six collaborators on the prob-
> lem of how to allow for the effects of economies of scale in the
> allocation of investment funds in underdeveloped countries. The
> papers represent a well-balanced integration of theory, empirical
> analysis, and policy generalizations using mathematical program-
> ming with sensitivity analysis for testing the validity of the as-
> sumptions. The industries analyzed include aluminum, caustic
> soda, cement, and nitrogenous fertilizer in the Indian economy.
> The papers are grouped under the following headings: industry
> studies, single producing area--further results, and multiple pro-
> ducing areas--further results. The authors are D. Erlenkotter,
> A.S. Manne, P.N. Radhakrishnan, R.M. Rao, T.N. Srinivasan,
> and A.F. Veinott, Jr. Portions of the book require knowledge
> of calculus and matrix algebra.

Manne, Alan S., and Markowitz, Harry M., eds. STUDIES IN PROCESS
ANALYSIS: ECONOMY-WIDE PRODUCTION CAPABILITIES. Foreword by
T.C. Koopmans. Cowles Foundation Monograph, no. 18. New York: Wiley,
1963. viii, 427 p. Appendix: Basic Concepts of Activity Analysis, by A.S.
Manne, pp. 417-22. Index, pp. 423-27.

> A collection of seventeen papers presented at a conference spon-
> sored by the Cowles Foundation and held at Yale University on
> April 24-26, 1961. The papers present methods for estimating the
> production capabilities of an economy or an industry, using engi-
> neering information and mathematical programming techniques.
> They are grouped under the following headings: process analysis
> (three papers), petroleum and chemical production: production,
> transportation, and plant location (three papers), food and agri-
> culture (two papers), and methods and methodology (seven papers).

The authors are A.C. Egbert, T. Fabian, E.O. Heady, K.A. Fox, H.M. Markowitz, T.A. Marschak, A.S. Manne, A.J. Rowe, and T. Vietorisz.

Martindale, Don, ed. FUNCTIONALISM IN THE SOCIAL SCIENCES: THE STRENGTH AND LIMITS OF FUNCTIONALISM IN ANTHROPOLOGY, ECONOMICS, POLITICAL SCIENCE, AND SOCIOLOGY. Monograph 5 in a Series Sponsored by the American Academy of Political and Social Science. Philadelphia: The American Academy of Political and Social Science, 1965. ix, 162 p. No index.

A collection of eight papers by nine authors on functionalism in the social sciences. The authors and their subjects are: anthropology (R.F. Spencer, I.C. Jarvie), economics (J.S. Chipman, S.R. Krupp), political science (R.T. Holt, Wm. Flanigan, E. Fogelman), and sociology (I. Whitaker, D. Martindale).

MATHEMATICAL MODEL BUILDING IN ECONOMICS AND INDUSTRY. New York: Hafner, 1968. viii, 165 p. No index.

A collection of eleven papers presented at a conference, Model Building in Economics and Industry, held in London July 4-6, 1967, and organized by C-E-I-R Ltd. The emphasis is on applications, and the following topics are discussed: econometric models, models of a company, macro model for the Dutch economy, traffic and marketing models, and economic fluctuations. The authors are R.J. Ball, C.A. Van den Beld, A.S.C. Ehrenberg, P.C. Haines, M.G. Kendall, W.J. Newby, A.S. Noble, A W. Phillips, G. Pompilj, J.R.N. Stone, and H.O.A. Wold. Assumes knowledge of operations techniques.

MATHEMATICAL THINKING IN BEHAVIORAL SCIENCES: READINGS FROM SCIENTIFIC AMERICAN. Introduction by D.M. Messick. San Francisco: Freeman, 1968. 231 p. Bibliographical Notes and Bibliographies, pp. 223-29. Index, pp. 230-31.

A collection of twenty-seven articles previously published in SCIENTIFIC AMERICAN and grouped under the following headings: the analysis of uncertainty, probability, communications and control, games and decisions, imitations of life, and recent computer applications. Among the authors are R.E. Bellman, L. Hurwicz, M. Kac, O. Morgenstern, A. Rapoport, and P. Suppes.

Mitra, Ashok, ed. ECONOMIC THEORY AND PLANNING: ESSAYS IN HONOUR OF A.K. DAS GUPTA. New York: Oxford University Press, 1974. xvi, 326 p. A.K. Das Gupta: A Bibliography, pp. xi-xiv. Index of Names, pp. 321-23. Subject Index, pp. 324-26.

Twenty-three heretofore unpublished essays presented to Professor A.K. Das Gupta on the occasion of his seventieth birthday. The essays deal with his fields of interest and include the following topics: preference in welfare and economic policy, capital theory, modes of production and exchange, population theory, development strategies, growth and economic structure, planning models, and fiscal, monetary, and employment policies. Some of the essays assume knowledge of elementary matrix algebra and calculus. The contributors include A. Bhaduri, V.V. Bhatt, P. Dasgupta, A.S. Guha, J.R. Hicks, L Lefeber, J.E. Meade, A. Sen, and S.R. Sen.

Morishima, Michio, et al., eds. THEORY OF DEMAND: REAL AND MONE-TARY. New York: Oxford University Press, 1973. xiii, 330 p. Bibliography, pp. 316–26. Index, pp. 327–30.

Twelve contributions (three previously published) on demand theory arranged in the following five parts: qualitative information, separability, consumption, money, and portfolio. The two contributions on qualitative economics and separable utilities are survey articles and make up one-third of the book. Other topics treated by the contributions include: separability and intrinsic complementarity, demand under uncertain expectation, consumer behavior and liquidity preference, Veblen effects and portfolio selection, transactions' demand for money, revealed preference theory, and portfolio diversification as optimal precautionary behavior. The models are presented in mathematical form with detailed economic implications provided. Assumes knowledge of calculus and probability theory. Among the authors are M.G. Allingham, P.T. Geary, M J. Martin, J.M. Parkin, and M. Morishima.

Moss, Milton, ed. THE MEASUREMENT OF ECONOMIC AND SOCIAL PER-FORMANCE. New York: Columbia University Press, 1973. x, 605 p. Index, pp. 593–605.

A collection of eleven papers on the following four topics: proposals for measurement of economic and social performance, household and business sector, public sector, and measuring the amenities and disamenities of economic growth. Assumes knowledge of calculus and probability.

National Bureau of Economic Research, ed. ECONOMIC GROWTH: FIFTIETH ANNIVERSARY COLLOQUIUM V. New York: Columbia University Press, 1972. xii, 92 p. No index.

The proceedings of the economic growth colloquium which was held at the Bank of America Center in San Francisco on December 10, 1970. The main paper in this monograph is: "Is Growth Obsolete?" by Wm. Nordhaus and J. Tobin with discussions by M. Abramovitz and R. Matthews.

Newmann, Peter K., ed. READINGS IN MATHEMATICAL ECONOMICS. Vol. 1: VALUE THEORY. Baltimore: Johns Hopkins Press, 1968. xii, 532 p. No index.

A collection of twenty-nine reprints of articles originally published in journals and elsewhere, arranged under the following five headings: mathematical tools, existence theory for static competitive equilibrium, adjustment processes in the theory of static equilibrium, the theory of individual behavior, and welfare economics and the theory of production. Among the authors are K.J. Arrow, G. Debreu, E. Eisenberg, D. Gale, H.W. Kuhn, O.R. Lange, H.B. Mann, and H. Nikaido. Assumes knowledge of linear algebra, mathematical analysis, and point-set topology.

_____. READINGS IN MATHEMATICAL ECONOMICS. Vol. 2: CAPITAL AND GROWTH. Baltimore: Johns Hopkins Press, 1968. ix, 358 p. Bibliography on Capital and Growth, pp. 329-58. No index.

Fifteen articles originally published in journals and elsewhere, grouped under the following headings: capital theory, classical and neoclassical growth models, and models of optimal growth. Among the authors are K.J. Arrow, M. Bruno, D.G. Champernowne, D. Gale, T.C. Koopmans, J. von Neumann, L.L. Pasinetti, R. Radner, R.M. Solow, T. Swan, and H. Uzawa.

Organization for Economic Co-operation and Development. QUANTITATIVE MODELS AS AN AID TO DEVELOPMENT ASSISTANCE POLICY. Report by the Expert Group on the Uses of Analytical Techniques. Paris: Organization for Economic Co-operation and Development, 1967. 78 p. Annexes, pp. 63-72. Bibliography, pp. 73-78.

A collection of eighteen papers on quantitative models by experts from several countries. They are nonmathematical and written for the layman.

Parkin, Michael, and Nobay, Areling Romeo, eds. ESSAYS IN MODERN ECONOMICS: THE PROCEEDINGS OF THE ASSOCIATION OF UNIVERSITY TEACHERS OF ECONOMICS: ABERYSTWYTH 1972. London: Longman, 1973. xii, 399 p. Index of Names, pp. 389-92. Index of Subjects, pp. 393-99.

Nineteen essays presented at the Conference of the Association of University Teachers of Economics held at the University of Wales, Aberystwyth, in March 1972. The papers are arranged under the following headings: microeconomics of the private sector (five papers), microeconomics of the public sector (six papers), on Keynes (one paper), and macro and monetary economics (seven papers). R.M. Solow presented the first Frank W. Paish lecture on urban economics. Some of the essays assume knowledge of mathematical economics and econometrics. Among the authors are M. Blaug, C.J. Bliss, F.H. Hahn, P.J. Hammond, P.N. Mathur, R.J. Nicholson, R. Rees, R.M. Solow, and N. Topham.

Pfouts, Ralph W., ed. ESSAYS IN ECONOMICS AND ECONOMETRICS: A VOLUME IN HONOR OF H. HOTELLING. Chapel Hill: University of North Carolina Press, for the School of Business Administration, University of North Carolina, 1960. xi, 240 p. Bibliography of H. Hotelling, pp. 233-40.

A collection of ten papers by different authors on mathematical economics and econometrics arranged under the following headings: mathematics of optimization, the theory of utility and demand, economic dynamics, the Edgeworth taxation paradox, and on econometric methods. The authors are K.J. Arrow, H.T. Davis, C.E. Ferguson, M. Friedman, R. Frisch, L. Hurwicz, L.R. Klein, R.W. Pfouts, G. Tintner, and Wm. Vickrey.

Prékopa, András, ed. COLLOQUIUM ON APPLICATIONS OF MATHEMATICS TO ECONOMICS. Budapest: Akademiai Kiado, 1965. 367 p. No index.

A collection of forty-one papers, arranged in alphabetical order of authors' names and presented at the colloquium on applications of mathematics in economics. Organized by the Janos Bolyar Mathematical Society and held in Budapest on June 18-22, 1963. Experts from eighteen countries participated in the colloquium where emphasis was placed on the mathematics of deterministic and stochastic models of operations research, with some papers dealing with applications. The topics include linear and dynamic programming, stochastic processes, design of experiments, sampling theory, the theory of graphs, inventory problems, polyhedral games, regional science, etc. Twenty-six of the papers deal with some aspect of mathematical programming. Twenty-four papers are printed in English, eleven in German, four in Russian, and two in French. Among the authors are E. Balas, R.E. Bellman, A. Bródy, A. Ghosal, M. Iosifescu, B. Ivanovic, I. Kadar, B. Krekó, C. Mack, A. Prékopa, J. Stahl, and G. Tintner.

Quirk, James P., and Zarley, Arvid M., eds. PAPERS IN QUANTITATIVE ECONOMICS. Lawrence: University Press of Kansas, 1968. 599 p. No index.

Twenty-two papers (and abstracts of seven additional papers) presented at the Kansas-Missouri joint seminar in theoretical and applied economics during the 1966-68 academic years. The papers are arranged under the following six headings: topics in economic theory (five papers); econometric theory and empirical studies (six papers); macroeconomics, growth theory and international trade (five papers); theory of the firm (three papers); and decision theory (three papers). Topics include the Kuhn-Tucker results in concave programming, differential growth among large U.S. cities, hypothesis formulation, a macro model with two interest rates, technical progress, risk aversion and bidding theory, and decision models for arms procurement. Among the authors are F.S.T. Hsiao, C.F. Menezes, J.C. Moore, J. Quirk, T. Rader, D.H. Richardson, and R. Saposnik.

Ramsey, F.P. FOUNDATIONS OF MATHEMATICS AND OTHER LOGICAL ESSAYS. Edited by R.B. Braithwaite. Paterson, N.J.: Littlefield, Adams, 1960. xviii, 292 p. Appendix, pp. 270–86. Epilougue, pp. 287–92.

A collection of published and unpublished papers by F.P. Ramsey, who is known for his contributions on the mathematical theory of saving.

Rao, Calyampudi Radhakrishna, et al., eds. ESSAYS ON ECONOMETRICS AND PLANNING. Foreword by C.D. Deshmukh. Elmsford, N.Y.: Pergamon Press, for the Statistical Publishing Society, 1963. 354 p. List of Scientific Papers Contributed by Professor P.C. Mahalanobis, pp. 329–42. Messages, pp. 343–54.

A collection of twenty-one essays presented to Professor P.C. Mahalanobis on the occasion of his seventieth birthday. The topics include the theory and philosophy of planning, planning experiences of various countries, and statistical problems of measurement. Among the contributors are R.L. Ackoff, R. Frisch, N. Georgescu-Roegen, R.W. Goldsmith, O.R. Lange, G. Myrdal, R. Stone, J. Tinbergen, and H.O.A. Wold.

Rosenstein-Rodan, P.N., ed. CAPITAL FORMATION AND ECONOMIC DEVELOPMENT. Studies in the Economic Development of India, no. 2. Cambridge, Mass.:' MIT Press, 1964. 164 p. Index, pp. 163–64.

A collection of ten papers on economic development with reference to the Indian economy using single equation and simultaneous equation models. It begins with an article by S. Chakravarty on the mathematical framework of the Third Five-Year Plan (1961–65) and considers in other papers alternative numerical models, a method for program evaluation, the use of shadow prices in program evaluation, capital formation, multisectoral intertemporal planning models, and an appraisal of alternative planning models. The other authors are R.S. Eckaus, L. Lefeber, and P.N. Rosenstein-Rodan. Assumes knowledge of elementary calculus and statistical estimation theory.

Roxin, Emilio O.; Liv, Pan-Tai; and Sternberg, Robert L., eds. DIFFERENTIAL GAMES AND CONTROL THEORY. Lecture Notes in Pure and Applied Mathematics, vol. 10. New York: Marcel Dekker, 1974. xi, 412 p. No index.

A collection of twenty-two papers presented at the Regional Research Conference on Differential Games and Control Theory held at the University of Rhode Island in Kingston, June 4–8, 1973, under the sponsorship of the National Science Foundation and the Conference Board of the Mathematical Sciences. Topics include application of control theory to population dynamics, pursuit games, existence of equilibrium points in n-person differential games, a dynamic version of Cournot's problem, search games with mobile hider, min-max feedback control of uncertain systems, symmetries and identification of nonautonomous control systems, and extended Isaacs equations for games of survival. Among the contributors are H.S. Blum, J. Flynn, O. Hajek, R. Isaacs, A. Kats, P. Kloeden, D.L. Lukes, R.J. Stern, D.R. Wakeman, and R.H. Worsham.

Ruggles, Nancy D., ed. THE ROLE OF THE COMPUTER IN ECONOMIC AND SOCIAL RESEARCH IN LATIN AMERICAN. A Conference Report of the National Bureau of Economic Research. New York: Columbia University Press, for the National Bureau of Economic Research, 1974. x, 399 p. List of Papers Presented, pp. 383–85. Index of Names, pp. 387–89. Subject Index, pp. 391–99.

A collection of twenty papers presented at a conference on the role of the computer in economic and social research in Latin American held at Cuernavaca, Mexico, October 25–29, 1971. Financial support for the conference was supplied by the IBM World Trade Corp., the National Science Foundation, and the National Bureau of Economic Research. The papers are concerned with the following areas: the computer and government statistical systems, data banks, and computer centers; computer simulation models; macroeconomic models; demography, manpower, employment, and education; international comparisons of income, consumption, and prices; and international trade patterns and commodity markets. A final paper by R. Ruggles summarizes the conference. Among the authors are B. Balassa, J.P.P. Castillo, L.R. Klein, R.E. Lipsey, J.V. Monteiro, T.H. Naylor, G.E. Peabody, J.A. Pechman, G. Sadowsky, and M. Shubik.

Samuelson, Paul A. THE COLLECTED SCIENTIFIC PAPERS OF PAUL A. SAMUELSON. 2 vols. Edited by Joseph E. Stiglitz. Cambridge, Mass.: MIT Press, 1966. 1813 p. Acknowledgments, pp. 763–71 (Vol. 1), and pp. 1783–91 (Vol. 2). Index, pp. 1793–1813.

Volumes 1 and 2 contain virtually all of Professor Paul A. Samuelson's contributions to economic theory through mid-1964. These contributions have been collected from economics journals, Festschrifts, and several books on current economic problems. A few are unpublished Rand memoranda, and others are lectures. The contributions of volume 1 are grouped under eight parts which are further arranged into the following two books: PROBLEMS IN PURE THEORY: THE THEORY OF CONSUMER'S BEHAVIOR AND CAPITAL THEORY, and TOPICS IN MATHEMATICAL ECONOMICS. The contributions in volume 2 are grouped under ten parts which are further arranged into the following three books: TRADE, WELFARE, AND FISCAL POLICY; ECONOMICS AND PUBLIC POLICY; and ECONOMICS--PAST AND PRESENT.

_____. THE COLLECTED SCIENTIFIC PAPERS OF PAUL A. SAMUELSON.
Vol. 3. Edited by Robert C. Merton. Cambridge, Mass.: MIT Press, 1972.
xii, 930 p. Contents of Volumes 1 and 2, pp. 901-7. Acknowledgments,
pp. 909-14. Index, pp. 915-30.

Volume 3 contains seventy-eight chapters (contributions) representing
all the scientific papers written by Paul A. Samuelson from mid-
1964 through 1970, and many of his papers for 1971. The contributors are arranged under nineteen parts, and they are grouped
within each part by their relationship to one another. The nine-
teenth part contains his first contributions to the theories of portfolio selection, warrant pricing, and speculative markets under uncertainty. The nineteen parts are arranged in five books as presented in volumes 1 and 2.

Sellekaerts, Willy, ed. ECONOMETRICS AND ECONOMIC THEORY: ESSAYS
IN HONOUR OF JAN TINBERGEN. White Plains, N.Y.: International Arts
and Science, 1974. viii, 298 p. Bibliography of Jan Tinbergen, pp. 285-96.
Index, p. 297.

A collection of thirteen hitherto unpublished papers by leading economists arranged under two groups: econometrics (seven papers),
and economic theory and miscellaneous (six papers). The topics
include autocorrelation in disturbances, estimation and prediction
in dynamic econometric models, aggregation theory, simulation,
quantitative economic policy making, theory of replacement and
depreciation, distribution of income among families, and demand
under multidimensional pricing. The authors are Wm. J. Baumol,
W.T. Dent, C. Hildreth, H.S. Houthakker, H.N. Johnston, P.W.
Jorgenson, G. Kadekodi, L.R. Klein, S. Kuznets, E. Kuh, A.P.
Lerner, L.H. Officer, P. Rao, M.V.R. Sastry, J.K. Sengupta,
K. Shinjo, G. Tintner, and A. Zellner.

_____. ECONOMIC DEVELOPMENT AND PLANNING: ESSAYS IN HON-
OUR OF JAN TINBERGEN. White Plains, N.Y.: International Arts and
Sciences, 1974. xxiv, 266 p. Appendix: Bibliography of Jan Tinbergen,
pp. 250-61. Index, pp. 263-66.

A collection of ten hitherto unpublished essays by noted economists
dealing with international trade, economic development and planning, econometrics, and economic theory. In addition, a reprint
of an article by B. Hansen appraising J. Tinbergen's contributions
to economics is included. The contributors are I. Adelman, B.
Balassa, V.S. Dadajan, J.C.H. Fei, K.A. Fox, A.C. Harberger,
H.G. Johnson, H. Leibenstein, J. Letiche, C.T. Morris, G.
Ranis, and H.W. Singer.

Shell, Karl, ed. ESSAYS ON THE THEORY OF OPTIMAL ECONOMIC GROWTH. Cambridge, Mass.: MIT Press, 1967. xi, 303 p. Selected Bibliography, pp. 295-99. Index, pp. 301-3.

A collection of fifteen essays based upon material presented during 1965-66 at a Massachusetts Institute of Technology seminar on the theory of optimal economic growth. Most of the essays are extensions of the optimal savings model of Ramsey incorporating technical progress, the impact of foreign trade and borrowing, and heterogeneous capital goods. One of the techniques used is Pontryagin's maximum principle. Among the authors are M. Bruno, N.G. Carter, E.S. Chase, M. Datta-Chaudhuri, S.A. Marglin, W.D. Nordhaus, H.E. Ryder, Jr., P.A. Samuelson, E. Sheshinski, K. Shell, and M.E. Yaari.

Shelly, Maynard W. II, and Bryan, Glenn L., eds. HUMAN JUDGEMENTS AND OPTIMALITY. New York: Wiley, 1964. xiii, 436 p. Table of Contributors, pp. v-vii. Author Index, pp. 419-24. Subject Index, pp. 425-36.

A collection of eighteen papers presented at the first symposium of the American Psychological Association held in St. Louis in September 1962, and at the second and third symposiums held December 26, 1962. The papers are arranged under the following headings: introduction: the relation of judgments to optimal decisions; judgmental differences between individuals; learning and optimal judgment-like processes; optimal decisions based on preferences; the fundamental role of judgment-like processes; and applications of judgments to decision-making programs. The authors include F.J. Anscombe, R.J. Aumann, C. West Churchman, H. Gulliksen, R.D. Luce, T.C. Koopmans, R. Radner, P. Suppes.

Shubik, Martin, ed. ESSAYS IN MATHEMATICAL ECONOMICS: IN HONOR OF OSKAR MORGENSTERN. Princeton, N.J.: Princeton University Press, 1967. xx, 475 p. The contribution of Oskar Morgenstern, pp. vii-viii. A Bibliography of Oskar Morgenstern, pp. ix-xviii. No index.

A collection of twenty-seven hitherto unpublished essays be well-known authors divided into the following groups: game theory (six essays), mathematical programming (three essays), decision theory (four essays), economic theory (five essays), management science (two essays), international trade (two essays), and econometrics (five essays). The topics treated include cooperative games with side payments, optimal programming, inventory theory, prior distributions, subjective utility, spectrum analysis of seasonal adjustment, and index numbers. Most of the papers assume knowledge of probability, statistical inference, calculus, mathematical analysis, and set theory. Among the authors are Wm. J. Baumol, C.W.J. Granger, M. Hatanaka, T. Whitin, and M. Suzuki.

Stogdill, Ralph M., ed. THE PROCESS OF MODEL-BUILDING IN THE BE-
HAVIORAL SCIENCES. Foreword by J.E. Corball, Jr. Columbus: Ohio
State University Press, 1970. viii, 181 p. Appendix: Problems in Model-
building, pp. 139-76. Notes on Contributors, pp. 177-78. Index, pp. 179-81.

A collection of nine papers presented at a symposium on the process of model-building in the behavioral sciences on April 20-21,
1967, at Ohio State University. The papers are nontechnical and
emphasize the factors that enter into the development of a model
rather than actual model construction. The authors are W.R.
Ashby, C. West Churchman, H. Guetzkow, R.D. Luce, J.G.
March, Wm. T. Morris, and R.M. Stogdill.

Stone, Richard, ed. MATHEMATICAL MODELS OF THE ECONOMY AND
OTHER ESSAYS. London: Chapman and Hall, 1970. xv, 335 p. A List
of Works Cited, pp. 330-35.

Nineteen essays, previously published by the author, sixteen of
which pertain to economics and economic statistics, and three to
sociodemographic studies. Of the sixteen economic essays, one
essay summarizes the Cambridge growth model; two deal with
foreign trade and the balance of payments; two essays are concerned with statistical techniques in econometric analysis; three
relate to the econometrics of consumers' behavior; and six essays
are devoted to national accounting--its origin, latest development,
and analytical uses. Assumes knowledge of matrix algebra, calculus,
and statistical estimation theory.

_____. MATHEMATICS IN THE SOCIAL SCIENCES AND OTHER ESSAYS.
Cambridge, Mass.: MIT Press, 1966. xiii, 291 p. A List of Works Cited,
pp. 283-91. No index.

Seventeen reprints of articles by the author on methodology in econometrics which can be used for supplementary readings in introductory econometrics and growth courses, and seminars. The topics
include demand theory and estimation, aggregate growth models,
regional and national social accounting systems, investment planning, savings functions of households and firms, and methodology.

Szegö, Giorgia P., and Shell, Karl, eds. MATHEMATICAL METHODS IN
INVESTMENT AND FINANCE. New York: American Elsevier, 1972. x,
665 p. Author Index, pp. 655-59. Subject Index, pp. 660-65.

Twenty-nine contributions presented at a symposium on mathematical methods in investment and finance held at the University of
Venice during September 1971. The papers are arranged under
the following headings: mathematical theories of portfolio allocation, some mathematical models of finance, applied portfolio
theories, optimal bank portfolio selection, forecasting models,
and application of stochastic models. Among the authors are O.
Castellino, K.J. Cohen, C.W.J. Granger, H.E. Leland, M.

Monti, K. Shell, J. Stiglitz, and G.P. Szegö. Assumes knowledge of differential and integral calculus, probability theory, and linear programming.

Thrall, Robert McDowell; Combs, C.H.; and Davis, R.L., eds. DECISION PROCESSES. New York: Wiley, 1954. viii, 332 p. Appendix A: List of Participants, p. 329. Appendix B: Papers Presented, pp. 331-32. No index.

> Nineteen papers presented at a seminar sponsored by the Ford Foundation on the design of experiments in decision processes and held in 1952 in Santa Monica, California. The papers are grouped under the following five headings: individual and social choice, learning theory, experimental studies, and theory and applications of utility. Assumes knowledge of set theory and calculus. Among the authors are C.H. Combs, G. Debreu, J. Marschak, F. Mosteller, J. Nash, E.D. Nering, R. Radner, H. Raiffa, and R.M. Thrall.

Tufte, Edward R., ed. THE QUANTITATIVE ANALYSIS OF SOCIAL PROBLEMS. Addison-Wesley Series in Behavioral Science: Quantitative Methods. Reading, Mass.: Addison-Wesley, 1970. 449 p.

> Nineteen papers, some reprints from journals and other sources and others appearing for the first time, on quantitative analysis of social problems grouped under the following six headings: statistical evidence and statistical criticism, experimental and quasi-experimental studies, economic and aggregate analysis, survey data, and data analysis and research design. The authors include R.E. Ayres, A.J. Coale, Wm. G. Cochran, C. Kaysen, L. Kish, F. Mosteller, A.M. Mood, J.W. Tukey, and D.E. Stokes.

Tuite, Matthew; Chisholm, Roger K.; and Radnor, Michael, eds. INTERORGANIZATIONAL DECISION MAKING. Chicago: Aldine, 1972. xvi, 298 p. Name Index, pp. 292-94. Subject Index, pp. 295-98.

> Seventeen papers grouped under the following headings: the concept of interorganizational decision making, a framework for solving joint decision problems, interorganizational decision making in a business context, decision making at the government-business interface, and interorganizational decision making in government. Some papers assume knowledge of calculus and matrix algebra.

Wold, Herman O.A. MODEL BUILDING IN THE HUMAN SCIENCES. Third Volume in a Series by the Centre International D'Etude des Problemes Humains. Monaco: Union Europeenne D'Editions, 1966. xii, 321 p. No index.

> Thirteen heretofore unpublished papers presented at an international conference held in Monaco in 1964. All of the papers are printed in English except for two, by E. Malinvaud, which are in French. The introduction and conclusion by H.O.A. Wold are in both lan-

guages. Topics include mathematical models of learning in education, causal models for qualitative attributes, some simple nonlinear models, a model for long-term planning in Poland, and forecasting models of the Netherlands economy. While higher mathematics is avoided, most papers assume knowledge of econometric model building. The authors are J.S. Coleman, P. de Wolff, L.R. Klein, E. Malinvaud, K. Porwit, J.J. Post, A. Prekopa, A. Robinson, P. Suppes, C.A. van den Beld, P.J. Verdoorn, P. Whittle, and H.O.A. Wold.

Young, Stanley, ed. MANAGEMENT: A DECISION-MAKING APPROACH. Foreword by D.R. Hampton. Dickenson Series on Contemporary Thought in Management. Belmont, Calif.: Dickenson, 1968. 146 p. Selected Bibliography, pp. 141–46.

Nine reprints of nontechnical articles on decision making. The authors are G.A.W. Boehm, K E. Boulding, B.V. Dean, H. Garfinkel, M. Shubik, W.B. Simon, G. Steiner, and S. Young.

Zarley, Arvid M., ed. PAPERS IN QUANTITATIVE ECONOMICS. Vol. 2. Lawrence: University Press of Kansas, 1971. 209 p. No index.

Volume 2 contains the following papers presented at the Kansas-Missouri, 1968–69 joint seminars in theoretical and applied economics: "The Compensation Principle in Welfare Economics," by J.S. Chipman and J.C. Moore; "Interrelated Consumer Preference and Voluntary Exchange," by T.C. Bergstrom; "Pareto Optimality and Competitive Equilibrium in a General Equilibrium Model of Economic Growth," by M. Ali El-Hodiri; "Spatial Price Equilibrium, Location Arbitrage, and Linear Programming," by S.A. Johnson; "Investment Decision, Uncertainty and the Incorporated Entrepreneur," by V.L. Smith; "The Abstract Transportation Problem," by A.M. Faden; and "International Trade and Development in a Small Country," by T. Rader.

Part III

OPERATIONS RESEARCH

Chapter 7

OPERATIONS RESEARCH AND MANAGEMENT SCIENCE

Abramowitz, Irving. PRODUCTION MANAGEMENT: CONCEPTS AND ANAL-
YSIS FOR OPERATION AND CONTROL. New York: Ronald Press, 1967.
vii, 362 p. Selected References, pp. 333-35. Appendixes, pp. 339-56.
Index, pp. 357-62.

A text for a course in production management for students with a
background in basic statistics and college algebra. It is divided
into the following parts: demand and capacity analysis, decision
alternatives, the production program, and information systems.
The problem-solving approach is used throughout. The appendixes
contain useful statistical tables, and the chapters conclude with
questions and problems.

Ackoff, Russell L., ed. PROGRESS IN OPERATIONS RESEARCH. Vol. 1.
Publications in Operations Research, no. 5. New York: Wiley, 1961. xii,
505 p. Author Index, pp. 493-500. Subject Index, pp. 501-5.

Volume 1 of a review series by the Operations Research Society
of America on the development of modeling techniques and the
applications of these techniques to problem solving. It contains
eleven contributions by practicing operations researchers in the
areas of inventories, linear and dynamic programming, sequencing
theory, replacement theory, simulation, and gaming. In addition,
a chapter is included on the foundations of operations research,
and decision and value theory. Can be followed by anyone familiar with basic operations research techniques.

Ackoff, Russell L., and Rivett, Patrick. A MANAGER'S GUIDE TO OPERA-
TIONS RESEARCH. New York: Wiley, 1963. 107 p. References, pp. 103-4.
Index, pp. 105-7.

A nontechnical guide to the role of operations research in a company. It includes a history of operations research, a concise
statement of problems and methods of operations research, how to
start operations research in a company, and sources of information
on education and experience in this field.

Ackoff, Russell [L.], and Sasieni, Maurice W. FUNDAMENTALS OF OPERA-
TIONS RESEARCH. New York: Wiley, 1968. ix, 455 p. Author Index,
pp. 449-52. Subject Index, pp. 453-55.

A text in operations research for students with knowledge of differential and integral calculus, and elementary probability theory.
It stresses computational rules, rather than algebraic abstractions,
and few rigorous proofs are provided. The following problems are
treated: linear and parametric programming, transportation, assignment, inventory, replacement, maintenance, dynamic programming,
queueing, networks, game theory, and search problems. Chapters
conclude with discussion topics, problems, and references.

Alberts, David S. A PLAN FOR MEASURING THE PERFORMANCE OF SOCIAL
PROGRAMS: THE APPLICATION OF OPERATIONS RESEARCH METHODOLOGY.
Foreword by O.A. Ornati. Labor Economics and Urban Studies. New York:
Praeger, 1970. xix, 159 p. Appendix, pp. 103-41. Bibliography, pp. 143-52.
Index, pp. 155-57. About the Author, p. 159.

Presents a framework in which the performance of a social action
can be measured. It deals with the problems of formulating social
goals from the individualistic and societal viewpoints; it attempts
to develop approximate measures of societal performance based on
individual and collective viewpoints; and it applies the methodology using data gathered by the School of Social Work of the
University of Pennsylvania. The appendix includes an explanation
of the Churchman-Ackoff measure of utility, procedures for estimating the level of satisfaction, portions of questionnaires used to
collect the data, and an illustration of a program's change in
value when the distribution of individual utilities is considered.
Some portions require knowledge of elementary integral calculus.

Anderson, David R.; Sweeney, Dennis J.; and Williams, Thomas A. AN IN-
TRODUCTION TO MANAGEMENT SCIENCE: QUANTITATIVE APPROACHES
TO DECISION MAKING. New York: West Publishing Co., 1976. xiv,
631 p. Appendixes, pp. 608-16. Index, pp. 617-31.

An applications-oriented text for a one-semester course in operations research for undergraduates who have knowledge of college
algebra, although some familiarity with basic statistics and differential calculus is also helpful. In each chapter a problem
situation is described in conjunction with the quantitative procedures being introduced. The problems at the end of each chapter
reinforce the concepts and illustrate the types of real life situations
in which the methods can be applied. Topics include linear programming and the simplex method; assignment and transportation
problems; decision theory; deterministic and probabilistic inventory
models; PERT and CPM (critical path method) models; waiting line
models; simulation; dynamic programming; and Markov chains. The
appendixes consist of tables and matrix notation and operations.

Aronofsky, Julius S., ed. PROGRESS IN OPERATIONS RESEARCH: RELA-TIONSHIP BETWEEN OPERATIONS RESEARCH AND THE COMPUTER. Vol. 3. Publications in Operations Research, no. 16. New York: Wiley, 1969. vii, 561 p. Author Index, pp. 551-57. Subject Index, pp. 559-61.

A collection of fourteen contributions on the use of computers in operations research. The papers fall into the following special areas: simulation, mathematical programming, integer programming, heuristic programming, the design of computer languages, and the management of information systems. Some of the contributions are descriptive and require no knowledge of mathematics; others assume knowledge of linear algebra and calculus. The contributors are R.L. Ackoff, J.S. Aronofsky, M.L. Balinski, R.E. Bargmar, E.M.L. Beale, Sir Stafford Beer, J.S. Bonner, G.B. Dantzig, J.C. Emery, D.B. Hertz, A. Powell, R.L. Sisson, K. Spielberg, J. Sussman, K.D. Tocher, and R.F. Wheeling.

Beer, Stafford. MANAGEMENT SCIENCE: THE BUSINESS USE OF OPERA-TIONS RESEARCH. Doubleday Science Series. Garden City, N.Y.: Double-day, 1968. 192 p. Suggested Reading, p. 7. Acknowledgments, p. 8. In-dex, pp. 109-92.

A nontechnical exposition of the role of science in the manage-ment of business based on examples from the author's experience as a practicing management scientist.

Bennion, Edward G. ELEMENTARY MATHEMATICS OF LINEAR PROGRAMMING AND GAME THEORY. Foreword by R.C. Henshaw, Jr. Michigan State Uni-versity Business Studies, 60. East Lansing: Bureau of Business and Economics Research, College of Business and Public Administration, Michigan State Uni-versity, 1960. xvi, 140 p. Bibliography, p. 138. Index, pp. 139-40.

A presentation of the following techniques of operations research using elementary mathematics: linear programming and the simplex method; degeneracy; duality; two-person, zero-sum game theory; and linear programming problems as games. The techniques are illustrated with numerical examples. This book is designed for students with minimal mathematics background who desire to gain an understanding of the mathematics of operations research rather than its application to practical problems.

Berczi, Andrew. PROBLEMS IN MANAGERIAL OPERATIONS RESEARCH. Vol. 1: NON-CALCULUS. Englewood Cliffs, N.J.: Prentice-Hall, 1969. x, 81 p. Appendix: Common Logarithms, pp. 80-81. No index.

A workbook containing problems for use by students in order to im-prove their skills in problem solving and by teachers who desire an additional source of problems for classroom, homework assignments, and examinations. The problems are based on actual experience of the author in applying operations research. The material is organized into eight chapters according to basic methodology applied, and within each chapter, problems are grouped according to the following functional areas of business: production, marketing, finance, personnel administration, purchasing, and miscellaneous. The eight chapter titles are "Utility Theory," "Decision Theory," "Game Theory," "Inventory Theory," "Sequencing," "Assignment," "Transportation," and "Linear Programming." Answers are given at the end of each chapter to problems marked by an asterisk, while answers to the remaining problems are available to teachers of the course.

Biegel, John E. PRODUCTION CONTROL: A QUANTITATIVE APPROACH. 2d ed., rev. Englewood Cliffs, N.J.: Prentice-Hall, 1971. xiii, 282 p. Exhibits, pp. 254-70. Bibliography, pp. 271-73. Index, pp. 277-82.

A text for a course in production control for students with a background in college algebra. Topics include forecasting, economic lot-size determination, inventory, linear programming, transportation problem, production planning using linear programming, PERT and CPM, and the use of computers in production control. Presentation is intuitive with many examples provided. Chapters conclude with problems.

Bierman, Harold, Jr.; Bonini, Charles P.; and Hausman, Warren H. QUANTITATIVE ANALYSIS FOR BUSINESS DECISIONS. 4th ed., rev. Irwin Series in Quantitative Analysis for Business. Homewood, Ill.: Richard D. Irwin, 1973. xii, 527 p. Appendix of Tables, pp. 500-518. Index, pp. 519-27.

An introduction to the application of mathematics to problems of business. It attempts to treat difficult problems with the simplest means of exposition, avoiding proofs and mathematical rigor. Familiarity with elementary probability and statistics is helpful, but not essential since the first third of the book is devoted to a review of probability and decision theory. The applications are to inventory problems, linear programming, utility theory, game theory, queueing theory, simulation, PERT, Markov processes, and dynamic programming. Chapters conclude with references and problems.

Bowman, Edward H., and Fetter, Robert B. ANALYSIS FOR PRODUCTION AND OPERATIONS MANAGEMENT. 3d ed., rev. Irwin Series in Quantitative Analysis for Business. Homewood, Ill.: Richard D. Irwin, 1967. xiii, 870 p. Appendixes, pp. 819-60. Index, pp. 863-70.

A combined revision of ANALYSIS FOR PRODUCTION MANAGEMENT (1957, 1961), and ANALYSES OF INDUSTRIAL OPERATIONS (1959). It is a text for graduates and undergraduates, and is divided into the following sections: orientation, mathematical programming, statistical analysis, economic analysis, simulation and heuristics, and case studies in production and operations management. Topics include linear and dynamic programming, statistical quality control, sampling inspection, analysis of variance, inventory models, waiting-line theory, maintenance policy, and regression analysis. The case studies include applications of linear programming to the oil, electric power, and aircraft industries; and production and inventory control in chemicals, Hawaiian pineapple, and electrical industries. The appendixes are devoted to statistical tables, glossary, and computer simulation languages. Chapters conclude with summaries, problems, and references.

Brabb, George J. INTRODUCTION TO QUANTITATIVE MANAGEMENT. New York: Holt, Rinehart, and Winston, 1968. xv, 576 p. Appendix A: Answers to Odd-numbered Exercises, pp. 501-5. Appendix B: Statistical Tables, pp. 507-66. Index, pp. 567-76.

A text for a course in quantitative analysis for graduates and undergraduates with a background in college algebra. Higher level mathematics, such as differential calculus, is introduced as needed. The book is divided into the following four sections: decision making under certainty, including classical constrained and unconstrained optimization and linear programming; decision making under risk and uncertainty, including basic statistics, Bayes' theorem, and regression; and decision making with mixed models, including queues and simulation. The chapters conclude with discussion questions, discussion problems, and exercises.

Buffa, Elwood Spencer. MODELS FOR PRODUCTION AND OPERATIONS MANAGEMENT. New York: Wiley, 1963. xii, 632 p. Appendixes, pp. 595-622. Index, pp. 623-32.

An introductory text for a course in production management written in a relatively nonmathematical style, with the appendixes devoted to the mathematical derivations, such as the economic lot size. It is divided into the following parts: introduction, models of flow and man-machine systems, statistical methods, waiting-line models, programming models, models of investment policy, inventory models, simulation models, and synthesis. Assumes knowledge of basic statistics and high school algebra. Chapters conclude with references and review questions, and some chapters also conclude with problems.

_____. MODERN PRODUCTION MANAGEMENT. 4th ed., rev. The Wiley Series in Management and Administration. New York: Wiley, 1973. xiii, 704 p. Appendix: Tables, pp. 687-92. Index, pp. 693-704.

A text for a course in production management for students with a high school algebra background. It is divided into the following five parts (part 6 consists of tables): introduction, analytical methods, design of production systems, operations planning and control, and synthesis. Techniques include linear programming, queueing theory, inventory management and control, machine maintenance, and quality control. Chapters conclude with summaries, problems, and references.

_____. OPERATIONS MANAGEMENT: PROBLEMS AND MODELS. 3d ed., rev. The Wiley Series in Management and Administration. New York: Wiley, 1972. xiii, 762 p. Appendix: Tables, pp. 725-51. Index, pp. 753-62.

A relatively nonmathematical text for courses in production and operations management for students with a background in basic statistics. It is divided into the following five parts: design of operational systems; operations planning and control-forecasting and inventories; linear programming models; operations planning and control-planning, scheduling, and controls; and synthesis. Chapters conclude with review questions, problems, and references.

Cabell, Randolph, and Phillips, Almarin. PROBLEMS IN BASIC OPERATIONS RESEARCH METHODS FOR MANAGEMENT. New York: Wiley, 1961. x, 110 p. Appendix: References to Instructional Material, pp. 107-10. No index.

A presentation of thirty-six statements of problems, without solutions, in the following areas of operations research: mathematical programming (eleven problems), inventory (eight problems), dynamic programming (two problems), queueing theory (five problems), replacement (two problems), analysis of variance (five problems), and miscellaneous (one each in sequencing, routing, and location of plant facilities). The problems are designed to acquaint managers with the types of situations to which operations research may be applied. The mathematics necessary for solving the problems consists of differential and integral calculus, elementary probability theory, and basic statistics.

Carlson, Phillip G. QUANTITATIVE METHODS FOR MANAGERS. New York: Harper and Row, 1967. viii, 181 p. Solution to Exercises, pp. 165-81. No index.

Considers first the business problem, and then the best alternative action based on quantitative analysis. The following problems are considered: lot size, assignment, transportation problem, simplex method, critical path, waiting line, games, statistical simulation, and replacement. Chapters conclude with exercises and references.

Carr, Charles R., and Howe, Charles W. INTRODUCTION TO QUANTITATIVE DECISION PROCEDURES IN MANAGEMENT AND ECONOMICS. New York: McGraw-Hill Book Co., 1964. xv, 383 p. Index, pp. 379-83.

A text for a course in operations research at the elementary level. Topics include the structure of decision problems, mathematical model building, classical optimization techniques, linear, integer, and nonlinear programming, dynamic programming and quadratic programming. Assumes knowledge of one year of college calculus. Chapters conclude with exercises, and notes and references.

Chorafas, Dimitris N. OPERATIONS RESEARCH FOR INDUSTRIAL MANAGE-
MENT. Foreword by D.E. Marlowe. New York: Reinhold, 1958. ix, 303 p.
Index, pp. 301-3.

> An early text in operations research which covers the following
> topics: simulation, games and decisions, linear programming, pro-
> duction scheduling and inventory control, transportation problem,
> and networks. The presentation utilizes mainly graphs and detailed
> examples, with mathematical derivations kept to a minimum. As-
> sumes knowledge of only high school algebra.

Chu, Kong. QUANTITATIVE METHODS FOR BUSINESS AND ECONOMICS.
Foreword by S. Dallas. An ITC Publication in Quantitative Methods. Scranton,
Pa.: International Textbook, 1969. xv, 373 p. Appendix I: Calculus for
Optimization, pp. 341-45. Appendix II: Probability, pp. 347-51. Appendix
III: Matrices and Matrix Operations, pp. 353-60. Appendix IV: Computer
Systems and Programming, pp. 361-70. Index, pp. 371-73.

> A text for a course in quantitative methods with a prerequisite of
> elementary calculus and matrix algebra. It stresses three areas:
> deterministic or mathematical programming models; probabilistic
> models with applications to queues and inventories; and computer
> applications in PERT and cost-benefit analysis. Useful to masters
> and doctoral candidates entering business or economic programs.
> Chapters conclude with summaries, problems, and references.

Churchman, Charles West; Ackoff, Russell L.; and Arnoff, E. Leonard. INTRO-
DUCTION TO OPERATIONS RESEARCH. New York: Wiley, 1957. x, 645 p.
Author Index, pp. 637-40. Subject Index, pp. 641-45.

> An introduction to operations research for students with a minimum
> mathematics background. It is divided into the following ten parts:
> introduction; the problem; the model; inventory models; allocation
> models; waiting-line models; replacement models; competitive models;
> testing, control, and implementation; and administration of opera-
> tions research.

Clough, Donald J. CONCEPTS IN MANAGEMENT SCIENCE. Englewood
Cliffs, N.J.: Prentice-Hall, 1963. 425 p. Appendixes: A, Elements of
Matrix Algebra; B, Statistical Tables, pp. 395-413. Selected References,
pp. 415-19. Index, pp. 420-25.

> A text for an undergraduate course in industrial engineering which
> is divided into the following three parts: decision-making and
> leadership concepts, economic concepts for executive decision
> making, and mathematical concepts for decision making. Topics
> include production activity analysis, in which classical production
> function models and linear programming techniques are discussed;
> probability and Markov chains; a statistical analysis and the de-
> sign of experiments; and simulation. Assumes knowledge of ele-
> mentary algebra, calculus, and economic theory. Chapters con-
> clude with problems and discussion topics.

Collcutt, R.H. THE FIRST TWENTY YEARS OF OPERATIONS RESEARCH. Old Woking, Surrey, England: Unwin Brothers, for the British Iron and Steel Research Association, 1965. 97 p. Charts, diagrams, and illustrations. No index.

A summary of the operations research accomplishments of the British iron and steel industry from 1945 to 1965.

Di Roccaferrera, Giuseppe M. Ferrero. OPERATIONS RESEARCH MODELS FOR BUSINESS AND INDUSTRY. Cincinnati, Ohio: South-Western, 1964. xxvi, 966, 126 p. Bibliography, pp. 915-36. Glossary of Symbols, pp. 937-66. Appendixes 1 to 7, pp. 1-22. Tables, pp. 23-112. Index of Names, pp. 113-15. Index of Subjects, pp. 117-26.

A book for managers--what managers should know about operations research, linear programming, and dynamic programming. It is divided into eight parts. Part 1, "Introduction," reviews basic concepts in economics and operations research. Part 2, "Curve Fitting and Forecasting," deals with regression, time series and forecasting, univariate, bivariate, and multivariate analyses, and probability distributions. Part 3, "Elements of Mathematics," takes up set theory, matrices, and vectors. Part 4, "Mathematical Models," treats Leontief's input-output analysis and inventory models. Part 5, "Linear Programming Techniques" (section A), is on the mathematical formulation of linear programming models, and the diet, transportation, and assignment problems, and sensitivity analysis. Part 6, "Linear Programming" (section B), deals with the simplex method and duality. Part 7, "Dynamic Models," takes up warehouse and caterer problems and production-smoothing problems. Part 8, "Queuing Problems," is on single and multichannel queues, and the Monte Carlo approach to queueing problems. The presentation is nonrigorous, and most of the mathematics is taken up as needed. Each of the thirty-three chapters concludes with problems. The seven appendixes are devoted to mathematical reviews, such as limits of functions and derivatives.

Duckworth, W. Eric. A GUIDE TO OPERATIONAL RESEARCH. 2d ed., rev. London: Methuen, 1974. 153 p. Appendix I: Rapid Statistical Tests for Use by Managers, pp. 137-46. Appendix II: Business Models Project Rules of Play, pp. 147-53. No index.

A nonmathematical introduction to the techniques of operations research for managers. The philosophy and techniques are illustrated by simple examples requiring knowledge of only high school algebra. Topics include linear programming, queueing theory, Monte Carlo and simulation, stock and production control models, decision theory, game theory, operational gaming, information theory, network analysis, and evolutionary operation. Each of the four chapters concludes with references.

Eck, Roger D. OPERATIONS RESEARCH FOR BUSINESS. Belmont, Calif.: Wadsworth Publishing Co., 1976. xvi, 757 p. Appendix A: Areas of the Standard Normal Density Function from the Mean to Z, pp. 721-23. Appendix B: Linear Programming Algorithm, pp. 724-27. Appendix C: Present Worth of $1, pp. 728-31. Appendix D: Integer Programming Algorithm, pp. 732-43. Appendix E: Capital Recovery Factors, pp. 744-47. Index, pp. 748-57.

A text for a one-year course in operations research for undergraduates who possess knowledge of college algebra and who have an exposure to the elements of probability and statistics. Emphasis is on model formulation and applications. Some computer programs and instructions for using the programs are included. The programs can be used on any medium or large computer usually found in academic institutions. Topics include linear programming with applications in marketing, accounting, finance, and the planning and coordination of corporate activities; duality and sensitivity analysis; goal programming, transportation, and other special-structure models for business planning; networks; integer and dynamic programming; deterministic inventory models; Markov chains; utility theory; Bayesian decision theory; queueing; simulation; and portfolio models. Each of the twenty chapters concludes with exercises which emphasize the application of the concepts to finding solutions to example problems.

Eddison, Roger T.; Pennycuick, K.; and Rivett, R.H.P. OPERATIONAL RESEARCH IN MANAGEMENT. New York: Wiley, 1962. xi, 330 p. Index, pp. 327-30.

A presentation of the techniques of operations research with emphasis on the underlying reasons, importance, and scope of applications. The book is mainly descriptive, with formulas used sparingly and mainly in the chapters on queueing applications and dynamic programming. Other topics include linear programming, simulation, data handling, distribution, and the measurement of efficiency and productivity. Knowledge of differential calculus helpful for some portions of the book. The chapters conclude with references.

Emory, William, and Niland, Powell. MAKING MANAGEMENT DECISIONS. Boston: Houghton Mifflin, 1968. ix, 306 p. Index, pp. 293-306.

A text for an introductory course in management at the undergraduate level or M.B.A. program. It is also useful for management training programs and for self-study by practicing executives. The first part develops a general model for making decisions, and considers such topics as the decision process, task delineation, and solution finding. The second part includes an explanation and application of the following techniques: linear programming, inventory theory, networks, queueing theory, decision theory, simulation, and game theory. Assumes knowledge of college algebra, but no prior knowledge of statistics is necessary.

Enrick, Norbert Lloyd. MANAGEMENT OPERATIONS RESEARCH. Modern Management Series. New York: Holt, Rinehart and Winston, 1965. xi, 320 p. Epilogue: Management in the Seventies, pp. 279-87. Appendix: Cases and Problems, pp. 289-315. Index, pp. 317-20.

> An elementary operations research text which stresses practical applications and which requires only a rudimentary knowledge of mathematics. Divided into the following four parts: the general nature of operations research as a management science; planning, programming, and program review; management of inventories and waiting lines; and sampling and statistical analysis. Numerous cases and illustrations are provided. The book concludes with twenty-five cases and problems designed as self-study exercises.

Fabrycky, Wolter J.; Ghare, P.M.; and Torgersen, Paul E. INDUSTRIAL OPERATIONS RESEARCH. Prentice-Hall International Series in Industrial and Systems Engineering. Englewood Cliffs, N.J.: Prentice-Hall, 1972. xiii, 578 p. Appendixes, pp. 531-68. Index, pp. 569-78.

> A revision of OPERATIONS ECONOMY: INDUSTRIAL APPLICATIONS OF OPERATIONS RESEARCH (Prentice-Hall, 1966) by the same authors. This book is a text in operations research for students with a background in college algebra and elementary calculus who have an understanding of industrial operations. It is divided into the following parts: introduction, economic models for production operations, models for production and project control, statistical models for quality assurance, procurement and inventory models, and programming models for operations. Topics include break-even analysis, time value of money (interest rate problems), equipment replacement, production scheduling, networks, quality control, inventory models (both deterministic and stochastic), waiting-line models (both deterministic and stochastic), and linear and dynamic programming. Emphasis is on examples to illustrate the concepts. Chapters conclude with questions and problems, and the appendixes consist of useful statistical and mathematical tables.

Farrar, Donald E., and Meyer, John R. MANAGERIAL ECONOMICS. Foundations of Modern Economics Series. Englewood Cliffs, N.J.: Prentice-Hall, 1970. x, 115 p. Selected Readings, pp. 105-6. Appendix: Tables, pp. 107-10. Index, pp. 111-15.

> A text for a short course in the economics of business decisions for students who have knowledge of elementary probability theory comparable to that covered in a basic statistics course. It can also be used, together with other books in the Foundations of Modern Economics Series, for a course in microeconomics. Topics include marginal analysis, including the inventory lot size problem; mathematical programming; capital budgeting, including multiple period planning and budget constraints; and uncertainty.

Fletcher, Allan, and Clarke, Geoffrey. MANAGEMENT AND MATHEMATICS: THE PRACTICAL TECHNIQUES AVAILABLE. New York: Gordon & Breach,

1964. xi, 235 p. Appendix: Electronic Computers and the Development of Data-processing Centers, pp. 225-30. Index, pp. 231-35.

> An applications-oriented reference for the nonmathematical executive. It covers network problems, PERT, inventory control, time series, regression theory, forecasting, queueing and simulation, replacement theory, dynamic programming, warehouse location problems, response surface analysis, and games. While mathematical derivations are kept to a minimum, rudimentary knowledge of calculus is helpful for some topics, especially queueing theory. The chapters conclude with bibliographies.

Gaver, Donald P., and Thompson, Gerald L. PROGRAMMING AND PROBABILITY MODELS IN OPERATIONS RESEARCH. Monterey, Calif.: Brooks/Cole, 1973. xiii, 683 p. Answers to Selected Exercises, pp. 635-75. Author Index, pp. 677-78. Subject Index, pp. 679-83.

> A comprehensive text for graduates and undergraduates who have studied calculus. It contains a development of linear algebra and probability theory for describing many types of mathematical models used in operations research. The approach is by means of definitions and theorems, followed by examples. Topics include linear and nonlinear programming, decision models, Markov chains, inventory theory, and simulation. Exercises follow the sections of the book, and the chapters conclude with references.

Giffin, W.C. INTRODUCTION TO OPERATIONS ENGINEERING. Irwin Series in Quantitative Analysis for Business. Homewood, Ill.: Richard D. Irwin, 1971. xi, 632 p. Tables, pp. 614-21. Index, pp. 623-32.

> A mathematically oriented text for engineers who have a background in calculus, Fourier analysis, and transform techniques. Topics include statistical quality control, waiting-line analysis, inventory models, mathematical programming, and simulation. Each chapter contains from eighteen and twenty-four problems and a reference list.

Gillett, Billy E. INTRODUCTION TO OPERATIONS RESEARCH: A COMPUTER-ORIENTED ALGORITHMIC APPROACH. McGraw-Hill Series in Industrial Engineering and Management Science. New York: McGraw-Hill Book Co., 1976. xvi, 617 p. Appendix A: Tables, pp. 593-97. Appendix B: Derivation of Queueing Formulas, pp. 598-602. Appendix C: Gauss-Jordan Method for Solving a System of Linear Equations, pp. 603-6. Index, pp. 609-17.

> A text for an introductory operations research course for undergraduates who possess knowledge of FORTRAN programming, calculus, and basic statistics. It differs from other texts in its overall computer-oriented approach to the formulation, construction, development, and solution of operations research models. The development of most methods of solution is preceded by one or more illustrative, step-by-step examples, many with printouts of computer

solutions run on an IBM 370/168 computer. The book is divided
into the following two parts: deterministic operations research
models, and probabilistic operations research models. Topics include linear, integer, and dynamic programming; deterministic
problems; regression analysis; game theory; PERT; queueing theory;
simulation; and Markov processes.

Goddard, Laurence Stanley. MATHEMATICAL TECHNIQUES OF OPERATIONAL
RESEARCH. Adiwes International Series of Monographs in Pure and Applied
Mathematics, vol. 38. New York: Pergamon Press, 1963. x, 230 p. Index,
pp. 228-30.

A monograph dealing with the following topics: linear programming, the transportation and assignment problems, queueing theory,
machine interference, and problems of stock control. A few applications to hypothetical problems are provided. Assumes knowledge of differential and integral calculus, and linear algebra.
Chapters conclude with references.

Goetz, Billy E. QUANTITATIVE METHODS: A SURVEY AND GUIDE FOR
MANAGERS. New York: McGraw-Hill, 1965. xxix, 541 p. Glossary of
Selected Technical Terms, pp. 502-8. Glossary of Symbols and Greek Alpha-
bet, pp. 509-12. Appendix, pp. 513-27. Index, pp. 529-41.

An introduction to the mathematical approach to managerial problems of planning, control, design, and operation of business en-
terprises. It is relatively nontechnical and nonmathematical and
requires knowledge of only high school algebra. It is divided
into the following three parts: the managerial problem, industrial
statistics, and mathematical models in managerial decisions where
linear programming and simulation are introduced. The sections
of the book conclude with problems, and the chapters conclude
with references. The appendix contains probability and interest
tables.

Griffith, John R. QUANTITATIVE TECHNIQUES FOR HOSPITAL PLANNING
AND CONTROL. Lexington, Mass.: D.C. Heath, 1972. xvii, 403 p.
Notes, pp. 385-96. Index, pp. 397-403.

A problem-oriented book for managers of hospitals and related
health care institutions, and for hospital research specialists interested in hospital applications. It provided a conceptual understanding of a variety of techniques (planning, forecasting, scheduling, cost control, and quality control) useful in hospital decisions, and a detailed description of the procedures for applying
the techniques. Topics include forecasting by multivariate analysis, determining population service areas and calculating use
rates, analysis of variance to evaluate demand rates, inventory
control, queueing and simulation models for resource allocation,
PERT and mathematical programming, cost-benefit analysis, statistical quality control, and information systems for hospital planning

and control. Assumes knowledge of elementary statistical and probability concepts and college algebra. Some chapters conclude with additional readings.

Groff, Gene K., and Muth, John F. OPERATIONS MANAGEMENT: ANALYSIS FOR DECISIONS. Irwin Series in Quantitative Analysis for Business. Homewood, Ill.: Richard D. Irwin, 1972. x, 572 p. Index, pp. 563-72.

A text for a course in operations management for students with a background in linear algebra, basic statistics, and computer programming (FORTRAN). It is divided into the following three parts: introduction, operations planning and control, and designing operations systems. The techniques employed include linear and nonlinear programming, the maximum principle, inventory models, stochastic control, least squares, and computer simulation. Topics include analysis for plant and equipment investment, project planning, reliability and maintenance, forecasting, decision rules for inventory control, operations scheduling, statistical quality control, labor performance, and work measurement. Chapters conclude with references and bibliographies.

Gue, Ronald L., and Thomas, Michael E. MATHEMATICAL METHODS IN OPERATIONS RESEARCH. New York: Macmillan, 1968. xi, 385 p. Appendix A: Matrix Algebra, pp. 351-61. Appendix B: Z Transforms, pp. 362-75. Index, pp. 379-85.

A text for a two-semester course for seniors or graduate students who have knowledge of calculus, matrix algebra, probability, and transform methods. Topics include constrained and unconstrained optimization of static systems, including linear programming and algorithms for optimization of nonlinear problems; optimization of dynamic systems using dynamic programming and the variational methods of the calculus of variations and the discrete maximum principle; queues, decision theory, game theory, and graphs with applications to modeling and optimization. The presentation is not rigorous, with most of the mathematical results stated without proof.

Gupta, Shiv. K., and Cozzolino, John M. FUNDAMENTALS OF OPERATIONS RESEARCH FOR MANAGEMENT: AN INTRODUCTION TO QUANTITATIVE METHODS. San Francisco: Holden-Day, 1974. viii, 405 p. Appendix A: Series, pp. 372-73. Appendix B: Derivatives, pp. 374-76. Appendix C: Integrals, pp. 377-78. Appendix D: Tables, pp. 379-402. Index, pp. 403-5.

A text for an introductory operations research course for students with a background in differential and integral calculus and probability. The book is concept and applications oriented, and puts mathematics in perspective as a tool. Topics include linear and dynamic programming, decision theory, game theory, stochastic processes, and simulation. Exercises follow the sections, and the chapters conclude with problems.

Hague, Douglas Chalmers. MANAGERIAL ECONOMICS: ANALYSIS FOR
BUSINESS DECISIONS. New York: Wiley, 1969. vii, 356 p. Index,
pp. 343-56.

A text for business schools and a reference for managers requiring
no college mathematics. About one half of the book is devoted
to microeconomics, and the remaining half to probability and decision theory such as break-even analysis, linear programming,
queueing theory, and inventory analysis. Chapters conclude with
references.

Hammer, Peter L., and Rudeanu, Sergiu. BOOLEAN METHODS IN OPERA-
TIONS RESEARCH AND RELATED AREAS. Foreword by P.L. Hammer. Econ-
ometrics and Operations Research, 8. New York: Springer-Verlag, 1968.
xv, 329 p. Appendix: Generalized Pseudo-Boolean Programming, by I.
Rosenberg, pp. 301-6. Bibliography, pp. 309-23. Author Index, pp. 324-26.
Subject Index, pp. 327-29.

Deals with the application of Boolean techniques to operations research using mainly Boolean matrix calculus, Boolean equations,
and pseudo-Boolean programming. Useful to individuals who are
interested in both operations research and Boolean algebra. Applications are to networks, sequencing problems, assignment and
transportation problems, timetable scheduling, and optimal plant
location. Integer programming problems can be reduced to pseudo-
Boolean programming problems, and various concepts of graph theory are cast in a Boolean framework. The book is self-contained
and assumes only a modest level of mathematical training.

Haner, Frederick Theodore, and Ford, James C. CONTEMPORARY MANAGE-
MENT. Columbus: Merrill, 1973. xii, 420 p. Appendixes, pp. 387-414.
Index, pp. 415-20.

An introductory text for a course in quantitative management. The
book is divided into four parts: foundations for decision making,
planning fundamentals and techniques (cash flow and profit analysis, probability, Markov processes, and investment analysis), qualitative and quantitative methods for organizing (linear programming
and network analysis), and implementation and control-important
concepts (games and simulation, and inventory control). The presentation is not rigorous, and requires knowledge of college algebra. The appendixes consist of interest and probability tables,
matrix operations, selected languages and computer programs, derivation of the EOQ equations, and answers to odd-numbered exercises. Chapters conclude with references and exercises.

Hanssmann, Fred. OPERATIONS RESEARCH TECHNIQUES FOR CAPITAL IN-
VESTMENT. New York: Wiley, 1968. vii, 269 p. Selected References,
pp. 260-66. Index, pp. 267-69.

An applied operations research book which emphasizes problem formulation and discussion of the solution, but does not emphasize algorithms. It is based on case studies of investment decisions, taken from both the published literature and the author's experiences, to illustrate concepts and models. The book is divided into five parts: introduction; criteria for the evaluation of investment proposals, with an illustration of planning an optimal structure in the French oil refining industry; decision models for typical investment situations, with case studies in the acquisition of property rights through competitive bidding and planning the composition of an air transport fleet; estimation and forecasting of investments and returns, with four illustrative cases; and conclusions. Assumes knowledge of probability, the Monte Carlo technique of simulation, and mathematical programming.

Harris, Roy D., and Maggard, Michael J. COMPUTER MODELS IN OPERATIONS MANAGEMENT: A COMPUTER-AUGMENTED SYSTEM. New York: Harper and Row, 1972. viii, 223 p. Index, pp. 221-23.

A collection of ten computer models for use in management analysis and decision making. The models are: FUTURE (a statistical forecasting model), INSYS (an inventory model), EOQ (a model for inventory ordering policy), DECIDE (a decision tree analysis model), CRIT (a project scheduling program), BALANCE (an assembly line balancing program), SQC (a model on statistical control charts), BEARSIM (maintenance policy simulation), QUESIM (a model for queueing systems), and UNISM (a one-product company simulation). For each model, the following is presented: an introduction to the theory and concepts, a simple hand-computed example, the correct input and the resulting output when the computer model is used to solve the simple problem, the solution to complex problems in order to illustrate additional uses of the computer model, and the actual computer program. Emphasis is on formulation of appropriate data inputs and interpretation of the outputs. A complete set of computer programs in FORTRAN and two types of data cards are available. One type is keypunched in the EBIDC code and fits the IBM 360 series of computers, while the other type is keypunched in the BCD code and fits most other computers. Knowledge of computer programming is not required.

Hein, Leonard W. THE QUANTITATIVE APPROACH TO MANAGERIAL DECISIONS. Englewood Cliffs, N.J.: Prentice-Hall, 1967. xxv, 386 p. Appendixes A to F, pp. 355-70. Index, pp. 373-86.

A text for an undergraduate course in quantitative techniques or introduction to operations research for students with a background in college algebra and basic statistics. The formulation of models is accomplished with as little mathematics as possible. Topics include linear programming, transportation problem, learning curves, probability, performance and cost evaluation techniques, PERT and CPM, and line of balance. The appendixes are devoted to a discussion of logarithms and basic statistical tables. Chapters conclude with discussion questions, problems, and references.

Hillier, Frederick S., and Lieberman, Gerald J. INTRODUCTION TO OP-
ERATIONS RESEARCH. 2d ed., rev. San Francisco: Holden-Day, 1974.
xii, 800 p. Appendixes, pp. 747-81. Answers to Selected Problems, pp. 782-89.
Index, pp. 791-800.

> A comprehensive text for undergraduate and graduate courses in
> operations research. It is divided into the following three parts:
> mathematical programming, requiring knowledge of high school al-
> gebra; probabilistic models, requiring knowledge of elementary
> probability theory; and advanced topics in mathematical program-
> ming, requiring knowledge of differential and integral calculus.
> The book covers all of the standard topics in operations research
> with emphasis on motivation and simplicity of explanation, rather
> than on rigorous proofs and technical details. The appendixes
> cover convexity, classical optimization methods, matrices, simul-
> taneous linear equations, and include statistical tables. Chapters
> conclude with exercises and references.

Holdren, Bob R.; Buckberg, Albert; and Loomba, N[arendra]. Paul. OPERA-
TIONS RESEARCH IN SMALL BUSINESS: A CASE STUDY PRIMER OF OPERA-
TIONS RESEARCH TECHNIQUES. Foreword by J.E. Horne. Ames, Iowa:
Bureau of Business and Economic Research, 1963. v, 69 p. Appendixes,
pp. 56-69. No index.

> A report on the feasibility of operations research in small business
> management based on a detailed study and analysis of one firm en-
> gaged in a Deere franchise operation. The following problems are
> analyzed: product-mix decisions, machine selection and optimum
> replacement period, allocation of equipment, and transportation
> problems. The conclusion is that the main need of small busi-
> nesses is to control inventories, while other techniques, such as
> queueing models, have little application. The appendixes are
> devoted to the derivation of formulas used in the report, and an
> algorithm for solving transportation problems. Assumes knowledge
> of college algebra.

Hopeman, Richard J. SYSTEMS ANALYSIS AND OPERATIONS MANAGE-
MENT. Columbus, Ohio: Charles E. Merrill, 1969. xiii, 346 p. Bibliography, pp. 305-40. Index, pp. 341-46.

> A descriptive exploration of the use of systems analysis in opera-
> tions management using few formulas. Part 1 is devoted to gen-
> eral concepts, the environmental setting of the firm, and the de-
> sign of systems. Part 2 applies systems analysis to operations
> management of the materials flow network, and includes inventory
> planning and control, logistics, inventory stock, and purchasing.
> Chapters conclude with summaries and conceptual problems.

Horowitz, Ira. AN INTRODUCTION TO QUANTITATIVE BUSINESS ANALY-
SIS. 2d ed., rev. New York: McGraw-Hill Book Co., 1972. x, 310 p.
Index, pp. 307-10.

A text for a one-semester course in quantitative analysis for students who have a background in college algebra and basic statistics. Topics include probability, Bayesian statistics, linear, integer, and dynamic programming, inventory and queueing models, simulation, and game theory. A review of matrix algebra is included, and no calculus is used. Each of the twelve chapters concludes with over one dozen problems, and a reference list.

Horton, Forest W., Jr., ed. REFERENCE GUIDE TO ADVANCED MANAGEMENT METHODS. New York: American Management Association, 1972. xii, 333 p. Glossary, pp. 306-20. Index, pp. 321-33.

A presentation of sixty-one tools of management science for use by managers and management technicians. Each tool is discussed in a separate section which includes a definition of the tool, the most common application areas, and possible constraints and pitfalls. Among the tools discussed are: break-even analysis, games, forecasting, linear programming, network analysis, profit analysis, scheduling, and simulation. References are provided.

Hough, Louis. MODERN RESEARCH FOR ADMINISTRATIVE DECISIONS. Englewood Cliffs, N.J.: Prentice-Hall, 1970. xix, 609 p. Appendixes, pp. 559-602. Index, pp. 605-9.

Concerned with techniques and problems of actual business research for individuals with backgrounds in statistics, industrial engineering, and cost accounting. It introduces: (1) the analytic methods used for handling uncertainty, (2) the subjective probabilities used in Bayesian statistics, (3) the research design that leads to administrative action, and (4) the system-analysis point of view. The techniques illustrated include inventory models, PERT, linear programming, and simulation. The appendixes deal with sources of data, mathematical rates of change, and Bayesian statistics. Chapters conclude with questions, problems, and references.

Houlden, B.T., ed. SOME TECHNIQUES OF OPERATIONAL RESEARCH. Foreword by D.G. Christopherson. London: English Universities Press, 1962. x, 154 p. Index, p. 154.

Each of the nine chapters was written by a different member of the operational research group of the National Coal Board. The book presents the basic mathematics and examples of the following techniques: linear programming and extensions; two-person, zero-sum game theory; queues; stock control; replacement; and the theory of search. Assumes knowledge of college algebra and matrix algebra, with differential calculus necessary for some portions. Chapters conclude with references.

Huysmans, Jan H.B.M. THE IMPLEMENTATION OF OPERATIONS RESEARCH: AN APPROACH TO THE JOINT CONSIDERATION OF SOCIAL AND TECH-

NOLOGICAL ASPECTS. Publications in Operations Research, no. 19. New York: Wiley, 1970. x, 234 p. Appendixes, pp. 117-201. References, pp. 202-6. Index, pp. 209-34.

Describes an approach to operations research implementation that focuses on the interdependence between social, psychological, and technical aspects of an OR project. The approach is general and schematic and does not rely on specific tools such as linear programming. It emphasizes the behavioral characteristics that determine if and how a specific proposal will be accepted by management. The presentation is based on an experiment of implementing operations research by management.

Jelen, F.C., ed. COST AND OPTIMIZATION ENGINEERING. New York: McGraw-Hill Book Co., 1970. xxii, 490 p. Appendixes, pp. 435-71. Answers to Problems, pp. 473-78. Name Index, pp. 479-82. Subject Index, pp. 483-90.

A text for an undergraduate course in cost and optimization written by the editor and sixteen other contributors. It is divided into four parts. Part 1 takes up the mathematics of cost comparisons and considers depreciation, taxes, interest and discounting, and profits. Part 2 deals with optimization, and includes such topics as break-even and minimum-cost analyses; probability, uncertainty, and simulation; learning curves; inventory problems; queueing problems; one variable and multivariable optimization; and linear and dynamic programming. Part 3 considers cost estimation and control and includes topics such as capital investment and operating-cost estimation, and critical path method. Part 4 is on cost accounting, social values, and other nonmathematical concepts. The main prerequisite is calculus. The appendixes contain interest and normal curve tables, values for exponential learning-curve tables, values for exponential learning-curve functions, and tables of random numbers.

Kaufmann, Arnold. METHODS AND MODELS OF OPERATIONS RESEARCH. Translated by Scripta Technica, Inc. Prentice-Hall International Series in Management. Englewood Cliffs, N.J.: Prentice-Hall, 1963. xi, 510 p. Tables, pp. 469-86. Bibliography, pp. 487-501. Index, pp. 503-10.

Translated from the French, this book presents operations research for engineers, accountants, managers, and others who have a background in high school algebra. Part 1 (methods and models) deals with linear programming, queueing, inventories, replacement, and maintenance of equipment. Part 2 contains mathematical reviews for those desiring to learn more about the mathematics of the models.

Kaufmann, Arnold, and Faure, R. INTRODUCTION TO OPERATIONS RESEARCH. Foreword by R. Bellman. Mathematics in Science and Engineering, vol. 47. New York: Academic Press, 1968. xi, 300 p. Bibliography, pp. 283-95. Index, pp. 297-300.

With the help of eighteen short studies of simplified problems, the authors explain the methods and procedures for setting up equations or constructing models for solving problems. The techniques illustrated are: dynamic programming, sequential decisions using Markov chains, waiting lines for tools in a factory, the Ford-Fulkerson algorithm for coffee shipments, assignment problem, investments and discounting, restocking, games of strategy, use of Boolean algebra, personnel management, linear programming, choice under uncertainty, Foulkes algorithm, Johnson's algorithm, theory of graphs, and the weighting of values using an index of poverty. Assumes knowledge of elementary calculus and matrix algebra. The bibliography is broken down by chapter.

King, John R. PRODUCTION PLANNING AND CONTROL: AN INTRODUCTION TO QUANTITATIVE METHODS. Omega Management Science Series. Elmsford, N.Y.: Pergamon Press, 1975. vii, 403 p. Appendix A: Complete Set of Conflict Resolution Tables for Example Scheduling Problem Discussed in Chapter 10, pp. 374-78. Appendix B: Tabulations of Machine Utilisation for the Ashcroft and Palm Models Discussed in Chapter 14, pp. 379-86. Answers to Example Problems, pp. 387-97. Index, pp. 399-403.

A problem-oriented text for a course in production and planning for graduate students who have knowledge of the rudiments of calculus, although the nonmathematically inclined should be able to follow and benefit from substantial sections of the text and the illustrative numerical examples. Topics include forecasting, linear programming, assignment problem, branch and bound methods, distribution, location and routing, scheduling of intermittent production, batch size determination, production smoothing, scheduling of continuous production, and machine supervision. Thirteen of the fourteen chapters conclude with problems and references.

Levin, Richard I., and Kirkpatrick, Charles A. QUANTITATIVE APPROACHES TO MANAGEMENT. 2d ed., rev. New York: McGraw-Hill Book Co., 1971. xiv, 461 p. Appendixes, pp. 425-55. Index, pp. 457-61.

A text for business students with a high school algebra background. Topics include probability with applications to decision theory, inventory and production models, linear programming, transportation and assignment problems, games, networks, queueing, and Markov analysis. Each of the sixteen chapters concludes with problems and a bibliography. The appendixes contain statistical tables and the derivation of economic lot size models. The book is also useful as a reference and self-study for managers.

Lientz, Bennet P. COMPUTER APPLICATIONS IN OPERATIONAL RESEARCH. Englewood Cliffs, N.J.: Prentice-Hall, 1975. xiv, 289 p. Appendix A: Linear Algebra and Matrices, pp. 243-57. Appendix B: Networks and Graph Theory, pp. 258-71. Appendix C: Tables of Statistics and Pseudo-random Numbers, pp. 272-82. Index, pp. 285-89.

A text for a one-year course at the undergraduate level. Purpose is to expose the student to operations research and statistics through computer applications in a variety of disciplines. Emphasis is on methods and the how, rather than on proofs and the why. Topics include data management and information systems, and applications in mathematical programming, networks, statistics, and numerical analysis. Assumes knowledge of set theory, algebra, calculus, and FORTRAN programming.

Luck, G.M.; Luckman, J.; Smith, G.W.; and Stringer, J. PATIENTS, HOSPITALS, AND OPERATIONAL RESEARCH. London: Tavistock Publications, 1971. xv, 210 p. References, pp. 201-4. Index, pp. 205-10.

A description of activities in a hospital which can be analyzed by operations research techniques with emphasis on queueing applications and simulation. The activities of main interest related to admissions of patients and their movement through the hospital, and to the deployment of resources around patients, such as medical and nursing staff, beds, operating theatres, and medical service departments. Useful to hospital managers and requires no mathematics background.

McKenney, James L., and Rosenbloom, Richard S. CASES IN OPERATIONS MANAGEMENT: A SYSTEMS APPROACH. Wiley Series in Management and Administration. New York: Wiley, 1969. xiii, 311 p. No index.

A supplement to a graduate course in operations research. It consists of twenty-one cases designed to assist the student to gain insights in the design and control of an operating system. The cases are grouped under the following two headings: system design and evaluation, and operations planning and control. There are no mathematics prerequisites.

McRae, Thomas W. ANALYTICAL MANAGEMENT. New York: Wiley-Interscience, 1970. xxi, 580 p. Selected Reading, pp. 569-71. Table of Random Numbers, p. 572. Author Index, pp. 573-74. Subject Index, pp. 575-80.

A simple introduction to a wide range of quantitative techniques in business with emphasis on their benefits and limitations rather than on developing competence in their use. It is divided into the following five parts: some basic concepts, information processing, problem solving, foretelling the future, and control. Topics include quantifying risk, statistical sampling methods, the structure of information systems, the elements of computer programming, break-even analysis, linear programming, inventory control, critical path method, queueing, simulation, and standard costing. Assumes knowledge of high school algebra. The chapters conclude with review questions, answer notes to review questions, and references.

Makower, Michael S., and Williamson, Eric. OPERATIONAL RESEARCH: PROBLEMS, TECHNIQUES AND EXERCISES. Teach Yourself Books. London: English Universities Press, 1967. vii, 264 p. Author Index, p. 259. Subject Index, pp. 261-64.

> A problem-oriented approach to operations research which treats the following problem or application areas: replacement, forecasting, stock control, queues, linear programming, games, networks, and dynamic programming. The prerequisite is high school algebra. Chapters conclude with exercises, references, and solutions to exercises.

Manne, Alan S. ECONOMIC ANALYSIS FOR BUSINESS DECISIONS. New York: McGraw-Hill Book Co., 1961. x, 177 p. Bibliography, pp. 169-74. Index, pp. 175-77.

> An introduction to the following techniques of analysis: linear and integer programming, dynamic programming, transportation models, sequencing problems, and inventory control. Assumes knowledge of high school algebra and elementary probability; only one chapter makes use of calculus. Chapters conclude with exercises.

Martin, Michael J.C., and Denison, Raymond A., eds. CASE EXERCISES IN OPERATIONS RESEARCH. Foreword by P. Rivett. New York: Wiley, 1971. xi, 210 p. No index.

> Presents a practical methodology for solving real problems. Nineteen authors discuss fifteen case studies which deal with plant location, cost estimation, stock and production control, political districting, aggregate manpower planning, and other topics. Each case, which may be assigned as a project in operations research, requires formulation of the problem, the construction of a model and the evaluation of alternative decisions or policies, and a written final report. Assumes knowledge of matrix algebra, and differential and integral calculus.

Menon, Palat Govind. MATERIALS MANAGEMENT AND OPERATIONS RESEARCH IN INDIA. New Delhi: MMJ Publications, 1968. iv, 347 p. Appendix, pp. 338-41. References and Acknowledgments, pp. 342-44. Index, pp. 345-47.

> Intended for the Indian layman, this book uses nontechnical language to describe materials management and operations research. Among the techniques treated are: economic order quantities, queueing theory, linear programming, learning curves, decision matrices, network techniques, and computer applications.

Mitchell, George H., ed. OPERATIONAL RESEARCH: TECHNIQUES AND EXAMPLES. London: English Universities Press, for the National Coal Board, 1972. ix, 275 p. Solutions to Examples, pp. 239-71. Index, pp. 273-75.

A collection of topics in operations research which formed the basis of study groups organized by the National Coal Board. Topics include linear programming and extensions, dynamic programming, decision theory and the theory of games, networks, queues, stock control, replacement, and simulation. Some chapters require knowledge of elementary calculus and probability theory. Chapters conclude with references and exercises.

Moore, Peter Gerald. BASIC OPERATIONAL RESEARCH. Topics in Operations Research. London: Sir Isaac Pitman, 1968. ix, 185 p. Appendix A: The Simplex Method of Linear Programming, pp. 167-74. Appendix B: Some Statistical Concepts, pp. 175-81. Index, pp. 183-85.

An introduction to operations research for managers and others who desire a survey before studying the higher level texts. Topics include linear programming, queueing, inventories, decision making, and networks. Emphasis is on case histories and problems in a real life setting. Assumes knowledge of high school algebra. Chapters conclude with references.

Morris, William T. MANAGEMENT SCIENCE: A BAYESIAN INTRODUCTION. Prentice-Hall International Series in Management. Englewood Cliffs, N.J.: Prentice-Hall, 1968. xiii, 226 p. Appendix A: Linear Normal Loss Integrals, pp. 219-20. Appendix B: Table of the Linear Normal Loss Integral, pp. 221-22. Appendix C: Glossary of Symbols, p. 223. Index, pp. 224-26.

Presents a balanced view of management science, and examines the possibility of using Bayesian logic as an integrating structure of the diverse ideas in management science. Useful as a supplement for management science and operations research courses, and requires knowledge of probability theory and elementary calculus. It is divided into three parts. Part 1, the basic logic of learning and decision, considers such topics as Bayes' theorem, the logic and psychology of decisions, scaling of preferences, and concepts encountered in decision theory. Part 2, applications, takes up the problems of decision making in capital investments, inventories, bidding among buyers and sellers, learning in binary data systems, and sequential decision making. Part 3, professional problems of management science, considers some general notions of models and the role of intuition in management science. Each chapter concludes with about one dozen exercises.

Morse, Philip McCord, and Kimball, George E. METHODS OF OPERATIONS RESEARCH. 1st ed., rev. New York: Wiley; Cambridge, Mass.: MIT Press, 1951. vii, 158 p. Tables, pp. 146-53. Bibliography, p. 154. Index, pp. 155-58.

The first publication of this work was in classified form and appeared just after World War II. It deals with some of the military applications of operations research, including measures of effectiveness of military equipment, tactical analysis, gunnery and bombardment problems, and operational experiments with equipment and tactics. Assumes knowledge of elementary calculus and probability theory.

Paik, C.M. QUANTITATIVE METHODS FOR MANAGERIAL DECISIONS. New York: McGraw-Hill Book Co., 1973. xiii, 403 p. Appendix A: Random Digits, p. 394. Index, pp. 395-403.

A text for a quantitative methods course in a master's degree program in business and public administration, and in upper level undergraduate programs in management. Topics include linear and nonlinear models, rates of change and derivatives, formulation and interpretation of linear programming, decision making under uncertainty, sampling, inventory models, queueing, and Monte Carlo simulation. Assumes no college level mathematics. A single exercise is presented after the discussion of a topic. Chapters conclude with summaries, questions, and problems.

Phillips, Don T.; Ravindran, A.; and Solberg, James J. OPERATIONS RESEARCH: PRINCIPLES AND PRACTICE. New York: Wiley, 1976. xvii, 585 p. Appendix A: Review of Linear Algebra, pp. 569-76. Appendix B: Table of Random Numbers, p. 577. Index, pp. 579-85.

A non-computer-oriented text for a one-year course in operations research for graduates and undergraduates who are familiar with calculus, linear algebra, and elementary probability theory. Extensive verbal explanations accompany the mathematical developments, and the emphasis is on problem formulation and interpretation with many examples provided. Topics include linear programming and the simplex method; parametric, integer, and dynamic programming; discrete and continuous probability distributions and their applications; Markov processes; networks; queueing processes; deterministic and probabilistic inventory models; simulation and simulation languages; and nonlinear programming. Special techniques such as Lagrangian optimization, separable programming, quadratic programming, and geometric programming are also treated. Each of the eleven chapters concludes with recommended readings, references, and exercises. The exercises stress theoretical derivations rather than practical applications.

Plane, Donald R., and Kochenberger, Gary A. OPERATIONS RESEARCH FOR MANAGERIAL DECISIONS. Irwin Series in Quantitative Analysis for Business. Homewood, Ill.: Richard D. Irwin, 1972. xi, 321 p. Appendix A: Zero-one Programming Routine, pp. 249-56. Appendix B: Linear Programming Routine, pp. 257-64. Appendix C: Lagrange Multiplier Program, pp. 265-73. Appendix D: Optimization of Functions of More than One Variable, pp. 274-80. Appendix E: Selected Topics in Nonlinear Programming, pp. 281-316. Appendix F: Random Number Table, pp. 317-18. Index, pp. 319-21.

A text for a one-semester introductory course in operations research at the senior or M.B.A. level. The mathematics prerequisite is college algebra, with calculus introduced as needed. Emphasis is on applications, and the examples and exercises at the ends of the chapters are designed to foster skills in formulating mathematical models. Topics include decision theory, classical optimization, linear programming and extensions, zero-one programming, analysis of waiting lines, and simulation with applications in queueing problems. The appendixes contain computer programs for zero-one, linear, and certain types of nonlinear programming problems.

Poage, Scott T. QUANTITATIVE MANAGEMENT METHODS FOR PRACTICING ENGINEERS. Professional Engineering Career Development Series. New York: Barnes and Noble, 1970. x, 186 p. Glossary, pp. 159-65. Selected Readings, pp. 166-81. Index, pp. 182-86.

Introduces the following tools for the practicing engineer: replacement theory, queueing, networks, inventory, linear programming, simulation, forecasting, and organization. Familiarity with elementary calculus and probability is assumed. Chapters conclude with references.

Rau, John G. OPTIMIZATION AND PROBABILITY IN SYSTEMS ENGINEERING. New York: Van Nostrand Reinhold, 1970. xi, 403 p. Index, pp. 401-3.

A text for a one-year course in the mathematics of operations research as applied to engineering problems, and a reference for operations research analysts, economists, management scientists, and applied mathematicians. The ten chapter titles are: "Mathematical Aspects of the Systems Engineering Process," "Calculus Techniques in Optimization," "Linear Programming," "Recursion Techniques," "System Reliability," "Systems with Repair," "System Availability and Dependability," "Systems with Spares," "Markov Techniques," and "Queueing Systems." The mathematics prerequisites include differential and integral calculus, linear algebra and matrices, differential equations, and probability theory. Chapters conclude with references and problems.

Reisman, Arnold. MANAGERIAL AND ENGINEERING ECONOMICS. The Allyn and Bacon Series in Mechanical Engineering. Boston: Allyn and Bacon, 1971. xix, 532 p. Appendix, pp. 422-515. Author Index, pp. 519-21. Subject Index, pp. 523-32.

A text for a graduate or undergraduate course for managerial economics majors in economics, engineering, and business. The book stresses the formulation of mathematical models with applications in engineering systems, capital budgeting, and corporate finance. Portions of the book require knowledge of calculus.

Richmond, Samuel B. OPERATIONS RESEARCH FOR MANAGEMENT DECI-
SIONS. New York: Ronald Press, 1968. xvi, 615 p. Appendixes, pp. 563-610.
Index, pp. 611-15.

A text for a one-year course in operations research for students
with only a high school algebra background. It is divided into
the following six parts: optimization models, including constrained
and unconstrained classical optimization; probability theory and
applications, including an introduction to integral calculus, continuous probability distributions, and applications to inventory,
maintenance, and replacement problems; allocation problems, including linear programming and related techniques; stochastic
models, including queueing theory and Markov chains; scheduling
problems, using dynamic programming, simulation and PERT; and
decision theory, including game theory and Bayes' theorem. Mathematical tools such as set theory, calculus, matrix algebra, series,
Lagrange multipliers, Newton's method of approximation, and other
topics are introduced as needed. The appendixes contain mathematical and statistical tables, and the Greek alphabet. Chapters
conclude with exercises and selected references.

Riggs, James L. ECONOMIC DECISION MODELS FOR ENGINEERS AND
MANAGERS. New York: McGraw-Hill Book Co., 1968. x, 401 p. Appendix: Compound Interest Factors, pp. 376-96. Index, pp. 397-401.

A text for an introductory course in economics for engineers and
managers with a background in calculus and probability theory.
Heavy reliance is placed on examples and explanation to develop
the rationale of solution methods. Topics include break-even analysis, inventory models, linear programming, networks, interest calculations, depreciation and replacement problems, investments,
queues, and games. Chapters conclude with problems and references.

Riggs, James L., and Inoue, Michael S. INTRODUCTION TO OPERATIONS
RESEARCH AND MANAGEMENT SCIENCE: A GENERAL SYSTEMS APPROACH.
McGraw-Hill Series in Industrial Engineering and Management Science. New
York: McGraw-Hill Book Co., 1975. xiii, 497 p. Appendix A: Conver-
sational Computer Programs for Linear Programming, pp. 452-72. Appendix B:
Mathematical Programming System: The Product-form Approach, pp. 473-84.
Annotated Bibliography, pp. 485-91. Index, pp. 492-97.

A text for courses in operations research and management science
which uses a new approach to problem formulation and solution
called Resource Planning and Management System (RPMS). The
traditional aspects of operations research are treated along with
the introduction of RPMS. Topics include problem formulation,
linear programming, PERT, dynamic programming, queueing models
and simulation, and RPMS: a tool for general systems analysis.
RPMS does not require knowledge of calculus or any special mathematics preparation; this method is illustrated with numerous solved
problems. Chapters conclude with exercises.

Rivett, Patrick. CONCEPTS OF OPERATIONAL RESEARCH. The New Thinker's Library. London: C.A. Watts, 1968. ix, 206 p. Selected Bibliography, pp. 202-3. Index, pp. 205-6.

A book for the layman on the methodology and application of models. The following techniques are illustrated: queueing, linear programming, games, simulation, and education and training. Knowledge of high school algebra and elementary probability is helpful, but not essential.

Saaty, Thomas L. MATHEMATICAL METHODS OF OPERATIONS RESEARCH. New York: McGraw-Hill Book Co., 1959. vii, 421 p. Index, pp. 413-21.

A text for a course in operations research for graduates and undergraduates who have a background in calculus and matrix algebra. It is divided into the following four parts. Part 1: scientific method—truth; mathematics and logic—validity; and outlines of some useful mathematical models. Part 2: optimization programming and game theory. Part 3: probability, statistics, and queueing theory. Part 4: an essay on the author's ideas on creativity in operations research. The presentation is not rigorous mathematically, and the models are described in detail. Parts 2, 3, and 4 conclude with problems and projects. Each of the twelve chapters concludes with a detailed reference list.

Sadowski, Wieslaw. THE THEORY OF DECISION-MAKING: AN INTRODUCTION TO OPERATIONS RESEARCH. Edited by H. Infeld and P. Knightsfield. Translated by E. Lepa. New York: Pergamon Press, 1965. viii, 292 p. Name Index, pp. 289-90. Subject Index, pp. 291-92.

This book is an exposition of operations research mainly for economists with a background in calculus and probability theory. No formal tools of linear algebra are used. The main topics are linear and dynamic programming with applications, and game theory. Chapters conclude with bibliographies.

Sargeaunt, Michael J. OPERATIONAL RESEARCH FOR MANAGEMENT. Heinemann Studies in Management. London: William Heinemann, 1965. vii, 157 p. Appendix I: Approach to Simplex Method of Linear Programming, pp. 140-47. Appendix II: Application of the Monte Carlo Method, pp. 148-49. Bibliography, pp. 150-53. Index, pp. 154-57.

A self-study book designed to meet the needs of executives faced with operations research for the first time. Topics include linear programming, inventories, queues, dynamic programming, networks production scheduling, games, search techniques, and simulation. High school algebra is the only mathematics prerequisite.

Sasieni, Maurice [W.]; Yaspan, Arthur; and Friedman, Lawrence. OPERATIONS RESEARCH: METHODS AND PROBLEMS. New York: Wiley, 1959. xi, 316 p. Appendixes: 1, Finite Differences; 2, Differentiation of Integrals; 3, Row

Operations, pp. 294-311. Index, pp. 312-16.

A text for a one-semester introductory techniques course for students who are familiar with differential and integral calculus. Each chapter presents the theory and techniques of a particular problem area, several solved problems, and problems for assignment. The areas considered are: inventory, replacement, waiting lines, competitive strategies, allocation sequencing, and dynamic programming. Chapters also conclude with references.

Schellenberger, Robert E. MANAGERIAL ANALYSIS. Irwin Series in Quantitative Analysis for Business. Homewood, Ill.: Richard D. Irwin, 1969. xv, 461 p. Epilogue, pp. 451-54. Index, pp. 457-61.

An introductory text for graduates and undergraduates covering the usual topics in operations research with special attention given to the needs of managers rather than specialists in operations research. Each chapter contains discussion topics, problems, and references.

Schweyer, Herbert E. ANALYTIC MODELS FOR MANAGERIAL AND ENGINEERING ECONOMICS. Reinhold Industrial Engineering and Management Science Textbooks Series. New York: Reinhold, 1964. xiii, 505 p. Selected Bibliography, p. 471. Appendixes, pp. 473-97. Index, pp. 499-505.

A text for a course in engineering economics for undergraduates with a background in calculus. The book demonstrates how simple mathematical models may be used in making practical business and production decisions. No proofs of theorems are included. Topics include interest calculations, amortization and depreciation, cash flow, pricing of economic goods, production cost curves, linear programming, profitability models, replacement models, inventory and cycle models, economic analysis and uncertainty, and capital outlay. The appendixes contain a glossary of selected terms, estimated life of equipment, interest tables, and tables of random numbers. Chapters conclude with summaries and problems.

Sengupta, Jati K., and Fox, Karl A. ECONOMIC ANALYSIS AND OPERATIONS RESEARCH: OPTIMIZATION TECHNIQUES IN QUANTITATIVE ECONOMIC POLICY. Studies in Mathematical and Managerial Economics, vol. 10. New York: American Elsevier, 1969. xvi, 478 p. Subject Index, pp. 468-73. Author Index, pp. 474-78.

Applications of linear and nonlinear programming, dynamic and variational programming, control methods, and probabilistic optimization techniques to microeconomic models to acquaint economists with operations research techniques. The book includes numerous results and concepts resulting from the authors' research, and it is therefore not a textbook. Other topics include sensitivity analysis, resource and planning in educational institutions, models of decomposition or decentralization in firm behavior and economic policy, and operations research and complex social sys-

tems. Assumes knowledge of advanced calculus and the calculus
of variations. Chapters conclude with references.

Shamblin, James E., and Stevens, G.T., Jr. OPERATIONS RESEARCH: A
FUNDAMENTAL APPROACH. New York: McGraw-Hill Book Co., 1974.
x, 404 p. Appendixes: Tables, pp. 393-400. Index, pp. 401-4.

A text for a course in operations research for graduates and undergraduates who are familiar with calculus and linear algebra. It
covers the standard topics, and relies on the intuitive approach
for understanding concepts. Chapters conclude with references
and problems.

Siemens, Nicolai; Marting, C.H.; and Greenwood, Frank. OPERATIONS RE-
SEARCH: PLANNING OPERATING AND INFORMATION SYSTEMS. New
York: The Free Press, 1973. vii, 450 p. Tables, pp. 429-43. Index,
pp. 445-50.

A text and reference for students, administrators, executives, and
others who are familiar with high school algebra. It is divided
into the following three sections: planning systems, operating
systems, and management information systems. It covers the standard topics in operations research. Chapters conclude with references, and some chapters conclude with exercises as well.

Sivazlian, B.D., and Stanfel, L.E. ANALYSIS OF SYSTEMS IN OPERA-
TIONS RESEARCH. Prentice-Hall International Series in Industrial and Systems Engineering. Englewood Cliffs, N.J.: Prentice-Hall, 1975. xii,
532 p. Appendix: The Laplace Transform, 520-28. Index, pp. 529-32.

An introduction to the main methods used in the study of systems
encountered in operations research for students in engineering,
operations research, and mathematics and who are familiar with
calculus and elementary differential equations. Topics include
reliability theory; queueing; inventories for single-commodity and
multicommodity systems, and single-installation and multiinstalla-
tion systems; replacement theory; information theory; and location
theory. The mathematical rigor is moderate, and emphasis is
placed on illustrations and examples. Chapters conclude with
references and exercises.

Stimson, David H., and Stimson, Ruth H. OPERATIONS RESEARCH IN HOS-
PITALS: DIAGNOSIS AND PROGNOSIS. Chicago: Hospital Research and
Educational Trust, 1972. ix, 110 p. Bibliography, pp. 85-110.

A critical assessment of the accomplishments and shortcomings of
operations research in hospitals over the past twenty years. The
studies that are included are grouped into the following categories: scheduling in outpatient clinics; staffing studies; admission,
discharge, and utilization of inpatient facilities; blood banking;
inventory control and menu planning; computer applications and

total hospital information systems; and models of hospitals. For each category there is first a brief review of the literature, then a discussion of the additional effort required of an administrator and his staff to implement the findings, and finally an attempt to evaluate the benefits and costs of implementation. Useful primarily to hospital administrators. There are no mathematics prerequisites.

Stoller, David S. OPERATIONS RESEARCH: PROCESS AND STRATEGY. Science Surveys: 2. Berkeley and Los Angeles: University of California Press, 1964. vii, 159 p. Remarks, pp. 153-54. Index, pp. 155-59.

Deals with two types of problems: (1) operations of systems that are characterized statistically and controlled by one decision maker, such as production lines, repair and service stations, and traffic flows, in which queueing theory provides the logical basis; and (2) systems controlled by several decision makers with conflicting interests in which game theory constitutes the basic model. Presentation is midway between a nontechnical form and a rigorous mathematical style. Assumes knowledge of elementary calculus, statistics, probability, and theory of equations. Chapters conclude with selected references.

Taha, Hamdy A. OPERATIONS RESEARCH: AN INTRODUCTION. 2d ed., rev. and enl. New York: Macmillan, 1976. xiv, 648 p. Appendix A: Review of Vectors and Matrices, pp. 601-13. Appendix B: Review of Basic Theorems in Differential Calculus, pp. 614-22. Appendix C: General Program for Computing Poisson Queueing Formulas, pp. 623-27. Appendix D: Answers to Selected Problems, pp. 628-36. Index, pp. 637-48.

A text for a one-year course in operations research for undergraduates who have knowledge of college algebra and elementary probability theory; calculus is used in only a few places. It is divided into the following three parts: linear, dynamic, and integer programming; probabilistic models; and nonlinear programming. Other topics include the simplex method, decision theory, network analysis, deterministic and probabilistic inventory models, queueing theory, and classical optimization techniques. There are over 480 problems, and almost every problem presents a new idea.

Teichroew, Daniel. AN INTRODUCTION TO MANAGEMENT SCIENCE: DETERMINISTIC MODELS. Huntington, N.Y.: Krieger, 1964. xix, 716 p. Appendix A: Miscellaneous Formulas, pp. 638-39. Appendix B: Interest Tables, pp. 640-89. Appendix C: Procedure for Reducing Polynomial Functions to Standard Forms, pp. 690-93. Answers to Selected Exercises, pp. 695-708. Author Index, pp. 709-10. Subject Index, pp. 711-13. Errata, pp. 714-16.

A text for a two-semester course in the application of mathematical models to business problems for students familiar with differential and integral calculus. It is divided into five parts: (1) formulation of business problems with applications to accounting, organization theory, and mathematics of finance; (2) optimization models of one variable; (3) optimization models of two variables; (4) linear systems of many variables, including linear programming; and (5) optimization models of many variables, including dynamic programming. Within each part, the mathematics is presented first, followed by examples, extensions, and proofs. Each chapter concludes with notes, references, and detailed exercises.

Theil, Henri; Boot, John C.G.; and Kloek, Teun. OPERATIONS RESEARCH AND QUANTITATIVE ECONOMICS: AN ELEMENTARY INTRODUCTION. Translated by Voorspellen en Beslissen. Economic Handbook Series. New York: McGraw-Hill Book Co., 1965. xiv, 258 p. Index, pp. 255-58.

Translated from the Dutch, this book is a survey for the layman of the methods and accomplishments of econometrics and operations research at a nontechnical level using few formulas and symbols. It stresses description of the type of problems that can be studied by given models and the formulation of problems, but not the solution. Topics include linear programming, critical path, input-output analysis, econometric macromodels, economic forecasts, probability and uncertainty, concept of a strategy, game theory, queues, simulation, management games, production and inventory systems, and the statistical specification of economic relations.

Thierauf, Robert J., and Klekamp, Robert C. DECISION MAKING THROUGH OPERATIONS RESEARCH. 2d ed., rev. Wiley Series in Management and Administration. New York: Wiley, 1975. xi, 650 p. Appendixes, pp. 601-44. Index, pp. 645-50.

A text for a course in operations research for students who have knowledge of matrix algebra, calculus, and basic statistics. It is divided into the following seven parts: overview of operations research, operations research models--probability and statistics, operations research models--matrix algebra, operations research models--calculus, operations research models--simulation techniques, operations research models--advanced topics, future of operations research. The appendixes contain a review of mathematical topics, tables, and answers to problems. Chapters conclude with questions, problems, and references.

Thompson, W.W. OPERATIONS RESEARCH TECHNIQUES. Merrill's Mathematics and Quantitative Methods Series. Columbus, Ohio: Charles E. Merrill, 1967. xiii, 157 p. Index, pp. 155-57.

Analytic, numerical, and simulation techniques of problem solving are discussed and illustrated primarily in reference to linear programming, transportation, and inventory models. Examples are used freely, and the mathematical prerequisites are minimal. Each chapter contains a reference list and about twelve problems for assignment.

Timms, Howard L. INTRODUCTION TO OPERATIONS MANAGEMENT. The Irwin Series in Operations Management. Homewood, Ill.: Richard D. Irwin, 1967. x, 159 p. Index, pp. 153-59.

An introduction to decision systems for managerial problems. Topics include management theory, decision theory, systems theory, and decision systems. Chapters conclude with review questions and references.

Trueman, Richard E. AN INTRODUCTION TO QUANTITATIVE METHODS FOR DECISION MAKING. Series in Quantitative Methods for Decision Making. New York: Holt, Rinehart and Winston, 1974. xvi, 624 p. Appendix A: Cumulative Standardized Normal Distribution Function, pp. 560-61. Appendix B: Unit Normal Linear Loss Integral, pp. 562-63. Appendix C: Cumulative Binomial Distribution, pp. 564-93. Appendix D: Cumulative Poisson Distribution Function, pp. 594-97. Appendix E: Random Variates, p. 598. Selected Answers to Exercises, pp. 601-7. Selected References, pp. 609-11. Index, pp. 615-24.

A text for a course in quantitative decision making for undergraduates and M.B.A. students. Topics include probability theory; decision making under certainty with emphasis on the Bayesian approach; and operations models including linear and dynamic programming, networks, inventories, queueing, and simulation. Emphasis is placed on problem formulation and illustrative examples.

Van der Veen, B. INTRODUCTION TO THE THEORY OF OPERATIONAL RESEARCH. Translated by W.D. Hoeksma. Philips Technical Library. New York: Springer-Verlag, 1967. viii, 204 p. Appendix, pp. 173-82. Problems, pp. 183-96. Answers, pp. 197-99. Bibliography, p. 200. Index, pp. 201-4.

Translated from the Dutch, this book is a text for a one-semester introductory course in operations research for graduates and undergraduates who are familiar with the rudiments of mathematical analysis, differential and integral calculus, and ordinary differential equations. The appendix contains a review of elementary probability theory. It stresses applications in the following areas: linear and dynamic programming, networks, inventories, queueing theory, and game theory. About forty problems, arranged by chapter at the end of the book, are designed to provide practice in applying the models.

Vollman, Thomas E. OPERATIONS MANAGEMENT: A SYSTEMS MODEL-BUILDING APPROACH. Reading, Mass.: Addison-Wesley, 1973. 716 p. Appendixes, pp. 687-707. Index, pp. 711-16.

A text for an introductory course in operations management for undergraduates or M.B.A. students who have a high school algebra background. It is divided into six parts. Part 1 explains the systems approach. Part 2 is concerned with models of cost and value, including PERT models. Part 3, quantitative models, presents an overview of deterministic and stochastic models and an introduction to mathematical programming and queueing models. Part 4 deals with operations research system design. Part 5 is on materials flow systems and includes a discussion of inventory models. Part 6 presents an overview of the book. The appendixes contain interest, normal curve, and random numbers tables. Chapters conclude with references, reviews, outlines, central issues, and assignments.

Wagner, Harvey M. PRINCIPLES OF MANAGEMENT SCIENCE WITH APPLICATIONS TO EXECUTIVE DECISIONS. Prentice-Hall International Series in Management. Englewood Cliffs, N.J.: Prentice-Hall, 1970. xx, 562 p. Appendix: Tables, pp. 547-49. Selected Readings, pp. 551-55. Index, pp. 557-62.

A text for an undergraduate course in management science. While only one chapter assumes knowledge of calculus, much of the text requires a mathematical maturity comparable to that acquired in a calculus course. The last five chapters assume knowledge of elementary probability theory. Topics include optimization with linear and dynamic programming techniques, sensitivity analysis and duality, networks, inventory scheduling using dynamic programming, integer and stochastic programming, waiting line models, and simulation. The aim of the book is to motivate the student to undertake further study in the field. Each chapter begins with a table of contents, and concludes with a large number of review exercises.

_____. PRINCIPLES OF OPERATIONS RESEARCH WITH APPLICATIONS TO MANAGERIAL DECISIONS. Prentice-Hall International Series in Management. Englewood Cliffs, N.J.: Prentice-Hall, 1969. xxii, 937 p. Appendixes, pp. A1-A78. Selected References, pp. R1-R31. Author Index, pp. 11-14. Subject Index, pp. 15-116.

A text of very broad coverage suitable for undergraduates and graduates, and a reference for administrators, consultants, executives, and others working in a variety of positions. Assumes knowledge of college calculus; topics requiring more advanced mathematics are contained in the appendixes. The aim of the text is to develop the student's skill in formulation and building models, and specifically, in translating a verbal description of a decision problem into an equivalent mathematical model. Topics include linear, dynamic, integer, and nonlinear programming, using both deterministic and stochastic models, Markov chains, inventories and waiting lines, network analysis, and simulation. Each chapter concludes with a large list of review exercises, formulation exercises, and mind-expanding exercises, and some chapters contain computational exercises. Chapter 15 on advanced techniques of nonlinear programming contains seventy-seven exercises.

Walldin, Knut-Erik. ASPECTS OF OPERATIONS RESEARCH IN THEORY AND PRACTICE. Uppsala, Sweden: Almqvist and Wiksells, 1969. 233 p. Index, pp. 232-33.

An attempt to bridge the gap between theory and application. It is divided into the following two parts: part 1 (chapters 2 to 6) describes five case studies in the application of operations research in different areas as follows: investment of liquid assets by a Swedish bank, dimensioning of a commercial bank's cash at the local offices, production and storage of welding electrodes, sampling inspection of vouchers, and queues in outpatient clinics; part 2 (chapters 7 to 9) contains theoretical contributions in the following areas: model building, stochastic processes, and variation-diminishing transformations. Assumes knowledge of differential and integral calculus and matrix algebra, and familiarity with operations research techniques. Some of the chapters conclude with references.

Ward, R.A. OPERATIONAL RESEARCH IN LOCAL GOVERNMENT. Preface by A.H. Marshall. London: George Allen and Unwin, 1964. xi, 96 p. Appendix: The Principle of Simulation, pp. 87-90. Bibliography, p. 92. Index, pp. 93-96.

A brief introduction to operations research techniques in the following areas of local government: buying policy and stock control, the checking of invoices, control of depot operations, refuse collection and disposal, and school transport. The techniques employed include simulation, linear programming, inventory models, queueing theory, and basic statistical techniques such as histograms and regression. The presentation is elementary and assumes knowledge of basic statistics.

White, Michael J. MANAGEMENT SCIENCE IN FEDERAL AGENCIES: THE ADOPTION AND DIFFUSION OF A SOCIO-TECHNICAL INNOVATION. Lexington, Mass.: D.C. Heath, 1975. xii, 111 p. Appendix A: Number and Types of Interviews in Forty-six Federal Civilian Agency Management Science Activities, 1967-72, pp. 87-89. Appendix B: Reproduction of the 1971 Questionnaire, pp. 91-100. Notes, pp. 101-8. Name Index, p. 109. Subject Index, p. 111.

Presents a largely descriptive model in terms of graphs and tables to describe the patterns of adoption or development of management science groups in federal civilian agencies. The model is used to explore the growth of actual management science activities in these agencies. The research is based on 1,100 pages of interview transcripts and 1,500 pages of questionnaires collected from forty-six groups engaged in management science activities in thirty-seven agencies. The histories of thirty-three management science activities are used in the model.

Woolsey, Robert E.D., and Swanson, Huntington S. OPERATIONS RESEARCH FOR IMMEDIATE APPLICATION: A QUICK AND DIRTY MANUAL. New York: Harper & Row, 1969. xvii, 204 p. Program Appendix, pp. 171-97. Index, pp. 199-204.

A supplementary text for courses in operations research, management science, and industrial engineering and a manual for practitioners. It brings together the best known methods for solving simple problems in production scheduling, inventory control, capital budgeting, Markov chains, networks, and engineering design. FORTRAN programs are provided for solving many of the problems. The appendix contains fourteen computer programs written in FORTRAN-IV for the Digital Equipment Corporation's PDP-10 computer.

Worms, Gerard. MODERN METHODS OF APPLIED ECONOMICS. New York: Gordon and Breach, 1970. xvi, 221 p. Appendix: Discount Tables, unpaged.

An introduction to economic calculation, operational research, and statistics with emphasis on mathematical reasoning rather than on mathematical rigor. Topics include interest and amortization, linear programming, game theory, regression and correlation. The final one-fourth of the book is devoted to case studies.

Zimmermann, Hans-Jürgen, and Sovereign, Michael G. QUANTITATIVE MODELS FOR PRODUCTION MANAGEMENT. Prentice-Hall International Series in Management. Englewood Cliffs, N.J.: Prentice-Hall, 1974. xvi, 650 p. Appendix A: Review of Microeconomics, pp. 573-80. Appendix B: Moving Averages and Exponential Smoothing for Forecasting, pp. 581-85. Appendix C: Duality in Constrained Optimization, pp. 586-99. Appendix D: Revised Simplex Method, pp. 600-601. Appendix E: Introduction to Queueing, pp. 602-23. Appendix F: Financial Tables, pp. 624-33. Appendix G: Statistical Tables, pp. 634-40. Index, pp. 641-50.

A text for a course in production management for graduates and undergraduates in business administration and industrial engineering who have knowledge of linear programming, basic statistics, probability theory, calculus, matrix algebra, and microeconomics. Emphasis is on decision making through models already developed for the common general problems in the production area: the size, location, design, cost, planning, scheduling, and maintenance of production processes. Topics include production functions, optimal long-run capacity planning with variable price, the multifacility firm, location theory, time studies, capital budgeting, inventory models, product sequencing, assignment problem, line balancing, and simulation and maintenance. Mathematical programming is used in nearly every chapter. The problems at the ends of the chapters apply the techniques discussed and also provide extensions of the text material. Chapters also conclude with references.

Chapter 8

GAME THEORY

Abt, Clark C. SERIOUS GAMES. New York: Viking Press, 1970. xvi, 176 p. Appendixes, pp. 135-76. No index.

A nonmathematical exploration of the activity among two or more independent decision makers seeking to achieve their objectives in some limiting context. It shows how games can be used in education, planning, and problem solving, and how to evaluate the cost-effectiveness of games. The appendixes contain exercises in games.

Aumann, Robert J., and Shapley, Lloyd S. VALUES OF NON-ATOMIC GAMES. Princeton, N.J.: Princeton University Press, for the Rand Corporation, 1974. xi, 333 p. Appendix A: Finite Games and Their Values, pp. 295-300. Appendix B: Epsilon-monotonicity, pp. 301-3. Appendix C: The Mixing Value of Absolutely Continuous Set Functions, pp. 304-14. References, pp. 315-20. Index of Special Spaces and Sets, p. 321. Index, pp. 323-33.

A presentation of the mathematical theory of multiperson games for use in economics and other fields. The author sets forth the axiomatic approach and treats separately the random order and asymptotic approaches. The relationship between the value and the core of games is taken up and applied to economic equilibrium. Assumes familiarity with the theory of measurable spaces and functional analysis, but no previous experience in game theory or mathematical economics is required.

Blackwell, David, and Girshick, M.A. THEORY OF GAMES AND STATISTICAL DECISIONS. New York: Wiley, 1954. xi, 355 p. References, pp. 337-45. Index, pp. 347-55.

A text for a course in decision theory for first-year graduate students in statistics. Topics include games in normal form, S games, utility and the principle of choice, classes of optimal strategies, fixed sample-size statistical games with a finite number of states of nature, games with a finite number of possible outcomes (multi-

decision games), sufficient statistics and the invariance principle
in statistical games, sequential games, Bayesian and minimax sequential procedures, perfect information games, and estimation procedures.
This book builds on earlier contributions to decision theory as contained
in STATISTICAL DECISION FUNCTIONS (Wiley, 1950), by A. Wald,
and GAMES AND ECONOMIC BEHAVIOR (Princeton, 1944), by O.
Morgenstern. Assumes knowledge of mathematical analysis (uniform
convergence, the Riemann integral, and the Heine-Borel theorem),
linear algebra, and intermediate probability theory. Sections of
the book conclude with problems.

Blaquière, Austin; Gérard, Francoise; and Leitmann, George. QUANTITATIVE
AND QUALITATIVE GAMES. Mathematics in Science and Engineering, vol. 58.
New York: Academic Press, 1969. xi, 172 p. Notation and Terminology,
p. xiv. Appendix, pp. 163-64. References, pp. 165-67. Subject Index,
pp. 169-71. Assumptions, Corollaries, Lemmas, Theorems, p. 172.

An introduction to the geometric theory of two-person games designed for further research and for applications in the physical,
biological, and social sciences. Chapter 1 is devoted to basic
concepts; chapters 2 to 4 are concerned with two-person, zero-
sum games, with perfect information of the state, called games
of degree, by R. Isaacs; and chapters 5 to 7 deal with qualitative two-player games, called games of kind by R. Isaacs. Assumes knowledge of advanced calculus and differential equations.
The appendix contains a proof of the axiom: an open set in the
i. t. is a subset of G which is open in E^n.

Burger, Ewald. INTRODUCTION TO THE THEORY OF GAMES. Translated
by J.E. Freund. Englewood Cliffs, N.J.: Prentice-Hall, 1959. vi, 202 p.
Appendix: Sperner's Lemma and Some of Its Consequences, pp. 194-98. Bibliography, pp. 199-200. Index, pp. 201-2.

Translated from the German, this book is an introduction to mathematical theory of games, with applications appearing mainly as
illustrations. The four chapter titles are: "The General Concept
of a Game," "The Non-cooperative Theory of Games," "Zero-
sum Two-person Games," and "The Cooperative Theory of Games."
The appendix consists of advanced topological concepts. This
book is useful as a text for a course in game theory for first-year
graduate students.

Danskin, John M. THE THEORY OF MAX-MIN AND ITS APPLICATIONS TO
WEAPONS ALLOCATION PROBLEMS. Econometrics and Operations Research
V. New York: Springer-Verlag, 1967. 126 p. Appendix: The Lagrange
Multiplier Theorem, pp. 123-25. Bibliography, p. 126. No Index.

A treatment of games in which standard game theory fails and the
concept of mixed strategy has no meaning. Specifically, it deals
with games in which a player cannot conceal his strategy from his

opponent, with applications to military situations in which the allocation of missiles cannot be concealed. The seven chapter titles are: "Introduction," "Finite Allocation Games," "The Directional Derivative," "Some Max-min Examples," "A Basic Weapons Selection Model," "A Model of Allocation of Weapons to Targets," and "On Stability and Max-min-max." Assumes knowledge of differential and integral calculus. Chapters conclude with exercises.

Davis, Morton D. GAME THEORY: A NONTECHNICAL INTRODUCTION. Foreword by O. Morgenstern. A Science and Discovery Book. New York: Basic Books, 1970. xii, 208 p. Appendix, pp. 197-98. Bibliography, pp. 199-204. Index, pp. 205-8.

A nonmathematical treatment of two-person zero-sum and non-zero-sum games, the n-person game, and von Neumann-Morgenstern utility theory. Liberal use is made of military, political, marketing, and economic examples.

Dresher, Melvin. GAMES OF STRATEGY: THEORY AND APPLICATIONS. Prentice-Hall International Series in Applied Mathematics. Englewood Cliffs, N.J.: Prentice-Hall, 1961. xii, 184 p. Bibliography, pp. 179-80. Index, pp. 181-84.

A text for a course in game theory for students who have knowledge of calculus. It considers both finite and infinite games, with emphasis on military applications, and concludes with a chapter on the moment space theory of the solution to infinite games.

Dresher, Melvin; Shapley, Lloyd S.; and Tucker, A.W., eds. ADVANCES IN GAME THEORY. Annals of Mathematical Studies, no. 52. Princeton, N.J.: Princeton University Press, 1964. x, 692 p. No Index.

A collection of twenty-nine papers on game theory. The first thirteen papers deal with two-person games, and the last sixteen are on n-person games.

Dresher, Melvin; Tucker, A.W.; and Wolfe, P., eds. CONTRIBUTIONS TO THE THEORY OF GAMES. Vol. 3. Annals of Mathematical Studies, no. 39. Princeton, N.J.: Princeton University Press, 1957. viii, 435 p. No Index.

A collection of twenty-three papers arranged under the following headings: moves as plays of other games (five papers); games with perfect information (six papers); games with partial information (four papers); games with a continuum of strategies (five papers); games with a continuum of moves (three papers). The authors are C. Berge, L.D. Berkovitz, L.E. Dubins, H. Everett, W.H. Fleming, D. Gale, D. Gillette, O. Gross, J.F. Hannan, J.C. Holladay, J.R. Isbell, S. Karlin, J.G. Kemeny, J. Milnor, J.C. Oxtoby, M.O. Rabin, R. Restrepo, H.E. Scarf, L.S. Shapley, M. Sion, G.L. Thompson, W. Walden, and P. Wolfe.

Duke, Richard D. GAMING: THE FUTURE'S LANGUAGE. New York: Wiley, Halsted Press, 1974. xvii, 223 p. References, p. 173. Appendixes, pp. 175–222. About the Author, p. 223.

A nonmathematical perspective on the nature of gaming in a context broad enough to include all serious gaming activity. A theoretical base is presented which provides insights into how gaming works and gaming as a language, and which can be used to set forth specifications for game design. There are no mathematical prerequisites, but it assumes that the reader has experienced the play of serious gaming/simulation exercises or will do so. The five chapter appendixes are devoted to specifications for game design, conceptual mapping, list of games cited, glossary, and gaming—the conceptual map.

Friedman, Avner. DIFFERENTIAL GAMES. Series in Pure and Applied Mathematics, vol. 25. New York: Wiley-Interscience, 1971. xi, 350 p. Bibliographical Remarks, pp. 335–38. Bibliography, pp. 339–45. Index, pp. 347–49. Index of Conditions, p. 350.

A text on the mathematical foundations of the theory of differential games, using existence theorems, for students in engineering, economics, and mathematics. Topics include games of fixed duration, pursuit and evasion, survival, values of games and saddle points, computational formulas for finding solutions, extensions of the theory to the case of restricted phase coordinates, capturability, delayed information, games with partial differential equations, and n-person games. The bibliography emphasizes Russian contributions to game theory. Chapters conclude with problems.

Isaacs, Rufus. DIFFERENTIAL GAMES: A MATHEMATICAL THEORY WITH APPLICATIONS TO WARFARE AND PURSUIT, CONTROL AND OPTIMIZATION. Foreword by F.E. Bothwell. The SIAM Series in Applied Mathematics. New York: Wiley, 1965. xii, 384 p. Widely Used Symbols and Formulas, pp. xxi–xxii. Appendix: A1, A Hit Probability Payoff; A2, The Fixed Battery Pursuit Game; A3, Optimal Trajectories of Guided Missiles; A4, An Illustration from Control Theory; A5, The Bomber and Battery Game, pp. 357–77. References, pp. 377–79. Index, pp. 379–84.

An attempt to integrate game theory, the calculus of variations, and control theory by emphasizing practical problems and their solutions. It makes use of a verification theorem which determines whether or not optimal strategies of a game are obtainable. This theorem is applied to the solution of problems characterized by transition surfaces, dispersal surfaces, and universal services with emphasis on applications to warfare. Assumes knowledge of advanced calculus and the calculus of variations. Problems requiring solutions are presented at intervals throughout the book.

Karlin, Samuel. MATHEMATICAL METHODS AND THEORY IN GAMES, PRO-
GRAMMING, AND ECONOMICS. Vol. 1: MATRIX GAMES, PROGRAM-
MING, AND MATHEMATICAL ECONOMICS. Addison-Wesley Series in Sta-
tistics. Reading, Mass.: Addison-Wesley, 1959. x, 433 p. Appendix A:
Vector Spaces and Matrices, pp. 362-96. Appendix B: Convex Sets and Con-
vex Functions, pp. 397-406. Appendix C: Miscellaneous Topics, pp. 407-13.
Bibliography, pp. 415-27. Index, pp. 431-33.

Volume 1 of a two-volume series on the synthesis of the concepts
of game theory and programming theory together with the concepts
and techniques of mathematical economics. It is divided into the
following two parts: the theory of games, and linear and nonlinear programming and mathematical economics. Topics include
min-max theorem, dimension relations for sets of optimal strategies, discrete games, linear and nonlinear programming, game
theory, and economic models, including the Arrow-Debreu equilibrium model of an expanding economy and the von Neumann
model of an expanding economy. Each chapter includes a fair
amount of advanced exposition, which is incorporated as starred
(*) sections. Elementary background material is presented in the
appendixes, and more advanced background material is incorporated directly in the text. The theory is applied to a variety
of simplified problems based on economic models, business decision, and military tactics. This book is useful as a text and
reference in these subjects.

_____. MATHEMATICAL METHODS AND THEORY IN GAMES, PROGRAM-
MING, AND ECONOMICS. Vol. 2: THE THEORY OF INFINITE GAMES.
Addison-Wesley Series in Statistics. Reading, Mass.: Addison-Wesley, 1959.
xi, 386 p. Notation, pp. x-xi. Solutions to Problems, pp. 279-318. Appendix A: Vector Spaces and Matrices, pp. 320-54. Appendix B: Convex
Sets and Convex Functions, pp. 355-64. Appendix C: Miscellaneous Topics,
pp. 365-70. Bibliography, pp. 371-83. Index, pp. 385-86.

Volume 2 in a two-volume series on the synthesis of games, programming, and mathematical economics. This volume treats infinite games (in volume 1 the strategy spaces are finite dimensional), and includes the following topics: structure of infinite
games, separable and polynomial games, games with convex kernels
and generalized convex kernels, games of timing of one action
for each player, infinite classical games, poker and general parlor
games, and miscellaneous games. The essential background material from volume 1 is reproduced here in its entirety (i.e.,
chapter 1 and the appendixes) in order that this volume may be
studied independently of volume 1.

Levin, Richard I., and DesJardins, Robert B. THEORY OF GAMES AND
STRATEGIES. International's Series in Management Science. Scranton, Pa.:
International Textbook, 1970. xi, 132 p. Bibliography, pp. 121-29. Index,
pp. 131-32.

A largely nonmathematical and intuitive presentation of game theory beyond the introductory level, requiring knowledge of elementary calculus in a few places. It treats the following games: 2x2 two-person, 2xM and Mx2, nonnegotiable, 3x3 and larger, non-zero-sum, and n-person.

Luce, R. Duncan, and Raiffa, Howard. GAMES AND DECISIONS: INTRODUCTION AND CRITICAL SURVEY. A Study of the Behavioral Models Project, Bureau of Applied Social Research, Columbia University. New York: Wiley, 1957. xix, 509 p. Appendixes, pp. 371-484. Bibliography, pp. 485-99. Index, pp. 501-9.

Attempts to communicate the ideas and results of game theory and related decision-making models unencumbered by their technical mathematical details (for example, almost no proofs are included). Topics include extensive, normal, and characteristic function forms of games; two-person zero-sum and non-zero-sum games; two-person cooperative games, n-person games with applications; and individual and group decision making under uncertainty. The presentation of the topics parallels the structuring given to the theory by von Neumann and Morgenstern in THEORY OF GAMES AND ECONOMIC BEHAVIOR. The main prerequisite is mathematical sophistication rather than specific courses in mathematics. The eight chapter appendixes deal with the more technical topics which arise in various parts of the book.

McKinsey, John Charles C. INTRODUCTION TO THE THEORY OF GAMES. New York: McGraw-Hill Book Co., for the Rand Corporation, 1952. x, 371 p. Bibliography, pp. 361-67. Index, pp. 369-71.

A text for a course in game theory for graduates and undergraduates who have knowledge of advanced calculus, including convergence, continuity, Rieman integrals, greatest lower and least upper bounds, and maxima and minima. Familiarity with classical algebra and matrix theory would be helpful for portions of the book. Other topics of mathematics, such as distribution functions and Stieltjes integrals, are introduced as needed. Topics include rectangular games, games in extensive form, continuous and separable games, games with convex payoff functions, linear programming, zero-sum, n-person games, games without the zero-sum restriction, and special problems, such as pseudo-games and games played over function spaces. Chapters conclude with historical and bibliographical remarks and from nine to twenty-two problems each.

Nicholson, Michael. OLIGOPOLY AND CONFLICT: A DYNAMIC APPROACH. Toronto: University of Toronto Press, 1972. ix, 228 p. Appendix: A Brief Analysis of the Extended Perfect Competition Model, pp. 215-21. Index, pp. 222-28.

An analysis of duopoly and oliogopoly behavior based on decision theory models, including game theory and a number of original models, such as the "conflict model" in which each participant in a two-person game possesses limited information concerning his opponent. Assumes knowledge of elementary calculus
and matrix algebra. Chapters conclude with mathematical appendixes.

Owen, Guillermo. GAME THEORY. Philadelphia: W.B. Saunders, 1968.
xii, 228 p. Appendix: Convexity; Fixed Point Theorems, pp. 211-15. Bibliography, pp. 217-22. Index, pp. 223-28.

Covers both two-person (including multistage) and n-person game
theory from a mathematical point of view. It considers the needs
of the social and management scientist and game theory's relationship to sociological phenomena by providing a heuristic interpretation of the mathematical argument whenever possible. This book
is suitable as a text for undergraduate and graduate students in
mathematics, economics, and management science. Chapters conclude with problems.

Parthasarathy, T., and Raghavan, T.E.S. SOME TOPICS IN TWO-PERSON
GAMES. Modern Analytic and Computational Methods in Science and Mathe-
matics, vol. 22. New York: American Elsevier, 1971. xii, 259 p. Author
Index, pp. 255-56. Subject Index, pp. 257-59.

Presents the mathematical theory of two-person games. After a
review of set topology and linear spaces (necessary for understanding minimax theorems), and theorems on convex sets (such as
the separation theorem and the fixed-point theorem) and convex
functions, the following topics are taken up: games in extensive
and normal form, including the minimax theorem due to von
Neumann; optimal strategies of finite games, including saddle
point theorems; an algorithm for finding a pair of optimal strategies,
including the Kuhn-Tucker theorem; minimax theorems with applications as tools to prove theorems in other branches of science;
non-zero-sum noncooperative two-person games; differential games;
n-person game solution theory due to von Neumann and Morgenstern; and stochastic games. Chapters conclude with selected
references.

Rapoport, Anatol. N-PERSON GAME THEORY: CONCEPTS AND APPLICA-
TIONS. Ann Arbor: University of Michigan Press, 1970. 331 p. Notes,
pp. 311-16. References, pp. 317-20. Index, pp. 321-31.

Covers the essential ideas of n-person games which are developed
in the original formulation of von Neumann and Morgenstern and
the subsequent extensions by the present generation of game theoreticians. It is aimed at readers with little mathematical background. Topics include individual and group rationality, the
Shapely value, the bargaining set, the kernel of an n-person

game, games in partition function form, and Harsanyi's bargaining
model. The applications are to small and large markets, legis-
latures, symmetric and quota games, and coalitions and power.

_____. TWO-PERSON GAME THEORY: THE ESSENTIAL IDEAS. Ann Arbor:
University of Michigan Press, 1966. 229 p. Notes, pp. 215-19. References,
pp. 220-21. Index, pp. 223-29.

A presentation of the essential ideas of game theory, including
its departures from the two-person zero-sum game, for the general
reader and for the social scientist using the barest minimum of
mathematical notation.

Saaty, Thomas L. MATHEMATICAL MODELS OF ARMS CONTROL AND DIS-
ARMAMENT: APPLICATIONS OF MATHEMATICAL STRUCTURES IN POLITICS.
Publications in Operations Research, no. 14. New York: Wiley, 1968. ix,
190 p. References, pp. 179-83. Index, pp. 185-90.

Presents models on arms control and disarmament based primarily
on game theory and Bayes' theorem. It is divided into four parts:
the objective of the book; stability of policies, optimal strategies
for coordinating policies, and negotiations for agreements; applications of agreements and the effectiveness of their enforcement;
and intermediate and long-range problems of arms control-analysis
of the growth of conflicts: ideas and perspectives. Assumes knowledge of probability theory and game theory.

Siegel, Sidney, and Fouraker, Lawrence E. BARGAINING AND GROUP DE-
CISION MAKING--EXPERIMENTS IN BILATERAL MONOPOLY. New York:
McGraw-Hill Book Co., 1960. x, 132 p. References, pp. 103-5. Appen-
dixes, pp. 107-27. Index, pp. 129-32.

A summary of three experiments in which bilateral monopoly transactions are simulated. In one experiment, each of the two partners was shown his own profits as a function of quantity and price;
in the second, one participant was given tables showing the profits of both bargainers; and in the third experiment, both participants knew the profits of both partners. Several hypotheses
were tested in these experiments. This book is useful to anyone
interested in the techniques of conducting laboratory experiments
on human beings, and to theoreticians who desire to test their
theories. Assumes knowledge of differential calculus. The appendixes contain the iso-profit tables used in the experiments.

Stahl, Ingolf. BARGAINING THEORY. Stockholm: Economic Research Institute, Stockholm School of Economics, 1972. 313 p. Literature Appendix,
pp. 213-52. Mathematical Appendix, pp. 253-88. General Appendix,
pp. 289-305. References, pp. 305-8. Name Index, pp. 309-10. Subject
Index, pp. 311-13.

A development of a theory and model for two-alternative bargaining games, followed by the presentation of a general model for n-alternative games and a special model for S-games. The behavioristic and institutional assumptions of the model are discussed and applications to merger, duopoly, and labor-management bargaining are presented. Most proofs of theorems are presented in the appendix, and few extensive mathematical derivations are included. Assumes knowledge of differential and integral calculus.

Telser, Lester G. COMPETITION, COLLUSION, AND GAME THEORY. Foreword by H.G. Johnson. Aldine Treatises in Modern Economics. Chicago: Aldine-Atherton, 1972. xix, 380 p. Appendixes, pp. 357-65. References, pp. 367-70. Name Index, p. 371. Subject Index, pp. 372-80.

Presents a new approach to the theory of competition and market structure called the theory of the core of the market. It is based on game theory as developed by von Neumann and Morgenstern and represents a shift in emphasis away from oligopoly and advertising and toward the influence on the outcome of market processes of such factors as number of traders on the two sides of the market, transactions costs, brokerage, the ways in which firms form their expectations of future demand, and the costs of collusion. The eight chapters fall into three parts: chapters 1 to 3 present the theory of the core; chapters 4 to 6 stress oligopoly; and chapters 7 and 8 relate industry structure to rates of return and other variables. Knowledge of advanced calculus, differential equations, and the theory of functions is helpful for some of the chapters.

Ventzel, Elena S. LECTURES ON GAME THEORY. International Monograph Series. Delhi, India: Hindustan Publishing Co., 1961. 67 p. No index.

An exposition of the elements of the theory of games and methods of solving matrix games. It contains no proofs, but illustrates with examples the main principles of the theory. Topics include pure and mixed strategies, general and approximate methods of the solution of games, and methods of solution of a few infinite games. Assumes knowledge of elementary probability mathematical analysis.

von Neumann, John, and Morgenstern, Oskar. THEORY OF GAMES AND ECONOMIC BEHAVIOR. 3d ed. Princeton, N.J.: Princeton University Press, 1953. 641 p. Appendix: The Axiomatic Treatment of Utility, pp. 617-32. Index of Figures, p. 633. Index of Names, p. 634. Index of Subjects, pp. 635-41.

A classic treatise on the theory of games as applied to economic analysis. It develops the mathematics and formulates the economic problems of the zero-sum two-person game, zero-sum three-person game, zero-sum n-person game, zero-sum four-person game, and non-zero-sum games.

Williams, John Davis. THE COMPLEAT STRATEGYST: BEING A PRIMER ON THE THEORY OF GAMES OF STRATEGY. With pictorial illustrations by C. Satterfield. The Rand Series. New York: McGraw-Hill Book Co., 1966. Appendix: Table of Random Digits, pp. 253-59. Solutions to Exercises, pp. 261-66. Index, pp. 267-68.

A primer on game theory for home study with arithmetic as the only prerequisite. It covers two-, three-, and four-strategy games, and an introduction to linear programming. Exercises follow the sections of the book.

Chapter 9

INVENTORY THEORY

Alfandary-Alexander, Mark. AN INQUIRY INTO SOME MODELS OF IN-
VENTORY SYSTEMS. Pittsburgh: University of Pittsburgh Press, 1962. vii,
108 p. Appendix, pp. 57-105. Bibliography, pp. 106-8. No index.

A development of four dynamic inventory systems using approaches
based on operational gaming, response function determination, and
symbolic logic. The systems are: multibin inventory, the (s,S)
system, base-stock level system, and fixed-order systems. The
optimal conditions for each policy function are determined, compared, and certain conclusions are drawn. The appendixes contain graphs and charts which permit the manager to choose the
optimal parameters of a policy by inspection. Very little mathematics is used and only a few formulas presented require knowledge of calculus.

Arrow, Kenneth J., et al. STUDIES IN THE MATHEMATICAL THEORY OF
INVENTORY AND PRODUCTION. Stanford Mathematical Studies in the
Social Sciences. Stanford, Calif.: Stanford University Press, 1958. x,
340 p. Bibliography of Inventory Theory, pp. 337-40. No index.

A collection of seventeen heretofore unpublished papers on mathematical and conceptual problems in the analysis of business decisions about inventories and production. The papers are grouped
under the following headings: introduction (three), optimal policies in deterministic inventory processes (four), optimal policies
in stochastic inventory processes (six), and operating characteristics of inventory policies (four). Most of the research upon which
papers are based was done at Stanford University during the period
1955-58 with the support of the Office of Naval Research. Assumes knowledge of differential and integral calculus and probability theory.

Barlow, Richard E., and Proschan, Frank. With contributions by L.C. Hunter.
MATHEMATICAL THEORY OF RELIABILITY. The SIAM Series in Applied Mathematics. New York: Wiley, 1965. xiii, 256 p. Appendix 1: Total Positivity,

pp. 227-31. Appendix 2: Test for Increasing Failure Rate, pp. 232-35. Appendix 3: Tables Giving Bounds on Distributions with Monotone Failure Rate, pp. 236-40. References, pp. 241-48. Index, pp. 249-56.

A monograph for mathematicians, scientists, and engineers on selected probabilistic models useful in solving reliability problems such as predicting, estimating, or optimizing the probability of survival, mean life, or, more generally, life distributions of components or systems. Topics include failure distributions, operating characteristics of maintenance policies, stochastic models for complex systems, redundancy optimization, and qualitative relationships for multicomponent structures. Statistical problems, such as parameter estimation, are not treated. Minimal assumptions are made so that the resulting mathematical deductions may be made about a large variety of commonly occurring reliability problems. Assumes knowledge of probability theory and the concept of total positivity.

Belsley, David A. INDUSTRY PRODUCTION BEHAVIOR: THE ORDER-STOCK DISTRIBUTION. Contributions to Economic Analysis, no. 62. New York: American Elsevier, 1969. xiii, 233 p. Appendix A: Sales Anticipations, pp. 159-64. Appendix B: The Time Horizon, pp. 165-67. Appendix C: Correction of Degrees of Freedom for Seasonal Adjustment, pp. 168-71. Appendix D: The Basic Regression Output, pp. 172-225. Bibliography, pp. 226-29. Index, pp. 230-33.

Highlights the distinction between production to stock and production to order in the investment behavior of the firm or industry. Models to explain production behavior for firms and industries producing exclusively to order and exclusively to stock, as well as for both order and stock, are developed and tested using monthly data from the Manufacturers' Shipments, Inventories, and Orders Series compiled by the Industry Division of the Bureau of Census, U.S. Department of Commerce.

Brown, Robert Goodell. DECISION RULES FOR INVENTORY MANAGEMENT. New York: Holt, Rinehart and Winston, 1967. x, 398 p. Index, pp. 393-98.

A practical application of the economic order quantity concept (EOQ) to the problem of inventory control in the Warmdot Company of Chicago. Topics include demand and forecast of sales, quantity discounts, shipping costs, shop scheduling, and other factors in inventory control. Sections of the book conclude with references. Assumes knowledge of college algebra.

Buffa, Elwood Spencer, and Taubert, William H. PRODUCTION-INVENTORY SYSTEMS: PLANNING AND CONTROL. Rev. ed. Homewood, Ill.: Richard D. Irwin, 1972. xiv, 616 p. Appendixes: A, Fourier Series Forecasting Program; B, Model and Dynamo Program for Repair Parts Supply System (RPSS); C, Adoptive Pattern Search Computer Programs, pp. 553-608. Index, pp. 609-16.

A text for a course in production control which is divided into the following seven parts: introduction; inventories; aggregate planning, concepts, and methods; planning and scheduling for high-volume standardized products; planning and scheduling for job shop systems; planning and scheduling for large-scale projects; and summary. Several techniques of operations research, such as the classical inventory models, linear programming, and network analysis, are discussed. The mathematics prerequisite is college algebra. Chapters conclude with questions, problems, and references.

Childs, Gerald L. UNFILLED ORDERS AND INVENTORIES: A STRUCTURAL ANALYSIS. Contributions to Economic Analysis, vol. 49. New York: American Elsevier, 1967. xiii, 142 p. Index, pp. 140-42.

An investigation of the role of orders as a determinant of production decisions using quadratic cost-linear decision models. The first half of the book is devoted to the derivation of decision rules and the remaining half to the estimation of decision rules by regression techniques in metals, fabricated metal products, nonelectrical machinery, electrical machinery, and the durable and nondurable aggregates. The empirical analysis utilizes posterior data subject to prior seasonal adjustment by standard Commerce Department procedures. Knowledge of calculus, probability, matrix algebra, and the Z transform is required.

Conway, Richard W.; Maxwell, William L.; and Miller, Louis W. THEORY OF SCHEDULING. Reading, Mass.: Addison-Wesley, 1967. x, 294 p. Bibliography, pp. 249-58. Appendixes, pp. 259-90. Index, pp. 291-94.

An attempt to define a theory of scheduling. It is organized according to the type of scheduling problem rather than the technique of solution. Topics include finite sequencing for a single machine, the general n/m and continuous-process job-shop problem, and single- and multiple-server queueing model. The appendixes contain experimental results. Assumes knowledge of advanced probability, and many theorems are stated and proved.

Cox, David Roxbee. RENEWAL THEORY. Methuen's Monographs on Applied Probability and Statistics. New York: Wiley, 1962. ix, 142 p. Appendix 1: Bibliographical Notes, pp. 125-27. Appendix 2: Exercises and Further Results, pp. 128-34. Appendix 3: References, pp. 135-38. Index, pp. 139-42.

A monograph on renewal theory with emphasis on formulas that can be used to answer specific problems rather than on proofs of theorems. Topics include the distribution of moments of the number of renewals, recurrence-times, superposition of renewal processes, alternating renewal processes, cumulative processes, probabilistic models of failure, and strategies of replacement. Assumes familiarity with calculus, probability theory, and the elementary properties of the Laplace transform. Useful for students and research workers in statistics, probability, and operations research whose work involves the application of probability.

Enrick, Norbert Lloyd. INVENTORY MANAGEMENT: INSTALLATION, OPERATION, AND CONTROL. QUANTITATIVE ASPECTS OF MANAGEMENT. San Francisco: Chandler Publishing Co., 1968. xviii, 121 p. Appendix: Case Problems and Assignments, pp. 99-118. Index, pp. 119-21.

Presents quantitative methods of analysis for various types of inventories, from incoming materials through final product and salable merchandise. Optimum inventory policies are developed for problems, such as determining proper lot sizes, providing for the right amount of safety or reserve stock, utilizing inventories to smooth out production under seasonal and other fluctuations, balancing risk factors in inventory decisions, and using price as a mechanism for inventory control. Use is made of intuitively understandable discussion and practically illustrated demonstrations, rather than mathematical derivations. Assumes knowledge of high school algebra.

Fabrycky, Wolter J., and Banks, Jerry. PROCUREMENT AND INVENTORY SYSTEMS: THEORY AND ANALYSIS. Reinhold Industrial Engineering and Management Sciences Series. New York: Reinhold Publishing Co., 1967. xii, 239 p. Appendix A: Progress Function Tables, pp. 215-18. Appendix B: Defining the Progress Function, pp. 219-24. Appendix C: Sensitivity Analysis, pp. 225-30. Appendix D: Some "Make or Buy" Decision Rules, pp. 231-36. Appendix E: Selected References, pp. 237-39. No index.

A text for graduates and undergraduates and a reference on the theory, structure, and decision-making procedures of procurement and inventory systems. The systems presented are: multi-item, single-source; single-item, multi-source; single-item, single-source; and multi-item, multi-source. The decision models are formulated for both deterministic and probabilistic situations. These models are based on calculus, linear programming, direct enumeration, Lagrangian multipliers, and dynamic programming. Numerous examples are used to illustrate the theories. Each chapter concludes with one dozen to two dozen problems oriented toward applications.

Fetter, Robert B., and Dalleck, Winston C. DECISION MODELS FOR INVENTORY MANAGEMENT. Irwin Series in Quantitative Analysis for Business. Homewood, Ill.: Richard D. Irwin, 1961. x, 123 p. Appendixes, pp. 99-118. Index, pp. 121-23.

A monograph which provides a guide to the theory of inventory problems which will lead to the development of ordering rules for effective inventory control. It contains three chapters and twelve appendixes. Chapter 1 develops graphical and mathematical models, which can be used to determine optimal inventory policy and the associated ordering rules. Chapter 2 is devoted to a discussion of the data requirements for the inventory models. Chapter 3 presents some simple illustrations of the numerical computations involved in using the models, followed by a complete case study demonstrating the application of the inventory material in a realistic situation. The appendixes contain graphs, tables, and formulas that are useful in the different stages of analyzing an inventory problem. Assumes knowledge of differential and integral calculus and probability theory.

Gertsbakh, Il'ía Borukhovich, and Kordonskiy, Kh. B. MODELS OF FAILURE. Translated by Scripta Technica. New York: Springer-Verlag, 1969. iv, 166 p. Appendix, pp. 157–61. Bibliography, pp. 162–66. No index.

Originally published in Russian in 1966, this book is an exposition of mathematical models of failure. It begins with a classification of failure of units, and then applies exponential, gamma, normal, logarithmic normal, and Weibull-Gnedenko probability distributions to the lifetime of units in the study of their failure. The estimation of the parameters of the theoretical distributions from empirical data is also taken up. The appendix contains a normal-curve table and table of logarithms. Assumes knowledge of frequency distributions of mathematical statistics.

Gnedenko, Boris V.; Belyayev, Yu. K.; and Solovyev, A.D. MATHEMATICAL METHODS OF RELIABILITY THEORY. Translated by Scripta Technica. Probability and Mathematical Statistics Series. New York: Academic Press, 1969. xi, 506 p. Appendix, pp. 454–99. Subject Index, pp. 501–6.

Translated from the Russian, this book is for engineers, mathematicians, economists, and industrial managers who have extensive preparation in mathematics and probability theory. Topics include testing reliability hypotheses, standby redundancy with and without renewal, statistical methods of quality control, and evaluation of reliability factors from experimental data. Graphs are used freely to illustrate concepts. Large numbers of tables appear in the appendix, many of which were prepared especially for this book. Chapters conclude with references.

Greene, James H. PRODUCTION AND INVENTORY CONTROL: SYSTEMS AND DECISIONS. Rev. ed. The Irwin Series in Management and the Behavioral Sciences. Homewood, Ill.: Richard D. Irwin, 1974. xv, 714 p. References, pp. 681–703. Index, pp. 707–14.

A text for undergraduate business and engineering students, and also for students in two-year technology and community colleges. Useful also by industrial personnel for solving day-to-day production and inventory problems. It is divided into six sections. Section 1 deals with production systems with emphasis on logic and on prepackaged software and on-line terminals. Section 2 develops all of the input information needed for a production and inventory control system. Section 3 is devoted to inventory control systems, including accounting and record keeping. Section 4 is devoted to decision making for production control, including techniques such as PERT and CPM. Section 5 deals with the design of production control systems, ranging from basic manual systems to complicated computerized systems. Section 6 is on the management of production and inventory systems. The main prerequisite is college algebra, but differential calculus is also used occasionally. Chapters conclude with summaries and questions.

————, ed. .PRODUCTION AND INVENTORY CONTROL HANDBOOK. Foreword by R.W. Van Cott. New York: McGraw-Hill Book Co., 1970. xx, irregularly paged. Index.

A reference on production and inventory control prepared under the supervision of the Handbook Editorial Board of the American Production and Inventory Control Society. It consists of thirty chapters by ninety contributors, all members of the society. The chapters are arranged in the following sections: organization for production and control, supporting systems for production and inventory control, planning for production control, production control operations, inventory control, systems for production and inventory control, techniques and tools for production and inventory control, and APICS-factory report. Each chapter will lead the reader from the basic concepts to the present state of the art and beyond.

Hadley, George F., and Whitin, T.M. ANALYSIS OF INVENTORY SYSTEMS. Prentice-Hall Quantitative Methods Series. Englewood Cliffs, N.J.: Prentice-Hall, 1963. xi, 452 p. Appendixes, pp. 433-47. Index, pp. 449-52.

A text for a one-semester course in the techniques of constructing and analyzing mathematical models of inventory systems. It may also be used for a course (or part of a course) in inventory theory, production, or operations research. The material is concerned almost exclusively with the determination of optimal operating procedures for systems consisting of a single stocking point and a single source of supply. The mathematical results are described in great detail and a large number of examples is provided. A unique feature of the book is the large number of problems at the ends of the chapters (two to three dozen per chapter). Assumes knowledge of differential and integral calculus and probability. The appendixes are devoted to Lagrange's multipliers, Newton's method, and tables of mathematical formulas.

Hanssmann, Fred. OPERATIONS RESEARCH IN PRODUCTION AND INVENTORY CONTROL. Huntington, N.Y.: Krieger, 1962. xii, 254 p. Mathematical Appendix, pp. 213-26. An Illustrative Case Study, pp. 227-35. Bibliography of Inventory Theory and Its Applications, pp. 237-49. Index, pp. 251-54.

A text for a two-semester course and a reference for the practitioner with emphasis on applications. It is divided into the following five parts: introduction; the single station, including determinants and probabilistic inventory processes, and dynamic inventory models; parallel stations; series of stations; and appendixes. The prerequisites include calculus, algebra, and probability, and knowledge of queueing theory and linear programming is helpful. Most of the mathematical concepts used in the text have been compiled without proof in the mathematical appendix. Chapters conclude with bibliographies.

Hillier, Frederick S. THE EVALUATION OF RISKY INTERRELATED INVESTMENTS. BUDGETING INTERRELATED ACTIVITIES-I. Studies in Mathematical and Managerial Economics, vol. 9. New York: American Elsevier, 1969. xi, 113 p. Appendix: Suggestions on Implementation, p. 87. Bibliography, pp. 101-10. Index, pp. 111-13.

An exploration of ways of developing systematic procedures for evaluating a set of large, risky, interrelated investments, using present value and expected utility of present value as criteria. A model is developed for determining the mean, variance, and the functional form of the probability distribution of present value. Other models are formulated for finding the expected utility of approving any particular combination of investments. Then a linear programming approach and an exact branch-and-bound algorithm is used for selecting the investments to be made. In addition, several chance-constrained programming models are formulated for maximizing expected present value subject to probabilistic constraints on the allowable risks. The presentation is mathematical, and requires knowledge of calculus and probability theory.

Hirsch, Albert A., and Lovell, Michael C. SALES ANTICIPATIONS AND INVENTORY BEHAVIOR. New York: Wiley, 1969. xiv, 256 p. Bibliography, pp. 245-51. Author Index, pp. 253-54. Subject Index, pp. 255-56.

An analysis of data provided by the Manufacturers' Inventory and Sales Expectations Survey of the Office of Business Economics, U.S. Department of Commerce, to determine the ways in which businesses form their expectations, the precision of their prediction, and the impact of forecasting errors, structure of expectations, and inventories and production. Assumes knowledge of statistical inference and regression analysis.

Holt, Charles C., et al. PLANNING PRODUCTION, INVENTORIES, AND WORK FORCE. Englewood Cliffs, N.J.: Prentice-Hall, 1960. xii, 419 p. Index, pp. 405-19.

A research report on the application of operations research in the operation of a factory warehouse system. Although the methods are developed in the context of a factory supplying a warehouse, they are applicable to decision problems of military and other governmental organizations. It is divided into the following five parts: overview for managers; decision rules for planning aggregate production and work force; order, shipment, production and purchase of individual products; design of decision systems; and generalization of decision methods. Exposure to elementary theory of system analysis and the techniques of operations research, such as dynamic programming and inventory theory, is desirable.

Jorgenson, Dale Weldeau; McCall, John Joseph; and Radner, Roy. OPTIMAL REPLACEMENT POLICY. Studies in Mathematical and Managerial Economics, vol. 8. New York: American Elsevier, 1967. xii, 225 p. Bibliography, pp. 215-19. Author Index, pp. 220-21. Subject Index, pp. 222-25.

A unified view of maintenance theory and its applications for stochastically failing equipment. Written on several levels of rigor in order to reach the largest possible audience--applied mathematicians, management scientists, operations research, and maintenance managers. Chapters 1 and 2 deal with the role of maintenance theory within economic theory and present a survey of the literature on stochastic maintenance policies; chapter 3 treats the theory of optimal replacement for stochastic problems. It provides proofs of theorems and is intended for the applied mathematician. Chapters 4 to 6 are concerned with the structure of optimal maintenance policy and are intended for the engineer and operations research analysts who have knowledge of calculus and probability theory. Chapters 7 to 9 describe applications of optimal replacement policies to military systems and can be followed by those with little mathematics background.

Killeen, Louis M. TECHNIQUES OF INVENTORY MANAGEMENT. New York: American Management Association, 1969. 175 p. Appendixes, pp. 137-60. Supplementary Reading, pp. 161-62. Index, pp. 165-75.

An introduction to the economic order quantity (EOQ) policy of inventory management under different assumptions such as fixed quantity/variable period and variable quantity/fixed period systems. It also considers other topics, such as safety stock, forecasting demand, the distribution system for finished goods, and inventory reduction techniques. It is written for the practitioner, and requires knowledge of high school algebra and basic statistics. The appendixes contain tables of the normal and Poisson distributions, square roots, and a review of the least squares procedure.

Lewis, Colin David. DEMAND ANALYSIS AND INVENTORY CONTROL. Lexington, Mass.: D.C. Heath, 1975. xv, 234 p. List of Figures, pp. vii-ix. List of Tables, pp. ix-xi. Appendix, pp. 219-24. Reference List, pp. 225-27. Bibliography, p. 229. Index, pp. 231-34.

Presents the main mathematical theories in demand analysis and inventory control with applications in day-to-day control of inventories. Useful to practitioners in operations research and management science. Can be used as a text for a course for practitioners extending over three days or ten evening sessions. Topics include short-term forecasting techniques in stationary and nonstationary demand conditions, adaptive forecasting and filtering, implementary forecasting techniques, stochastic reorder level and cyclical inventory models, inventory policies for situations where demand is known in advance, simulation of inventory problems, and practical problems of implementing inventory control. The techniques are presented mathematically and also illustrated with graphs and numerical examples. Assumes knowledge of basic statistics and elementary inventory analysis. Chapters conclude with exercises.

_____. SCIENTIFIC INVENTORY CONTROL. Operational Research Series. New York: American Elsevier, 1970. vii, 209 p. Appendix A: Summary of Equations, pp. 194-201. Appendix B: The Normal Distribution, p. 202. Appendix C: The Cumulative Poisson Distribution, p. 204. Index, pp. 205-9.

A reference on probabilistic inventory models with a minimum of mathematical and statistical theory. Topics include separate and joint calculation of reorder levels and replenishment order quantities, cyclical policies, inventory queues, multiproduct inventory systems, simulation of inventory situations, and inventory control problems. A complete glossary and collection of equations are provided in the appendixes. Assumes knowledge of elementary calculus and probability theory. Chapters conclude with problems and references.

Lloyd, David K., and Lipow, Myron. RELIABILITY: MANAGEMENT, METHODS, AND MATHEMATICS. Prentice-Hall Space Technology Series. Englewood Cliffs, N.J.: Prentice-Hall, 1962. xxii, 528 p. Appendix, pp. 485-508. Index of Authors, pp. 509-11. Subject Index, pp. 512-28.

Presents the concepts and methodology of reliability for a mixed group, including engineers, managers, and mathematicians. It is divided into three sections as follows: Section 1: management, organization, and communication; section 2: the mathematics of reliability; and section 3: examples of reliability evolution and demonstration programs. The emphasis is on probability theory and statistics, and knowledge of mathematical statistics is assumed. Applications are in space industry, such as solid propellant rocket engines.

Magee, John F., and Boodman, David M. PRODUCTION PLANNING AND INVENTORY CONTROL. 2d ed. New York: McGraw-Hill Book Co., 1968. x, 397 p. Appendix A: Derivation of Economical Order-quantity Formulas, pp. 349-59. Appendix B: Production Control Rules, pp. 360-64. Appendix C: Techniques of Seasonal Planning, pp. 365-73. Glossary, pp. 375-93. Index, pp. 395-97.

Designed to introduce operating executives, students of business, and engineers of planning and control systems to the concepts and methods of inventory control and costs. Mathematical formulation of concepts and techniques are de-emphasized, while extensive use is made of graphs and numerical examples. Chapters conclude with problems and discussion topics, and references.

Moran, Patrick Alfred Pierce. THE THEORY OF STORAGE. Methuen's Monographs on Applied Probability and Statistics. New York: Wiley, 1959. 111 p. Bibliography, pp. 106-10. Index, p. 111.

Describes probability problems which arise in the theory of storage. The problems are grouped under the following types: (1) the inventory problem with a random demand and a rule of restocking; and (2) queueing problems with a random input and a rule of release. About one-half of the book is devoted to the theory of dams in which inputs occur at discrete points of time and are independent, or occur continuously in the form of an additive homogeneous process. Monte Carlo simulation is described as a technique for constructing a theory of dams, under more complicated situations, such as when inputs are not independent over time. The mathematical results are either stated without proof, or intuitive proofs are provided. Assumes knowledge of the calculus of probability and calculus.

Moriguchi, Chikashi. BUSINESS CYCLES AND MANUFACTURERS' SHORT-TERM PRODUCTION DECISIONS. Contributions to Economic Analysis, no. 52. New York: American Elsevier, 1967. 152 p. Appendix: Notes on Autocorrelation Bias in Least Squares Estimates, pp. 127-45. Bibliography, pp. 147-49. Index, pp. 151-52.

An analysis of nonlinear relationships between sales fluctuations and production decisions by means of standard least squares regression techniques with extensive use of dummy variables. One set of dummy variables allows the demand stock-sales ratio to depend upon the season of the year. Additional dummy variables make the speed with which firms plan to adjust inventory toward the equilibrium level depend upon whether capacity is fully or partially utilized. The empirical applications are in the cement, paper, and lumber industries for the postwar period, and much attention is given to the nature of autocorrelation bias. Assumes knowledge of econometrics.

Naddor, Eliezer. INVENTORY SYSTEMS. New York: Wiley, 1966. xiv, 341 p. References and Bibliography, pp. 329-32. Index, pp. 333-41.

A text for a one-semester course in inventory systems for graduates and undergraduates, and a reference for practitioners engaged in solving inventory problems. Useful for a mixed group, including engineers, operations research analysts, economists, business students,

and others, as well as departments in industry handling scheduling,
inventories, market research, forecasting, etc. It is divided into
four parts. Part 1, introduction, surveys properties of inventory
systems and requires knowledge of algebra. Part 2, deterministic
systems, analyzes inventory systems in which demand is known
with certainty. In part 3, probabilistic scheduling-period systems,
the probability distribution of demand is known and replenishment
occurs at equal intervals. Part 4, probabilistic reorderpoint systems,
treats cases where replenishments occur when the reorder point is
reached. Part 2 requires knowledge of calculus, and parts 3 and
4 assume knowledge of mathematical statistics and Markov chains.
Contains numerous examples, and chapters conclude with problems.

Neuschel, Richard F., and Johnson, H. Tallman. HOW TO TAKE PHYSICAL
INVENTORY. New York: McGraw-Hill Book Co., 1946. vii, 159 p. Index, pp. 155-59.

A practical business guide on the various activities that are necessary in taking an inventory in a firm.

Niland, Powell. PRODUCTION PLANNING, SCHEDULING, AND INVEN-
TORY CONTROL: A TEXT AND CASES. New York: Macmillan, 1970.
xiv, 553 p. Appendixes A to E, pp. 491-541. Index, pp. 543-53.

A text for graduates in business who have knowledge of algebra,
basic statistics, and probability. It is divided into two parts:
text and cases. The text considers process organization, finished
goods inventories, and production control; planning and scheduling;
forecasting; production scheduling with linear programming; and
elementary inventory control. The cases are scheduling and work
in process; inventory control; and planning, scheduling, and inventory control. The appendixes are devoted to network analysis,
linear programming, simulation models, and background information
on Europe and Korea. Chapters conclude with questions and bibliographies.

Prichard, James W., and Eagle, Robert H. MODERN INVENTORY MANAGE-
MENT. New York: Wiley, 1965. xii, 419 p. Tables, pp. 381-90. Glossary, pp. 391-96. Development of Order Quantity, Review Cycle, and Reorder Point Formulas, pp. 397-405. Answers to Exercises, pp. 406-13. Index,
pp. 415-19.

A book for individuals with little mathematics background who are
entering the field of inventory management. Chapters conclude
with a few practical problems.

Proschan, Frank. POLYA TYPE DISTRIBUTIONS IN RENEWAL THEORY: WITH
AN APPLICATION TO AN INVENTORY PROBLEM. Englewood Cliffs, N.J.:
Prentice-Hall, 1960. x, 36 p. Bibliography, p. 36. No index.

A thesis on the use of Polya-type distributions arising in renewal theory.

Reisman, Arnold; Dean, Burton V[ictor].; Salvador, Michael S.; and Oral, Muhittih. INDUSTRIAL INVENTORY CONTROL. Studies in Operations Research. New York: Gordon and Breach, 1972. xi, 180 p. Appendix A: Probability Theory, pp. 127-42. Appendix B: Theoretical Development of the Field of Warehouse Replenishment Model, pp. 143-56. Glossary of Terms, pp. 157-67. Index, pp. 169-80.

> Describes a major industrial inventory control study, from its for-
> mulation to completion and implementation, of an anonymous com-
> pany. It provides a systems description of the company; perfor-
> mance evaluation criteria used in the firm's inventory decision
> rules are discussed; and the decision models and policies are de-
> veloped. The relevant cost parameters are described, and com-
> puter simulation is used to evaluate and compare performance in
> terms of costs and service for three sets of decision rules. Dis-
> cusses the methods used to establish the costs and the probabilities
> associated with each alternative. Intended mainly for students
> and practitioners in inventory control and logistics management.
> Knowledge of college algebra and probability theory is assumed.
> Chapters conclude with references.

Roberts, Norman H. MATHEMATICAL METHODS IN RELIABILITY ENGINEER-
ING. New York: McGraw-Hill, 1964. xiii, 300 p. Appendixes, pp. 249-95.
Index, pp. 297-300.

> A text for a course in reliability engineering for students who have
> the mathematical maturity comparable to that obtained in a stan-
> dard freshman course in mathematical analysis. The necessary
> statistics and probability is reviewed in the beginning chapters.
> The seven appendixes consist of reviews of certain mathematical
> concepts, such as inequalities and the Laplace transform, statis-
> tical tables, and a selected bibliography. Chapters conclude
> with problems.

Starr, Martin K., and Miller, David W. INVENTORY CONTROL: THEORY
AND PRACTICE. Englewood Cliffs, N.J.: Prentice-Hall, 1962. ix, 354 p.
Appendix, pp. 331-36. Tables, pp. 337-39. Glossary, pp. 340-43. Bibliog-
raphy, pp. 344-46. Index, pp. 347-54.

> An intermediate level text for students with a background in cal-
> culus and elementary probability theory. It is divided into two
> parts. Part 1, inventory theory, covers both static and dynamic
> inventory models under risk and uncertainty; part 2, implemen-
> tation phase: the theory of practice, covers methods for the anal-
> ysis of the control aspects of different kinds of inventory systems
> and methods for estimating the advantages and disadvantages of
> various systems. The appendix is devoted to rigorous derivations,
> such as convolutions of demand distributions. Chapters conclude
> with problems.

Tijms, H.C. ANALYSIS OF (s,S) INVENTORY MODELS. Mathematical Centre Tracts, 40. Amsterdam: Mathematical Centre, 1972. i, 149 p. References, pp. 145–49. No index.

> A presentation of periodic review and continuous review inventory models with a single product and a single stocking point. Emphasis is given to optimality of (s,S) policies in the infinite period inventory model and with the determination of a number of characteristics for certain (s,S) inventory systems. Renewal theory is widely used in the analysis. Assumes knowledge of probability theory and mathematical analysis. The presentation is rigorous from a mathematical point of view, with proofs of theorems provided.

Van Hees, R.N.; Monhemius, A.; with Muyen, A.R.F. AN INTRODUCTION TO PRODUCTION AND INVENTORY CONTROL. Philips Technical Library. New York: Macmillan, 1972. xiv, 146 p. Notation, pp. xiii–xiv. No index.

> An English edition of a Dutch book first published in 1964. It is an introduction to production planning and management of stocks, using numerous formulas. Section 1 takes up the essence of planning and the irregularities in consumption of a product and the reasons why production and consumption processes differ from each other. Section 2 investigates the behavior of stocks and waiting times, the existing system of stock management, methods of calculating the average quantity of stock, and chains of stocks. In section 3 a practical example of the stock management of electronic components is discussed.

_____. PRODUCTION AND INVENTORY CONTROL: THEORY AND PRACTICE. Philips Technical Library. New York: Macmillan, 1972. xvi, 370 p. Appendixes, pp. 351–62. Bibliography, pp. 363–66. Index, pp. 367–70.

> An English edition of a Dutch book first published in 1969 and a continuation of AN INTRODUCTION TO PRODUCTION AND INVENTORY CONTROL (see above). It is intended for specialists in production management and stock control who are charged with the supervision or execution of practical problems. It is divided into the following eight sections: background; statistical inventory control; forecasting (plans and prediction); calculations associated with reorder systems and production batches: intermittent supply; calculation of production level (continuous supply); calculation of capacity; some examples; and approach to a problem. Elementary calculus is necessary to understand parts of the book. The appendixes consist of probability tables and a mathematical derivation of the optimum batch, with due regard to risk of obsolescence.

Wagner, Harvey M. STATISTICAL MANAGEMENT OF INVENTORY SYSTEMS. Publications in Operations Research, no. 6. New York: Wiley, 1962. xiv, 235 p. References, pp. 229-30. Index, pp. 231-35.

A research study (not a text) on the mathematical approach to problems of management control of complex inventory systems, using methods based on statistical aggregates and indexes. The mathematical analysis is limited to propositions capable of numerical calculation, and numerous examples are given to illustrate the computational schemes. The presentation is based on a class of (s, S) policies which utilize stationary probability analysis. Assumes knowledge of the basic concepts of inventory theory, probability, mathematical statistics, and sequential decision theory.

Chapter 10

MATHEMATICAL PROGRAMMING

Adams, William J.; Gewirtz, Allan; and Quintas, Louis V. ELEMENTS OF
LINEAR PROGRAMMING. New York: Van Nostrand Reinhold, 1969. iv,
186 p. Answers to Selected Exercises, pp. 173-81. Index, pp. 183-86.

> An exposition of linear programming and two-person zero-sum games
> for students with minimal mathematics background. It can be used
> as a text in business administration, social sciences, and education.
> Numerous examples are used to illustrate the concepts, and the
> chapters conclude with exercises.

Angel, Edward, and Bellman, Richard E. DYNAMIC PROGRAMMING AND
PARTIAL DIFFERENTIAL EQUATIONS. Mathematics in Science and Engineering, vol. 88. New York: Academic Press, 1972. xi, 204 p. Appendix:
Computer Programs, pp. 182-200. Author Index, pp. 201-2. Subject Index,
pp. 203-4.

> An attempt to show that dynamic programming and invariant im-
> bedding are useful techniques for the solution of linear elliptic
> and parabolic partial differential equations over regular and ir-
> regular regions. It presents a number of algorithms for obtaining
> numerical solutions, and the appendix contains four sample com-
> puter programs. Chapters conclude with exercises and references.

Aris, Rutherford. DISCRETE DYNAMIC PROGRAMMING: AN INTRODUC-
TION TO THE OPTIMIZATION OF STAGED PROCESSES. New York: Blaisdell,
1964. x, 148 p. Index, pp. 146-48.

> An introduction to the method of optimization by dynamic program-
> ming as it applies to a process with discrete stages for the under-
> graduate or interested layman who has knowledge of elementary
> algebra and differential calculus. The latter half of the book
> contains applications in economics, communications and information
> theory, reliability theory, growth and production, and Jacobi
> matrices. The chapters conclude with references and problems.

Barsov, A.S. WHAT IS LINEAR PROGRAMMING? Topics in Mathematics Series. Lexington, Mass.: D.C. Heath, 1964. vi, 110 p. Bibliography, p. 110. No index.

> An elementary exposition of the general problem dealt with by linear programming and a discussion of the simplex procedure and the combinatorial method for solving it. The transportation problem is treated thoroughly, and the use of digital computers for solving linear programming problems is treated. Useful as an introductory text in linear programming for students of mathematics, engineering, economics, and business.

Beale, Evelyn Martin Landsdowne. MATHEMATICAL PROGRAMMING IN PRACTICE. Topics in Operational Research Series. London: Sir Isaac Pitman, 1968. xi, 195 p. Appendix: Note on the LP/90/94 system, pp. 184–88. References, pp. 189–91. Author Index, p. 193. Subject Index, pp. 194–95.

> A brief assessment of the actual and potential fields of application of mathematical programming and the economics of these applications. Part 1 covers the conventional material on linear programming, including the simplex and inverse matrix methods. Part 2 is on the organization of linear programming calculations. Part 3 considers special procedures, such as Lagrange multipliers and the Kuhn-Tucker conditions; quadratic, separable, integer, and stochastic programming; and the decomposition approach to the solution of large linear programming problems. The appendix describes a computer program for solving linear programming problems with up to 1,125 rows and an arbitrary number of columns on an IBM 7090 or 7094 computer. Assumes knowledge of elementary algebra and matrix notation.

Beckmann, Martin J. DYNAMIC PROGRAMMING OF ECONOMIC DECISIONS. Econometrics and Operations Research, vol. 9. New York: Springer-Verlag, 1968. xii, 143 p. Author Index, pp. 136–37. Subject Index, pp. 138–43.

> An introduction to the basic ideas of dynamic programming followed by applications to economic analysis, operations research, and decisions in general. Topics include discrete and continuous sequences and decision variables; certainty, risk, and uncertainty; and applications to automobile replacement, inventory control, adaptive programming, machine care, and the maximum principle. It omits combinatorial problems, stopping rules, statistical decision theory, and control theory. Assumes knowledge of advanced calculus and probability theory. Chapters conclude with references.

Bellman, Richard E. DYNAMIC PROGRAMMING. Princeton, N.J.: Princeton University Press, for the Rand Corporation, 1957. xxii, 340 p. Index of Applications, p. 338. Name and Subject Index, pp. 339–40.

> An introduction to the mathematical theory of deterministic and stochastic, discrete and continuous, multistage decision processes

for mathematicians, economists, statisticians, engineers, and operations research analysts who have knowledge of advanced calculus or real analysis. The processes are treated by both dynamic programming and the calculus of variations. Topics include existence and uniqueness theorems for dynamic programming processes, the optimal inventory equation, bottlenecks, multistage games, and Markovian decision processes. Proofs of some theorems are included, and each of the eleven chapters concludes with a lengthy list of exercises and research problems, as well as a bibliography and comments.

_____. SOME VISTAS OF MODERN MATHEMATICS: DYNAMIC PROGRAMMING, INVARIANT IMBEDDING, AND THE MATHEMATICAL BIOSCIENCES. Lexington: University of Kentucky Press, 1968. viii, 141 p. Index, pp. 139-41.

An expository presentation based on lectures given by the author at the University of Kentucky in 1966. Assumes knowledge of advanced calculus. Each of the three chapters concludes with references and comments.

Bellman, Richard E., and Dreyfus, Stuart E. APPLIED DYNAMIC PROGRAMMING. Princeton, N.J.: Princeton University Press, 1962. xxii, 363 p. Appendix 1: On a Transcendental Curve, by O. Gross, pp. 337-39. Appendix 2: A New Approach to the Duality Theorem of Mathematical Programming, by S. Dreyfus and M. Freimer, pp. 340-47. Appendix 3: A Computational Technique Based on Successive Approximations in Policy Spaces, by S. Dreyfus, pp. 348-53. Appendix 4: On a New Functional Transform in Analysis: The Maximum Transform, by R. Bellman and W. Karush, pp. 354-56. Appendix 5: The Rand Johnniac Computer, by S. Dreyfus, p. 357. Name Index, pp. 359-61. Subject Index, pp. 362-63.

A development of the field of dynamic programming which attempts to translate problems from several fields into mathematical formulas with the aim of obtaining solutions. The applications are from engineering, economics, industry, and military areas. Topics include one- and multidimensional allocation processes, optimal search techniques, the calculus of variations, optimal trajectories, input-output models, steady-state growth, feedback control processes, Markovian decision processes, and numerical analysis in dynamic programming. Only deterministic models are considered. Assumes knowledge of advanced calculus or real analysis. Each of the chapters concludes with two-page comments and bibliography section.

Bellman, Richard E., and Kalaba, Robert E. DYNAMIC PROGRAMMING AND MODERN CONTROL THEORY. New York: Academic Press, 1965. xi, 112 p. Author Index, pp. 107-8. Subject Index, pp. 109-12.

An introduction to the mathematical theory of processes to assist the reader in developing realistic models that will permit meaningful computations. Topics include multistage decision processes, computational aspects, analytic results in control and communications theory, and adaptive control processes. Assumes knowledge of the calculus of variations and the main tools of operations research. Sections of the book conclude with exercises, and chapters conclude with references and comments.

Berge, Claude, and Ghouila-Houri, A. PROGRAMMING, GAMES AND TRANSPORTATION NETWORKS. Translated by M. Merrington and C. Ramanujacharyulu. New York: Wiley, 1965. ix, 260 p. Index, pp. 247-59. Index of Notation, p. 260.

First published in French in 1962, this book is divided into two parts: general theory of convex programming, by A. Ghouila-Houri, and problems of transportation and of potential, by Claude Berge. Assumes knowledge of differential and integral calculus, and matrix algebra.

Berman, Abraham. CONES, MATRICES AND MATHEMATICAL PROGRAMMING. Lecture Notes in Economics and Mathematical Systems, vol. 79. New York: Springer-Verlag, 1973. v, 96 p. References, pp. 85-94. Glossary of Notations, pp. 95-96.

A presentation of the basic theory of cones with applications to mathematical programming and matrix theory. Knowledge of advanced calculus and matrix algebra is helpful.

Bertele, Umberto, and Brioschi, Francesco. NONSERIAL DYNAMIC PROGRAMMING. New York: Academic Press, 1972. xii, 235 p. Appendix A: A Review of Graph Theory, pp. 219-21. Appendix B: Some Set-theoretical Definitions, pp. 222-23. Appendix C: Combinatorial Aspects in the Solution of Linear Systems by Gaussian Elimination, pp. 224-28. References, pp. 229-32. Subject Index, pp. 233-35.

A presentation of algorithms for the solution of discrete, deterministic optimization problems (constrained and unconstrained, parametric and nonparametric, serial and nonserial) with the following aims: (1) to present a standard mathematical formulation for each class of problems; (2) to define three different classes of dynamic programming procedures; (3) to provide a simple procedure for measuring the computational effort needed for each procedure; and (4) to construct an efficient algorithm. This book is useful to anyone desiring more efficient algorithms for solving dynamic programming problems.

Boot, John C.G. QUADRATIC PROGRAMMING: ALGORITHMS-ANOMALIES-APPLICATIONS. Studies in Mathematical and Managerial Economics, vol. 2. New York: American Elsevier, 1964. xvii, 213 p. Bibliography, pp. 208-10. Index, pp. 211-13.

A self-contained book on a number of algorithms used in the portion of quadratic programming which deals exclusively with the maximization of a quadratic function subject to linear inequalities. The Theil–Van de Panne combinatorial method and the Houthakker capacity procedure are discussed at length. The main application is to the problem of surplus milk in the Netherlands. Assumes knowledge of matrix algebra.

Boudarel, René J.; Delmas, Jacques; and Guichet, Pierre. DYNAMIC PROGRAMMING AND ITS APPLICATION TO OPTIMAL CONTROL. Translated with a foreword by R.N. McDonough. Mathematics in Science and Engineering, vol. 81. New York: Academic Press, 1971. xiv, 252 p. Appendix: Filtering, pp. 233–47. References, pp. 249–50. Index, pp. 251–52.

Originally published in French in 1968, this book presents a complete treatment of optimal control theory for dynamic systems, both linear and nonlinear, discrete and continuous, and deterministic and stochastic, using only simple calculus, matrix notation, and the elements of probability theory. The presentation is based on dynamic programming, rather than the calculus of variations, and emphasis is placed on illustrating concepts by nontrivial-examples problems. It is divided into five parts: discrete deterministic processes, discrete random processes, numerical synthesis of the optimal controller for a linear process, continuous processes, and applications. Assumes knowledge of advanced calculus, matrix algebra, and an introduction to both differential equations and systems theory.

Bracken, Jerome, and McCormick, Garth P. SELECTED APPLICATIONS OF NONLINEAR PROGRAMMING. New York: Wiley, 1968. xii, 110 p. Index, pp. 107–10.

Presents selected applications of nonlinear programming in detail as an aid to those interested in acquiring facility in building mathematical programming models. Each chapter presents a brief summary of the problem, a mathematical formulation of a nonlinear model, and one or two examples, based on both real-world data and hypothetical data. The applications are weapons assignment, bid evaluation, alkylation process optimization, chemical equilibrium, structural optimization, launch vehicle design and costing, parameter estimation and curve fitting, stochastic linear programming, and stratified sampling on several variates. Linear programming problems are not considered. This book is useful as a supplement in courses in mathematical programming and as a reference for practitioners. Assumes knowledge of linear programming. Chapters conclude with references.

Brown, John P. THE ECONOMIC EFFECTS OF FLOODS: INVESTIGATIONS OF A STOCHASTIC MODEL OF RATIONAL INVESTMENT BEHAVIOR IN THE FACE OF FLOODS. Lecture Notes in Economics and Mathematical Systems, vol. 70. New York: Springer-Verlag, 1972. 87 p. References, pp. 76–81.

Appendix: FORTRAN Computer Program, pp. 82-87. No index.

A monograph on the application of dynamic programming to one aspect of the economics of floods, namely the choice of land use by a single landowner. Bayesian statistical analysis is employed in determining the optimal investment policy, and the appendix contains a computer program to simulate investment behavior in a flood plain.

Calman, Robert F. LINEAR PROGRAMMING AND CASH MANAGEMENT/CASH ALPHA. Foreword by P.H. Cootner. Cambridge, Mass.: MIT Press, 1968. xi, 154 p. Appendixes A-J, pp. 76-152. Index, pp. 153-54.

An application of linear programming to optimum decisions in banking based on monthly short-run forecasts. The linear program, CASH ALPHA, incorporates the price of specific services performed, forms of compensation for those services, and company policies. The appendixes contain printouts of programs and examples of company data sheets. Knowledge of computer programming is helpful.

Carrillo-Arronte, Ricardo. AN EMPIRICAL TEST ON INTER-REGIONAL PLANNING: A LINEAR PROGRAMMING MODEL FOR MEXICO. Foreword by J. Tinbergen. Rotterdam University Press Economic Series, vol. 8. Rotterdam: Rotterdam University Press, 1970. xi, 213 p. Statistical Appendix and Annexes, pp. 159-204. Selected Bibliography, p. 205. List of Maps and Tables, pp. 206-9. Subject Index, pp. 210-13.

Four versions of the Netherlands Economic Institute's interregional planning models are applied to the Mexican economy using 1960 census data in an attempt to integrate the regional and the sectoral subdivisions of a development plan. It is divided into the following four parts: theoretical framework of the model; background, data and objectives of the Mexican economy; applications of the models and their solutions; summary, conclusions, and appendixes. The main mathematical tool used is linear programming. The data and results are presented in over 100 tables. Chapters conclude with references.

Carsberg, Bryan V. AN INTRODUCTION TO MATHEMATICAL PROGRAMMING FOR ACCOUNTANTS. London: George Allen and Unwin, 1969. xii, 108 p. Index, pp. 107-8.

A comparison of the mathematical programming approach to the problem of product selection in a firm with some of the traditional approaches of cost accounting. Assumes knowledge of college algebra.

Charnes, Abraham, and Cooper, William Wager. MANAGEMENT MODELS AND INDUSTRIAL APPLICATIONS OF LINEAR PROGRAMMING. VOLUME I. New York: Wiley, 1961. xxiii, 467 p. Glossary of Symbols, pp. xxii-xxiii.

Bibliography, pp. xxv-lvii. Index, pp. 1-4 at end of book.

A text for graduates and undergraduates and a reference on the managerial applications of linear programming based on the author's research experiences. The emphasis is on assembling and evaluating information for decision making by linear programming. By means of examples, interpretations, and suggested points of view, areas of past and potential managerial applications are pointed out. Topics include the stepping-stone method for transportation-type models; an input-output example; simplex method; dual theorem; basic existence theorems and goal programming; Tucker's existence theorem; geometry of the functional; theory and computations for delegation models of activity analysis; horizons and surrogates in a dynamic model for production scheduling; and the Kuhn-Tucker theorem and some applications to storage, inventory, and functional efficiency. The main prerequisite is college algebra; other mathematics such as matrix theory is taken up as needed. There are six chapter appendixes that are devoted to statements and proofs of theorems.

_____. MANAGEMENT MODELS AND INDUSTRIAL APPLICATIONS OF LINEAR PROGRAMMING. VOLUME II. New York: Wiley, 1961. xxi, pp. 469-859. Glossary of Symbols, pp. xx-xxi. Final Appendix, pp. 809-26. Bibliography, pp. 827-59.

Presents general techniques of linear programming, including the modified and dual methods, and the revised simplex code for electronic calculators, transportation models, the double-reverse method of linear programming, dyadic models and subdual methods, networks and models of incidence type, convex and quadratic programming, the relations between the theory of games and linear programming, and extensions and applications of game theory. The final appendix is devoted to linear transformations in expanding or contracting a "unit" volume in R^n. The organization and presentation of volume 2 is similar to that of volume 1.

Chung, An-min. LINEAR PROGRAMMING. Columbus, Ohio: Charles E. Merrill, 1963. xiii, 338 p. Bibliography, pp. 329-33. Index, pp. 335-38.

An introduction to linear programming which assumes only a knowledge of elementary algebra. Topics include the simplex method, duality, and the transportation problem. Advanced topics such as integer and quadratic programming are taken up briefly in the final chapter. Numerous examples are used to illustrate concepts, and the explanation is clear and concise. Each chapter concludes with exercises.

Cooper, Leon, and Steinberg, David. METHODS AND APPLICATIONS OF LINEAR PROGRAMMING. Philadelphia: W.B. Saunders, 1974. ix, 434 p. Index, pp. 431-34.

An intermediate-level text for graduates and undergraduates which deals rigorously with the theory, computational procedures, and selected applications of linear programming. Topics include the simplex method, the dual simplex and primal-dual algorithms, the transportation and assignment problems, network flows, and integer programming: models and algorithms. The computational techniques discussed include the decomposition principle of Dantzig and Wolfe, upper bound constraints, and generalized upper bound techniques. Assumes knowledge of linear algebra.

Cooper, William Wager; Henderson, A.; and Charnes, A[braham]. AN INTRODUCTION TO LINEAR PROGRAMMING. New York: Wiley, 1953. ix, 74 p. No index.

An introduction to linear programming in two parts: introduction, and mathematical development. Most of the mathematics is introduced as needed. This book is one of the early publications in the field.

Daellenbach, Hans G., and Bell, Earl J. USER'S GUIDE TO LINEAR PROGRAMMING. Englewood Cliffs, N.J.: Prentice-Hall, 1970. xii, 226 p. No index.

An attempt to provide the user of linear programming with a background sound enough to enable him to recognize problems that can be solved by this technique, formulate linear programming problems, set them up for computer solution using prepared computer programs, and correctly interpret and evaluate the solution without having first acquired a thorough knowledge of the mathematical theory. The applications include product mix, the diet problem, fuel blending, transportation, and assignment problems. The final chapter is devoted to advanced computational features of linear programming codes and considers problems such as tolerances, errors, changes in constraints and variables, and continuous parameter variation. Assumes knowledge of college algebra. Chapters conclude with exercises and references.

Danø, Sven. LINEAR PROGRAMMING IN INDUSTRY: THEORY AND APPLICATIONS. 3d ed., rev. New York: Springer-Verlag, 1965. viii, 120 p. Appendix, pp. 104-13. Numerical Exercises, pp. 114-15. Bibliography, pp. 116-18. Index, pp. 119-20.

An elementary text for students of economics and operations research who are interested in the theory of production and cost and its practical applications. It employs only simple linear relations, and the appendix contains proofs of theorems and an explanation of the simplex method.

_____. NONLINEAR AND DYNAMIC PROGRAMMING: AN INTRODUCTION. New York: Springer-Verlag, 1975. vii, 164 p. Answers to Exercises, pp. 156-59. References, pp. 160-61. Index, pp. 163-64.

An introductory text of nonlinear and dynamic programming which emphasizes practical applications. It is aimed at students of economics, operations research, and engineering, as well as at managers responsible for planning industrial operations. The mathematical development is kept to a minimum, and it should be understandable to anyone with a background in calculus and linear programming techniques. Topics include nonlinear programming theory, quadratic programming, multistage problem solving by dynamic programming, decision and state variables, infinite-stage problems, and dynamic programming under risk. Numerous examples and applications to problems of production, inventory, and investment planning are provided. Each chapter contains exercises. The principles of forward and backward recursion procedures are described in the appendix (chapter 10).

Dantzig, George B. LINEAR PROGRAMMING AND EXTENSIONS. Princeton, N.J.: Princeton University Press, for the Rand Corporation, 1963. xvi, 625 p. Bibliography, pp. 592-610. Subject Index, pp. 611-18. Name Index, pp. 619-21. Other Rand Books, pp. 623-25.

Presents the theory and solutions of linear inequality systems by simplex algorithms. Topics include proof of simplex method algorithm and duality theorem, the geometry of linear programming, the simplex method using multipliers, variants of the simplex algorithm, the price concept in linear programming, games and linear programs, transportation and network problems, primal-dual method for transportation problems, a decomposition principle for linear programs, convex programming, uncertainty, and discrete variable extremum problems. The book concludes with two chapters on applications to Stigler's diet problem and on the allocation of aircraft to routes under uncertain demand. Assumes knowledge of matrix algebra. Chapters conclude with references and problems.

Day, Richard H. RECURSIVE PROGRAMMING AND PRODUCTION RESPONSE. Contributions to Economic Analysis, vol. 30. Amsterdam: North-Holland, 1963. xiv, 226 p. Appendixes, pp. 149-220. References, pp. 221-26. No index.

The development and testing of a dynamic general interdependence model of agriculture production using recursive programming which incorporates a sequence of optimizing decisions, rather than a single optimizing decision which is characteristic of dynamic programming. It is divided into the following three parts: introduction, recursive programming, and dynamic production model. The model is applied in production response analyses of field crop production in the Mississippi Delta for the period 1940-57. It consists of thirty-seven structural equations, 103 variables, hundreds of parameters, nine output price variables and forty-five input price variables. Estimates are made for acreage and yields. The appendixes are for chapters 3 to 7 and consist of explicit solutions of recursive models, data used for analysis, and the numerical estimates derived from the model.

Dennis, Jack Bonnell. MATHEMATICAL PROGRAMMING AND ELECTRICAL NETWORKS. Foreword by J.A. Stratton. Technology Press Research Monographs. New York: Wiley, 1959. vi, 186 p. Appendixes A-H, pp. 118-80. Bibliography, pp. 181-83. Index, pp. 184-86.

An application of linear programming to electrical networks for solving network flow problems. The eight appendixes are devoted to reviews of mathematics. Assumes knowledge of matrix algebra and calculus. This book is a report on research, rather than a text.

Di Roccaferrera, Giuseppe M. Ferrero. INTRODUCTION TO LINEAR PROGRAMMING PROCESSES: AN ABRIDGEMENT OF OPERATIONS RESEARCH MODELS FOR BUSINESS AND INDUSTRY. Cincinnati, Ohio: South-Western, 1967. xix, 672 p. Bibliography, pp. 641-53. Glossary of Symbols, pp. 655-72. Appendixes and Tables, pp. 1-32. Index, pp. 33-39 at end of book.

A detailed and relatively elementary treatment of linear programming and the simplex method, as well as univariate, bivariate, and multivariate statistical analysis, for both managers and students. Detailed examples are used to illustrate the concepts.

Dorfman, Robert. APPLICATION OF LINEAR PROGRAMMING TO THE THEORY OF THE FIRM, INCLUDING AN ANALYSIS OF MONOPOLISTIC FIRMS BY NON-LINEAR PROGRAMMING. Berkeley: University of California Press, 1951. ix, 98 p. References, pp. 95-98. No index.

A comparison of the marginal analysis and linear programming approaches to the theory of the firm in competition and also monopoly. Assumes knowledge of matrix algebra and differential calculus. Reprints of this book are available from University Microfilms, Ann Arbor, Michigan.

Dorfman, Robert; Samuelson, Paul A.; and Solow, Robert M. LINEAR PROGRAMMING AND ECONOMIC ANALYSIS. The Rand Series. New York: McGraw-Hill Book Co., for the Rand Corp., 1958. ix, 525 p. Appendix A: Chance, Utility, and Game Theory, pp. 465-69. Appendix B: The Algebra of Matrices, pp. 470-506. Bibliography, pp. 507-12. Index, pp. 513-25.

Develops theory of linear programming for economists with a background in college algebra. The main applications are in game theory, welfare economics, general equilibrium, and the statical Leontief system. Useful as a reference.

Dreyfus, Stuart E. DYNAMIC PROGRAMMING AND THE CALCULUS OF VARIATIONS. Mathematics in Science and Engineering, vol. 21. New York: Academic Press, 1965. xix, 248 p. Bibliography, pp. 242-44. Author Index, p. 245. Subject Index, pp. 246-48.

A development of the relationships between the calculus of variation and the dynamic programming approaches to optimization.
One objective of this book is to reveal the simple and intuitive
results from the calculus of variations when they are deduced from
the dynamic programming viewpoint, and another objective is to
familiarize the reader with the applications of dynamic programming methods to classical problems. The theory is developed for
minimization problems only. The book concludes with applications of dynamic programming to stochastic and adaptive optimization problems. Assumes knowledge of advanced calculus and
intermediate probability theory.

Driebeek, Norman J. APPLIED LINEAR PROGRAMMING. Reading, Mass.:
Addison-Wesley, 1969. viii, 230 p. Bibliography, pp. 223-26. Index,
pp. 227-30.

This book, written by an employee of Arthur D. Little, is different from most books on linear programming in that algorithm development and formal mathematical model development are de-
emphasized in favor of detailed descriptions of operational procedures used in solving a variety of industrial problems. Applications are in natural gas and petroleum, oil tanker operations,
and inventory control. Assumes knowledge of computer programming. Some chapters contain three or four discussion topics and
problems.

Duffin, Richard J.; Peterson, Elmor L.; and Zener, Clarence. GEOMETRIC
PROGRAMMING: THEORY AND APPLICATION. New York: Wiley, 1967.
xi, 278 p. Appendix A: The Farkas Lemma, pp. 244-46. Appendex B: The
Perturbation of Optimized Parameters Due to Variations in the Posynomial Coefficients, pp. 247-54. Appendix C: Chemical Equilibrium Treated by Geometric Programming, pp. 255-64. Appendix D: Expressing Linear Programs
as Geometric Programs, pp. 265-68. Index, pp. 269-78.

Explains the mathematical theory of geometric programming and illustrates its application to engineering design. Topics include
linear and convex programming, the duality theorem of geometric
programming by way of Lagrange multipliers, a refined duality
theory, and extensions of geometric programming to functions that
are not necessarily of the posynomial (positive polynomial) form.
Assumes knowledge of calculus, mathematical programming, and
systems engineering, although the latter topic is reviewed in chapter 2. Chapters conclude with detailed exercises and references.

Evers, J.J.M. LINEAR PROGRAMMING OVER AN INFINITE HORIZON. Tilburg Studies on Economics, 8. Groningen, the Netherlands: Tilburg University Press, 1973. 187 p. References, p. 184. List of Symbols, pp. 185-86.
Subject Index, p. 187.

An application of linear programming theory to growth models over

an infinite horizon. Topics include mathematical formulation of
the linear programming system, partial objective functions, optimality, parametric properties, paths of equilibrium, semiequili-
brium paths, and equivalent linear programming problems over a
finite horizon. Assumes knowledge of advanced calculus, mathematical analysis, and linear algebra.

Fiacco, Anthony V., and McCormick, Garth P. NONLINEAR PROGRAMMING:
SEQUENTIAL UNCONSTRAINED MINIMIZATION TECHNIQUES. New York:
Wiley, 1968. xiv, 210 p. Symbols and Notation, pp. xi-xii. References,
pp. 196-201. Index of Theorems, Lemmas, and Corollaries, pp. 203-4. Author
Index, pp. 205-6. Subject Index, pp. 207-10.

A unified theory on the methods of transforming a constrained minimization problem into a sequence of unconstrained minimzations
of certain auxilliary functions which define "interior point" and
"exterior point" methods, depending on whether or not the constraints are strictly satisfied by the minimizing sequence. Topics
include interior and exterior point unconstrained minimization,
extrapolation of unconstrained minimization techniques, and computational aspects of unconstrained minimization algorithms. It is
useful to anyone concerned with mathematical programming theory
or computations and with the evolution, theory, and computational
implementation of auxiliary function sequential unconstrained methods
and recent advances in these methods. Assumes knowledge of advanced calculus and mathematical analysis.

Ficken, F.A. THE SIMPLEX METHOD OF LINEAR PROGRAMMING. New York:
Holt, Rinehart and Winston, 1961. vi, 58 p. Bibliography, p. 36. Appen-
dix 1: Prerequisites, pp. 37-52. Appendix 2: Theorems on Existence and
Duality, pp. 53-58. No index.

A brief account of the simplex method for specialists in engineering
and business. It stresses both rigorous mathematical development
and examples, and employs graphs for illustrations. Appendix 1
contains mathematical reviews for those without previous training
in linear algebra.

Frazer, J. Ronald. APPLIED LINEAR PROGRAMMING. Englewood Cliffs, N.J.:
Prentice-Hall, 1968. ix, 174 p. No index.

Concentrates on the use of linear programming as a problem-solving
technique rather than on the mathematics of this technique. About
thirty pages are devoted to the graphic solution of a simple linear
program; about ninety pages are devoted to the simplex method
with emphasis on setting up the first table of a simplex linear
programming solution; and the remaining forty pages consist of
the solution of the transportation problem. A FORTRAN program
for the simplex algorithm is included. Chapters conclude with
exercises.

Garfinkel, Robert S., and Nemhauser, George L. INTEGER PROGRAMMING. Series in Decision and Control. New York: Wiley, 1972. xiv, 427 p. Bibliography, pp. 392-414. Author Index, pp. 415-19. Subject Index, pp. 421-27.

A comprehensive treatment of the theory, methodology, and application of integer programming which can be used both as a reference and a text in a one- or two-semester course in operations research, management science, applied mathematics, or industrial engineering. The style of presentation is rigorous with theorems and proofs provided, but every technique that is presented is illustrated by at least one numerical example. Topics include enumeration methods, including branch and bound techniques, cutting plane methods, the Knapsack problems, integer programming over cones, the set covering and partitioning problems, and approximate methods. The book concludes with a chapter on the computational experience of various integer programming algorithms. The chapters conclude with exercises and notes on the historical development of the material presented. Assumes knowledge of linear algebra, set theory, and elementary abstract algebra.

Garvin, Walter W. INTRODUCTION TO LINEAR PROGRAMMING. New York: McGraw-Hill Book Co., 1960. xiv, 281 p. References, pp. 271-74. Index, pp. 275-81.

A text for a one-semester course in linear programming requiring knowledge of only college algebra. It is divided into three parts: the general linear programming problem and sensitivity analysis; the transportation problem and its variants; and special methods, such as stochastic linear programming, the revised simplex method, parametric programming, and duality. The chapters conclude with problems.

Gass, Saul I. LINEAR PROGRAMMING: METHODS AND APPLICATIONS. 4th ed., rev. New York: McGraw-Hill Book Co., 1975. x, 406 p. Bibliography of Linear Programming Applications, pp. 371-81. References, pp. 383-401. Index, pp. 402-6.

A text for an introductory course in linear programming for students who possess knowledge of linear algebra and calculus, and a reference for research workers and practitioners. It is divided into three parts: theoretical, computational, and applied. It assists the student in recognizing potential linear-programming problems, formulating such problems as linear-programming models, and employing computational techniques to solve these problems. Topics include simplex and revised simplex methods, duality, degeneracy procedures, parametric linear and integer programming, nonlinear and quadratic programming, sensitivity analysis, decomposition of large-scale systems, and bounded-variable problems. The applications are to transportation, production-scheduling and inventory control, interindustry analysis, diet problems, networks, and game theory. Each of the twelve chapters concludes with remarks and exercises.

Geary, Robert Charles, and McCarthy, M.D. ELEMENTS OF LINEAR PROGRAMMING WITH ECONOMIC APPLICATIONS. Griffin's Statistical Monographs and Courses, no. 15. New York: Hafner, 1964. 126 p. Appendixes, pp. 97-123. Index, pp. 125-26.

An elementary treatment of linear programming with emphasis on the simplex method of solution. Applications are to transportation, industrial firms, interindustry analysis, and agriculture. Assumes knowledge of only high school algebra. The appendixes take up the discrete variable in linear programming, the use of the simplex method in nonlinear programming, and the primal and the dual solution of a linear program.

Glicksman, Abraham M. AN INTRODUCTION TO LINEAR PROGRAMMING AND THE THEORY OF GAMES. New York: Wiley, 1963. 131 p. References, pp. 127-28. Subject Index, pp. 129-31.

An introduction to linear programming and the theory of games in an elementary yet mathematically sound manner. Even though this book can be used as a high school text, it contains proofs of theorems, and relatively deep concepts such as the extreme point theorem for convex polygons, Dantiz's simplex method, the fundamental duality theorem, and von Neumann's minimax theorem. Chapters conclude with exercises.

Gluss, Brian. AN ELEMENTARY INTRODUCTION TO DYNAMIC PROGRAMMING: A STATE EQUATION APPROACH. Boston: Allyn and Bacon, 1972. xxi, 402 p. Glossary of Symbols, pp. xx-xxi. Bibliography, pp. 380-81. Answers to Exercises, pp. 382-94. Index, pp. 395-402.

A text for an introductory course in dynamic programming for students who have knowledge of elementary differential and integral calculus and probability theory. Both deterministic and stochastic models are covered, and emphasis is given to the construction and solution, both numerical and analytical, of typical functional equations of dynamic programming. Applications include inventories, adaptive multistage decision processes, sequential analysis problems, optimal trajectories and control theory, and simulation. The last two chapters take up special solution methods, such as the approximate techniques. This book would be useful to individuals in a variety of fields, including economics and business. Each chapter concludes with about one dozen problems.

Gol'steĭn, Evgeniĭ Grigor'evich. THEORY OF CONVEX PROGRAMMING. Translated by K. Makowski. Translations of Mathematical Monographs, vol. 36. Providence, R.I.: American Mathematical Society, 1972. v, 57 p. Bibliography, p. 57. No index.

Originally published in Russian under the title VYPUKLOE PRO-
GRAMMIROVANIE in 1970, this book presents the theory of convex
programming for finite-dimensional problems with emphasis on the
duality theorem. The seven section topics are the dual problem;
the theorem of antagonistic games and its generalizations; duality
theorems; dual problems, problems of finding saddle points, and optimality criteria; quasiconvex problems generalized duality theorems;
and stability and marginal values. Assumes knowledge of real analysis.

Greenberg, Harold. INTEGER PROGRAMMING. Mathematics in Science and
Engineering, vol. 76. New York: Academic Press, 1971. xii, 196 p.
Author Index, p. 193. Subject Index, pp. 194-96.

An introduction to the study of integer programming, beginning
with a review of linear programming. The variables of the integer program are expressed in terms of parameters which are varied
until a solution is obtained. The enumerative procedure is included, and a branch and bound scheme that eliminates the necessity for large computer storage capacity is also included.
Chapters conclude with problems and references.

Greenwald, Dakota Ulrich. LINEAR PROGRAMMING: AN EXPLANATION
OF THE SIMPLEX METHOD. New York: Ronald Press, 1957. vii, 75 p.
References, p. 75. No index.

An exposition of the simplex algorithm as used in hand-computed
solutions of linear programming problems. It is accessible to anyone having a college algebra background, and sample problems
are included.

Hadley, George F. LINEAR PROGRAMMING. Addison-Wesley Series in Industrial Management. Reading, Mass.: Addison-Wesley, 1962. xii, 520 p.
Subject Index, pp. 517-20.

Presents a detailed mathematical treatment of the theory of linear
programming with emphasis on computational procedures. There
are more than 200 pages on the use of the simplex method for
the solution of the transportation, transshipment, and game theory
problems. Assumes knowledge of linear algebra comparable to
that contained in LINEAR ALGEBRA (Reading, Mass.: Addison-
Wesley, 1961), also by Hadley. Each chapter contains references and problems.

_____. NONLINEAR AND DYNAMIC PROGRAMMING. Addison-Wesley
Series in Management Science. Reading, Mass.: Addison-Wesley, 1964. xi,
484 p. Index, pp. 481-84.

Presents the theory and computational aspects of nonlinear programming. Topics include classical optimization and the properties of
convex functions, using Lagrange multipliers, stochastic program-

ming, Kuhn-Tucker theory, approximate methods for solving problems involving separable functions, quadratic and integer programming, gradient methods, dynamic programming, and algorithms for
solving nonlinear programming problems. Chapter 2 summarizes
the most important mathematical background needed and introduces
the notation used throughout the book. The presentation is rigorous and assumes knowledge of linear programming, linear albegra,
calculus, and probability theory. Chapters conclude with references and problems (one chapter concludes with fifty-four problems).

Haley, Keith Brian. MATHEMATICAL PROGRAMMING FOR BUSINESS AND
INDUSTRY. Studies in Management. New York: St. Martin's Press, 1967.
viii, 156 p. References, pp. 142-43. Bibliography, pp. 144-46. Exercises,
pp. 147-54. Index, pp. 155-56.

Presents the following models based on typical industrial applications: linear programming, integer and parametric programming,
transportation and transshipment, queueing, and game theory. The
mathematical results are presented without proof and illustrated by
example, making this book accessible to those with only a high
school algebra background. The exercises at the end of the book
are arranged by chapter, and the references are arranged into
three categories: direct references to papers, theoretical references to four standard texts, and those pertaining to practical
applications.

Hastings, N.A.J. DYNAMIC PROGRAMMING (WITH MANAGEMENT APPLI-
CATIONS). New York: Crane, Russack and Co., 1973. viii, 173 p. Appendix: Dynacode: A Dynamic Programming Software Package, p. 166. References, pp. 167-68. Index, pp. 169-73.

A text for a course in dynamic programming for engineers and
operations research analysts. It begins at an elementary level,
deals thoroughly with problem formulation, contains numerous
worked examples and exercises, and deals with stochastic problems in reasonable depth. Topics include the value iteration algorithm, with applications to production planning, equipment replacement, product assortment, and resource allocation; finite
stage Markov programming; infinite stage Markov programming;
and continuous time probabilistic decision problems. Assumes
knowledge of linear algebra and probability theory. The chapters
conclude with exercises and answers.

Heady, Earl O., and Candler, Wilfred. LINEAR PROGRAMMING METHODS.
Ames: Iowa State University Press, 1958. vii, 597 p. Index, pp. 591-97.

An introduction to linear programming for teachers, research workers,
and extension specialists with emphasis on the economic interpretation of programming results. Topics include input-output models,

game theory, risk, and nonlinear programming. Requires no mathematics beyond arithmetic. Chapters conclude with selected references.

Himmelblau, David M. APPLIED NONLINEAR PROGRAMMING. New York: McGraw-Hill Book Co., 1972. xi, 498 p. Appendix A: Nonlinear Programming Problems and Their Solutions, pp. 393-431. Appendix B: Computer Codes (in FORTRAN) That Are Not Available Commercially, pp. 433-68. Appendix C: Matrices, pp. 469-77. Appendix D: Standard Timing Program, pp. 475-77. Appendix E: Notation, pp. 478-85. Name Index, pp. 489-91. Subject Index, pp. 493-98.

A presentation of nonlinear programming algorithms that have been demonstrated in practice to be relatively effective. The algorithms are compared using the following criteria: reliability (success in obtaining a solution), speed of solution, preparation time by the user, accuracy of the solution, and degree of satisfaction of the constraints. The models are deterministic, and all algorithms treated are operational only with the aid of large-sized, high-speed digital or hybrid computers. Part 1 is the introduction; part 2 treats unconstrained nonlinear programming algorithms; and part 3 describes algorithms for constrained optimization. Assumes knowledge of calculus, matrix algebra, and methods of solution of linear programs. Chapters conclude with summaries, references, and problems.

Hinderer, K. FOUNDATIONS OF NON-STATIONARY DYNAMIC PROGRAMMING WITH DISCRETE TIME PARAMETER. Lecture Notes in Operations Research and Mathematical Systems, vol. 33. New York: Springer-Verlag, 1970. vi, 160 p. Literature, pp. 153-58. Index of Definitions and Notation, pp. 159-60.

A rigorous foundation of stochastic dynamic programming for nonstationary models with no applications. Chapter 1 treats models with countable state space (and arbitrary action space), and chapter 2 presents models in which the state space and action space are Borel subsets of complete separable metric spaces.

Howard, Ronald A. DYNAMIC PROGRAMMING AND MARKOV PROCESSES. MIT Press Books in Operations Research. Cambridge, Mass.: MIT Press, 1960. viii, 136 p. Appendix: The Relationship of Transient to Recurrent Behavior, pp. 127-31. References, p. 133. Index, pp. 135-36.

A supplementary text for courses in operations research and systems analysis which deals mainly with dynamic programming as a method for the solution of sequential problems. Topics include Markov processes, solution of the sequential process by value iteration and policy-iteration methods, sequential decision processes with discounting, and continuous time decision and Markov processes. Applications are to taxicab operation and automobile replacement. Assumes knowledge of mathematical analysis and probability theory.

Hu, T[e]. Chiang. INTEGER PROGRAMMING AND NETWORK FLOWS: Reading, Mass.: Addison-Wesley, 1969. xii, 452 p. List of Notations, pp. xi-xii. Appendix A: Smith's Normal Form, pp. 377-81. Appendix B: An Alternative Proof of Duality, pp. 382-84. Appendix C: Tree Search Type of Algorithms, pp. 385-88. Appendix D: Faces, Vertices, and Incidence Matrices of Poly-hedra, pp. 389-432. References, pp. 433-45. Index, pp. 446-52.

> A text which is divided into three parts: linear programming, net-work flows, and integer programming. The first part can be used for a one-semester course in linear programming; the second part is useful for a course on network flows; and the third part can be used for a course on integer programming. The book deals mainly with algorithms for solving problems, and three chapters present in great detail Gamory's cutting plane algorithms and the parti-tioning algorithms of Benders for solving integer programming prob-lems. The presentation is fairly intuitive, and assumes knowledge of linear algebra. Chapters conclude with exercises and suggested readings.

Hughes, Ann J., and Grawoig, Dennis E. LINEAR PROGRAMMING: AN EM-PHASIS ON DECISION MAKING. Reading, Mass.: Addison-Wesley, 1973. xv, 414 p. Index, pp. 413-14.

> A text for an introductory course in linear programming designed to give the student an intuitive, in-depth comprehension of the subject, including the simplex method. Linear programming is presented as a tool for teaching the broad concepts of modeling, and the coverage is extended to include both goal and integer programming. Most of the eighteen chapters conclude with exer-cises designed to provide practice in model building. Assumes knowledge of high school algebra.

Jacobs, O.L.R. AN INTRODUCTION TO DYNAMIC PROGRAMMING: THE THEORY OF MULTISTAGE DECISION PROCESSES. London: Chapman and Hall, 1967. xi, 126 p. Bibliography, p. 124. Index, pp. 125-26.

> A text for an introductory course in dynamic programming for un-dergraduates, and an introduction for graduates in mathematics-based subjects. It presents dynamic programming as a theory of multi-stage decision processes, and it is organized with respect to the following types of processes: discrete-time deterministic, continous-time deterministic, stochastic, and adaptive. Assumes knowledge of differential and integral calculus, probability theory, and ele-mentary mathematical programming.

Jacobson, David H., and Mayne, David Q. DIFFERENTIAL DYNAMIC PRO-GRAMMING. Modern Analytic and Computational Methods in Science and En-gineering, vol. 24. New York: American Elsevier, 1970. 208 p. Appendix A: Second-order and Successive Sweep Algorithms: A Comparison, pp. 199-202. Appendix B: Error Analysis for Bang-bang Algorithms, pp. 203-4. Author In-dex, pp. 205-6. Subject Index, pp. 207-8.

A presentation of algorithms in optimal control for the applied scientist. Topics include algorithms for the solution of continuous-time and bang-bang control problems, discrete-time systems, and stochastic systems with discrete-valued and continuous disturbances. Assumes knowledge of advanced calculus.

Kantorovich, Leonid Vitalevich. THE BEST USE OF ECONOMIC RESOURCES. Edited by G. Morton. Translated by P.F. Knightsfield. Cambridge, Mass.: Harvard University Press, 1965. xxxiii, 349 p. Appendix I: Mathematical Formulation of the Problem of Optimal Planning, pp. 262–301. Appendix II: Numerical Methods for the Solution of Problems of Optimal Planning, pp. 302–42. References to Appendixes I and II, pp. 343–44. Author Index, p. 345. Subject Index, pp. 347–49.

Translated from the Russian, this book presents linear programming as a tool for economic planning in the USSR. The exposition is based on selected sample problems which are developed in a step-by-step manner. It is written for nonmathematicians.

Kaufmann, Arnold. GRAPHS, DYNAMIC PROGRAMMING, AND FINITE GAMES. Translated by H.C. Sneyd. Mathematics in Science and Engineering, vol. 36. New York: Academic Press, 1967. Bibliography, pp. 473–79. Subject Index, pp. 481–84.

Originally published in French in 1964, this book is an introduction to the theory of graphs with applications to dynamic programming and games. It is divided into two parts. Part 1 is devoted to simple and concrete examples in networks, dynamic programming, decisions, and games and assumes knowledge of college algebra. Part 2 develops these examples in a theoretical framework based on knowledge of calculus and elementary mathematical analysis. Proofs are provided for the most important theorems. The bibliography is arranged under three headings: theory of graphs and their applications, dynamic programming, and the theory of games of strategy.

Kaufmann, Arnold, and Cruon, R. DYNAMIC PROGRAMMING: SEQUENTIAL SCIENTIFIC MANAGEMENT. Foreword to the French edition by H.P. Galliher. Translated by H.C. Sneyd. Mathematics in Science and Engineering, vol. 37. New York: Academic Press, 1967. xv, 278 p. List of Principal Symbols, pp. xiii–xv. Bibliography, pp. 269–75. Subject Index, pp. 277–78.

Originally published in French in 1965, this book treats sequential problems which are discrete in time for both deterministic and probabilistic models. The problem of convergence with an infinite horizon is thoroughly treated and the influence of an introduced rate of change is studied in detail. Markov decision models with a finite number of states are also treated. Proofs of theorems are provided, and many examples and graphs are used to illustrate the concepts. Assumes knowledge of advanced calculus or real analysis, and probability theory.

Mathematical Programming

Kendrick, David A. PROGRAMMING INVESTMENT IN THE PROCESS INDUS-
TRIES: AN APPROACH TO SECTORAL PLANNING. MIT Monographs in Eco-
nomics. Cambridge, Mass.: MIT Press, 1967. xiii, 160 p. Appendix A:
A Brief Description of the Production of Flat Steel Products, pp. 101-5. Ap-
pendix B: Specific Inputs and Unit Costs of Inputs, pp. 106-17. Appendix
C: The Single-period Matrix, pp. 118-26. Appendix D: Transportation and
Cost Data for the Mixed-integer Programming Model, pp. 127-32. Appendix
E: Evidence on Equipment Cost, pp. 133-41. Appendix F: Structure of the
Matrix and the Right-hand Side for the Multiperiod Model, pp. 142-52. Se-
lected Bibliography, pp. 153-56. Index, pp. 157-60.

> Presents both single period (for studying the operation of the exist-
> ing system) and multiperiod (for analyzing investment problems)
> linear programming models in order to assess the optimality and
> feasibility of investment alternatives over time. The applications
> are to steel production and distribution in Brazil. This analysis
> can be used as a foundation for a more extensive analysis of
> intersectoral programming for a developing economy. The appen-
> dixes are devoted to a description of steel products and a presen-
> tation of the numerical results of the linear programming matrices.

Kim, Chaiho. INTRODUCTION TO LINEAR PROGRAMMING. New York:
Holt, Rinehart and Winston, 1971. xviii, 556 p. Bibliography, pp. 528-44.
Index, pp. 545-56.

> A text for students in economics, business, and related disciplines
> who have limited mathematics background and who wish to study
> the geometric and computational aspects of linear programming in
> some depth. Part 1, consisting of 100 pages, reviews the neces-
> sary mathematics, including matrix algebra and convex sets in
> n-dimensional spaces. Part 2 takes up linear programming with
> applications to transportation and transshipment problems. Other
> topics include parametric programming, sensitivity analysis, and
> duality. Omissions include networks, integer programming, and
> risk and uncertainty. Chapters conclude with problems.

Kornai, János. MATHEMATICAL PLANNING OF STRUCTURAL DECISIONS.
Translated by Pál Morvay and József Hatvany. Contributions to Economic Analysis, vol. 45. New York: American Elsevier, 1967. xxvi, 526 p. Appendixes, pp. 429-506. References, pp. 507-17. Author Index, pp. 519-21.
Subject Index, pp. 523-26.

> Originally published in Hungarian in 1965, this book deals with
> mathematical programming for a planned economy and compares
> planned and market economies from the standpoints of function
> and administration. It presents a system of programming models
> to generate optimum allocations and shadow prices for socialistic
> economies. Topics include constraint structure, social performance
> criteria, data problems, forecasts of exogenous foreign prices,
> technological alternatives, imports vs. domestic production, econ-
> omies of scale, pricing of labor and capital, risk, risk aversion,

sensitivity analysis, parametric programming, exchange rates, and optimal central and sector plans.

Krekó, Béla. LINEAR PROGRAMMING. Translated by J.H.L. Ahrens and C.M. Safe. New York: American Elsevier, 1968. xii, 355 p. Appendix: The Hungarian Method, pp. 327-44. Bibliography, pp. 345-50. Index, pp. 351-55.

Originally published in German in 1962, this book deals with the method and theory of linear programming in the following three parts: the techniques of linear programming, the mathematical foundations of linear programming, and practical applications. Topics include parametric and concave programming, the relationship between linear programming and game theory, and applications to transportation, machine and production scheduling, warehouse problems, and extensions to the models of Kantorovich and the Hungarian paper industry. The appendix treats the Hungarian method and provides a solution to the transportation problem. The mathematics beyond college algebra is taken up as needed.

Künzi, Hans Paul; and Krelle, Wilhelm; in collaboration with Werner Oettli. NONLINEAR PROGRAMMING. Translated by Frank Levin. A Blaisdell Book in the Pure and Applied Sciences. Waltham, Mass.: Blaisdell Publishing Co., 1966. xiv, 240 p. Bibliography, pp. 227-36. Index, pp. 237-40.

Originally published in German in 1962, this book is useful as a text for a course in nonlinear programming and as a reference. Emphasis is on theory, and the following nonlinear algorithms are treated: the methods of Hildreth and D'Esopo, Theil and Van de Panne, Beale, Wolfe, Barankin and Dorfman, and Frank and Wolfe; Rosen's gradient projection method; Frisch's multiplex method; Zoutendijk's method of feasible directions; and Houthakker's capacity method. For each method the important theorems are proved and an example is given. The main prerequisites are linear algebra, and linear, convex, and quadratic programming.

Kwak, N.K. MATHEMATICAL PROGRAMMING WITH BUSINESS APPLICATIONS. New York: McGraw-Hill Book Co., 1973. xiv, 334 p. Appendixes: 1, Determinants and Matrix Algebra; 2, Notes on Concavity and Convexity; 3, Review of Differential Calculus; 4, Review of Logarithms; 5, Logarithmic Tables, pp. 273-329. Index, pp. 330-34.

A text for an introductory course in mathematical programming for advanced undergraduates and graduates in business, as well as a supplementary text for courses in operations research, management science, and quantitative analysis. Topics include linear programming and the simplex method, transportation problem, assignment problem, integer, nonlinear and dynamic programming, and game theory. Assumes knowledge of matrix algebra and elementary differential calculus. Rigorous proofs are avoided, and the chapters conclude with summaries, problems, and references.

Kwang, Ching-Wen, and Wu, Yuan-Li. MATHEMATICAL PROGRAMMING AND ECONOMIC ANALYSIS OF THE FIRM: AN INTRODUCTION. Scranton, Pa.: Intext Educational Publishers, 1971. x, 486 p. Selected Bibliography, pp. 471-78. Index, pp. 479-86.

> A reexamination of the theory of the firm from the viewpoint of mathematical programming. Topics include the classical theory of the firm, factor productivity and demand for inputs, and production and investment planning.

Laidlaw, Charles D. LINEAR PROGRAMMING FOR URBAN DEVELOPMENT PLAN EVALUATION. Praeger Special Studies in U.S. Economic and Social Development. New York: Praeger, 1972. xiv, 283 p. Selected References, pp. 279-83. No index.

> Presents five case study demonstrations in Baltimore and Jersey City to show how linear programming may be used to evaluate urban development planning alternatives. The applications are: urban renewal programming, intensive housing development evaluation, large-scale community development evaluation, regional airport system programming, and urban renewal design translation. Assumes knowledge of linear programming. Chapters conclude with notes and references.

Land, Ailsa H., and Powell, S. FORTRAN CODES FOR MATHEMATICAL PROGRAMMING: LINEAR, QUADRATIC AND DISCRETE. New York: Wiley, 1973. xiii, 249 p. Appendixes, pp. 200-248. References, p. 249. No index.

> Presents several computer programs for use by research workers to test their ideas about mathematical programming, such as alternative branching rules, and to experiment with the versions of so-called "intersection" cuts. The purpose and method of each algorithm is described briefly before the FORTRAN programs and routines are presented. Topics include the modifications needed to adopt the programs for different computers; the setting of tolerance levels; how to change the programs to solve programs for adding and deleting constraints other than those discussed in the book; and a brief record of the performance of two different integer programming algorithms on some published problems. The appendixes serve to index the subroutines and common variables.

Lange, Oskar Richard. OPTIMAL DECISIONS: PRINCIPLES OF PROGRAMMING. Translated by Irena Dobosz. New York: Pergamon Press, 1971. x, 292 p. Bibliography, pp. 285-87. Index, pp. 289-92.

> Translated from the Polish, this book is an exposition of the theory of linear programming with solutions by the simplex method. Almost the entire second half of the book deals with linear programming under conditions of uncertainty. Topics include linear and marginal programming, activity analysis, programming under uncertainty,

dynamic programming of purchases under certainty and uncertainty, and dynamic programming of production under certainty and uncertainty. Assumes knowledge of calculus and probability theory.

Larson, Robert E. STATE INCREMENT DYNAMIC PROGRAMMING. Modern Analytic Computational Methods in Science and Engineering. New York: American Elsevier, 1968. xvi, 256 p. Author Index, p. 251. Subject Index, pp. 253-56.

Presents a complete treatment of the theory of dynamic programming and the standard computational algorithms with applications to optimization problems arising in several fields, including guidance and control of aerospace vehicles, natural gas pipeline networks, and airline scheduling. It contains flow charts for constructing computer programs. Assumes knowledge of probability theory and mathematical analysis. Chapters conclude with references.

Lee, Sang M. GOAL PROGRAMMING FOR DECISION ANALYSIS. Auerbach Management and Communications Series. Philadelphia: Auerbach Publishers, 1972. xiv, 387 p. Notes, pp. 365-82. Index, pp. 383-87.

A book for both students and practitioners of decision analysis with emphasis on application of goal programming to real-world problems (more than one-half of the book is devoted to examples and applications). It is divided into the following four parts: decision analysis and goal programming, methods and processes of goal programming, applications of goal programming, and final remarks. The applications are in production planning, financial and marketing decisions, corporate and academic planning, medical care, and government. It is useful as a text for courses in management science, operations research, or decision theory. A simplex algorithm and FORTRAN program for goal programming are included. Assumes knowledge of linear programming.

Levin, Richard I., and Lamone, Rudolph P. LINEAR PROGRAMMING FOR MANAGEMENT DECISIONS. Irwin Series in Quantitative Analysis for Business. Homewood, Ill.: Richard D. Irwin, 1969. xi, 308 p. Appendix, pp. 291-98. Bibliography, pp. 299-306. Index, pp. 307-8.

Presents linear programming for students and managers with modest mathematical background. Numerous examples are provided, and each chapter concludes with exercises. The appendix consists of a simplex algorithm written in FORTRAN for solving linear programming problems.

Llewellyn, Robert W. LINEAR PROGRAMMING. New York: Holt, Rinehart and Winston, 1964. x, 371 p. Appendix, pp. 358-67. Index, pp. 369-71.

An intermediate mathematical treatment of linear programming with applications to the transportation, assignment, traveling salesman, and production scheduling problems, and game theory. Emphasis is placed on computational procedures using the simplex method and its revisions. Assumes knowledge of finite mathematics. The appendix consists of a case problem, and the chapters conclude with references and exercises.

Loomba, Narendra Paul. LINEAR PROGRAMMING: AN INTRODUCTORY ANALYSIS. New York: McGraw-Hill Book Co., 1964. xvi, 284 p. Appendix 1: The Meaning of Linearity, pp. 241-49. Appendix 2: A Note on Inequalities, pp. 250-52. Appendix 3: A System of Linear Equations Having a Unique Solution, pp. 253-55. Appendix 4: A System of Linear Equations Having No Solution, pp. 256-58. Appendix 5: A System of Linear Equations Having an Infinite Number of Solutions, pp. 259-61. Exercises, pp. 263-78. Selected Bibliography, pp. 279-80. Index, pp. 281-84.

An introductory text on linear programming for business and economics students with emphasis on the solution of problems by the simplex method. It makes no attempt to explain the theory or to show how practical problems can be formulated as linear programming problems. The exercises at the end of the book are arranged according to chapter.

Loomba, Narendra Paul, and Turban, Efraim. APPLIED PROGRAMMING FOR MANAGEMENT. New York: Holt, Rinehart and Winston, 1974. xvi, 475 p. Appendix A: General Problems, pp. 448-52. Appendix B: Optimization of Unconstrained Functions, pp. 453-57. Appendix C: Convexity and Concavity, pp. 458-60. Appendix D: Lagrange Multipliers and Kuhn-Tucker Conditions (constrained optimization), pp. 461-65. Author Index, pp. 467-69. Subject Index, pp. 471-75.

A presentation of various linear, integer, nonlinear, and dynamic programming methods for managers with each method illustrated with examples and applications. Topics include transportation and assignment problems, inventory control, production scheduling, portfolio selection, decision trees and input-output analysis. This textbook may be used in an introductory course in management science, operations research, or mathematical programming for the M.B.A. student. All algebra and higher mathematics that is required to comprehend the models is contained in the book. Chapters conclude with bibliographies, remarks, and applications.

Luenberger, David G. INTRODUCTION TO LINEAR AND NONLINEAR PROGRAMMING. Reading, Mass.: Addison-Wesley, 1973. xii, 356 p. Appendix A: Mathematical Reviews, pp. 324-31. Appendix B: Convex Sets, pp. 332-39. Appendix C: Gaussian Elimination, pp. 340-43. Bibliography, pp. 344-49. Index, pp. 353-56.

A text on optimization techniques for students in system analysis, operations research, numerical analysis, management science, and related fields. Topics include simplex method, constrained and unconstrained optimization, and penalty and barrier methods. Assumes knowledge of linear algebra. Chapters conclude with exercises and references.

McMillan, Claude. MATHEMATICAL PROGRAMMING: AN INTRODUCTION TO THE DESIGN AND APPLICATION OF OPTIMAL DECISION MACHINES. Wiley Series in Management and Administration. New York: Wiley, 1970. xi, 496 p. Appendixes A-I, pp. 413-92. Index, pp. 493-96.

A text for a course in mathematical programming with emphasis on nonlinear and discrete aspects of management science, operations research, and economics. Topics include classical optimization of functions of one and of many variables with and without constraints; gradient methods of optimization; simplex method; branch and bound algorithms; and the following types of programs: geometric, dynamic, linear, integer, binary (zero-one), discrete (integer nonlinear), and heuristic. Assumes knowledge of one year of calculus. The nine appendixes contain FORTRAN programs, a review of matrix algebra, and the Newton-Raphson method for identifying the extrema of an unconstrained function. Chapters conclude with exercises and references.

Mangasarian, Olvi L. NONLINEAR PROGRAMMING. McGraw-Hill Series in Systems Science. New York: McGraw-Hill Book Co., 1969. xiii, 220 p. Appendix A: Vectors and Matrices, pp. 177-81. Appendix B: Resume of Some Topological Properties of R^n, pp. 182-90. Appendix C: Continuous and Semi-continuous Functions, Minima and Infima, pp. 191-99. Appendix D: Differentiable Functions, Mean-value and Implicit Function Theorems, pp. 200-204. References, pp. 205-12. Name Index, pp. 215-16. Subject Index, pp. 217-20.

A text for a one-semester course on nonlinear programming for graduates and undergraduates in engineering, computer science, and operations research who have backgrounds in advanced calculus and real analysis. It gives a rigorous account of the theory with every result either proved or stated precisely with adequate references cited. Topics include saddlepoint optimality criteria of nonlinear programming with and without differentiability, duality, and optimality and duality in the presence of nonlinear equality constraints. Chapters conclude with problems.

Massé, Pierre. OPTIMAL INVESTMENT DECISIONS: RULES FOR ACTION AND CRITERIA FOR CHOICE. Translated by Scripta Technica. Editor's introduction by W. Allen Spivey. Prentice-Hall Quantitative Methods Series. Englewood Cliffs, N.J.: Prentice-Hall, 1962. xvi, 500 p. Author Index, pp. 493-96. Subject Index, pp. 497-500.

Originally published in French (Paris: Dunod, 1959) as LE CHOIX DES INVESTISSMENTS CRITERES ET METHODES, this book analyzes the problem of an optimum investment program for an economy. It seeks to develop rules for action and criteria for choice. The first part of the book studies investment by individual economic agents in the economy (the firm); the second part introduces uncertainty into decisions; and the third part takes up the problem of socially optimum investment. The tools employed include linear programming and certain concepts of stochastic processes (expectation, for example), and almost every aspect of optimal investment decisions is touched upon, including the replacement problem, the choice of equipment, and inventory management. The applications are in hydroelectric power, agriculture, and oil field exploration. Assumes knowledge of calculus, matrix algebra, and probability theory. Chapters conclude with references.

Meisels, Kurt. A PRIMER OF LINEAR PROGRAMMING. Foreword by D.S. Philipps. New York: New York University Press, 1962. x, 103 p. Bibliography, p. 102. Index, p. 103.

An introduction to linear programming for the manager and practitioner. It begins with a presentation of basic mathematical vocabulary, followed by a graphical and then algebraic solution of a linear program. The simplex method is introduced and used to solve a trim-loss reduction problem and a blending problem. The book concludes with a discussion of the transportation problem. The treatment is not rigorous mathematically, but it is thorough and can be understood by anyone with a college algebra background.

Metzger, Robert W. ELEMENTARY MATHEMATICAL PROGRAMMING. New York: Wiley, 1963. ix, 246 p. Bibliography, pp. 240-44. Index, pp. 245-46.

An introduction to mathematical programming for managers and a text for a one-semester introductory course. The applications are in production planning, material handling, scheduling, and job and salary evaluation. Assumes knowledge of high school algebra.

Miller, Ronald E. DOMESTIC AIRLINE EFFICIENCY: AN APPLICATION OF LINEAR PROGRAMMING. Foreword by W. Isard. The Regional Science Studies Series. Cambridge, Mass.: MIT Press, 1963. xiv, 174 p. Appendix A: Balanced Competition, pp. 139-44. Appendix B: Extensions of the Model, pp. 145-52. Appendix C: Airline Demand Forecasting, pp. 153-65. Appendix D: Data and Sources, pp. 166-70. Index, pp. 171-74.

An application of linear programming to the study of the efficiency of passenger traffic between a sample of U.S. city pairs during 1957. Although there is no large discrepancy between the results of the author's research and empirical observation, the techniques used would be of interest and hold promise in further multiregional linear programming developments.

Müller-Merbach, H. ON ROUND-OFF ERRORS IN LINEAR PROGRAMMING. Lecture Notes in Operations Research and Mathematical Systems, vol. 37. New York: Springer-Verlag, 1970. vi, 48 p. Appendix, pp. 35-46. References, pp. 47-48. No index.

An investigation of the error accumulation (increase in round-off errors) in linear programming procedures. Three different algorithms were used for this investigation: the normal simplex method, the revised simplex method with explicit form of the inverse, and the symmetric revised simplex method. The appendix shows graphically how the round-off errors increase in several test runs from the first to the last simplex iteration.

Näslund, Bertil. DECISIONS UNDER RISK: ECONOMIC APPLICATIONS OF CHANCE-CONSTRAINED PROGRAMMING. Stockholm: The Economic Research Institute at the Stockholm School of Economics, 1967. 188 p.

A presentation of chance-constrained programming with applications to linear static and dynamic models of portfolio selection and other investment situations. It includes brief introductions to other methods of probabilistic programming, such as stochastic linear programming and the two-stage programming under certainty.

Naylor, Thomas H.; Byrne, Eugene T.; and Vernon, John M. INTRODUCTION TO LINEAR PROGRAMMING: METHODS AND CASES. Belmont, Calif.: Wadsworth, 1971. xii, 229 p. Bibliography, pp. 166-69. Cases in Linear Programming, pp. 171-225. Index, pp. 226-29.

An introduction to linear programming for individuals in business administration, economics, industrial management, and industrial engineering, with emphasis on solving problems by the simplex method and also on applications in business and industry. It also includes one chapter on each of the following areas: integer programming, input-output analysis, and game theory. Assumes knowledge of linear algebra. Fifteen representative problems which can be solved by linear programming techniques are included. The chapters conclude with problems.

Nemhauser, George L. INTRODUCTION TO DYNAMIC PROGRAMMING. Series in Decision and Control. New York: Wiley, 1967. xiii, 256 p. References, pp. 248-51. Index, pp. 253-56.

An introductory text on the theory and computational aspects of dynamic programming for operations researchers, management scientists, statisticians, engineers, and social scientists. The models contain many decision variables and have a mathematical structure which is such that calculations of the optimal decisions can be done sequentially. Extensive applications are made in inventory control, allocation problems, control theory, search theory, and chemical engineering design. Two chapters, consisting of about

100 pages, deal with computational aspects using computer programming with emphasis on construction of flow charts. The relationship between dynamic programming and the calculus of variations is revealed. Assumes knowledge of differential and integral calculus, and probability theory. The chapters conclude with exercises.

Nijkamp, Peter. PLANNING OF INDUSTRIAL COMPLEXES BY MEANS OF GEOMETRIC PROGRAMMING. Foreword by J.H.P. Paelinck. Rotterdam: Rotterdam University Press, 1972. x, 146 p. References, pp. 140–44. Index, pp. 145–46.

A contribution to economic planning consisting of several development programs that are solvable by geometric programming algorithms. It attempts to operationalize the concept of growth poles which may be defined "as an ensemble of economic forces with high interlinkages that are able to transmit growth impulses to all sectors." The five chapter titles are: "Growth Poles and Industrial Complex Analysis," "Geometric Programming: Mathematical Programming of Posynomial Inequalities," "A Minimum Investment Model," "Alternative Objectives and Dynamics," and "Conclusions." Assumes knowledge of elementary mathematical analysis and linear programming. Posynomial is a term used to designate a particular function called a positive polynomial.

Norman, John Malcolm. HEURISTIC PROCEDURES IN DYNAMIC PROGRAMMING. Manchester, England: Manchester University Press, 1972. vii, 95 p. References, p. 79. Appendixes, pp. 81–95. No index.

A presentation of some methods of reducing the computational burden in dynamic programming. Topics include reductions in grid size, problem embedding, path restriction, the use of expectations in stochastic dynamic programming, state reduction by identifying problem characteristics, and practical use of the procedures. The appendixes are devoted to additional mathematical results and computational problems. Assumes knowledge of dynamic programming.

Nugent, Jeffrey B. PROGRAMMING THE OPTIMAL DEVELOPMENT OF THE GREEK ECONOMY 1954–1961: FORMULATION OF A LINEAR PROGRAMMING MODEL IN EVALUATING ECONOMIC PLANNING AND PERFORMANCE OF THE GREEK ECONOMY. Preface by G. Coutsoumaris. Research Monograph Series, no. 15. Athens, Greece: Center of Planning and Economic Research, 1966. 171 p. Appendix A: Input–output Analysis, pp. 151–53. Appendix B: Linear Programming, pp. 157–60. Appendix C: Tables (separate volume). Bibliography, pp. 163–71. No index.

A nonmathematical monograph on the application of linear programming to problems of economic planning with reference to the Greek economy. It is based primarily on an input–output interindustry model of the structure of the Greek economy.

296

Orchard-Hays, William. ADVANCED LINEAR-PROGRAMMING COMPUTING TECHNIQUES. New York: McGraw-Hill Book Co., 1968. xi, 355 p. Appendix A: Elementary Transformations, pp. 309-19. Appendix B: Design Criteria for a Complete Mathematical Programming System, pp. 321-40. Appendix C: Selected Bibliography, pp. 341-47. Index, pp. 349-55.

Presents a number of algorithms for solving linear programs based on the author's experiences. Among the classes of algorithms presented are: simple, dual, crashing techniques, ranging procedures, parametric algorithms, and decomposition techniques. Flow charts for computer programs for some of these algorithms are provided, and knowledge of computer programming is not essential. Assumes knowledge of linear programming. This book is useful as a supplementary text in operations research courses and as a reference for practitioners.

Plane, Donald R., and McMillan, Claude. DISCRETE OPTIMIZATION: INTEGER PROGRAMMING AND NETWORK ANALYSIS FOR MANAGEMENT DECISIONS. Englewood Cliffs, N.J.: Prentice-Hall, 1971. xii, 251 p. Appendixes, pp. 177-247. Index, pp. 249-51.

An introduction to integer programming for students of managerial analysis which can be used as a text or supplementary text for undergraduate courses in industrial engineering. The seven chapter titles are: "Management Science and Mathematical Programming," "Problem Formulation for Zero-one Programming," "Solving Zero-one Problems via Implicit Enumeration," "Integer Linear Programming," "Special Applications of Integer Programming," "Network Optimization," and "An Integer Programming Case Study." The main prerequisite is college algebra. Several FORTRAN programs are provided for some of the algorithms. The presentation is intuitive with proofs omitted and mathematical results brief and simple. Chapters conclude with references and exercises.

Reinfeld, Nyles V., and Vogel, William R. MATHEMATICAL PROGRAMMING. Englewood Cliffs, N.J.: Prentice-Hall, 1958. viii, 274 p. Appendix A: General Notation Used in Mathematics, pp. 243-54. Appendix B: A Discussion of the Theory of the Simplex Method, pp. 255-58. Bibliography, pp. 259-63. Organizations, p. 264. Index, pp. 265-74.

A practical approach to linear programming using examples from everyday management problems. Topics include the distribution method for solving mathematical programming problems, the modified distribution method, Vogel's approximation method, the simplex method, relaxation method, case studies in machine allocation, seasonal sales, railraod and tractor-trailer movements, and marketing and sales. The book concludes with a chapter on the use of computers.

Rust, Burt W., and Burrus, Walter R. MATHEMATICAL PROGRAMMING AND THE NUMERICAL SOLUTION OF LINEAR EQUATIONS. Modern Analytic and Computational Methods in Science and Mathematics, no. 38. New York: American Elsevier, 1972. x, 218 p. Author Index, p. 213. Subject Index, pp. 215-18.

This book deals with the solutions to the linear vector matrix Ax=y+e where the system is poorly conditioned. The main conclusion is that useful solutions can be obtained to undetermined problems. The six chapter headings are: "Ill-conditioned Linear Systems," "Linear Estimation," "A Thought Experiment," "Constrained Linear Estimation," "Mathematical Programming," and "Applications and Generalizations of the Constrained Estimation Technique." Assumes knowledge of matrix algebra, mathematical analysis, and statistical decision theory. Chapters conclude with references.

Saaty, Thomas L. OPTIMIZATION IN INTEGERS AND RELATED EXTREMAL PROBLEMS. New York: McGraw-Hill Book Co., 1970. xv, 295 p. Name Index, pp. 285-88. Subject Index, pp. 289-95.

A text for a course in optimization for students in mathematics, science, engineering, social science, and operations research who have advanced calculus and linear algebra prerequisites. The book covers optimization in integers in its entirety, and includes a wide variety of applications in both geometric and algebraic settings. Topics include optimization subject to diophantine constraints, and geometric optimization. The sections conclude with exercises, and the chapters conclude with references.

Sakarovitch, M. NOTES ON LINEAR PROGRAMMING. New York: Van Nostrand, 1971. vii, 195 p. Aide Memoire, pp. 171-72. References, p. 173. Index, pp. 174-75.

This set of notes was used in an undergraduate course on linear programming for engineers and students of operations research who had one year of calculus. The text is computer oriented in that the algorithms and theoretical results are presented in a format which makes their coding on a digital computer simple. Proofs of theorems are provided. Each of the eight chapters concludes with exercises.

Salkin, Harvey M. INTEGER PROGRAMMING. Reading, Mass.: Addison-Wesley, 1975. xx, 537 p. List of Commonly Used Symbols, pp. xi-xiii. Name Index, pp. 529-32. Subject Index, pp. 533-37.

A text for a course in integer programming for students who possess knowledge of matrix algebra and the primal and dual simplex methods. It is a unified, easy-to-read exposition of integer (linear) programming techniques and applications, and it is divided into the following four parts: cutting plane techniques; enumerative techniques; partitioning and group theory; and particular problems, specialized algorithms, and applications. Each chapter is structured so that introductory and background material are presented first, followed by the chapter's principal content, and then by more detailed points and mathematical proofs. Problems and references (except for the last chapter) are at the end of each chapter. Topics include dual fractional integer and mixed integer programming; dual and primal all-integer programming; branch and bound enumeration; search enumeration; partitioning in mixed integer programming; group theory in integer programming; the set covering problem; the knapsack problem; the plant location problem; and how to do integer programming in the real world.

Scott, Allen J. COMBINATORIAL PROGRAMMING, SPATIAL ANALYSIS AND PLANNING. London: Methuen, 1971. vii, 204 p. Bibliography, pp. 166-95. Index, pp. 196-204.

An introductory survey of modern combinatorial programming methods and their application in urban and regional analysis with emphasis on spatially structured problems. Topics include heuristic and exact methods of solving combinatorial programming problems; spatially structured combinatorial programs for the assignment, transportation, and transshipment problems; combinatorial programs for network analysis; location-allocation systems; and special grouping and partitioning problems. It makes extensive use of mathematical and symbolic methods of exposition, but rigorous axiomatic methods of mathematical analysis are avoided. Assumes knowledge of linear algebra and linear programming.

Sengupta, Jati K. STOCHASTIC PROGRAMMING: METHODS AND APPLICATIONS. New York: American Elsevier, 1972. xi, 313 p. References, pp. 303-13. No index.

A presentation of the theory and applications of stochastic (probabilistic) programming. The five chapter titles are: "Stochastic Programming Methods"; "Stochastic Programming with Chance Constraints"; "Nonlinear and Dynamic Models of Stochastic Programming"; "Applications to Reliability Programming, Approximation Models, and Sensitivity Analysis"; and "Future Trends and Problems." It also presents computational techniques, such as sequential unconstrained minimization, for solving stochastic programming problems. Assumes knowledge of mathematical analysis. This book is useful to research workers in fields such as applied mathematics, operations research, and mathematical economics.

Simmons, Donald M. LINEAR PROGRAMMING FOR OPERATIONS RESEARCH. San Francisco: Holden-Day, 1972. xii, 288 p. References, pp. 282-83. Index, pp. 284-88.

A text for a one-semester course in linear programming for students of varied backgrounds (economics, business, engineering, mathematics) with emphasis on the art of formulating linear programming problems and the science of solving them. Topics include duality theorem, simplex method and revised simplex method, the dual simplex method and postoptimality problems, and the transportation problem. Assumes knowledge of calculus; linear algebra is also used although it can be taken concurrently with the linear programming course. To meet the needs of the mathematically oriented student, care is taken to state theorems explicitly and to prove them clearly. The chapters conclude with exercises.

Simonnard, Michael. LINEAR PROGRAMMING. Translated by Wm. S. Jewell. Prentice-Hall Series in Management. Englewood Cliffs, N.J.: Prentice-Hall, 1966. xxiv, 430 p. Appendixes, pp. 349-405. List of Symbols and Notation, pp. 406-8. English-French Glossary, pp. 409-12. Bibliography, pp. 413-22. Index, pp. 423-30.

Originally published in French in 1962, this book is a text on the mathematical theory of linear programming at the intermediate level which is divided into the following four parts: the theory of linear programming and the general methods of calculation (chapters 1 to 7), integer linear programming (chapters 8 and 9), special structures devoted primarily to transportation networks (chapters 10 to 15), and the appendixes which consist of reviews of linear algebra, the theory of convex polyhedrons, and the theory of graphs. This book strikes a balance between theory and practical procedures for calculations. Assumes knowledge of college algebra, and exposure to matrix algebra is helpful.

Singleton, Robert R., and Tyndall, William F. GAMES AND PROGRAMS: MATHEMATICS FOR MODELING. San Francisco: Freeman, 1974. xxiii, 304 p. Answers to Exercises, pp. 259-93. Suggestions for Further Reading, pp. 295-99. Index, pp. 301-4.

Presents linear programming and game theory as related and complementary theories as bases for models rather than simply as mathematics. The book concludes with a chapter on the use of the duality theorem to prove the fundamental minimax theorem of matrix games. It is suitable for a one-semester course in economics, business, political science, psychology, and related disciplines. Assumes knowledge of only high school algebra. The chapters conclude with exercises.

Smythe, William R., and Johnson, Lynwood A. INTRODUCTION TO LINEAR PROGRAMMING WITH APPLICATIONS. Englewood Cliffs, N.J.: Prentice-Hall, 1966. viii, 221 p. Reference, p. 211. Answers to Selected Exercises, pp. 213-18. Index, pp. 219-21.

Develops the elementary mathematical theory of the simplex algorithm and describes some of the applications of linear programming to maximal flow, transportation, and transshipment problems. Assumes knowledge of algebra and analytic geometry. The sections conclude with exercises, and numerous examples are used to illustrate the concepts.

Spivey, W. Allen. LINEAR PROGRAMMING: AN INTRODUCTION. New York: Macmillan, 1963. viii, 184 p. Bibliography, pp. 180-81. Index, pp. 182-84.

Presents a reasonably self-contained discussion of the mathematical model of linear programming together with a variety of applications. Assumes knowledge of matrix algebra, geometry of vector spaces, and convexity. The sections conclude with exercises.

Spivey, W. Allen, and Thrall, Robert McDowell. LINEAR OPTIMIZATION. New York: Holt, Rinehart and Winston, 1970. xii, 530 p. Appendix A: Foundations, Sets, and Functions, pp. 375-429. Appendix B: Linear Algebra, pp. 430-502. Appendix C: Flow Charts for the Graves-Thrall Algorithm, pp. 503-15. References, pp. 517-20. Index, pp. 523-30.

A treatment of the simplex method in linear optimization presented at three levels: (1) linear programming with applications; (2) a development of the simplex method with illustrations; and (3) a development of the schema approach of Tucker, Gomory, and Balinski. Other topics include game theory, capacitated transportation problem, and decomposition and upper-bound constraints. Assumes knowledge of linear algebra.

Stockton, R.S. INTRODUCTION TO LINEAR PROGRAMMING. Homewood, III.: Richard D. Irwin, 1971. ix, 150 p. Appendix: An Alternative Computational Procedure for Revision of Simplex Tables, pp. 132-37. Answers for Checking Your Comprehension, pp. 138-50.

An introduction to linear programming methods within the broader context of decision making. Each section contains self-checking questions and problems. Assumes knowledge of only high school algebra.

Strum, Jay E. INTRODUCTION TO LINEAR PROGRAMMING. San Francisco: Holden-Day, 1972. xi, 404 p. Index, pp. 401-4.

A text for an introductory course in linear programming with emphasis on the algebraic approach using row operations. Topics include the simplex method, duality, game theory, transportation problem, and introduction to networks. Assumes knowledge of college algebra and matrix algebra. The sections of the book conclude with exercises.

Tabak, Daniel, and Kuo, Benjamin C. OPTIMAL CONTROL BY MATHEMATICAL PROGRAMMING. Prentice-Hall Instrumentation and Control Series. Englewood Cliffs, N.J.: Prentice-Hall, 1971. xiii, 237 p. Author Index, pp. 231-33. Subject Index, pp. 235-37.

A unified presentation of the application of mathematical programming to optimal control problems. Mathematical programming is employed to solve linear and nonlinear continuous-time systems, linear and nonlinear discrete-time systems, stochastic systems, and distributed-parameter systems. Assumes knowledge of calculus, differential equations, matrix algebra, and probability theory. Chapters conclude with references.

Thompson, Gerald L. LINEAR PROGRAMMING: AN ELEMENTARY INTRODUCTION. New York: Macmillan, 1971. vii, 384 p. Index, pp. 377-84.

A text for a one-semester course in introductory linear programming for students with only a high school algebra background. Topics include linear, integer, and convex programming, and linear programming under uncertainty. Emphasis is placed on geometric solutions and illustrations. The chapters conclude with references and problems.

Throsby, C.D. ELEMENTARY LINEAR PROGRAMMING. Introduction by E.J. Mishan. Quantitative Techniques for Economists Series. New York: Random House, 1970. xiv, 242 p. Index, pp. 239-42.

An introductory text in linear programming with emphasis on setting up linear programs and finding solutions by computer programs. It is divided into the following three parts: setting up simple linear programming problems, setting up more complex linear programming problems, and computational aspects. The computer programming portion of the book is limited to the establishment of flow charts and the interpretation of computer solutions. Assumes knowledge of high school algebra. The chapters conclude with exercises.

Vajda, Steven. AN INTRODUCTION TO LINEAR PROGRAMMING AND THE THEORY OF GAMES. London: Methuen, 1960. 76 p. Appendix 1: Proof of the Fundamental Theorem of Duality, pp. 45-48. Appendix 2: Proof of the Main Theorem, pp. 50-74. Index, pp, 75-76.

A brief presentation of two closely related topics: linear programming and game theory. Linear programming forms the basis for the numerical treatment of game theory. The examples are simple and schematic, and very little mathematics beyond high school algebra is required.

_____. MATHEMATICAL PROGRAMMING. Addison-Wesley Series in Statistics. Reading, Mass.: Addison-Wesley, 1961. ix, 310 p. Appendix, pp. 252-55. Solutions to Exercises, pp. 256-300. Bibliography, pp. 301-7. Index, pp. 309-10.

A text on linear, nonlinear, and dynamic programming with emphasis on computational methods. The theoretical foundations are laid in chapters 2 to 4 which deal with the algebra, the theory of graphs, and combinatorial theory. Chapters 5 and 6 are concerned with special and general algorithms, and chapters 7 and 8 take up applications. The remaining chapters are devoted to special topics such as parametric and stochastic linear programming, dynamic and nonlinear programming. The appendix is devoted to a review of matrix algebra. Assumes knowledge of calculus in some portions, but college algebra is sufficient for most of the text. The chapters conclude with exercises.

_____. PLANNING BY MATHEMATICS. 2d ed., rev. Topics in Operational Research. London: Sir Isaac Pitman, 1973. vii, 142 p. Appendix, pp. 130–39. Bibliography, pp. 140–42. No index.

This book is a completely revised version of READINGS IN MATHE-MATICAL PROGRAMMING (New York: Wiley, 1962) which itself was the second edition of READINGS IN LINEAR PROGRAMMING (New York: Wiley, 1958). It is a collection of twenty-one simplified problems (one problem per chapter) which are formulated as linear programs and solved by one or more of the following techniques: simplex, dual simplex, the inverse matrix method, decomposition and integer routines, dynamic programming, fractional programming, and branch and bound methods. The problems include blending aviation gasoline, minimum cost diet, production scheduling, the knapsack problem, personnel assignment, nutrition, critical path scheduling, and lot manufacturing. Assumes knowledge of college algebra.

_____. PROBABILISTIC PROGRAMMING. Probability and Mathematical Statistics Series, vol. 9. New York: Academic Press, 1972. ix, 127 p. Appendix 1: Linear Programming in Various Fields (references), pp. 115–17. References, pp. 118–24. Index, pp. 125–27.

A fairly nonrigorous exposition of probabilistic programming with emphasis on two-stage decision problems. The three chapter titles are: "Stochastic Programming," "Decision Problems," and "Chance Constraints." Algorithms are not discussed in detail. Familiarity with linear programming and the simplex method, discussed briefly in appendix 1, is assumed.

_____. THE THEORY OF GAMES AND LINEAR PROGRAMMING. Methuen's Monographs on Physical Subjects. New York: Wiley, 1956. 106 p. References, pp. 101–3. Index of Definitions, pp. 104–5. Directory of Games, p. 105. Directory of Linear Programming Problems, p. 106.

An elementary presentation of the theory of games as a special case of linear programming. It deals with the computational aspect thoroughly, but without specific attention to the problem formulation. Assumes knowledge of college algebra and analytic geometry.

_____. THEORY OF LINEAR AND NON-LINEAR PROGRAMMING. London: Longman, 1974. viii, 118 p. References, pp. 112-15. Index, pp. 117-18.

A text for a two-semester course in mathematical programming for mathematicians, engineers, and other practitioners of operations research who have knowledge of linear algebra, calculus, and topology. It concentrates on theory, rather than algorithms, and considers linear and quadratic programming, but omits integer, geometric, dynamic and infinite programming. Topics include theorems of alternatives, duality theorems, theorems of sufficiency and of necessity, saddle point theorems, and the generalized Kuhn-Tucker theory.

Van de Panne, C. LINEAR PROGRAMMING AND RELATED TECHNIQUES. New York: American Elsevier, 1971. ix, 364 p. Bibliography, pp. 361-62. Index, pp. 363-64.

An introductory text for students in economics and business administration who have a high school algebra background. It is divided into the following four parts: basic linear programming and the simplex method, sensitivity analysis, anticycling and upper-bound methods, and parametric programming; applications in production planning, sales, inventory control, capital budgeting, demand analysis, monopoly, and games; the transportation problem and its variants and networks; and decision tree methods, including branch and bound techniques for solving integer programming problems. Linear programming is explained without the use of linear algebra, and the transportation problem, networks, and integer programming are treated in a manner which follows naturally from linear programming. Each chapter concludes with about six exercises.

_____. METHODS FOR LINEAR AND QUADRATIC PROGRAMMING. Studies in Mathematical and Managerial Economics, vol. 17. New York: American Elsevier, 1975. xii, 477 p. Bibliography, pp. 470-75. Index, pp. 476-77.

A detailed exposition of the most important methods of linear and nonlinear programming for a varied audience using numerical examples throughout. A large number of techniques in both linear and quadratic programming are related to each other. Topics include simplex method, duality, sensitivity analysis and parametric programming, quadratic forms, simplex and dual method for quadratic programming, parametric methods for quadratic programming, multiparametric linear and quadratic programming, concave quadratic programming, linear complementarity problems, and termination for copositive plus matrices. This book is useful as a text in mathematical programming for graduates and undergraduates who have knowledge of matrix algebra.

Vazsonyi, Andrew. SCIENTIFIC PROGRAMMING IN BUSINESS AND INDUSTRY. New York: Wiley, 1958. xix, 474 p. Author Index, pp. 469-70. Subject Index, pp. 471-74.

A text on programming for those interested in applications in business and industry. Topics include simplex method, convex and dynamic programming, theory of games, and programming in production and inventory control.

Waverman, Leonard. NATURAL GAS AND NATIONAL POLICY: A LINEAR PROGRAMMING MODEL OF NORTH AMERICAN NATURAL GAS FLOWS. Toronto: University of Toronto Press, 1973. xiv, 112 p. References, pp. 119-22. No index.

Presents estimates of additional costs paid by final consumers of natural gas because of restrictions on trade between Canada and the United States, and also estimates of the benefits of trade restrictions to producers. The following two models based on linear programming are used: a free trade model which allocates supplies to demands so as to minimize the costs to final consumers, assuming no restrictions to trade, and a constrained model which incorporates restrictions which prevent American penetration of Canadian markets. Each model has an associated dual model which yields information on the prices (shadow values) at supply points and at markets. In the solution to the constrained model, the actual natural gas flow pattern of 1966 is simulated. Assumes knowledge of the rudiments of linear programming.

Webb, Michael. TRANSPORTING GOODS BY ROAD. London School of Economics Monograph no. 10. London: Weidenfeld and Nicolson, for the London School of Economics and Political Science, 1972. 435 p.

Deals with the problems of road transport in the United Kingdom with emphasis on policies aimed at increasing efficiency. It consists of nine parts which are subdivided into forty-two chapters. Topics include description of the road goods transport expenditures in the United Kingdom, government policy, economies of scale from the point of view of large manufacturers and postal services, and journey planning and the estimation of journey duration and planned idle time. Part 8 of the book presents the calculations of journey plans and is the most interesting to operations researchers. It shows that the pigeon-holing method is inadequate for planning wholesale delivery journeys, and that the mathematical optimization techniques, such as linear programming and dynamic programming, can only be applied with difficulty to simplified journey planning problems. Useful to those in government transport agencies and industry transportation departments.

Weingartner, H. Martin. MATHEMATICAL PROGRAMMING AND THE ANALYSIS OF CAPITAL BUDGETING PROBLEMS. Chicago: Markham, 1967. xiv, 265 p. Bibliography, pp. 195-200. Criteria for Programming Investment Project Selection, pp. 201-12. Capital Budgeting of Interrelated Projects: Survey and Synthesis, pp. 213-44. Methods for the Solution of the Multi-dimensional 0/1 Knapsack Problem, pp. 245-65. No index.

Mathematical Programming

This book considers the problem of which investment project to undertake when there is no access to external sources of capital and when investment planning is done under capital rationing and under imperfect competition. It is formulated as an integer programming problem, and the major portion discusses matters such as the nature of the solution when the integer constraints are dropped, the nature of the dual, and variations for cases such as mutually exclusive projects. The Markham edition includes the original publication (Prentice-Hall, 1963) and sixty-five additional pages consisting of an abstract of recent research by the author and reprints of two articles originally published in MANAGEMENT SCIENCE and OPERATIONS RESEARCH. Assumes knowledge of matrix algebra.

White, Douglas John. DYNAMIC PROGRAMMING. Mathematical Economics Texts, no. 1. San Francisco: Holden-Day, 1969. vii, 181 p. Appendixes: Proof of Some Results, pp. 157-78. References, pp. 179-80. Index, p. 181.

A text for a one-semester course in dynamic programming for graduates and undergraduates and a reference for workers in varied fields, including economics. It presents some of the main contributions in dynamic programming and the relative advantages and disadvantages of other approaches to optimization, such as the calculus of variations and Pontryagin's maximum principle. It treats deterministic, stochastic, and adaptive processes with applications in equipment replacement, operation of a multistand rolling mill, multistage chemical processes, investment valuations, inventory control, pricing problems, capacity expansion, and search processes. Assumes knowledge of calculus, real analysis, and probability theory. The appendixes consist of proofs of some of the mathematical results.

Williams, N. LINEAR AND NON-LINEAR PROGRAMMING IN INDUSTRY. Topics in Operational Research. London: Sir Isaac Pitman, 1967. x, 182 p. No index.

Applies mathematical programming to a variety of problems, including mixing-type and transportation, with emphasis on the techniques of expressing a problem in a form suitable for calculation and on the use of computer programs in obtaining the solution. No attempt is made to describe in detail the steps involved from problem formulation to the solution. While no computer programs are included, one chapter is devoted to a discussion of the desirable features of good programs. Assumes knowledge of college algebra.

Wolfe, Carvel S. LINEAR PROGRAMMING WITH FORTRAN. Glenview, Ill.: Scott, Foresman, 1973. vii, 232 p. References, pp. 201-3. Appendix A: FORTRAN subroutines for the Primal-dual Algorithm, pp. 204-8. Appendix B: FORTRAN Subroutines for Gomory's Algorithm, pp. 209-11. FORTRAN Glossary, pp. 212-30. Index, pp. 231-32.

An introduction to linear programming with emphasis on the use of computers in solving problems. Many of the concepts are stated in the form of algorithms which lead naturally to computer routines that perform the necessary computations. Topics include graphical methods, convexity, simplex method, duality, integer and parametric programming, and the transportation problem. The chapters conclude with two types of problems: those that can be done by hand while the student is learning techniques, and longer problems that are expected to be run on a digital computer.

Yudin, David Borisovich, and Gol'stein, Evgeniĭ Grigor'evich. LINEAR PRO-GRAMMING. New York: Daniel Dovey, 1965. x, 509 p. Appendix: Mathematical Principles of Linear Programming, pp. 463-99. References, pp. 500-505. Subject Index, pp. 506-9.

Presents the mathematical theory of linear programming and three computational methods which provide an exact solution over a finite number of steps: simplex, dual simplex, and Hungarian methods. This book is aimed at students and practitioners in a variety of fields, including economics. Many theorems are proved, and the chapters conclude with exercises.

Zangwill, Willard I. NONLINEAR PROGRAMMING: A UNIFIED APPROACH. Prentice-Hall International Series in Management. Englewood Cliffs, N.J.: Prentice-Hall, 1969. xvi, 356 p. Appendix: Mathematical and Linear Programming Review, pp. 322-31. Bibliography, pp. 332-45. Index, pp. 347-56.

A presentation of the theoretical aspects of potentially practical algorithms for solving nonlinear programming problems defined on finite dimensional spaces. Among the algorithms discussed are: steepest ascent, Newton, cyclic coordinate ascent, Fibonacci, Bolzano, conjugate gradient, simplex, convex simplex, manifold suboptimization, method of centers, Lagrangian, penalty barrier feasible direction, epsilon-perturbation, and three cutting plane algorithms. The Kuhn-Tucker theory and saddle point theory and their relevance to discrete optimal control, geometric programming, regression, and quadratic programming are also discussed. Little attention is given to restrictions on the number of iterations in reaching the solution, and no attention is given to computer-related problems of round-off errors, computation time, and memory requirements.

Zeleny, Milan. LINEAR MULTIOBJECTIVE PROGRAMMING. Lecture Notes in Economics and Mathematical Systems, vol. 95. New York: Springer-Verlag, 1974. xii, 220 p. Bibliography, pp. 183-85. Appendixes, pp. 187-220. No index.

A development of the theory and algorithms of linear programming problems involving multiple noncommensurable objective functions. The single optimal solution is replaced by a whole set of nondominated solutions. The last section points out extensions to nonlinearity and other topics. The appendixes contain a note on the redundancy among linear constraints, printouts of FORTRAN codes for multicriteria simplex methods, and examples of printouts.

Zionts, Stanley. LINEAR AND INTEGER PROGRAMMING. International Series in Management. Englewood Cliffs, N.J.: Prentice-Hall, 1974. xiv, 514 p. References, pp. 492-507. Index, pp. 508-14.

A comprehensive text on mathematical programming (linear and integer) which emphasizes both theory and applications. Topics include constrained optimization and the simplex method, duality and its significance, network flow methods, applying linear programming to problems, and integer programming. This book is useful for graduates and undergraduates in a variety of fields, including economics, engineering, and mathematics. Assumes knowledge of matrix algebra. The chapters conclude with problems.

Zoutendijk, G. METHODS OF FEASIBLE DIRECTIONS: A STUDY IN LINEAR AND NON-LINEAR PROGRAMMING. New York: American Elsevier, 1960. vi, 126 p. References, pp. 121-26.

A monograph on the mathematical and computational aspects of linear, quadratic, and convex programming. The main results are: (1) the development of a new computational algorithm for the simplex method, called the revised product-form algorithm; (2) the development of several algorithms for convex programming, called the method of feasible directions; and (3) the discovery of an equivalence between many existing methods for linear, quadratic, and convex programming and the method of feasible directions.

Zukhovitskiy, Semen Izrailevich, and Avdeyeva, L.I. LINEAR AND CONVEX PROGRAMMING. Translated by Scripta Technica. Edited and with a foreword by B.R. Gelbaum. Philadelphia: W.B. Saunders, 1966. viii, 286 p. References, pp. 281-83. Index, pp. 285-86.

Originally published in Russian in 1964, this text for a course in linear programming stresses both theory and illustrative examples. Topics include linear, integer, quadratic, and convex programming with applications to games, transportation, networks and maximal flow, flight scheduling, agriculture, military problems, shortest route, and production planning. Rigorous proofs are omitted, and Chebyshev approximations are used in convex (nonlinear) programming problems. Assumes knowledge of linear algebra.

Chapter 11
NETWORKS

Antill, James M., and Woodhead, Ronald W. CRITICAL PATH METHODS IN
CONSTRUCTION PRACTICE. 2d ed., rev. New York: Wiley-Interscience,
1970. xii, 414 p. Appendix A: CPM and Linear Graph Theory, pp. 369-80.
Appendix B: Specification for Planning and Project Control, pp. 381-84. Appendix C: Answers to Problems, pp. 385-403. Index, pp. 405-14.

The application of critical path methods to problems arising in
management, administration, and construction. Emphasis is on use
of diagrams, and no mathematics beyond arithmetic is required.
Topics include network diagrams and calculations, practical planning with critical path methods, project control, evaluation of
work changes and delays, and computer processing. Chapters conclude with problems.

Archibald, Russell D., and Villoria, Richard L. NETWORK-BASED MANAGE-
MENT SYSTEMS (PERT/CPM). Information Sciences Series. New York: Wiley,
1967. xiv, 508 p. Appendixes: A, Glossary of Terms; B, Precedence Dia-
gramming--An Alternative Method of Network Construction; C, An Analytical
Study of the PERT Assumptions; D, Suggested Course Outline; E, Selected Bibliography; F, DOD and NASA Guide--PERT/Cost Output Reports, pp. 433-504.
Index, pp. 505-8.

A text for a course in network analysis based on PERT and CPM,
and a reference for on-the-job use. It is divided into the following five parts: network planning: what it is, how it works,
implementing the system, case studies, pitfalls and potentials, and
appendixes. The presentation is based on extensive experience in
the use of network planning for industrial and governmental work
and in teaching the subject to a wide range of persons, including
managers and undergraduate students. Knowledge of college algebra is helpful, but not necessary. Chapters conclude with summaries.

Ashour, Said. SEQUENCING THEORY. Lecture Notes in Economics and
Mathematical Systems, vol. 69. New York: Springer-Verlag, 1972. v,
133 p. No index.

A supplementary text for a course in operations research with a prerequisite of college algebra, elementary matrix algebra, and basic statistics. It consists of an introduction to sequencing theory and a treatment of the combinatorial aspects of sequencing problems. The presentation is not rigorous mathematically, the concepts are described and illustrated in great detail. Chapters conclude with references.

Ashton, Winifred Dianna. THE THEORY OF ROAD TRAFFIC FLOW. Methuen's Monographs on Applied Probability and Statistics. New York: Wiley, 1966. viii, 178 p. References, pp. 169-74. Subject Index, pp. 175-76. Author Index, pp. 177-78.

A comprehensive account of the theory of traffic flows for statisticians, applied mathematicians, and engineers specializing in the traffic field. Topics include dynamical and kinematic theories of traffic flow, traffic cybernetics, stochastic approach to traffic problems, simulation of traffic problems, and an analysis of accident statistics. Emphasis is on description and application of theories, rather than on rigorous proofs. Assumes knowledge of differential and integral calculus, statistical distributions, and queueing theory.

Baker, Bruce N., and Eris, Rene L. AN INTRODUCTION TO PERT-CPM. Homewood, Ill.: Richard D. Irwin, 1964. x, 85 p. Appendix: Monthly Progress Report Format, pp. 75-78. Glossary of Symbols, pp. 79-83. Bibliography, p. 85. No index.

An introduction to the fundamentals and uses of PERT and CPM for planning and scheduling work that involves a high degree of uncertainty. Few formulas are used, and emphasis is placed on the use of charts, tables, and numerical examples. Topics include basic elements of PERT, network preparation, data structuring processing, analysis and management action, applications of PERT, evaluation of PERT-CPM and other network planning techniques, such as COMET (computer operated management evaluation techniques) and IMPACT (implementation, planning, and control techniques).

Baker, Kenneth R. INTRODUCTION TO SEQUENCING AND SCHEDULING. New York: Wiley, 1974. ix, 305 p. Appendix A: Sixteen Test Problems for the T Problem, p. 289. Appendix B: A Dynamic Programming Code for Sequencing Problems, pp. 290-91. Glossary, p. 292. FORTRAN Code, pp. 293-94. Appendix C: A Design Strategy for a Branch and Bound Code, pp. 295-99. Index, pp. 301-5.

A text for a course on sequencing and scheduling for graduates and undergraduates in operations research, industrial engineering, and management science. It is also a reference on models and methodologies in the field. Topics include single-machine sequencing and methodologies, parallel-machine methods, flow shop and job shop scheduling, simulation studies of a dynamic job shop, network methods for project scheduling, and resource-constrained project scheduling. The problem-solving techniques include branch and bound, integer programming, dynamic programming, and random sampling and discrete search. Examples are used freely to illustrate concepts. Chapters conclude with three types of exercises: computational exercises requiring simple calculations; conceptual exercises involving the construction of counterexamples and building models; and algorithmic exercises requiring the design and implementation of computerized solution methods. Chapters conclude with summaries and references. Assumes knowledge of probability and statistics and computer programming.

Barnetson, Paul. CRITICAL PATH PLANNING: PRESENT AND FUTURE TECHNIQUES. London: Newnes-Butterworths, 1968. 102 p. Bibliography, pp. 98-99. Index, pp. 101-2.

A nontechnical presentation of the basic techniques of CPM for the practitioner. Chapter 1 explains basic concepts, such as how to draw and alter a network. Chapter 2 covers network cost control. Chapter 3 considers the problem of resource allocation-- the problem of scheduling a limited number of men on a complex project with as little delay as possible. Chapter 4 lists some of the newer methods of CPM being developed.

Battersby, Albert. NETWORK ANALYSIS FOR PLANNING AND SCHEDULING. 3d ed., rev. New York: Wiley, 1970. ix, 332 p. Appendix 1: The PERT Statistics, pp. 261-69. Appendix 2: Standard Symbols, pp. 270-71. Appendix 3: Types of Float, pp. 272-75. Appendix 4: Probabilistic Networks, pp. 276-81. Appendix 5: Estimating Durations and Costs, pp. 282-86. Appendix 6: A Note on the Use of Modulex, pp. 287-88. Bibliography, pp. 289-94. Solutions to Exercises, pp. 295-324. Index, pp. 325-32.

An introduction to the techniques and applications of network analysis (primarily PERT) for the practitioner. Little formal mathematics training is assumed. It incorporates practical applications in a wide variety of British industries, and it includes three case studies in installing network analysis: the construction of liquid oxygen tankers (a real life study), launching a new product (a hypothetical case), and prevention and detection of crime (hypothetical case). Topics include arrow diagrams; critical paths; the method of potentials; variable costs and durations; analogue and analytical methods; heuristic methods; capital, materials and shortage costs; and monitoring, control, and project organization. Numerous charts, diagrams, and tables are included as aids to applications. Chapters conclude with questions and exercises to test the reader's skill in the application of the concepts.

Benes, Vaclav E. MATHEMATICAL THEORY OF CONNECTING NETWORKS AND TELEPHONE TRAFFIC. Mathematics in Science and Engineering, vol. 17.

New York: Academic Press, 1965. xiv, 319 p. Suggested Readings, p. 314. Subject Index, pp. 317-19.

An introduction to telephone traffic theory emphasizing three kinds of problems: (1) combinatorial problems of network design; (2) probabilistic problems of traffic analysis; and (3) variational problems of routing telephone traffic in networks. It is based almost entirely on research papers written by the author at Bell Telephone Laboratories during the period 1954 to 1964. One chapter on non-blocking networks is based on a research paper by C. Clos. The prerequisites include advanced calculus, mathematical analysis, probability theory, and topology. Chapters conclude with references.

Brandon, Dick H., and Gray, Max. PROJECT CONTROL STANDARDS. New York: Brandon/Systems Press, 1970. vii, 204 p. Appendix: Software Packages for Project Control, pp. 183-89. Bibliography, pp. 193-204. No index.

A guide to the principles, techniques, and development of project control. It is divided into four sections: general principles of project control; procedures on project selection, personnel assignment, scheduling, progress reporting, and cost allocation; outline of recommended documents; and steps in developing and implementing a project control system. The bibliography is divided into the following three parts: recommended references, PERT and CPM bibliography, and project control bibliography. The appendix describes several computer program packages which perform some of the functions of project control.

Burman, P.J. PRECEDENCE NETWORKS FOR PROJECT PLANNING AND CONTROL. New York: McGraw-Hill Book Co., 1972. xi, 374 p. Appendixes, pp. 305-68. References, p. 368. Index, pp. 369-74.

A nonmathematical book on project planning for the layman, managers, and others responsible for implementing network projects. It presents a form of analysis designed to take the guesswork out of project management. Emphasis is placed on the use of examples, charts, and diagrams.

Elmaghraby, Salah Eldin. SOME NETWORK MODELS IN MANAGEMENT SCIENCE. Lecture Notes in Operations Research and Mathematical Systems, vol. 29. New York: Springer-Verlag, 1970. iii, 176 p. No index.

An expository treatment of four network models: shortest path problems, maximum flow models, signal flow graphs, and activity networks. Emphasis is placed on the relevance and applicability of each model to management science and operations research, rather than on mathematical rigor. Each of the four chapters concludes with references. Assumes knowledge of calculus.

Evarts, Harry F. INTRODUCTION TO PERT. The Allyn and Bacon Series in Quantitative Methods for Business and Economics. Boston: Allyn and Bacon, 1964. xi, 112 p. Glossary, pp. 94-96. Selected Bibliography, pp. 97-98. Appendix: IBM 709 PERT Computer Operations, pp. 99-112. No index.

A nonmathematical introduction to PERT with emphasis on understanding commonly used management techniques.

Ford, Lester Randolph, Jr., and Fulkerson, D.R. FLOWS IN NETWORKS. Princeton, N.J.: Princeton University Press, for the Rand Corporation, 1962. xii, 194 p. Index, pp. 193-94.

A monograph on the linear theory of network flows with emphasis on proofs of theorems that lead to computationally efficient algorithmic procedures. It is restricted to flow problems for which the assumption of integral data implies the existence of an integral solution. Combinatorial methods of solution are developed, and the simplex method of solution for network flow problems is omitted. The notion of a spanning subtree of a network is not introduced until the last chapter. The four chapter titles are: "Static Maximal Flow," "Feasibility Theorems and Combinatorial Applications," "Minimal Cost Flow Problems," and "Multi-terminal Maximal Flows." Assumes knowledge of linear algebra. Chapters conclude with references.

Horowitz, Joseph. CRITICAL PATH SCHEDULING: MANAGEMENT CONTROL THROUGH CPM AND PERT. New York: Ronald Press, 1967. v, 254 p. Bibliography, pp. 213-15. Glossary, pp. 217-22. Solutions to Problems, pp. 223-49. Index, pp. 251-54.

A book for business and industrial managers as well as for architects and engineers on critical path methods (CPM) and its companion network method PERT (program evaluation and review technique). Emphasis is placed on the use of examples and illustrations, and no knowledge of special mathematical techniques is assumed. Chapters conclude with problems.

Iannone, Anthony L. MANAGEMENT PROGRAM PLANNING AND CONTROL WITH PERT, MOST AND LOB. Englewood Cliffs, N.J.: Prentice-Hall, 1967. xvii, 202 p. Appendixes, pp. 169-93. Bibliography, p. 194. Index, pp. 195-202.

Presents the following three techniques on planning, scheduling, and control in production, plant engineering, construction, research and development, as well as other areas: program evaluation and review technique (PERT), management operation system technique (MOST), and line of balance (LOB). The appendixes consist of pictorial diagrams and charts, glossary, and a list of abbreviations. No formal mathematics is used.

Iri, Masao. NETWORK FLOW, TRANSPORTATION, AND SCHEDULING: THEORY AND ALGORITHMS. Mathematics in Science and Engineering, vol. 57. New York: Academic Press, 1969. xi, 316 p. Appendixes, pp. 256-306. References, pp. 307-9. Index, pp. 311-16.

> An exposition of the theory of networks and a development of practical algorithms for solving network problems. Aimed at a mixed audience, including applied mathematicians, students of operations research, and engineers. Emphasis is on the mathe- matical programming formulation of problems and, to some extent, on the graph-theoretic approach. Topics include topological prop- erties of networks, general network flow problems, and linear network flow problems. Assumes knowledge of linear algebra, differential and integral calculus, set theory and elementary topol- ogy. Convex sets and functions and the topological properties of linear graphs are reviewed in chapter 2. The appendixes con- tain a survey of topics and research problems in related fields, as well as mathematical supplements and suggestions on digital and analog computations.

Jurecka, Walter. NETWORK PLANNING IN THE CONSTRUCTION INDUS- TRY. Translated by F.L. Carvalho and F.H. Turner. London: Maclaren and Sons, 1969. vii, 112 p. No index.

> First published in German in 1967, this book is a practical guide to network planning for the businessman. It includes examples in road construction, sewer tunnel construction, multistory building, and pipe-laying activities. There are no mathematics prerequi- sites, and a large number of network charts are included in the cover packet.

Kaufmann, Arnold, and Desbazeille, G. THE CRITICAL PATH METHOD: AP- PLICATION OF THE PERT METHOD AND ITS VARIANTS TO PRODUCTION AND STUDY PROGRAMS. Preface by E. Ventrua. New York: Gordon and Breach, 1969. x, 167 p. Appendix 1: Beta Distribution, pp. 141-43. Ap- pendix 2: The Critical Path Mean Value in a Program Graph, pp. 144-52. Appendix 3: Subdivision of Jobs and Realization of a Schedule, pp. 153-59. Appendix 4: Application of the Method of Potentials, pp. 160-64. Bibliog- raphy, pp. 165-67. No index.

> A reference on the application of PERT, mainly for planners. It begins by describing the techniques of graphs and sequencing and the steps in establishing a program of research or production, and then considers examples of PERT, including stochastic models. The last chapter takes up the problem of optimizing a program using the Fulkerson algorithm. Knowledge of elementary calcu- lus would be helpful, especially for the appendixes.

Lang, Douglas W. CRITICAL PATH ANALYSIS: TECHNIQUES, EXERCISES, AND PROBLEMS. Teach Yourself Books. London: English Universities Press, 1970. xi, 174 p. No index.

A practical guide to critical path analysis using scarcely any mathematics. Included are two variants of critical path analysis: activity-on-node, and line of balance, which is applicable to fluctuating batch manufacturing. The last two chapters consist of two case studies: installing a computer, and building a V.H.F. transmitter. Chapters conclude with exercises.

Levin, Richard I., and Kirkpatrick, Charles A. PLANNING AND CONTROL WITH PERT/CPM. New York: McGraw-Hill Book Co., 1966. x, 179 p. Appendixes, pp. 147-73. Index, pp. 175-79.

Presents PERT and CPM for practitioners who have knowledge of only high school algebra. Topics include PERT fundamentals, work breakdown schedules, networking principles, network replanning and adjustment, probability concepts, and the use of the computer in PERT applications. The appendixes consist of other methods of project planning and control, bibliographies, and tables.

Lockyer, K.G. AN INTRODUCTION TO CRITICAL PATH ANALYSIS. New York: Pitman Publishing Corp., 1964. viii, 111 p. Selected Reading, pp. 105-7. Index, pp. 109-11.

Presents critical path analysis for those who wish to apply the techniques. Requires no mathematical expertise, and makes liberal use of graphs and tables to explain concepts.

Lowe, Cecil William. CRITICAL PATH ANALYSIS BY BAR CHART: A NEW ROLE OF JOB PROGRESS CHARTS. New York: Brandom/Systems Press, 1966. x, 188 p. Glossary, pp. 180-81. Bibliography, pp. 182-84. Index, pp. 185-88.

Presents a simple procedure for critical path analysis using job progress charts without the use of networks, computers, and mathematics. A job progress chart describes a collection of related jobs as a network while at the same time it defines the time limitations in a time chart. This method is illustrated by actual examples taken from the operations of Monsanto Chemicals Limited.

McLaren, K.G., and Buesnel, Eric Leonard. NETWORK ANALYSIS IN PROJECT MANAGEMENT: AN INTRODUCTORY MANUAL BASED ON UNILEVER EXPERIENCE. Cassell Management Studies. London: Cassell and Co., 1969. vii, 219 p. Appendix A: Alternative Symbols, Types of Float, Errors in Float Calculation, Three-time Estimates, Constructing a Project Cost Curve, pp. 182-91. Appendix B: Glossary of Terms in Network Analysis, pp. 192-94. Appendix C: Answers to and Comments on the Exercises, pp. 195-212. Appendix D: Bibliography, pp. 213-15. Index, pp. 216-19.

A manual for self-instruction in the application of network analysis in Unilever, Ltd., of the United Kingdom. Based on the authors' experiences in teaching and applying network analysis to various projects in that company. Topics include time analysis of the diagram, activity times and floats, developing networks in practice, examples in the use of network analysis, management implications, and using the computer. Chapters conclude with exercises.

Martino, R.L. CRITICAL PATH NETWORKS. New York: Gordon and Breach, 1967. 157 p. Appendix 1: Probability, pp. 147-53. Appendix 2: Computer Programs, pp. 154-55. About the Author, p. 157.

A primer on the nature, scope, and applications of critical path networks in the technical environment in which they must be applied. Presents universal techniques independent of any particular profession, discipline, or configuration of computer, and assumes knowledge of only high school algebra. Topics include network formulation, modeling a real project, using master nets, job boundaries and float, and critical path--its nature and use in scheduling, and event-oriented networks. Each of the ten chapters concludes with summaries and exercises.

Maxwell, Lee M., and Reed, Myril B. THE THEORY OF GRAPHS: A BASIS FOR NETWORK THEORY. Pergamon Unified Engineering Series. New York: Pergamon Press, 1971. xv, 164 p. Table of Symbols, pp. xiv-xv. Index, pp. 162-64.

A presentation of the development of connection aspects of systems for use by students in a wide variety of disciplines who have knowledge of college algebra and introductory set theory. Topics include: the path and circuit, parts and connected subgraphs, separable and nonseparable parts, the tree and co-tree, the seg and co-seg, the circ, and network theory. Chapters conclude with problems and references.

Mayeda, Wataru. GRAPH THEORY. New York: Wiley-Interscience, 1972. 588 p. Bibliography, pp. 559-77. Symbols, pp. 579-81. Nomenclature, pp. 583-88.

An introduction to linear graph theory and applications which will bring the reader far enough into the field to enable him to embark on a research problem of his own. Topics include incidence set and cut-set, matrix representation of linear graphs and trees, planar graphs, special cut-set and pseudo-cut, topological analysis of passive networks, and signal flow graphs. The applications are confined to electrical network theory, switching theory, communication net and transportation theory, and system diagnosis. Linear algebra and elementary point set topology are the main mathematics prerequisites. Chapters conclude with problems.

Miller, Robert W. SCHEDULE, COST AND PROFIT CONTROL WITH PERT: A COMPREHENSIVE GUIDE FOR PROGRAM MANAGEMENT. New York:

McGraw-Hill Book Co., 1963. xvi, 227 p. Appendix 1: Simplified Derivation of PERT Equations, pp. 199-201. Appendix 2: Representative Bibliography, pp. 203-6. Appendix 3: Glossary of Some Representative Management Systems Terms, pp. 207-15. Appendix 4: Glossary of PERT Terminology,
pp. 217-22. Index, pp. 223-27.

> A historical treatment of PERT techniques with emphasis on spe-
> cial purpose or one-time-through programs, in contrast to continous
> production operations. It also stresses application of PERT techniques
> to areas of time, cost, and product performance, representing an inte-
> grated management approach to all three of these factors. The pre-
> sentation is strongly applied, and may be of little interest from a re-
> search point of view. Chapters conclude with references.

Moder, Joseph J., and Phillips, Cecil R. PROJECT MANAGEMENT WITH
CPM AND PERT. 2d ed., rev. New York: Van Nostrand Reinhold, 1970. xviii,
360 p. Nomenclature, pp. xvii-xviii. Solution to Exercises, pp. 344-52.
Index, pp. 353-60.

> A text and reference on the theory and application of CPM and
> PERT. Part 1 deals with basic topics, such as how to develop
> a network, time estimates, basic scheduling computations, and
> use of computer. Useful for a two- or three-day training course
> for industrial personnel. Part 2 takes up specialized topics, in-
> cluding a variety of networking schemes, resource allocation, and
> network cost control. Part 2 is suitable for college-level courses
> in business administration, industrial engineering, civil engineer-
> ing, and systems engineering. Certain portions of the text re-
> quire some knowledge of mathematical statistics and linear pro-
> gramming. Some chapters contain appendixes with computer pro-
> grams, statistical tables, and construction project methods and
> reports. Chapters conclude with summaries, references, and exer-
> cises.

Morris, L.N. CRITICAL PATH: CONSTRUCTION AND ANALYSIS. Elmsford,
N.Y.: Pergamon, 1967. vii, 114 p. Tables and Charts. No index.

> A self-teaching book on the construction and analysis of networks
> with no college mathematics required. Problems with detailed
> solutions make up the content of the last chapter. Topics include
> the logic of networks, construction of networks, network analysis-
> project duration, network analysis-float, allocation of resources,
> and the impact of critical path analysis on supervisory management.

Mulvaney, John Edward. ANALYSIS BAR CHARTING: A SIMPLIFIED CRITI-
CAL PATH ANALYSIS TECHNIQUE. International Scientific Series. Cleveland: Chemical Rubber Co. Press, 1969. 100 p. Index, pp. 97-100.

> Presents the analysis of bar charting (ABC) method of critical path
> analysis for managers in marketing, production, construction, and
> administration. It presents simple procedures, in the form of

charts and graphs, for scheduling and controlling projects, and it sets up the training procedures required in the use of ABC. There are no mathematics prerequisites.

Peart, Alan Thomas. DESIGN OF PROJECT MANAGEMENT SYSTEMS AND RECORDS. London: Goiver Press, 1971. xiii, 189 p. Appendix 1: Glossary of Terms, pp. 169-81. Appendix 2: Bibliography and Suppliers, pp. 182-83. Index, pp. 185-89.

A book for managers of minor or isolated projects which is based upon practical experience in the management of both major and minor projects. Assumes knowledge of critical path and network analyses.

Price, W.L. GRAPHS AND NETWORKS: AN INTRODUCTION. Philadelphia: Auerbach Publishers, 1971. 108 p. Index, pp. 105-8.

An introduction to graphs and networks for graduates and undergraduates in operations research and related fields. A mathematical review is presented in chapter 2, and chapters 3 and 4 deal with distance and flow networks, including algorithms for calculating shortest paths and maximum flows. Chapter 5 is concerned with activity networks, PERT, and CPM. Each chapter contains worked examples, exercises and answers, and selected references.

Reutlinger, Shlomo. TECHNIQUES FOR PROJECT APPRAISAL UNDER UNCERTAINTY. Foreword by R.S. McNamara. World Bank Staff Occasional Papers, no. 10. Baltimore: Johns Hopkins Press, for the International Bank for Reconstruction and Development, 1970. xiii, 95 p. Annex Tables, pp. 92-93. Bibliography, pp. 94-95. No index.

A presentation of a cost-benefit method for evaluating the riskiness of investment projects with emphasis on methodology and problems of measurement. Part 1 of the book deals with probability analysis, and part 2 takes up hypothetical case illustrations of a highway project and irrigation project. Costs and benefits are evaluated in terms of probability distributions to yield a probability distribution of the rate of return, or net worth. Assumes knowledge of elementary calculus and probability theory.

Riggs, James L., and Heath, Charles O. GUIDE TO COST REDUCTION THROUGH CRITICAL PATH SCHEDULING. Englewood Cliffs, N.J.: Prentice-Hall, 1966. 221 p. Appendix 1: Illustrative Problems, pp. 171-86. Appendix 2: Problem Solutions, pp. 187-214. Selected References, pp. 215-16. Index, pp. 217-21.

A self-teaching text with numerous examples and step-by-step instructions. Topics include setting up the project plan, determining the critical path, reading and constructing time charts, analyzing and controlling project costs by critical path scheduling (CPS), implementing the CPS program, computerizing CPS applications, and decision making with CPS.

Robertson, D.C. PROJECT PLANNING AND CONTROL: SIMPLIFIED CRITICAL PATH ANALYSIS. Cleveland: Chemical Rubber Co. Press, 1967. 105 p. Index, pp. 103-5.

Presents a practical and simple approach to critical path analysis based on diagrams and tables. The five chapter titles are: "Introduction to the Problems," "Project Planning and Scheduling," "Resource Leveling," "Project Control," and "Critical Path Analysis in Practice." There are no mathematics prerequisites.

Shaffer, Louis Richard; Ritter, J.B.; and Meyer, W.L. THE CRITICAL PATH METHOD. New York: McGraw-Hill Book Co., 1965. xi, 212 p. References, pp. 191-93. Appendix: Solution to Problems, pp. 195-212. No index.

A self-teaching text for those who have not had the opportunity to become familiar with linear programming and the other mathematical principles upon which CPM is based. Emphasis is on application without detailed calculation, and use is made of words and graphs, rather than formulas. One chapter is devoted to calculations by computer programs.

Simms, Alfred G., and Britten, John R. PROJECT NETWORK ANALYSIS AND CRITICAL PATH. Modern Aids to Project Management. London: Machinery Publishing Co., 1969. 98 p. References, p. 95. Index, pp. 96-98.

An introduction to network analysis for practitioners. Topics include network diagrams, duration of stages, project timing, project control, the cost of time, and critical path analysis and the computer. Assumes knowledge of high school algebra.

Staffurth, C., ed. PROJECT COST CONTROL USING NETWORKS. London: The Operational Research Society and The Institute of Cost and Works Accountants, 1969. 92 p. References, p. 87. Appendix: Glossary of Terms, pp. 89-92. No index.

An attempt to improve the communication between the accountant, who may not be familiar with operations research terminology, and the engineer who is well versed in the techniques of network analysis. The book is a cooperative undertaking of eleven individuals representing varied interests and experiences. Topics include the fundamentals and applications of project cost control, management problems, use of computers, advantages of project cost control over traditional methods, and the cost of implementation. There are no mathematics prerequisites.

Whitehouse, Gary E. SYSTEMS ANALYSIS AND DESIGN USING NETWORK TECHNIQUES. Englewood Cliffs, N.J.: Prentice-Hall, 1973. xii, 500 p. Appendix, pp. 491-93. Index, pp. 495-500.

> The network techniques discussed include PERT, CPM, decision trees, network flows, flowgraphs, and GERT (graphic evaluation and review techniques). Useful as a text in schools of business, industrial engineering, operations research, and computer science. Assumes knowledge of probability theory. Chapters conclude with lengthy exercises and references.

Wiest, Jerome D., and Levy, Ferdinand K. A MANAGEMENT GUIDE TO PERT/CPM. Englewood Cliffs, N.J.: Prentice-Hall, 1969. v, 176 p. Bibliography, pp. 137-46. Exercises, pp. 147-67. Index, pp. 168-70.

> A guide and basic reference to the techniques of PERT and CPM for actual users in industry and for graduate students in management-oriented courses in college. Advanced methods in PERT/CPM techniques requiring mathematics and computer programs are discussed in the appendixes to the chapters. Knowledge of finite mathematics is helpful, and graphs are used freely to illustrate the concepts.

Woodgate, Harry Samuel. PLANNING BY NETWORKS: PROJECT PLANNING AND CONTROL USING NETWORK TECHNIQUES. London: Business Publications, 1964. xvii, 338 p. Appendix: Glossary of Terms, pp. 323-27. Index, pp. 328-38.

> A nonmathematical presentation of planning by networks based on the experience of International Computers and Tabulators in more than 200 different applications worldwide. Topics include multi-level and sectionalized networks, time estimates for network activities, analysis of networks, presentation of results and replanning of networks, planning for minimum cost, and the use of a computer for project planning and control.

Chapter 12

OPTIMIZATION AND CONTROL

Adby, P.R., and Dempster, Michael Alan Howarth. INTRODUCTION TO OPTIMIZATION METHODS. Chapman and Hall Mathematics Series. New York: Wiley, Halsted Press, 1974. x, 204 p. References, pp. 187-95. Further Reading, pp. 196-98. Index, pp. 199-204.

A text for a course in nonlinear methods of optimization for undergraduates and graduates in mathematics, the physical and social sciences, and engineering. The first half covers basic optimization techniques, including linear search methods, steepest descent, least squares, and the Newton-Raphson method, all illustrated with worked examples. The second half describes algorithms such as variate metric methods for unconstrained problems and penalty function methods for constrained problems. The mathematical notation is rigorous, but the presentation is relatively simple and interdisciplinary. No flow charts are included and no knowledge of computer programming is necessary. Assumes knowledge of differential and integral calculus. Chapters conclude with exercises.

Akhiezer, Naum I. THE CALCULUS OF VARIATIONS. Translated by A.H. Frink. A Blaisdell Book in the Pure and Applied Sciences. New York: Blaisdell, 1962. viii, 247 p. Appendix: Supplementary Topics and Exercises, pp. 161-239. Answers, pp. 239-41. Bibliography, pp. 242-43. Index, pp. 245-47.

Translated from the Russian, this book is a text for a course in the calculus of variations for graduates and undergraduates who possess knowledge of differential and integral calculus, and mathematical analysis. Topics include the necessary conditions of Weierstrauss and Legendre, the Hamilton-Jacobi theorem, and the Lagrange problem, and concludes with a chapter on direct methods and proofs of the theorems of Hilbert and Tonelli. Each of the four chapters concludes with up to one dozen exercises. Additonal exercises are contained in the appendix.

Optimization and Control

Albrecht, Felix. TOPICS IN CONTROL THEORY. Lecture Notes in Mathematics, vol. 63. New York: Springer-Verlag, 1968. 65 p. References, p. 65. No index.

Concerned with the Pontryagin maximum principle, which states a necessary condition for a control to be optimal. Included is the proof that the Pontryagin principle holds for a larger class of controls which are characterized by geometric rather than variational properties. Knowledge of Banach spaces is essential.

Anderson, Brian D.O., and Moore, John B. LINEAR OPTIMAL CONTROL. Prentice-Hall Network Series. Englewood Cliffs, N.J.: Prentice-Hall, 1971. xiv, 399 p. Appendix A: Brief Review of Some Results of Matrix Theory, pp. 376-88. Appendix B: Brief Review of Several Results of Linear Systems, pp. 389-92. Index, pp. 393-99.

A text for a graduate course in optimal control for students who are familiar with elementary control theory, introduction to the state-space description of linear systems, and linear algebra. It is divided into the following five parts: introduction, basic theory of the optimal regulator, properties and application of the optimal regulator, extensions to more complex problems, and computational aspects. Restricted to continuous, time-invariant systems which do not employ the Pontryagin maximum principle. The presentation is fairly rigorous with the mathematical results either stated or derived, but few theorems are proved. The sections of the book conclude with a few problems, and the chapters conclude with references.

Aoki, Masanao. INTRODUCTION TO OPTIMIZATION TECHNIQUES: FUNDAMENTALS AND APPLICATIONS OF NONLINEAR PROGRAMMING. Macmillan Series in Applied Computer Sciences. New York: Macmillan, 1971. xvi, 335 p. Appendixes, pp. 269-309. List of Algorithms, pp. 311-13. References, pp. 315-20. Flow charts, pp. 321-29. Index, pp. 331-35.

A text for an undergraduate course in nonlinear programming with applications to design optimization problems. It presents the following techniques for locating extrema (minima or maxima) of a function of several variables: small- and large-step gradient methods, the Newton-Raphson and Davidson-Fletcher-Powell methods, and direct search methods. Function extremization subject to linear and nonlinear constraints are also discussed. Knowledge of calculus and vector spaces is assumed.

Astrom, Karl J. INTRODUCTION TO STOCHASTIC CONTROL THEORY. Mathematics in Science and Engineering, vol. 70. New York: Academic Press, 1970. xi, 299 p. Index, pp. 295-99.

A presentation of stochastic control theory from the viewpoints of analysis, parametric optimization, and optimal stochastic control with emphasis on applications. Limited to linear systems with

quadratic criteria. Each chapter includes both the discrete and
continuous time versions. Topics include minimal variance control
strategies, with an industrial application in the paper industry in
Sweden, and prediction and filtering problems. Most of the mathematical results are stated without proof. Assumes knowledge of
mathematical analysis, probability, and dynamical systems. Useful as a text for a short applications-oriented course, for an introductory course in discrete time stochastic control, or for a
course in stochastic control. Many sections of the book conclude
with exercises, and the chapters conclude with additional exercises, references, and comments.

Balakrishnan, A.V. INTRODUCTION TO OPTIMIZATION THEORY IN A
HILBERT SPACE. Lecture Notes in Operations Research and Mathematical
Systems, vol. 42. New York: Springer-Verlag, 1971. iv, 153 p. Bibliog-
raphy, p. 153. No index.

Presents basic concepts and techniques of functional analysis relevant to optimization problems in control, communications, and
other areas in system science. Chapter titles are: "Basic Properties of Hilbert Spaces"; "Functions, Transformations, Operators";
"Semigroups of Linear Operators"; and "Probability Measures on
a Hilbert Space." Assumes knowledge of Hilbert space theory.

Barnett, Stephen. MATRICES IN CONTROL THEORY WITH APPLICATIONS
TO LINEAR PROGRAMMING. New York: Van Nostrand Reinhold, 1971.
xiv, 221 p. Appendixes, pp. 165-77. Solutions to Exercises, pp. 178-213.
Subject Index, pp. 215-18. Author Index, pp. 219-21.

Presents some of the more interesting and useful applications of
matrices to control theory. Topics include polynomials and polynomial matrices, rational and unimodular matrices, stability and
inertia, matrix Riccati equations, and generalized inverses. Proofs
of theorems are given only where they can be presented in a
simple and clear manner. Assumes knowledge of matrix algebra,
elementary control theory, and linear programming. The appendixes contain brief summaries of mathematical developments. Useful as a self-study for graduates and undergraduates. Exercises
follow the sections of the book, and the chapters conclude with
references.

Becker, Martin. THE PRINCIPLES AND APPLICATIONS OF VARIATIONAL
METHODS. Foreword by J.A. Stratton. Research Monograph no. 27. Cam-
bridge, Mass.: MIT Press, 1964. vii, 120 p. Appendixes, pp. 85-115.
References, pp. 116-17. Index, pp. 119-20.

Presents methods based on the calculus of variations which combine
"trial functions" (guesses about the form of the solution) into a
satisfactory approximate solution. One method is called the simulated boundary-value approach, and another is the least-squares

variational approach, which is applied to the problem of fuel depletion in a nuclear reactor. The appendixes contain a review of the adjoint function, the elements of the calculus of variations, appropriate functionals, and eigenfunctions and error interpretations. Assumes knowledge of differential and integral calculus.

Bellman, Richard E. ADAPTIVE CONTROL PROCESSES: A GUIDED TOUR. Princeton, N.J.: Princeton University Press, for the Rand Corporation, 1961. xvi, 255 p. Index, pp. 251-55.

An expanded version of lectures given by the author at the Hughes Aircraft Company on the dynamic programming approach to optimal control. Topics include computational aspects of dynamic programming, the Lagrange multiplier, two-point boundary value problems, sequential machines, Markovian decision processes, quasilinearization, learning models, games and pursuit processes, adaptive processes, communications theory, and successive approximation. Useful as a reference for mathematicians, engineers, physicists, operations analysts, and mathematical economists. Assumes knowledge of differential and integral calculus, mathematical analysis, and probability theory.

_____. INTRODUCTION TO THE MATHEMATICAL THEORY OF CONTROL PROCESSES. Vol. 1: LINEAR EQUATIONS AND QUADRATIC CRITERIA. Mathematics in Science and Engineering, vol. 40-I. New York: Academic Press, 1967. xvi, 245 p. Author Index, pp. 241-42. Subject Index, pp. 243-45.

A text for an introductory course in control theory for students in applied mathematics, engineering, biology, economics, operations research, and related fields who are familiar with undergraduate calculus, ordinary differential equations, and matrix algebra. After reviewing second-order linear differential equations, attention is given to a rigorous treatment of the problem of minimizing a quadratic functional by obtaining the Euler equations. Then dynamic programming is applied to discrete control problems and is shown to yield results obtainable from the calculus of variations approach. Finally, using matrix notation, multidimensional optimization problems are treated using both dynamic programming and the calculus of variations. The book concludes with an account of variational problems by functional analysis using Hilbert space theory. Some sections of the book conclude with exercises, and the chapters conclude with additional exercises and references.

_____. INTRODUCTION TO THE MATHEMATICAL THEORY OF CONTROL PROCESSES. Vol. 2: NONLINEAR PROCESSES. Mathematics in Science and Engineering, vol. 40-II. New York: Academic Press, 1971. xix, 306 p. Author Index, pp. 299-303. Subject Index, pp. 304-6.

An introduction to the analytic and computational aspects of deterministic optimization models involving nonlinear differential equations and functionals. The discrete version is treated by the theory of dynamic programming, and the continuous version by the calculus of variations. Applications are avoided, and numerical solutions are motivated by questions which arise in numerical investigations. Three chapters are devoted specifically to numerical solutions of control problems. Useful to students in the applied fields, including economics. Sections of the book contain exercises, and the chapters conclude with miscellaneous exercises, bibliographies, and comments. Assumes knowledge of partial differential equations.

Bellman, Richard E., and Cooke, Kenneth L. DIFFERENTIAL-DIFFERENCE EQUATIONS. Mathematics in Science and Engineering, vol. 6. New York: Academic Press, for the Rand Corporation, 1963. xvi, 462 p. Author Index, pp. 457-59. Subject Index, pp. 460-62.

A reference for advanced topics in control theory. Topics include the Laplace transform, various types of first-order linear differential-difference equations, renewal equations, asymptotic behavior of linear and nonlinear differential-difference equations and stability theories of these equations, and asymptotic location and stability properties of the zeros of exponential polynomials. Chapters conclude with exercises and research problems, bibliographies, and comments.

Bellman, Richard E.; Cooke, Kenneth L.; and Lockett, Jo Ann. ALGORITHMS, GRAPHS, AND COMPUTERS. Mathematics in Science and Engineering, vol. 62. New York: Academic Press, 1970. xv, 246 p. Author Index, pp. 239-41. Subject Index, pp. 243-46.

An introduction to the structure, operation, and control of large systems for those desiring a general view of the methodologies of solutions and specific information on mathematical and computational solution methods. Emphasis is on computer programming for solving problems, but since languages vary widely and change quickly, no programs are provided. One problem considered is that of traversing a network of streets in such a way as to minimize the time to go from one point to another. Assumes a maturity comparable to that obtained in a calculus course. Sections contain exercises, and the chapters conclude with miscellaneous exercises, and references.

Beltrami, Edward J. AN ALGORITHMIC APPROACH TO NONLINEAR ANALYSIS AND OPTIMIZATION. Mathematics in Science and Engineering, vol. 63. New York: Academic Press, 1970. xiv, 235 p. Appendix: Computer Program for the Solution of Two-point Boundary Value Problems, pp. 204-28. Bibliography, pp. 229-30. Author Index, pp. 231-32. Subject Index, pp. 233-35.

An introductory text in nonlinear functional analysis and its applications for engineers and workers in operations research. Emphasis is on computationally feasible algorithms and on constructive methods for the analysis and optimization of nonlinear systems. The approach is one of reducing a nonlinear problem to a sequence of linear problems, or of replacing a constrained optimization problem by a sequence of unconstrained problems, or of interpreting an infinite-dimensional problem in terms of a sequence of finite-dimensional ones, and so on. The techniques employed are contraction mapping, Newton's method, the penalty function concept of Courant, and Frechet differentials. Assumes knowledge of linear analysis and abstract analysis. Sections conclude with exercises.

Bensoussan, Alain; Hurst, E. Gerald, Jr.; and Näslund, Bertil. MANAGEMENT APPLICATIONS OF MODERN CONTROL THEORY. Studies in Mathematical and Managerial Economics, vol. 18. New York: American Elsevier, 1974. xvii, 346 p. Appendix A: Optimal Deterministic Control Theory, pp. 229-80. Appendix B: Optimal Stochastic Control Theory, pp. 281-320. References, pp. 321-33. Author Index, pp. 335-38. Subject Index, pp. 339-46.

A fairly comprehensive presentation of optimal control theory with applications in management. The first main part of the book takes up management problems for which control theory is appropriate, and reviews introductory control theory. The second part treats the following applications aspects of the management of an enterprise: inventory and production control, theory of the firm, optimal maintenance and life of machines, finance, marketing, and control of natural resources. For each application a simple example and its related model are introduced, and suggestions for expanding the model are made. The third part summarizes and presents examples of the computational techniques available for control problems, from simple parameter solution through full dynamic simulation. The fourth part consists of a review of the mathematics necessary for the appendixes, and two appendixes, in which the main theorems and their proofs in both deterministic and stochastic control theory are included. This book is useful to undergraduates and graduates in management, engineering, and applied mathematics. It should also be useful to industrial planning and operations research groups. Assumes knowledge of calculus, elementary differential equations, and mathematical programming. The stochastic models also assume knowledge of probability theory.

Beveridge, Gordon S.G., and Schechter, Robert Samuel. OPTIMIZATION: THEORY AND PRACTICE. McGraw-Hill Chemical Engineering Series. New York: McGraw-Hill Book Co., 1970. xix, 773 p. Appendix 1: Some Notes on Determinants, pp. 727-31. Appendix 2: Optimization of Nonlinear Objective Functions Subject to Equality and Inequality Constraints--A Comparison, pp. 732-34. Appendix 3: References, pp. 735-56. Name Index, pp. 757-62. Subject Index, pp. 763-73.

An introductory text of broad coverage on optimization using techniques ranging from differential calculus through direct search and mathematical programming to the more specialized game theory and decision theory approaches. The calculus of variation method is not considered. It is divided into the following three parts: the organization for optimization, optimization techniques, and optimization in practice. Assumes knowledge of calculus. Chapters conclude with problems for assignment.

Bliss, Gilbert A. LECTURES ON THE CALCULUS OF VARIATIONS. Chicago: University of Chicago Press, 1946. ix, 296 p. Appendix: Existence Theorems for Implicit Functions and Differential Equations, pp. 257-83. A Bibliography for the Problem of Bolza, pp. 285-91. Index, pp. 293-96.

Presents selected problems in the calculus of variations, rather than a complete account of the theory. Part 1, consisting of six chapters, deals with the following problems: the necessary conditions of Weierstrauss, Legendre, and Jacobi; sufficient conditions for a minimum; fields and the Hamilton-Jacobi theorem; problems in the plane and in higher space; problems in parametric form; and problems with variable end-points. Part 2, consisting of three chapters, takes up the necessary and sufficient conditions of the problem of Bolza.

Boltvanskiĭ, Vladimir Gregor'evich. MATHEMATICAL METHODS OF OPTIMAL CONTROL. Edited by I. Tarnove. Translated by K.N. Trirogoff. Balskrishnan-Neustadt Series. New York: Holt, Rinehart and Winston, 1971. xiv, 272 p. References, p. 268. Index, pp. 269-72.

A Russian book published in 1965, enhanced by the inclusion of two additional sections on optimal control in nonoscillatory systems and nonlinear oscillatory systems of the second order. It presents the theory of optimal control, as developed by L.S. Pontryagin, and includes the following topics: the time-optimal problem, the maximum principle, the method of dynamic programming and sufficient conditions for optimality, linear time-optimal processes, the problem with variable endpoints, the maximum principle for nonautonomous systems, and the isoperimetric problems with fixed time. It is useful as a graduate level text in optimal control and assumes knowledge of the Lebesgue integral, differential equations, and weak compactness theorem.

Bolza, Oskar. LECTURES ON THE CALCULUS OF VARIATIONS. 2d ed. New York: Chelsea Publishing Co., 1960. ix, 271 p. Addenda, pp. 265-66. Index, pp. 269-71.

A reprint of the first edition in which a number of corrections and improvements in notation have been made. Presents that part of the calculus of variations in which the function under the integral sign depends upon a plane curve and involves no derivatives higher than the first. Topics include the Kneser's

Optimization and Control

theory, which is based upon an extension of certain theorems on geodesics to extremals in general, and Hilbert's existence proof for an extremum of a definite integral. Emphasis is on definitions of concepts, examples, and rigorous proofs.

Burley, David Michael. STUDIES IN OPTIMIZATION. New York: Wiley, Halsted Press, 1974. xii, 228 p. Answers and Comments on Problems, pp. 215-22. References, pp. 223-24. Index, pp. 225-28.

A text in optimal control for undergraduates who have knowledge of elementary partial differentiation and vector calculus. Emphasis is on the Euler equation, with dynamic programming as the main tool for solving problems. After a review of classical optimization, the following topics are treated: hill climbing with and without constraints, approximate methods in variational problems, the Euler equation, classical and physical applications, differential equations and eigenvalues, Rayleigh-Ritz method, dynamic programming, control problems, and related minimization problems. The book concludes with a chapter on problems for further study. The presentation is concerned more with application of the theory, rather than with proofs of theorems.

Butkovskii, Anatolii Gregor'evich. DISTRIBUTED CONTROL SYSTEMS. Edited by G.M. Kranc. Translated by Scripta Technica. Modern Analytic and Computational Methods in Science and Mathematics, vol. 20. New York: American Elsevier, 1969. xv, 446 p. Appendixes, pp. 428-35. References, pp. 436-42. Subject Index, pp. 443-46.

Originally published in Russian in 1965, this book develops the mathematical theory of optimal control for systems whose state is characterized by one or several parameters distributed in space. Using functional analysis, both necessary and sufficient conditions for the optimality of such systems are derived, and approximate methods of solving the equations by analog or digital computer are presented. Emphasis is given to applications of the theory to the optimization of thermal processes. The problem of modeling a broad class of systems with distributed parameters is also discussed. Assumes knowledge of functional analysis.

Canon, Michael D.; Cullum, Clifton D., Jr.; and Polak, Elijah. THEORY OF OPTIMAL CONTROL AND MATHEMATICAL PROGRAMMING. McGraw-Hill Series in Systems Science. New York: McGraw-Hill Book Co., 1970. xii, 285 p. Appendixes: A, Convexity; B, Constrained Minimization Problems in Finite-dimensional Spaces, pp. 231-76. Glossary of Symbols, pp. 277-80. Index, pp. 281-85.

Combines the disciplines of optimal control and nonlinear programming in order to achieve three objectives: (1) a presentation of a unified theory of optimization; (2) an introduction to nonlinear programming algorithms for the control engineer; and (3) an intro-

duction of optimal control to the nonlinear programming expert.
Optimal control and nonlinear programming problems are shown to
be equivalent to a simple canonical form of mathematical programming. It is also shown how certain linear and nonlinear algorithms can be used for the solution of discrete optimal control
problems. The appendixes are devoted to the mathematics of convexity and the Pontryagin maximum principle. The book omits
discussions of dynamic programming, stochastic optimal control,
and stochastic mathematical programming. Assumes knowledge of
advanced calculus and linear algebra. Chapters conclude with
selected references.

Citron, Stephen J. ELEMENTS OF OPTIMAL CONTROL. New York: Holt,
Rinehart and Winston, 1969. xiii, 266 p. Index, pp. 263-66.

An introductory text for graduate and undergraduate engineering
students on the development and application of techniques applicable for the solution of optimal control problems. Topics include
classical calculus of variations, the maximum principle of Pontryagin,
problems with control variable and state variable inequality constraints, techniques for solving two-point boundary value problems,
direct methods of steepest descent, dynamic programming, and the
relationship between dynamic programming and the calculus of
variations. The solution of optimal feedback control problems by
the Riccati transformation technique is also treated. Theorems
are illustrated with numerical examples from engineering. Chapters conclude with references.

Connors, Michael M., and Teichroew, Daniel. OPTIMAL CONTROL OF DY-
NAMIC OPERATIONS RESEARCH MODELS. International Textbooks in Man-
agement Science. Scranton, Pa.: International Textbook, 1967. ix, 118 p.
Bibliography, pp. 95-114. Selected Readings, p. 115. Index, pp. 117-18.

A self-contained introduction to dynamic optimization techniques
by means of the calculus of variations and the maximum principle.
Useful as a reference and as a supplement to operations research
texts in the treatment of dynamic models. Assumes familiarity
with the calculus of variations and mathematical analysis.

Converse, Alvin O. OPTIMIZATION. Huntington, N.Y.: Krieger, 1970.
xvi, 295 p. Pedagogical Notes, pp. xi-xii. Index, pp. 291-95.

A text for a one-semester course in optimization with a prerequisite of digital computer programming, calculus, and elementary
differential equations. It is divided into two parts: problem formulation and problem solution. Topics include Lagrange multipliers, discrete and continuous sequential problems, linear programming, and quadratic and geometric programming. Several
computer programs in Dartmouth BASIC are included. Chapters
conclude with exercises which are extensions of worked problems.

Cooper, Leon, and Steinberg, David. INTRODUCTION TO METHODS OF OPTIMIZATION. Philadelphia: W.B. Saunders, 1970. vii, 381 p. Answers to Selected Exercises, pp. 369-76. Index, pp. 377-81.

A text for an introductory course in the methods and techniques available for the solution of optimization problems for engineers, economists, and social scientists who are familiar with elementary differential and integral calculus. It omits advanced topics such as the calculus of variations, the maximum principle, and various topics in stochastic optimization. Topics treated include linear, nonlinear, and integer programming. Although the mathematical rigor is not high, many proofs are included. Chapters conclude with references and problems.

De La Barrière, R. Pallu. OPTIMAL CONTROL THEORY: A COURSE IN AUTOMATIC CONTROL THEORY. Edited and with a foreword by B.R. Gelbaum. Translated by Scripta Technica. Saunders Mathematics Books. Philadelphia: W.B. Saunders, 1967. xii, 412 p. Bibliography, pp. 405-8. Index, pp. 409-12.

Originally published in French in 1966, this book is divided into two parts. Chapters 1 through 7 deal with the mathematical background of the Fourier and Laplace transforms, Markov chains, and second-order stationary random processes. Chapters 8 through 18 deal with automatic control theory and include treatments of linear servo systems; filtering, prediction, and anticipation; discrete systems; convex sets; linear and dynamic programming; Markov control systems; and Pontryagin's principle. The main prerequisite is a senior-level course in mathematical analysis.

Dem'yanov, Vladimir F., and Rubinov, Aleksandr M. APPROXIMATE METHODS IN OPTIMIZATION PROBLEMS. Edited by G.M. Kranc. Translated by Scripta Technica. Modern Analytic and Computational Methods in Science and Mathematics. New York: American Elsevier, 1970. ix, 256 p. References, pp. 249-54. Index, pp. 255-56.

Originally published in Russian in 1968, this book presents methods of investigating and solving certain nonlinear extremal problems using functional analysis. Topics include necessary conditions for an extremum, algorithms for testing for extrema, and solutions of optimal control problems. Chapter 5, the last chapter, treats finite-dimensional problems. Many important mathematical results are stated without proof. Assumes knowledge of functional analysis.

Denn, Morton M. OPTIMIZATION BY VARIATIONAL METHODS. New York: McGraw-Hill Book Co., 1969. xvi, 419 p. Name Index, pp. 411-14. Subject Index, pp. 415-19.

A text for a course in optimization for graduates and undergraduates in engineering, statistics, and computer science who possess knowledge of calculus (through Taylor series) and ordinary differential equations. Topics include optimization with differential calculus including computational procedures such as the Newton-Raphson method, Fibonacci search, and the method of steep descent; calculus of variations; continuous systems including linear feedback control, optimal-yield problems, a bang-bang control problem, and nonlinear time-optimal control; a Pontryagin type minimum principle; staged systems; optimal and feedback control; nonserial processes; distributed-parameter systems; and dynamic programming and Hamilton-Jacobi theory. Omitted are such advanced topics as the existence of optimal solutions, the Kuhn-Tucker theorem, the control-theory topics of observability and controllability, and optimization under uncertainty. Matrix notation is not used and for this reason the analysis in several chapters is limited to two-variable processes. Some of the problems at the end of each chapter supplement the theoretical developments, while others present further applications. The concepts are illustrated with applications to simple but typical process design and control situations.

Dyer, Peter, and McReynolds, Stephen R. THE COMPUTATION AND THEORY OF OPTIMAL CONTROL. Mathematics in Science and Engineering, vol. 65. New York: Academic Press, 1970. x, 242 p. Appendix: Conjugate Points, pp. 235-40. Index, pp. 241-42.

An advanced text on the theory and application of optimal control with emphasis on the development of numerical algorithms for solutions to practical problems. Topics include fundamental concepts of parametric optimization, the Newton-Raphson method, multistage optimization based on dynamic programming, and continuous and discontinuous optimal control problems. Many worked problems are provided, but computer algorithms are omitted. Assumes knowledge of advanced calculus and ordinary differential equations. Chapters conclude with problems and references.

Edelen, Dominic G.B. NONLOCAL VARIATIONS AND LOCAL INVARIANCE OF FIELDS. Modern Analytic and Computational Methods in Science and Mathematics. New York: American Elsevier, 1969. xvi, 197 p. Appendix: Stationarization with Constraints, pp. 186-90. List of Frequently Used Symbols, pp. 191-92. Subject Index, pp. 193-97.

A self-contained book for second-year graduate students in mathematics which deals with certain extensions of classical variational methods based on functionals of functionals and which lead to Euler's equations that are integro-differential in nature. This approach is useful for the quantification and description of physical systems. The four chapter titles are "Variations with One Dependent Function," "Variational Calculus for Several Dependent Functions," "Geometric Objects and Lie Derivatives," and "Invariance Considerations." Assumes knowledge of functional analysis. Chapters conclude with problems.

El-Hodiri, Mohamed A. CONSTRAINED EXTREMA: INTRODUCTION TO THE DIFFERENTIABLE CASE WITH ECONOMIC APPLICATIONS. Lecture Notes in Operations Research and Mathematical Systems, vol. 56. New York: Springer-Verlag, 1971. 130 p. References, pp. 128-30. No index.

A supplement for a course in optimization for students who are familiar with advanced calculus. Part 1 deals with constrained extrema of differentiable functionals on finite and not-so-finite dimensional problems. Part 3 is devoted to variational problems leading to a discussion of some optimal control examples.

El'sgol'ts, Lev Ernestovich. CALCULUS OF VARIATIONS. Adiwes International Series in Mathematics, vol. 19. Reading, Mass.: Addison-Wesley, 1962. 178 p. Solutions to Problems, pp. 173-76. Index, pp. 177-78.

Translated from the Polish, this book provides engineers and students of colleges of technology with the basic concepts of the calculus of variations, including direct methods of solution. Each chapter is illustrated with a large number of examples. Topics include variational problems with fixed boundaries and movable boundaries, sufficient conditions for an extremum, variational problems of constrained extrema, and direct methods of Euler, Ritz, and Kantorovic. Each of the five chapters concludes with six to twenty problems.

Ewing, George M. CALCULUS OF VARIATIONS WITH APPLICATIONS. New York: W.W. Norton, 1969. xii, 343 p. Bibliography, pp. 331-40. Index, pp. 341-43.

An introduction to the subject requiring knowledge of advanced calculus for the first half of the book and modern real analysis, topology, and differential equations for the second half. Topics include necessary and sufficient conditions for an extremum, principles of Hamilton and Bolza, parametric problems, direct methods, variational theory in terms of the Lebesgue integral, and the Hamilton-Jacobi theory. The presentation is rigorous, with proofs of theorems provided. Exercises follow the sections of the book.

Fan, Liang-Tseng. THE CONTINUOUS MAXIMUM PRINCIPLE: A STUDY OF COMPLEX SYSTEMS OPTIMIZATION. Huntington, N.Y.: Krieger, 1966. xiii, 411 p. Appendix 1: Computational Procedures for Simple Continuous Processes, pp. 353-71. Appendix 2: Optimization of Complex Multistage Processes by the Dynamic Programming Technique, pp. 372-86. Nomenclature, pp. 387-403. Author Index, pp. 405-7. Subject Index, pp. 409-11.

A sequel to THE DISCRETE MAXIMUM PRINCIPLE, by Fan and Wang (see below). About half of the book presents maximum principle algorithms, and the remaining half deals with applications to the design and control of industrial and process engineering systems.

The concluding chapter explains the relationships existing among dynamic programming, the calculus of variations, and the maximum principle. Assumes knowledge of the calculus of variations. Other authors who contributed in the writing of the book are Y.K. Ahn, T.C. Chen, S.J. Chen, L.E. Erickson, Y.C. Ko, C.L. Hwang, and C.S. Wang.

Fan, Liang-Tseng, and Wang, Chiu-Sen. THE DISCRETE MAXIMUM PRINCIPLE: A STUDY OF MULTISTAGE SYSTEMS OPTIMIZATION. New York: Wiley, 1964. ix, 158 p. Appendix 1: Pontryagin's Maximum Principle, pp. 127-30. Appendix 2: Dynamic Programming, pp. 131-33. Appendix 3: The Kth Best Policy, pp. 134-44. Appendix 4: Comments on the Necessary and Sufficient Condition for Optimality, pp. 145-47. Nomenclature, pp. 149-54. Author Index, pp. 155-56. Subject Index, pp. 157-58.

Presents control theory techniques for the maximization of an objective function composed of a finite number of separate and interconnected stages. Applications are to design problems encountered in engineering, such as cross-current extraction, grain dryers, multistage reactors, and multistage rockets. While only one economic example is included, the theory and applications are useful in optimization problems in economics. Assumes knowledge of advanced calculus and linear analysis. Chapters conclude with references.

Fel'dbaum, A.A. OPTIMAL CONTROL SYSTEMS. Translated by A. Kraiman. Mathematics in Science and Engineering, vol. 22. New York: Academic Press, 1965. x, 452 p. Bibliography, pp. 428-42. Author Index, pp. 443-46. Subject Index, pp. 447-52.

Originally published in Russian in 1963, this book takes up the mathematical methods applicable to the theory of optimal systems, such as dynamic programming and the maximum principle, and applies the theory to the following systems: systems with complete information about the controlled object, systems with maximal partial information, systems with independent (passive) storage of information about the object, and systems with active information storage. Assumes knowledge of advanced calculus.

Gelfand, Izrail Moiseevich, and Fomin, S.V. CALCULUS OF VARIATIONS. Edited and translated by R.A. Silverman. Rev. English ed. Selected Russian Publications in the Mathematical Sciences. Englewood Cliffs, N.J.: Prentice-Hall, 1963. vii, 232 p. Appendix 1: Propagation of Disturbances and the Canonical Equations, pp. 208-17. Appendix 2: Variational Methods in Problems of Optimal Control, pp. 218-26. Bibliography, p. 227. Index, pp. 228-32.

Translated from the Russian, this book is a text for a graduate course in the calculus of variations with a prerequisite of advanced calculus and differential equations. About two-thirds of the book is devoted to the necessary and sufficient conditions for weak and

strong extrema, while the remainder is devoted to variational
methods for a system with infinitely many degrees of freedom and
a brief statement of direct methods used in the calculus of variations. The problems at the conclusion of the chapters were written specifically for the English edition.

Gottfried, Byron S., and Weisman, Joel. INTRODUCTION TO OPTIMIZA-
TION THEORY. Prentice-Hall International Series in Industrial and Systems
Engineering. Englewood Cliffs, N.J.: Prentice-Hall, 1973. xviii, 571 p.
Nomenclature, pp. xiv-xviii. Appendixes, pp. 535-61. Index, pp. 562-71.

An introductory text for graduate or undergraduate courses in optimization with prerequisites of calculus and differential equations.
Topics include classical optimization using calculus, linear and
nonlinear programming, integer programming and the method of
decomposition, optimization of functionals, dynamic programming
and the discrete maximum principle, and optimization under risk
and uncertainty. The appendixes contain guides to selection of
optimization techniques, and a review of matrix algebra. Chapters conclude with references and problems.

Gould, Sidney Henry. VARIATIONAL METHODS FOR EIGENVALUE PROB-
LEMS: AN INTRODUCTION TO THE WEINSTEIN METHOD OF INTERME-
DIATE PROBLEMS. 2d ed., rev. and enl. Mathematical Expositions, no. 10.
Toronto: University of Toronto Press, 1966. xiv, 275 p. Bibliography, pp. 265-69.
Index, pp. 271-75.

The first edition presented the Weinstein method of intermediate
problems for calculating lower bounds of eigenvalues. The second
edition is an extension of this method, and, in particular, it presents both necessary and sufficient conditions for maximum-minimum
values of eigenvalues. Applications are to linear operators in a
Hilbert space; problems of the string, rod, membrane, and pole;
and to differential equation problems. Assumes knowledge of advanced calculus, differential equations, mathematical analysis,
and Hilbert space theory. Exercises follow some of the sections.

Gumowski, Igor, and Mira, C. OPTIMIZATION IN CONTROL THEORY AND
PRACTICE. New York: Cambridge University Press, 1968. viii, 242 p. Ap-
pendix: Some Elementary Definitions Used in Functional Analysis, pp. 224-25.
References, pp. 226-37. Index, pp. 239-42.

An attempt to bridge the gap between designers of control systems
and theoreticians in the field. Topics include the fundamental
properties of extremal problems; equivalent formulation of extremal
problems in terms of boundary value problems associated with partial differential equations; the relationship among the calculus
of variations, dynamic programming, and the maximum principle;
and approximate solutions of optimization problems, either by means
of direct methods of the calculus of variations or by numerical

algorithms associated with ordinary and partial boundary value problems. The mathematical results are explained in detail, usually without proofs, and illustrated with examples from control engineering. Useful as a text (or supplementary text) in control theory courses, and as a reference for practitioners.

Hadley, George F., and Kemp, Murray C. VARIATIONAL METHODS IN ECONOMICS. Advanced Textbooks in Economics, vol. 1. New York: American Elsevier, 1971. ix, 378 p. Appendixes, pp. 364-73. References, pp. 374-76. Index, pp. 377-78.

A comprehensive treatment of classical calculus of variations for students in economics and related disciplines who are familiar with multivariable calculus, linear algebra, differential equations, and the methods of ordinary linear and nonlinear optimization. It is primarily a mathematics book, with applications to economic growth. Topics include the Ramsey model, the problem of Bolza, Samuelson's catenary turnpike theorem, the necessary and sufficient conditions of Weierstrauss and Legendre, the problem of Lagrange, and two-sector models of optimal economic growth. The appendixes include statements and proofs of existence and uniqueness theorems. Useful for courses in mathematical economics, control theory, and the calculus of variations.

Hestenes, Magnus R. CALCULUS OF VARIATIONS AND OPTIMAL CONTROL. Applied Mathematics Series. New York: Wiley, 1966. xii, 405 p. Appendix: Embedding Theorems for Differential Equations, pp. 375-400. Bibliography, pp. 401-2. Index, pp. 403-5.

A text for a two-semester course in the calculus of variations which presents variational theory from three points of view: the classical view (chapters 1 to 3), in which a relatively complete account is given for the standard fixed endpoint problems with extensions to variable end points and isoperimetric problems; the variational approach (chapters 4 through 8), which includes optimal control with fixed initial and terminal states, bounded state variables, and the theory of optimal fields; and an approach which is based on the first-order necessary conditions for a minimum for general variational and optimal control and includes the problems of Bolza and Mayer (chapter 7). Sufficiency theorems are discussed for only special cases, and existence theorems and computational methods are omitted. The treatment is rigorous, and knowledge of calculus, differential equations, and mathematical analysis is assumed.

Hsu, Jay C., and Meyer, Andrew U. MODERN CONTROL PRINCIPLES AND APPLICATIONS. New York: McGraw-Hill, 1968. xix, 769 p. Appendix A: Vectors and Matrices, pp. 691-706. Appendix B: The Z-transform and the Advanced Z-transform, pp. 707-22. Appendix C: Mathematical Background of Chapters 10 and 11, pp. 723-43. Bibliography, pp. 745-58. Index, pp. 759-69.

A text and reference on modern control theory with the mathematics introduced as needed. It is divided into the following three parts: introduction and basic techniques, emphasizing the state-space approach to systems analysis; system-stability analysis, with stress on the frequency-domain techniques in nonlinear system analysis; and optimum system performance analysis where attention is given to the calculus of variations and optimal control, the maximum principle of Pontryagin, dynamic programming, abnormal and singular control problems, and applications and potentialities of optimal control. Exposure to advanced calculus, mathematical analysis, and the calculus of variations is helpful. Chapters conclude with summaries, references, and problems.

Intriligator, Michael D. MATHEMATICAL OPTIMIZATION AND ECONOMIC THEORY. Prentice-Hall Series in Mathematical Economics. Englewood Cliffs, N.J.: Prentice-Hall, 1971. xix, 508 p. Appendix A: Analysis, pp. 450-75. Appendix B: Matrices, pp. 476-500. Index, pp. 501-8.

A self-contained introduction to and survey of static and dynamic optimization techniques with applications to economic theory. It is divided into the following five parts: introduction, static optimization, application of static optimization, dynamic optimization, and applications of dynamic optimization. Topics include linear and nonlinear programming; general equilibrium as illustrated in input-output models; the von Neumann model of an expanding economy, optimal control theory using the calculus of variation, dynamic programming, and the maximum principle; differential games; and optimal growth, including neoclassical growth models and two-sector growth models. Most of the mathematical results are stated without proof, and the analysis and matrix algebra are reviewed in the appendixes. Useful as a text in mathematical economics, mathematical optimization, and as a supplement in courses in economic theory and operations research. Chapters conclude with teaching and research problems, footnotes, and references.

Jacoby, Samuel L.; Kowalik, Janusz S.; and Pizzo, Joseph T. ITERATIVE METHODS FOR NONLINEAR OPTIMIZATION PROBLEMS. Prentice-Hall Series in Automatic Computation. Englewood Cliffs, N.J.: Prentice-Hall, 1974. ix, 274 p. Appendix: Auxilliary Techniques, pp. 233-50. Notation, pp. 251-52. Glossary, pp. 253-63. Annotated Bibliography, pp. 265-67. Index, pp. 269-74.

Presents in algorithmic form the currently available techniques of mathematical programming and unconstrained optimization for practitioners. These techniques include methods for optimization along a line, direct search techniques, descent techniques, and transformation methods. The final chapter deals with methods for constrained optimization. Assumes familiarity with mathematical programming and mathematical analysis. Chapters conclude with references.

Kalman, Rudolf Emil; Falb, P.L.; and Arbib, Michael A. TOPICS IN MATH-
EMATICAL SYSTEM THEORY. International Series in Pure and Applied Mathematics. New York: McGraw-Hill Book Co., 1969. xiv, 358 p. Appendixes, pp. 325-39. References, pp. 341-48. Author Index, pp. 349-51.
Subject Index, pp. 353-58.

A text for a one-semester course in optimization, automata theory,
or algebraic theory of linear systems. It is also useful as a self-
study for individuals who have a background in advanced calculus,
mathematical analysis, and abstract algebra. The four parts are
elementary control theory from the modern point of view; optimal
control theory, including the maximum principle of Pontryagin and
numerical techniques; automata theory, including loop-free decomposition of finite automata; and advanced theory of linear systems.
The presentation is rigorous, and omits such topics as stability,
qualitative theory of differential equations, algebraic linguistics,
practical control theory, and switching circuits. The appendixes
contain a review of module theory, partial realization of an input/
output map (scalar case), proof of uniqueness theorem of canonical
realizations, and index of notations.

Kirk, Donald E. OPTIMAL CONTROL THEORY: AN INTRODUCTION. Prentice-
Hall Network Series. Englewood Cliffs, N.J.: Prentice-Hall, 1970. ix,
452 p. Appendixes, pp. 428-42. Index, pp. 443-52.

An introduction to three facets of optimal control theory: dynamic programming, Pontryagin's maximum principle, and techniques
for trajectory optimization. Useful as a text for a graduate course
with a prerequisite of advanced calculus and matrix algebra. The
appendixes present matrix algebra, difference equation representation of linear sampled data systems, Euler equations, and answers
to selected problems. Chapters conclude with references and problems.

Koppel, Lowell B. INTRODUCTION TO CONTROL THEORY WITH APPLICA-
TIONS TO PROCESS CONTROL. Prentice-Hall International Series in the
Physical and Chemical Engineering Sciences. Englewood Cliffs, N.J.: Prentice-
Hall, 1968. xi, 466 p. Appendixes: A, Existence and Uniqueness Theorems;
B, Vectors and Matrices; C, Classical Methods for Sampled-data Systems; D,
Controllability and Observability; E, Filtering, Estimation, Differentiation, and
Prediction, pp. 383-461. Index, pp. 463-66.

A text for a graduate course in automatic control for students familiar with classical control theory and matrix algebra. Topics
include discrete and continuous systems, state variables, Lyapunov
stability, and optimization based on Pontryagin's minimum principle
and Bellman's dynamic programming. The applications to process
control are designed to be as simple as possible and do not require a computer for solution. The book is oriented toward chemical engineering aspects of control theory. Chapters conclude with
problems and references.

Kowalik, Janusz J., and Osborne, M.R. METHODS FOR UNCONSTRAINED OPTIMIZATION PROBLEMS. Modern Analytic and Computational Methods in Science and Engineering. New York: American Elsevier, 1968. xii, 148 p. Appendix 1: Summary of Matrix Formulas and Notation, pp. 135–37. Appendix 2: Some Results and Definitions Concerning Convexity, pp. 138–41. Notes on Recent Developments, pp. 142–44. Author Index, pp. 145–46. Subject Index, pp. 147–48.

> Presents the following three classes of methods for unconstrained optimization for students with a background in matrix algebra and advanced calculus: direct search, descent, and minimizing sums of squares. A brief discussion of constrained optimization techniques is also included. The book concludes with a chapter on the numerical results of the methods presented. Most of the mathematical results are stated without proof, and graphs are used freely to illustrate the concepts. Each of the four chapters concludes with references.

Künzi, Hans Paul; Tzschach, H.G.; and Zehnder, C.A. NUMERICAL METHODS OF MATHEMATICAL OPTIMIZATION WITH ALGOL AND FORTRAN PROGRAMS. Translated by Werner C. Rheinbolt and Cornelie J. Rheinbolt. Computer Science and Applied Mathematics Series. New York: Academic Press, 1968. x, 171. List of Existing Computer Programs, pp. 154–61. Bibliography, pp. 162–68. Index, pp. 169–71.

> Originally published in German in 1967, the first part of this book (chapters 1 and 2) is devoted to the mathematical theory of linear and nonlinear optimization while the second part (chapters 3 and 4) presents sixteen computer programs and their numerical application of optimization theory. Assumes knowledge of linear algebra and differential calculus.

Kushner, Harold J. INTRODUCTION TO STOCHASTIC CONTROL. New York: Holt, Rinehart and Winston, 1971. xvii, 390 p. Appendix: Stochastic Processes: Continuous Range Spaces, pp. 363–74. Bibliography, pp. 375–83. Index, pp. 385–90.

> A text for a one-semester course in stochastic control theory for students in engineering, economics, operations research, and applied mathematics who are familiar with probability comparable to the first half of Wm. Feller, PROBABILITY THEORY AND ITS APPLICATIONS, volume 1 (New York: Wiley, 1968), and also a reference for practitioners. Most of the development is for systems that can be modeled by a finite-state Markov chain or embedded in such a model. However, continuous-state space and continuous-time Markov models are also considered, since one of the objectives is to introduce the reader to continuous-time control processes by starting with Markov chains. Other topics include stochastic stability, and quadratic cost problems, both with and without filtering. The presentation is fairly rigorous, with most of the proofs provided. Examples and graphs are used to illustrate the theory. Chapters conclude with problems.

_____ . STOCHASTIC STABILITY AND CONTROL. Mathematics in Science and Engineering, vol. 33. New York: Academic Press, 1967. xiv, 161 p. References, pp. 153-58. Author Index, p. 159. Subject Index, pp. 160-61.

Develops the Liapunov function approach to optimality of control systems with emphasis on continuous time parameters. The five chapter titles are: "Introduction," "Stochastic Stability," "Finite Time Stability and First Exit Time," "Optimal Stochastic Control," and "Design of Controls." The presentation is rigorous with proofs provided, and assumes knowledge of supermartingales, weak infinitesimal generators, and Ito processes.

Lack, Geoffrey N.T. OPTIMIZATION STUDIES WITH APPLICATIONS TO PLANNING IN THE ELECTRIC POWER INDUSTRY AND OPTIMAL CONTROL THEORY. Report CCS-5. Palo Alto, Calif.: Stanford University, Institute of Engineering-Economic Systems, 1965. ix, 205 p. Appendixes, pp. 178-203. References, pp. 204-5. No index.

An application of an optimization procedure, using functional analysis in conjunction with Hurwicz's generalization of the Kuhn-Tucker theorem, to the problem of long-range planning of generation and transmission capacity installation in the electric power industry where demand diversity is a significant factor. A second application is to a general problem of optimal control theory in which there may be inequality constraints on the control and state variables.

Lasdon, Leon S. OPTIMIZATION THEORY FOR LARGE SYSTEMS. Macmillan Series in Operations Research. New York: Macmillan, 1972. xi, 523 p. Appendix 1: Convex Functions and Their Conjugates, pp. 493-501. Appendix 2: Subgradients and Directional Derivatives of Convex Functions, pp. 502-13. List of Symbols, p. 515. Index, pp. 517-23.

A text for a one-semester graduate course in operations research. Topics include the Dantzig-Wolfe decomposition principle, partitioning and relaxation procedures in linear programming, compact inverse methods, partitioning procedures in nonlinear programming, duality and decomposition in mathematical programming, and decomposition by right-hand-side allocation. The techniques are developed in a clear and concise manner from a set of basic ideas and principles, and many are illustrated with examples. Chapters conclude with problems and references.

Lee, Ernest Bruce, and Markus, L. FOUNDATIONS OF OPTIMAL CONTROL THEORY. The SIAM Series in Applied Mathematics. New York: Wiley, 1967. x, 576 p. Appendix A: Steepest Descent and Computational Techniques for Optimal Control Problems, pp. 481-520. Appendix B: Bibliography on Optimal Control Processes Governed by Ordinary and Partial Functional-differential Systems, pp. 521-31. References, pp. 533-36. Bibliography, pp. 537-68. Index, pp. 569-76.

Presents deterministic control theory problems definable in terms of ordinary differential systems. The exposition is in the mathematical style of definitions, theorems, and proofs. Topics include necessary and sufficient conditions for optimal control, control system properties, and optimal controls for some basic nonlinear control processes. Assumes knowledge of differential equations and advanced calculus; knowledge of real analysis and linear control systems is also desirable. Sections conclude with problems and extensions of the text material.

Lee, Robert C.K. OPTIMAL ESTIMATION, IDENTIFICATION, AND CONTROL. Research Monograph, no. 28. Cambridge, Mass.: MIT Press, 1964. viii, 152 p. Appendix A: Vector Differentiation, pp. 142-44. Appendix B: Basic Computation Program Equations, pp. 145-46. References, pp. 147-51. Index, p. 152.

Presents the similarities and limitations of the various known techniques in optimal control. Topics include optimal-control theory, optimal estimation, identification, optimal closed-loop control systems, and recommendations. Assumes knowledge of optimal control theory, statistics, and probability.

Lee, Tong Hun; Adams, G.E.; and Gaines, W.M. COMPUTER PROCESS CONTROL: MODELING AND OPTIMIZATION. New York: Wiley, 1968. xiii, 386 p. Bibliography, pp. 350-58. Problems, pp. 359-80. Index, pp. 381-86.

A text for graduates and undergraduates who are familiar with calculus, basic statistics, and matrix algebra. Topics include regression models, Lagrange multiplier techniques, linear programming, adaptive control, physical process models, evolutionary optimization, and dynamic programming. Useful in various disciplines, such as instrumentation, control theory, computer design and programming, statistics, operations research, and mathematical optimization.

Lefschetz, Solomon. STABILITY OF NONLINEAR CONTROL SYSTEMS. Mathematics in Science and Engineering, vol. 13. New York: Academic Press, 1965. xi, 150 p. Appendix A: An Application of Multiple Feedback Control, pp. 139-41. Appendix B: An Example from the Theory of Nuclear Power Reactors, by K. Meyer, pp. 142-43. Bibliography, pp. 144-47. Index, pp. 149-50.

A monograph on control stability in the method of Liapunov. Useful mainly to mathematicians, physicists, and engineers. Assumes knowledge of linear algebra, advanced calculus, and mathematical analysis.

Leitmann, George. AN INTRODUCTION TO OPTIMAL CONTROL. McGraw-Hill Series in Modern Applied Mathematics. New York: McGraw-Hill Book Co., 1966. xiv, 163 p. Appendix A: An Illustrative Example, pp. 141-48.

Appendix B: Existence of Optimal Control, pp. 149-56. Bibliography, pp. 157-59. Index, pp. 161-63.

A text for a brief course on variational calculus and optimal control for graduate students having the mathematical ability of an undergraduate in engineering. Restricted to deterministic dynamical systems whose behavior is governed by ordinary differential equations. Useful in engineering and the applied sciences, including the social and biological sciences. The presentation is primarily geometric in nature.

_____, ed. OPTIMIZATION TECHNIQUES WITH APPLICATIONS TO AEROSPACE SYSTEMS. Mathematics in Science and Engineering, vol. 5. New York: Academic Press, 1962. xiii, 453 p. Author Index, pp. 447-49. Subject Index, pp. 450-53.

Consists of fourteen chapters, each by a different author, on methods for solutions of problems in optimization and optimal control. Chapter titles are: "Theory of Maxima and Minima," "Direct Methods," "Extremization of Linear Integrals by Green's Theorem," "The Calculus of Variations in Applied Aerodynamics and Flight Mechanics," "Variational Problems with Bounded Control Variables," "Methods of Gradients," "Pontryagin's Maximum Principle," "On the Determination of Optimal Trajectories via Dynamic Programming," "Computational Considerations for Some Deterministic and Adaptive Control Processes," "General Imbedding Theory," "Impulsive Transfer between Elliptical Orbits," "The Optimum Spacing of Corrective Thrusts in Interplanetary Navigation," "Propulsive Efficiency of Rockets," and "Topics in Nuclear Rocket Optimization." Contributors are R. Bellman, R.W. Bussard, T.N. Edelbaum, F.D. Faulkner, R. Kalaba, C.M. Kashmar, H.J. Kelley, R.E. Kopp, D.F. Lawden, G. Leitmann, A. Miele, and E.L. Peterson.

Leondes, Cornelius T., ed. ADVANCES IN CONTROL SYSTEMS: THEORY AND APPLICATIONS, VOL. 1. New York: Academic Press, 1964. x, 365 p. Author Index, pp. 359-61. Subject Index, pp. 363-65.

Six heretofore unpublished essays on the theory and application of automatic control as follows: "On Optimal and Suboptimal Policies in Control Systems," by M. Aoki; "The Pontryagin Maximum Principle and Some of Its Applications," by J.S. Meditch; "Control of Distributed Parameter Systems," by P.K.C. Wang; "Optimal Control for Systems Described by Difference Equations," by H. Halkin; "An Optimal Control Problem with State Vector Measurement Errors," by P.R. Schultz; and "On Line Computer Control Techniques and Their Application to Re-entry Aerospace Vehicle Control," by F.H. Kishi. Each contribution concludes with references. Assumes knowledge of advanced mathematics and probability theory. Useful as a supplement in courses concerned with control systems.

_____. ADVANCES IN CONTROL SYSTEMS: THEORY AND APPLICATIONS, VOL. 2. New York: Academic Press, 1965. x, 313 p. Author Index, pp. 309-11. Subject Index, pp. 312-13.

Five heretofore unpublished essays on the theory and application of automatic control as follows: "The Generation of Liapunov Functions," by D.G. Schultz; "The Application of Dynamic Programming to Satellite Intercept and Rendezvous Problems," by F.T. Smith; "Synthesis of Adaptive Control Systems by Function Space Methods," by H.C. Hsieh; "Singular Solutions in Problems of Optimal Control," by C.D. Johnson; and "Several Applications of the Direct Method of Liapunov," by R.A. Nesbit. Assumes knowledge of advanced mathematics, including functional analysis and complex variables.

_____. ADVANCES IN CONTROL SYSTEMS: THEORY AND APPLICATIONS, VOL. 3. New York: Academic Press, 1966. x, 346 p. Author Index, pp. 341-43. Subject Index, pp. 344-46.

Six heretofore unpublished essays on the theory and application of automatic control. The titles of the essays are: "Guidance and Control of Reentry and Aerospace Vehicles," by T.L. Gunckel II; "Two-Point Boundary-Value-Problem Techniques," by P. Kenneth and R. McGill; "The Existence Theory of Optimal Control Systems," by W.W. Schmaedeke; "Application of the Theory of Minimum-Normed Operators to Optimum-Control-System Problems," by J.M. Swiger; "Kalman Filtering Techniques," by H.W. Sorenson; "Application of State-Space Methods to Navigation Problems," by S.F. Schmidt. This third volume continues the purpose of the series in bringing together diverse information on important progress in the field of automatic control.

_____. ADVANCES IN CONTROL SYSTEMS: THEORY AND APPLICATIONS, VOL. 4. New York: Academic Press, 1966. xiv, 320 p. Author Index, pp. 317-18. Subject Index, pp. 319-20.

Six heretofore unpublished essays on the theory and application of automatic control. The titles of the essays are: "Algorithms for Sequential Optimization of Control Systems," by D. Isaacs; "Stability of Stochastic Dynamical Systems," by H.J. Kushner; "Trajectory Optimization Techniques," by R.E. Kopp and H.G. Moyer; "Optimum Control of Multidimensional and Multilevel Systems," by R. Kulikowski; "Optimal Control of Linear Stochastic Systems with Complexity Constraints," by D.E. Johansen; "Convergence Properties of the Method of Gradients," by D.E. Johansen.

_____. ADVANCES IN CONTROL SYSTEMS: THEORY AND APPLICATIONS, VOL. 5. New York: Academic Press, 1967. xii, 426 p. Author Index, pp. 421-24. Subject Index, pp. 425-26.

Six heretofore unpublished essays on recent progress in the field of control and systems theory and applications. The titles of the essays are: "Adaptive Optimal Steady State Control of Nonlinear Systems," by A.E. Pearson; "An Initial Value Method for Trajectory Optimization Problems," by D.K. Scharmack; "Determining Reachable Regions and Optimal Controls," by D.R. Snow; "Optimal Nonlinear Filtering," by J.R. Fisher; "Optimal Control of Nuclear Reactor Systems," by D.M. Wiberg; and "On Optimal Control with Bounded State Variables," by J. McIntyre and B. Paiewonsky.

_____. ADVANCES IN CONTROL SYSTEMS: THEORY AND APPLICATIONS, VOL. 6. New York: Academic Press, 1968. xiii, 321 p. Author Index, pp. 317-19. Subject Index, pp. 320-21.

Four heretofore unpublished essays on recent progress in the theory and applications of control systems. The titles of the essays are: "The Application of Techniques of Artificial Intelligence to Control System Design," by J.M. Mendel and J.J. Zapalac; "Controllability and Observability of Linear, Stochastic, Time-discrete Control Systems," by H.W. Sorenson; "Multilevel Optimization Techniques with Application to Trajectory Decomposition," by E.J. Bauman; and "Optimal Control Theory Applied to Systems Described by Partial Differential Equations," by Wm. L. Brogan.

_____. ADVANCES IN CONTROL SYSTEMS: THEORY AND APPLICATIONS, VOL. 7. New York: Academic Press, 1969. xiii, 314 p. Author Index, pp. 311-12. Subject Index, pp. 313-14.

Six heretofore unpublished essays on progress in the field of control and systems theory and applications, as achieved and discussed by leading researchers. The titles of the essays are: "Computational Problems in Random and Deterministic Dynamical Systems," by M.M. Connors; "Approximate Continuous Nonlinear Minimal-variance Filtering," by L. Schwartz; "Computational Methods in Optimal Control Problems," by J.A. Payne; "The Optimal Control of Systems with Transport Lag," by R.R. Bate; "Entropy Analysis of Feedback Control Systems," by H.L. Weidemann; and "Optimal Control of Linear Distributed Parameter Systems," by E.I. Axelband.

_____. ADVANCES IN CONTROL SYSTEMS: THEORY AND APPLICATIONS, VOL. 8. New York: Academic Press, 1971. xiv, 259 p. Author Index, pp. 253-56. Subject Index, pp. 257-59.

Seven heretofore unpublished essays on progress in the field of control and systems theory and applications as achieved and presented by leading contributors. Emphasis in this volume is on applications to large-scale systems and decision making. The titles of the essays are: "Method of Conjugate Gradients for Optimal Control Problems with State Variable Constraint," by T.S. Fong and C.T. Leondes; "Final Value Control Systems," by C.E. Seal and A. Stubberud; "Singular Problems in Linear Estimation and Control," by K. Simon and A. Stubberud; "Discrete Stochastic Differential Games," by K.B. Bley and E.B. Stear; "Optimal Control Applications in Economic Systems," by L.F. Buchanan and F.E. Norton; "Numerical Solution of Nonlinear Equations and Nonlinear, Two-point Boundary-value Problems," by A. Miele, S. Naqvi, A.V. Levy, and R.R. Iyer; and "Advances in Process Control Applications," by C.H. Wells and D.A. Wismer.

———. CONTROL AND DYNAMIC SYSTEMS: ADVANCES IN THEORY AND APPLICATIONS, VOL. 9. New York: Academic Press, 1973. xvi, 514 p. Author Index, pp. 507-10. Subject Index, pp. 511-14.

A continuation of the series ADVANCES IN CONTROL SYSTEMS: THEORY AND APPLICATIONS (see above). Vol. 9 consists of five heretofore unpublished contributions on important progress in the field of control and systems theory and applications as achieved and presented by leading contributors. The titles of the contributions are: "Optimal Observer Techniques for Linear Discrete Time Systems," by L.M. Novak; "Application of Sensitivity Constrained Optimal Control to National Economic Policy Formulation," by D.L. Erickson and F.E. Norton; "Modified Quasilinearization Method for Mathematical Programming Problems and Optimal Control Problems," by A. Miele, A.V. Levy, R.R. Iyer, and K.H. Well; "Dynamic Decision Theory and Techniques," by Wm. R. Osgood and C.T. Leondes; and "Closed Loop Formulations of Optimal Control Problems for Minimum Sensitivity," by R.N. Crane and A.R. Stubberud. Assumes knowledge of advanced mathematics, including differential equations and linear algebra.

———. CONTROL AND DYNAMIC SYSTEMS: ADVANCES IN THEORY AND APPLICATIONS, VOL. 10. New York: Academic Press, 1973. xvii, 527 p. Author Index, pp. 519-24. Subject Index, pp. 525-27.

Seven heretofore unpublished essays on progress in the field of control and systems theory and applications as achieved and presented by leading contributors. The titles of the essays are: "The Evaluation of Suboptimal Strategies Using Quasilinearization," by R.G. Graham and C.T. Leondes; "Aircraft Symmetric Flight Optimization," by M. Falco and H.J. Kelley; "Aircraft Maneuver Optimization by Reduced-order Approximation," by H.J. Kelley; "Differential Dynamic Programming--A Unified Approach to the Optimization of Dynamic Systems," by D.Q. Mayne; "Estimation of Uncertain Systems," by J.O. Pearson; "Application of Modern Control and Optimization Techniques to Transportation Systems," by D. Tabak; and "Integrated System Identification and Optimization," by Y.Y. Haimes.

_____. CONTROL AND DYNAMIC SYSTEMS: ADVANCES IN THEORY AND APPLICATIONS, VOL. 11. New York: Academic Press, 1974. xv, 516 p. Subject Index, pp. 515-16.

Six heretofore unpublished contributions on progress in the field of control and systems theory and applications as achieved and presented by leading contributors. The titles of the contributions are: "Fitting Multistage Models to Input/Output Data," by P.L. Smith; "Computer Aided Control Systems Design Techniques," by J.A. Page and E.B. Stear; "Multilevel Optimization of Multiple Arc Trajectories," by R.D. Sugar; "Nonlinear Smoothing Techniques," by J.B. Peller; "Toward the Synthesis of Solutions of Dynamic Games," by L.C. Westphal; and "A Survey of Soviet Contributions to Control Theory," by A. Ya. Lerner.

_____. CONTROL AND DYNAMIC SYSTEMS: ADVANCES IN THEORY AND APPLICATIONS, VOL. 12. New York: Academic Press, 1976. xx, 627 p. Subject Index, pp. 625-27.

Nine heretofore unpublished contributions on progress in the field of control and systems theory as achieved and presented by leading contributors. The titles of the contributions are: "An Overview of Filtering and Stochastic Control in Dynamic Systems," by H.W. Sorenson; "Linear and Nonlinear Filtering Techniques," by G.T. Schmidt; "Concepts and Methods in Stochastic Control," by Y. Bar-Shalom and E. Tse; "The Innovations Process with Applications to Identifications," by W.C. Martin and A.R. Stubberud; "Discrete-time Optical Stochastic Observers," by L.M. Novak; "Discrete Riccati Equations: Alternative Algorithms, Asymptotic Properties, and System Theory Interpretations," by L.M. Silverman; "Theory of Disturbance-accommodating Controllers," by C.D. Johnson; "Identification of the Noise Characteristics in a Kalman Filter," by H.W. Brewer; and "Adaptive Minimum Variance Estimation in Discrete-time Linear Systems," by R.F. Ohap and A.R. Stubberud.

Lions, Jacques Louis. OPTIMAL CONTROL OF SYSTEMS GOVERNED BY PARTIAL DIFFERENTIAL EQUATIONS. Translated by S.K. Mitter. New York: Springer-Verlag, 1971. xi, 396 p. Bibliography, pp. 384-94. Subject Index, pp. 395-96.

Translated from the German, this book develops optimal control theory for a model described by a family of partial differential equations. The chapter titles are: "Minimization of Functions and Unilateral Boundary Value Problems"; "Control of Systems Governed by Elliptic Partial Differential Equations"; "Control Systems Governed by Parabolic Partial Differential Equations"; "Control of Systems Governed by Hyperbolic Partial Differential Equations or by Equations Well-posed in the Sense of Petrovsky"; and "Regularization, Approximation and Penalization." Each chapter begins with an outline and concludes with bibliographical notes

and indications of problems which are unsolved or aspects which
have not been considered in the book. Assumes knowledge of
mathematical analysis and functional analysis.

_____. SOME ASPECTS OF THE OPTIMAL CONTROL OF DISTRIBUTED PA-
RAMETER SYSTEMS. Regional Conference Series in Applied Mathematics.
Philadelphia: Society for Industrial and Applied Mathematics, 1972. vi,
92 p. References, pp. 88-92. No index.

Ten lectures given by the author at the National Science Foundation Regional Conference on control theory, August 23-27, 1971,
at the University of Maryland in College Park. Topics include
examples of variational inequalities arising in mechanics, optimal
control for linear distributed parameter systems, a nonlinear system in biochemistry, existence theorems for geometrical problems
of optimization, applications of boundary layers in singular perturbations, and numerical techniques. Assumes knowledge of
mathematical analysis and differential equations.

Liusternik, Lazar' Aronovich. SHORTEST PATHS VARIATIONAL PROBLEMS.
Translated by P. Collins and R.B. Brown. Popular Lectures in Mathematics,
vol. 13. New York: Macmillan, 1964. x, 102 p. No index.

Originally published in Russian in 1955 as KRATCHAISCHIE LINII,
this book investigates a number of simple problems, including
shortest paths on simple surfaces, geodesic paths, problems connected with the potential energy of a stretched string, the isoperimetric problem, and Fermat's principle. Graphs are used
freely, and the only prerequisite is high school algebra. Useful
to anyone planning to undertake a more advanced study of the
mathematical theory of the calculus of variations.

_____. THE TOPOLOGY OF THE CALCULUS OF VARIATIONS IN THE
LARGE. Translated by J.M. Danskin. Translations of Mathematical Monographs, vol. 16. Providence, R.I.: American Mathematical Society, 1966.
vii, 96 p. Bibliography, pp. 94-96.

Originally published in Russian in 1947, this book deals with the
homological properties of functional spaces as they relate to variational problems. Assumes knowledge of functional analysis and
topology.

Luenberger, David G. OPTIMIZATION BY VECTOR SPACE METHODS. Series
in Decision and Control. New York: Wiley, 1969. xiii, 326 p. Bibliography, pp. 312-19. Symbol Index, p. 321. Subject Index, pp. 323-26.

A text for a graduate course in optimization for students in engineering, mathematics, operations research, and related disciplines
who are familiar with mathematical analysis and linear algebra.
Purpose is to show that optimization can be unified by a few geo-

metric principles of linear vector space theory. Topics include an
introduction to functional analysis, dual spaces, least-squares estimation, linear operators and adjoints, optimization of functionals,
global and local theories of constrained optimization, and iterative
methods of optimization. The problems which follow the chapters
are of two kinds: miscellaneous mathematical problems and proofs
which extend and supplement the theory, and optimization problems which illustrate further areas of application. Chapters conclude with references.

McCausland, Ian. INTRODUCTION TO OPTIMAL CONTROL. New York:
Wiley, 1969. xiii, 258 p. Notation and Symbols, pp. xii-xiii. Bibliography,
pp. 229-31. Appendixes, pp. 233-54. Index, pp. 255-58.

A text for a graduate course in optimal control with a prerequisite of matrix algebra, elementary probability theory, and linear
system theory, including the Laplace transform. Topics include
the Wiener and Kalman filtering theories, the maximum principle,
and dynamic programming. Among the computational methods
discussed are: Newton-Raphson method of maximization, methods
which do not involve derivatives, hill-climbing methods, and direct computation of an optimal control input. The appendixes
take up Parseval's theorem, the bilateral Laplace transform, autocorrelation functions and spectral densities, table of integrals,
and answers to problems. The presentation does not rely on rigorous proofs, and the chapters conclude with references and problems.

Meditch, J.S. STOCHASTIC OPTIMAL LINEAR ESTIMATION AND CONTROL.
McGraw-Hill Series in Electronic Systems. New York: McGraw-Hill Book
Co., 1969. xiv, 394 p. Name Index, pp. 385-86. Subject Index, pp. 387-94.

Presents optimal estimation and control for processes which can be
modeled as discrete-time (sampled-data) or continuous-time linear
dynamic systems. The linear theory which is developed is used to
solve practical optimization problems in navigation, guidance, control, postflight data analysis for aircraft and spacecraft, the control of
large-scale production and chemical processes, and the processing
of seismological and biological data. The book is intended for a
wide audience, including the beginning graduate student in engineering and applied science. The principle results are summarized
in the form of theorems and corollaries, and the examples are
purposely kept simple. Assumes knowledge of matrix algebra, ordinary differential equations, and probability theory. Chapters
conclude with problems and references.

Merriam, Charles Wolcott III. OPTIMIZATION THEORY AND THE DESIGN
OF FEEDBACK CONTROL SYSTEMS. McGraw-Hill Electronics Sciences Series.
New York: McGraw-Hill Book Co., 1964. xv, 391 p. Appendix A: Basic
Nomenclature, pp. 305-9. Appendix B: References, pp. 310-13. Appendix

C: Statistical Signals and Linear Systems, pp. 314-22. Appendix D: State-determined Dynamic Processes, pp. 323-26. Appendix E: Case Study for Linear Optimum Controls--The Design of an Aircraft Landing System, pp. 327-51. Appendix F: Numerical Integration of Ordinary Differential Equations with Discontinuous Time Functions, pp. 352-62. Appendix G: Expansion of the Hamiltonian Function, pp. 363-70. Appendix H: Problems, pp. 371-87. Index, pp. 389-91.

> Suitable as a text for a two-semester course in feedback control for engineers, and as a reference for anyone interested in optimal control theory. It is divided into the following four parts: a review of the chronological developments in optimization theory applied to control-system design (three chapters), introductory mathematical background required for control-system optimization (two chapters), linear optimum controls (two chapters), and nonlinear control systems (three chapters). The problems in appendix H are grouped according to chapter.

Mickle, Marlin H., and Sze, T.W. OPTIMIZATION IN SYSTEMS ENGINEERING. The Intext Series in Circuits, Systems, Communications, and Computers. Scranton, Pa.: Intext Educational Publishers, 1972. xiii, 312 p. Appendix A: The Impulse Function of Finite-state, Discrete-time Systems, pp. 293-301. Appendix B: Subroutines, pp. 302-5. Appendix C: List of Symbols, pp. 306-8. Index, pp. 309-12.

> A text for a graduate course in optimization for engineering students with the usual mathematics preparation, including linear algebra and advanced calculus. Intended as an introduction to methodology and as a basis for further work in optimization. Topics include linear and nonlinear systems of equations; simultaneous inequalities; linear and dynamic programming; the maximum principle; and finite-state systems. Chapters conclude with summaries, exercises, and references.

Minorsky, Nicolai. THEORY OF NONLINEAR CONTROL SYSTEMS. McGraw-Hill Electrical and Electronic Engineering Series. New York: McGraw-Hill Book Co., 1969. xx, 331 p. Appendix: The New Mexico Test, pp. 319-24. Index, pp. 325-31.

> Presents both linear and nonlinear control systems. Topics include fundamentals of nonlinear control theory, nonlinearities in control systems, stability in the sense of Liapunov, stability of nonlinear control systems, the general theory of relay systems, method of harmonic linearization, piecewise linear methods, point transformation method, and functional transformers and analogs. Chapters conclude with references.

Mohler, Ronald R. BILINEAR CONTROL PROCESSES WITH APPLICATIONS TO ENGINEERING, ECOLOGY, AND MEDICINE. Mathematics in Science and Engineering, vol. 106. New York: Academic Press, 1973. xi, 224 p. Author Index, pp. 215-18. Subject Index, pp. 219-24.

A reference and a text on the control of a class of nonlinear systems which are coined bilinear, that is, systems described by ordinary differential equations which are linear in state, linear in
control, but not jointly linear in both. Chapters 1 to 3 establish
a theoretical and computational base for bilinear control systems,
including sufficient conditions for complete controllability and
optimal regulation of bilinear systems with bounded control. In
chapter 4, bilinear models for nuclear fission and for heat transfer are studied, and several biochemical, physiological, and ecological variable structures, which can be modified by bilinear systems of equations, are analyzed in chapter 5. Chapter 6, the
final chapter, applies bilinear modeling to socioeconomic systems,
including a case study of the system prevailing in mainland China.
Assumes knowledge of advanced calculus and introductory systems
theory. Useful as a text or supplementary text for control courses,
and in seminars in computer science, engineering, mathematics,
economics, and medicine. Chapters conclude with exercises and
references.

Morrey, Charles B., Jr. MULTIPLE INTEGRALS IN THE CALCULUS OF VA-
RIATIONS. New York: Springer-Verlag, 1966. ix, 506 p. Bibliography,
pp. 494-503. Index, pp. 504-6.

Deals with the existence and differentiability of the solutions of
variational problems involving multiple integrals. Physical applications to engineering, and so forth, are not discussed, but mathematical applications to such objects as the theory of harmonic
integrals are taken up. Knowledge of Banach and Hilbert spaces
is essential. The presentation is rigorous and takes the form of
difinitions, assumptions, theorems, and remarks.

Mufti, I.H. COMPUTATIONAL METHODS IN OPTIMAL CONTROL PROB-
LEMS. Lecture Notes in Operations Research and Mathematical Systems,
vol. 27. New York: Springer-Verlag, 1970. 45 p. References, pp. 43-45.
No index.

A report on computational methods for the solution of optimal control problems based on the maximum principle of Pontryagin. Topics
include necessary conditions for a maximum of the control problem
of Bolza, the method of steepest descent, the successive sweep
method, the Newton-Raphson method, and the generalized Newton-
Raphson method.

Nicholson, T.A.J. OPTIMIZATION IN INDUSTRY. VOLUME I. OPTIMI-
ZATION TECHNIQUES. London Business School Series. New York: Aldine-
Atherton, 1971. xiii, 222 p. Answers to Exercises, pp. 178-220. Index,
pp. 221-22.

A text for a graduate course which gives a detailed review of the
available search techniques for solving optimization problems.

Each technique is illustrated by numerical and graphical examples. Topics include calculus and Lagrange multipliers, linear and dynamic programming, branch and bound methods, permutation procedures, and heuristic procedures. Chapters conclude with exercises and references. Assumes knowledge of calculus and matrix algebra.

_____. OPTIMIZATION IN INDUSTRY. VOLUME II. INDUSTRIAL APPLICATIONS. London Business School Series. New York: Aldine-Atherton, 1971. xiii, 252 p. Answers to Exercises, pp. 208-50. Index, pp. 251-52.

A detailed review of the application of optimization techniques in industry. Examples and case studies are taken from a wide range of industrial contexts, and each application is formulated up to the point where it can be solved directly by one of the techniques described in volume 1. Assumes knowledge of finite mathematics. Applications are to chemical process control, machine sequencing and resource utilization, stock control and production scheduling, design optimization, plant renewal and location, transportation, and advertising. Chapters conclude with references and exercises.

Oǧuztöreli, M. Namik. TIME-LAG CONTROL SYSTEMS. Mathematics in Science and Engineering, vol. 24. New York: Academic Press, 1966. xii, 323 p. Bibliography, pp. 276-314. Author Index, pp. 315-19. Subject Index, pp. 320-23.

A monograph on the theory of ordinary delay-differential equations and control processes with time delay. Topics include existence, uniqueness, and construction of solutions; nature of solutions in relation to the initial functions, initial moments, parameters, and right-hand side of the equation; piecewise continuous delay-differential equations; existence of optimal policies and sufficient conditions for their existence; optimality using dynamic programming, and time optimal problems and optimal pursuit problems. Presentation is rigorous, consisting primarily of proofs of theorems, and knowledge of functional analysis, differential and integral equations, and dynamic programming is assumed.

Panik, Michael J. CLASSICAL OPTIMIZATION: FOUNDATIONS AND EXTENSIONS. Studies in Mathematical and Managerial Economics, vol. 16. New York: American Elsevier, 1975. xi, 312 p. Bibliography, pp. 307-8. Subject Index, pp. 309-12.

A text for a two-semester course in classical optimization (optimization based on calculus) for undergraduates and graduates in business, economics, and operations research who have a background in elementary calculus. Three chapters are devoted to a review of the topics in advanced calculus and linear algebra; five chapters deal with unconstrained classical optimization; two chapters take up optimization involving equality functional constraints; and three chapters are devoted to extensions of classical optimization to cover the case where nonnegative conditions along with equality and/or inequality functional constraints are present. Emphasis is on proofs of theorems, and several chapters have appendixes devoted to advanced mathematical topics. Short numerical examples are used to illustrate the concepts. Topics include global and local extrema of real-valued functions; polynomial approximation of real-valued functions, inflection points of real-valued functions, convex and concave real-valued functions, the technique of Lagrange, and the Kuhn-Tucker theory with m inequality constraints and with mixed constraints.

Pars, L.A. AN INTRODUCTION TO THE CALCULUS OF VARIATIONS. New York: Wiley, 1962. xi, 350 p. Examples, pp. 338-45. Bibliography, p. 347. Index, pp. 349-50.

An introductory text on the calculus of variations based on the Rieman integral, rather than on the Lebesgue integral, with applications in geometry, dynamics, and physics. The chapter titles are: introduction, fundamental theory, illustrative examples, variable end-points, the fundamental sufficiency theorem, the isoperimetrical problem, curves in space, the problem of Lagrange, the parametric problem, and multiple integrals. The last chapter on multiple integrals, which is largely oncerned with Dirichlet's principle, assumes knowledge of point-set theory, harmonic functions, and Fourier series.

Pervozyvanskiĭ, Anatoliĭ A. RANDOM PROCESSES IN NONLINEAR CONTROL SYSTEMS. Foreword by R. Bellman. Mathematics in Science and Engineering, vol. 15. New York: Academic Press, 1965. xv, 341 p. Appendixes, pp. 293-331. Bibliography, pp. 332-37. Author Index, pp. 339-40. Subject Index, p. 341.

A reference for design engineers and research scientists whose work is concerned with control systems based on simple techniques of approximation. The chapter titles are: "Introduction," "Nonlinear Transformations without Feedback," "Nonlinear Transformations with Feedback: Stationary States," "Nonlinear Transformations with Feedback: Nonstationary State," and "Extremal Systems." The appendixes are devoted to a review of mathematical topics.

Petrov, Lu. P. VARIATIONAL METHODS IN OPTIMUM CONTROL THEORY. Translated by M.D. Friedman, with H.J. ten Zeldam. Mathematics in Science and Engineering, vol. 45. New York: Academic Press, 1968. ix, 216 p. Appendix 1: Historical Survey, pp. 194-200. Appendix 2: Glossary, pp. 201-3. References, pp. 204-12. Author Index, pp. 213-14. Subject Index, pp. 215-16.

Translated from the Russian, this book presents the variational methods underlying optimum control theory and applications to technical problems primarily in engineering. Topics include the simplest problem of the calculus of variations, application of Euler's equation, necessary and sufficient conditions for an extremum, examples of variational methods in electrical and mechanical engineering.

Pierre, Donald A. OPTIMIZATION THEORY WITH APPLICATIONS. Series in Decision and Control. New York: Wiley, 1969. xv, 612 p. Appendixes, pp. 555-91. Author Index, pp. 593-98. Subject Index, pp. 599-612.

A text for a course in optimization for undergraduates and graduates and a reference for practicing engineers and system designers. It emphasizes concepts from both classical and modern developments, and certain detailed proofs are passed up in favor of examples that help to clarify the theory. The sequence of presentation is as follows: Classical necessary and sufficient conditions for relative extrema, including classical calculus of variations; modern mathematical programming; and dynamic programming, which provides insights into the maximum principle. Assumes knowledge of differential equations and matrix theory. Some of the required mathematics, such as the Laplace transform, is reviewed in the appendixes. Chapters conclude with appendixes and problems.

Pindyck, Robert S. OPTIMAL PLANNING FOR ECONOMIC STABILIZATION: THE APPLICATION OF CONTROL THEORY TO STABILIZATION POLICY. Contributions to Economic Analysis, vol. 81. New York: American Elsevier, 1973. xii, 167 p. Appendix: State-variable Representation of a Dynamic System, pp. 155-59. References, pp. 161-64. Index, pp. 165-67.

Presents optimal control as a tool for the design and analysis of economic stabilization policies based on Pontryagin's minimum principle. Most of the mathematical derivations are contained in the first two chapters, with the remaining five chapters devoted to applications and experiments based on the quarterly model of the U.S. economy. Assumes knowledge of elementary calculus of variations.

Plant, John B. SOME ITERATIVE SOLUTIONS IN OPTIMAL CONTROL. Foreword by H.W. Johnson. Research Monograph, no. 44. Cambridge, Mass.: MIT Press, 1968. xiii, 218 p. Appendixes A to I, pp. 147-212. References, pp. 213-15. Index, pp. 216-18.

Presents algorithms for using optimal control to improve and compare optimal control systems without having to build complete systems. Applications are to: time-optimal, terminal-cost, minimum-fuel control, and minimum-effort control. Assumes knowledge of advanced calculus, linear algebra, and mathematical analysis. Appendixes A to E are devoted to some essential mathematical

concepts and theorems, and appendixes F to I contain FORTRAN
programs for solving problems.

Polak, Elijah. COMPUTATIONAL METHODS IN OPTIMIZATION: A UNI-
FIED APPROACH. Mathematics in Science and Engineering, vol. 77. New
York: Academic Press, 1971. xvii, 329 p. Conventions and Symbols,
pp. xv-xvii. Appendix A: Further Models for Computational Methods; B:
Properties of Continuous Functions; C: A Guide to Implementable Algorithms,
pp. 283-316. References, pp. 317-21. Index, pp. 323-29.

A graduate text and reference which presents a large number of
optimization, boundary value and root-solving algorithms on convergence and implementation in optimal control and nonlinear
programming. The methods considered are: gradient and quasi-
Newton methods in R^n, conjugate methods in R^n, penalty function methods, methods of centers, methods of feasible directions,
second-order methods of feasible directions, gradient projection
methods, and decomposition and variate metric algorithms. Special topics include convex optimal control problems and rate of
convergence of algorithms. Assumes knowledge of advanced calculus, real analysis, mathematical analysis, and linear and quadratic programming. Appendix C assists the reader in the choice
of algorithms.

Pontryagin, L[ev]. S[emenovitch].; Boltvanskiĭ, V[ladimir]. G[rigor'evich].;
Gamkrelidze, R.V.; and Mishchenko, E.F. THE MATHEMATICAL THEORY
OF OPTIMAL PROCESSES. Edited by L.W. Neustadt. Translated K.N.
Trirogoff. New York: Wiley-Interscience, 1962. viii, 360 p. References,
pp. 354-56. Index, pp. 357-60.

Translated from the Russian, this book treats control processes
using the maximum principle. Processes considered are: time-
optimal, control processes involving retardation, pursuit processes,
stochastic control processes, and applications to approximation
theory. The relationship between dynamic programming and the
calculus of variation is described.

Postnikov, Mikhail M. THE VARIATIONAL THEORY OF GEODESICS. English
edition edited and with a foreword by Bernard R. Gelbaum. Translated by
Scripta Technica. Saunders Mathematics Books. Philadelphia: W.B. Saunders,
1967. viii, 200 p. Index, pp. 199-200.

Originally published in Russian in 1965, this book presents the
fundamental aspects of differential geometry and the calculus of
variations, and the basic tools for the study of Morse theory.
The five chapters are: "Smooth Manifolds," "Spaces of Affine
Connection," "Riemannian Spaces," "The Variational Properties
of Geodesics," and "A Reduction Theorem." Assumes thorough
familiarity with calculus.

Optimization and Control

Pun, Lucas. INTRODUCTION TO OPTIMIZATION PRACTICE. New York:
Wiley, 1969. x, 309 p. General References, pp. 303–6. Index, pp. 307–9.

A reference for graduate students and practicing engineers on deterministic optimization techniques with special emphasis on industrial applications. Topics include static optimization techniques,
extremum seeking techniques, dynamical optimization techniques,
and dynamical suboptimization techniques. Assumes knowledge of
linear algebra and differential calculus. References follow the
chapters.

Rockafellar, R. Tyrrell. CONJUGATE DUALITY AND OPTIMIZATION.
Philadelphia: Society for Industrial and Applied Mathematics, 1974. vi,
74 p. References, pp. 73–74. No index.

An introduction to conjugate duality in both finite- and infinite-
dimensional problems. Examples are taken from nonlinear (including nonconvex) programming, stochastic programming, the calculus of variations, and optimal control. Proofs of theorems are
provided, and knowledge of mathematical analysis and functional
analysis is helpful.

Rund, Hanno. THE HAMILTON-JACOBI THEORY IN THE CALCULUS OF VARI-
ATIONS: ·ITS ROLE IN MATHEMATICS AND PHYSICS. The New University
Mathematics Series. Huntington, N.Y.: Krieger, 1966. xi, 404 p. Appen-
dix, pp. 366–84. Bibliography, pp. 385–95. Index, pp. 397–404.

A self-contained presentation of the Hamilton-Jacobi theory. Topics
include congruences of curves and families of subspaces related
to the theory, multiple integrals, and the problem of Lagrange.
Emphasis is on properties of concepts and their applications, and
intuitive proofs for many of the mathematical results are provided.
The appendix contains advanced mathematical concepts, such as
tensor analysis, Riemannian geometry, and the general divergence
theorem.

Russell, David L. OPTIMIZATION THEORY. Mathematics Lecture Note
Series. New York: W.A. Benjamin, 1970. xiv, 405 p.

A set of lecture notes suitable for an undergraduate course in
optimization given at the University of Wisconsin. Limited to
processes that can be described by functions of finitely many
variables. Topics include existence theory, one-dimensional block
search techniques, least squares, differentiation and Newton's
method, convexity, eigenvectors, and gradient methods. Assumes
knowledge of calculus and linear algebra. Chapters conclude
with exercises.

Sage, Andrew P. OPTIMUM SYSTEMS CONTROL. Prentice-Hall Networks
Series. Englewood Cliffs, N.J.: Prentice-Hall, 1968. ixv, 562 p. Appen-

dix: The Algebra, Calculus, and Differential Equations of Vectors and Matrices, pp. 545-58. Index, pp. 559-62.

A text for an introductory graduate course in optimization theory as applied to the control of systems. Written from an engineer's point of view and emphasizes basic concepts of various techniques and the relationships, similarities, and limitations among the concepts. It is divided into the following three parts: optimal control with deterministic inputs, state estimation and combined estimation and control, sensitivity and computational techniques in systems control. A unique feature is the use of numerous examples to illustrate the concepts. Prerequisites include advanced calculus, although some exposure to linear control system design, numerical analysis, variational calculus, modern analysis, stochastic processes and state space techniques would be helpful. Chapters conclude with references and problems.

Sawaragi, Yoshikazu; Sunahara, Yoshifumi; and Nakamizo, Takayoshi. STATISTICAL DECISION THEORY IN ADAPTIVE CONTROL SYSTEMS. Mathematics in Science and Engineering, vol. 39. New York: Academic Press, 1967. xiii, 216 p. Author Index, pp. 213-14. Subject Index, pp. 215-16.

Combines control theory with statistical decision theory in an effort to overcome the lack of modern control theory in practical applications. The first half consists of a review of stochastic and other processes, and statistical decision theory. Then, attention is turned to sequential and nonsequential approaches in adaptive control systems, and adaptive adjustments of parameters of nonlinear control systems. The final chapter presents some problems that may arise in the future in applications of statistical decision theory to control processes. Emphasis is on examples and applications, rather than on rigorous proofs. Assumes knowledge of calculus, mathematical analysis, and probability theory.

Schechter, Robert Samuel. THE VARIATIONAL METHOD IN ENGINEERING. McGraw-Hill Chemical Engineering Series. New York: McGraw-Hill Book Co., 1967. ix, 287 p. Index, pp. 283-87.

An introductory text for undergraduates and graduates in engineering and related disciplines on variational calculus. Concepts and methods are illustrated by examples rather than abstract arguments in the hope that the reader will translate from the examples to forms suitable to his purpose. One category of application consists of problems that lead directly to variational formulations which utilize the Euler-Lagrange equations to yield new differential equations. The second category consists of synthetic problems in which the differential equation is known and an extremal principle is selected to match the particular differential equation. For the latter category, various approximate methods of solution are presented, including the Ritz method. Other topics include the macroscopic balance equations, applications to transport problems, and hydrodynamic stability. Assumes knowledge of calculus and an introduction to differential equations and the calculus of variations. Some sections conclude with problems.

Schultz, Donald G., and Melsa, James L. STATE FUNCTIONS AND LINEAR CONTROL SYSTEMS. McGraw-Hill Series in Electronic Systems. New York: McGraw-Hill Book Co., 1967. xi, 435 p. Index, pp. 427-35.

The application of control theory to the control of linear systems. It is largely a self-teaching book for the control engineer in industry and for graduate students. Sections conclude with exercises and answers, and the chapters conclude with general problems and references.

Smith, Donald R. VARIATIONAL METHODS IN OPTIMIZATION. Englewood Cliffs, N.J.: Prentice-Hall, 1974. xv, 378 p. Appendixes A1 to A10, pp. 352-68. Subject Index, pp. 369-74. Author Index, pp. 375-78.

An elementary exposition of the use of differentiation to solve a wide range of optimization problems in engineering, astronautics, mathematics, physics, economics, and operations research. It takes up many problems in the calculus of variations, such as problems with fixed endpoints, variable endpoints, isoperimetric constraints, and certain types of global inequality constraints, along with other optimization problems customarily handled by the methods of optimal control. The main tool used is the Euler-Lagrange multiplier theorem of differential calculus. Applications are in economics, business, engineering, and the physical sciences. Useful as a text, and assumes knowledge of a first course in calculus. Sections conclude with exercises. The appendixes are devoted to reviews of mathematical derivations and concepts, such as the Cauchy and Schwarz inequalities, and the Du Bois-Reymond's derivation of the Euler-Lagrange equation.

Strauss, Aaron. AN INTRODUCTION TO OPTIMAL CONTROL THEORY. Lecture Notes in Operations Research and Mathematical Economics, vol. 3. New York: Springer-Verlag, 1968. vi, 153 p. References, pp. 143-48. List of Symbols, pp. 149-51. Index for Definitions, p. 153.

A study of control processes that can be described by a system of ordinary differential equations of the form $dx/dt=f(t,x,u)$ where x is a vector summarizing the state of the process at time t, and u is a vector of variables subject to control. The necessary and sufficient conditions for optimal control are discussed, including a statement of Pontryagin's maximum principle. Theory, rather than applications, are emphasized, and knowledge of ordinary differential equations and measure theory is assumed. Useful to anyone interested in the mathematical theory of control.

Sworder, David. OPTIMAL ADAPTIVE CONTROL SYSTEMS. Mathematics in Science and Engineering, vol. 25. New York: Academic Press, 1966. xi, 187 p. Appendixes A to H, pp. 139-83. Author Index, p. 185. Subject Index, pp. 186-87.

> An exposition of optimal control processes with incomplete mathematical descriptions with emphasis on the properties of these processes. It is divided into three parts: formulation of the control problem, recurrence formulas for a class of optimal control policies, and examples which clarify the meaning of optimization and identification in adaptive control systems. Assumes knowledge of state space description of dynamical systems and elementary statistical decision theory. The appendixes are devoted to mathematical reviews.

Tse, Edison Tack-Shuen. ON THE OPTIMAL CONTROL OF LINEAR SYSTEMS WITH INCOMPLETE INFORMATION. Detroit: Management Information Services, 1970. vii, 285 p. Appendix A: On the Pseudo-inverse of a Matrix, p. 277. Appendix B: Weiner-Kopf Equation, p. 278. Appendix C: Equation for Error Process (continuous time case), pp. 279-80. Bibliography, pp. 281-85. No index.

> A study of optimal control problems in which future effects cannot be predicted exactly, that is, control problems with incomplete information. The cases treated first assume that the system is linear (either discrete time or continuous time). Then, problems of controlling a linear system with known dynamics is taken up, followed by systems whose poles are known but whose zeros are unknown. Finally, the control of discrete time linear systems with unknown gain parameters, and control of third order systems with unknown zeros, are treated. Assumes knowledge of optimal control theory.

Varaiya, P.P. NOTES ON OPTIMIZATION. New York: Van Nostrand Reinhold, 1972. viii, 202 p. References, pp. 196-200. Index, pp. 201-2.

> A set of notes for a short course in mathematical programming and optimal control for students having diverse backgrounds and interests. Assumes knowledge of advanced calculus, linear algebra, and linear differential equations. Topics include optimization over sets defined by equality and inequality constraints, linear and nonlinear programming, sequential decision problems: discrete- and continuous-time optimal control, and dynamic programming. The presentation is rigorous, consisting mainly of theorems and proofs, with some explanation and application.

Weinstock, Robert. CALCULUS OF VARIATIONS: WITH APPLICATIONS TO PHYSICS AND ENGINEERING. International Series in Pure and Applied Mathematics. New York: McGraw-Hill Book Co., 1952. x, 326 p. Bibliography, p. 319. Index, pp. 321-26.

A text for a course in the calculus of variations with applications in engineering and physics for graduates and undergraduates who have knowledge of advanced calculus. Topics include isoperimetric problems, Fermat's principle, the vibrating string problem, the Sturm-Lioville eigenvalue-eigenfunction problem, the vibrating membrane problem, theory of elasticity, quantum mechanics, and electrostatics. Problems of sufficiency and existence, second variation, and the conditions of Legendre, Jacobi, and Weierstrass are not treated. Expansion theorems for the eigenfunctions associated with certain boundary value problems are stated without proof. Assumes knowledge of advanced calculus. Chapters conclude with exercises.

Whittle, Peter. OPTIMIZATION UNDER CONSTRAINTS: THEORY AND APPLICATIONS OF NONLINEAR PROGRAMMING. Wiley Series in Applied Probability and Statistics. New York: Wiley-Interscience, 1971. ix, 241 p. References, p. 238. Index, pp. 239-41.

An account of Lagrangian theory in constrained optimization problems, including applications of physical, technological and economic interest. Topics include linear programming, linear models of economic development, geometric programming, control problems, economic development programs with nonlinear recursions, chemical equilibrium, game theory, and decision theory. The relationship of the Pontryagin maximum principle to Lagrangian theory is considered. Assumes knowledge of advanced calculus and probability theory.

Wilde, Douglass J. OPTIMUM SEEKING METHODS. Prentice-Hall International Series in the Physical and Chemical Engineering Sciences. Englewood Cliffs, N.J.: Prentice-Hall, 1964. xiii, 202 p. Index, pp. 195-202.

A guide for developing strategies for seeking the optimum of any function about which full knowledge is not available. After an introduction to the search problem, the book considers single variable search procedures, the geometry of multidimensional response surfaces, tangents and gradients, acceleration along a ridge, and experimental error. The methods include Kiefer's "Fibonacci" technique, steepest ascent, contour tangent elimination, parallel tangents, pattern search, and several schemes designed to follow ridges. In the final chapter on experimental error, Dvoretzky's work on stochastic convergence occupies the central position.

Wilde, Douglass J., and Beightler, Charles S. FOUNDATIONS OF OPTIMIZATION. Englewood Cliffs, N.J.: Prentice-Hall, 1967. xiv, 480 p. Index, pp. 471-80.

A text for a one-year (or more) course in optimization for graduates and advanced undergraduates, and a supplementary text for other courses, such as system design. The main prerequisite is differen-

tial calculus; matrix algebra is deliberately avoided in most cases.
Proofs of theorems are passed up in favor of detailed industrial
examples based on the authors' experiences as practicing engineers and consultants. Topics include indirect methods of optimization, such as Newton-Raphson, geometric programming, constrained derivatives, and sensitivity analysis; direct elimination
(for example, Fibonacci and Bolzano search methods) and direct
climb (least squares, pattern search, and quadratic programming);
inequality constraints (Wolfe's algorithm and Beale's method);
polynomial inequalities; partial optimization of multistage systems,
including networks; and optimal control by policy improvements.
Each of the ten chapters concludes with a bibliography and eight
chapters conclude with exercises.

Wonham, W. Murray. LINEAR MULTIVARIABLE CONTROL: A GEOMETRIC
APPROACH. Lecture Notes in Economics and Mathematical Systems, vol. 101.
New York: Springer-Verlag, 1974. x, 344 p. References, pp. 328-36.
Index, pp. 337-44.

A monograph on the geometric approach, using the state space
representation, to the structural synthesis of multivariable control
systems that are time-invariant, linear, and of finite dynamic
order. It is addressed to graduate students specializing in control,
engineers engaged in control system research and development,
and to mathematicians with some previous acquaintance with control problems. The presentation is primarily theoretical, with
nearly all numerical examples located in the exercises at the end
of each chapter. The computational procedures are programmed
in APL (A Programming Language). It presents a reasonably complete structure theory for two control problems: regulation, and
noninteraction. The two closing chapters (chapters 12 and 13)
deal with quadratic optimization. Assumes knowledge of advanced
calculus or real analysis. Chapters conclude with notes and references, as well as exercises.

Yakowitz, Sidney J. MATHEMATICS OF ADAPTIVE CONTROL PROCESSES.
Modern Analytic and Computational Methods in Science and Engineering,
vol. 14. New York: American Elsevier, 1969. xv, 158 p. Appendixes,
pp. 93-150. List of Symbols, pp. 151-53. Author Index, p. 155. Subject
Index, pp. 157-58.

A rigorous, unified, and inclusive systems theory for multistage
decision processes. Its aim is to demonstrate the generality of
adaptive control processes and to show the common underlying
mathematical structure of information theory, sequential decision
theory, and dynamic programming. Detailed investigations in
two-armed bandit and pattern recognition problems are carried out.
Useful as a text for a course in optimization for seniors and graduates who have knowledge of advanced calculus, mathematical
analysis, and the calculus of probability. The nine appendixes

are devoted to theorems and computer programs for two-armed bandit problems.

Young, Laurence Chisholm. LECTURES ON THE CALCULUS OF VARIATIONS AND OPTIMAL CONTROL THEORY. 2 vols. Foreword by W.H. Fleming. Philadelphia: W.B. Saunders, 1969. xi, 331 p. References, pp. 323-26. Index, pp. 327-31.

Divided into the following two volumes: volume 1, LECTURES ON THE CALCULUS OF VARIATIONS; and volume 2, OPTIMAL CONTROL THEORY. Volume 1 is concerned with the unconstrained variational problem where existence and sufficiency theories are stressed, and volume 2 deals mainly with time-optimal control problems using the maximum principle. The main prerequisite is advanced calculus. Useful to economists specializing in optimal economic growth.

Zahradinik, Raymond L. THEORY AND TECHNIQUES OF OPTIMIZATION FOR PRACTICING ENGINEERS. Professional Engineering Career Development Series. New York: Barnes and Noble, 1971. xi, 326 p. Bibliography, pp. 320-21. Index, pp. 323-26.

A text for a course in optimization for engineers which attempts to give highlights, key points, and workable procedures of optimization, omitting most of the history, theory, and aesthetic appreciation of the subject. It is divided into two parts. Part 1 deals with parameter optimization in which values are sought for a set of independent variables (static optimization). Part 2 is devoted to trajectory optimization in which the path of the independent variables is sought. In part 1, linear, quadratic, geometric, and discrete dynamic programming, together with direct and indirect optimization techniques, are considered. In part 2, dynamic programming, as well as direct and indirect methods of solution, are taken up. Assumes knowledge of the calculus of variations. Chapters conclude with references.

Chapter 13

QUEUEING THEORY

Beckmann, Peter. INTRODUCTION TO ELEMENTARY QUEUEING THEORY AND TELEPHONE TRAFFIC. Boulder, Colo.: Golem Press, 1968. 144 p. Table 1: Erland Loss Probabilities, pp. 126-33. Table 2: Erland Delay Probabilities, pp. 134-41. Index, pp. 141-44.

An introduction to queueing theory for beginning students who have knowledge of calculus and elementary probability theory. Topics include review of probability, Poisson and exponential distributions, single-server systems, multiple service with and without queueing, delay times, first come first served, and random order of service. Chapters conclude with problems.

Benes, Vaclav E. GENERAL STOCHASTIC PROCESSES IN THE THEORY OF QUEUES. Addison-Wesley, 1963. viii, 88 p. References, pp. 84-85. Index, pp. 87-88.

A monograph on the theory of delays in queueing systems with one server presented in a way that meets the needs of both the mathematician, who insists on rigorous mathematical presentations, and the engineer, whose interest is how to apply the theory. The six chapter titles are: "Virtual Delay," "Delay Formulas: A Direct Approach," "Delay Formulas: An Approach Using Transforms," "Weak Stationarity: Preliminary Results," "Weak Stationarity: Convergence Theorems," and "Weak Markov Assumptions." Assumes knowledge of calculus, probability, and mathematical analysis.

Bhat, U. Narayan. A STUDY OF THE QUEUEING SYSTEMS M/G/1 and GI/M/1. Lecture Notes in Operations Research and Mathematical Economics, vol. 2. New York: Springer-Verlag, 1968. v, 78 p. Bibliography, pp. 72-78. No index.

A presentation of two queueing systems based on queue length process which utilize transforms only to derive steady state results. The three chapter titles are: "The Queue M/G/1 with Group Arrivals," "The Queue GI/M/1 with Group Service," and "Queueing

Systems in Discrete Time." This book represents an attempt to bridge the gap between theory and real life problems which can be represented by queueing systems.

Cohen, Jacob William. THE SINGLE SERVER QUEUE. North-Holland Series in Applied Mathematics and Mechanics, vol. 8. New York: Wiley-Interscience, 1969. xiv, 657 p. Appendix, pp. 622-28. Notes on Literature, pp. 629-38. References, pp. 639-47. Author Index, pp. 648-49. Index of Notations, pp. 650-52. Subject Index, pp. 653-57.

A description of the mathematical techniques useful for investigating the single server model, and an extensive analysis of the single server queue and its most important variants. It is divided into three parts. Part 1 deals with stochastic processes; part 2 is devoted to the single-server model and several of its analytical techniques, including the Pollaczek's approach combined with renewal theory; and part 3 discusses several variants of the single server model. Assumes knowledge of advanced probability, the Laplace-Stieltjes transform, and the theory of functions. Chapters begin with a short review of the literature.

Cooper, Robert B. INTRODUCTION TO QUEUEING THEORY. New York: Macmillan, 1972. x, 277 p. Appendix A: Engineering Curves for Systems with Poisson Input, pp. 253-66. Appendix B: References, pp. 267-72. Index, pp. 273-77.

A text for a course in queueing theory for graduates and undergraduates in operations research and engineering who have knowledge of calculus, differential equations, and probability theory. The presentation is informal with emphasis on the probabilistic reasoning underlying various approaches. Topics include review of probability theory, birth-and-death queueing models, imbedded Markov chain queueing models, and waiting times. Chapters conclude with exercises which are mainly outlines of derivations not covered in the chapter, or different derivations of old results by important or interesting methods. Many exercises require insights, rather than long calculations. Many examples are taken from the telephone industry.

Cox, David Roxbee, and Smith, Walter L. QUEUES. Methuen's Monographs on Statistical Subjects. New York: Barnes and Noble, 1961. xii, 180 p. Appendixes, pp. 160-75. Author Index, p. 176. Subject Index, pp. 177-80.

A monograph for operations research workers on the introduction to queueing theory with emphasis on applications. The main topics are: introduction, simple queues with random arrivals, queues with many servers and with priorities, machine interference, simulation and Monte Carlo methods, and series of queues. The mathematical techniques are illustrated by examples. The appendixes are devoted to bibliographical notes, exercises, and reviews of probability theory and Laplace-Stieltjes transforms. Assumes knowledge of probability theory and advanced calculus.

Cruon, R., ed. QUEUEING THEORY: RECENT DEVELOPMENTS AND APPLICATIONS. New York: American Elsevier, 1967. xiv, 224 p. Appendix: Terminology of Queueing Theory, pp. 219-24. No index.

Proceedings of a conference on the theory of queues and their applications organized under the auspices of the NATO Advisory Panel on Operational Research and held in Lisbon from September 27 to October 1, 1965. The twenty-two papers of the proceedings are grouped under the following headings: introduction (one paper, "The Application of Queueing Theory in Operational Research," by P. Morse), point processes (three papers), mathematical methods in queueing theory (three papers), nearly saturated queues and transient behavior (three papers), inventories and maintenance (two papers), statistical problems (three papers), case histories and simulation (five papers), and conclusions (two papers). Applications are in the areas of traffic control, open hearth furnace repairs, petroleum refining, electrical circuits, and atomic weapons research. The preface is written in both French and English, and some of the papers appear in French. Among the contributors are J. Auberger, B.W. Conolly, R.R.P. Jackson, L. Kosten, T. Lewis, P. Passau, T.L. Saaty, D.S. Stroller, and J. Teghem.

Descloux, A. DELAY TABLES FOR FINITE- AND INFINITE-SOURCE SYSTEMS. New York: McGraw-Hill Book Co., 1962. vii, 440 p.

Provides tables of values for a delay system having a finite number of source inputs. The infinite case is included for completeness. The thirteen-page introduction includes the derivation of formulas most commonly used in queueing theory and methods used to compute the values given in tables, examples, and a bibliography. The assumptions upon which the tables are based are: (1) N identical and independent sources, or requests, with full availability of servers; and (2) the service time of a request is the amount of time that a server needs to handle that request, but it does not include the time spent waiting, if any, before the service starts. Service times are distributed according to the negative exponential distribution. Useful to anyone involved in finite queueing studies.

Ghosal, Amitava. SOME ASPECTS OF QUEUEING AND STORAGE SYSTEMS. Lecture Notes in Operations Research and Mathematical Systems, vol. 23. New York: Springer-Verlag, 1970. iv, 93 p. Bibliography, pp. 88-93. No index.

A development of queueing and storage problems from a unified point of view. The chapter titles are: "A Unified Treatment of

Queueing and Storage Problems--Probability Distributions," "First Passage Problems and Duality Relations," "Cybernetic Queueing and Storage Systems," and "Optimal Capacity of a Storage System." Assumes knowledge of calculus and probability theory.

Gnedenko, Boris V., and Kovalenko, I.N. INTRODUCTION TO QUEUEING THEORY. Edited by D. Louvish. Translated by E.R. Kondor. Jerusalem: Israel Programs for Scientific Translation, 1968. ix, 281 p. Bibliography, pp. 270-79. Explanatory List of Abbreviations, p. 279. Subject Index, pp. 280-81.

Originally published in Russian in 1966. It is a text for an advanced course in queueing theory for mathematicians and others interested in the theory of the subject. Topics include various limit theorems for incoming customer streams, special types of stochastic processes (semi-Markov, L-processes, piecewise-linear), embedded Markov chains, applications of the processes, and methods of solution of queueing problems, including the Monte Carlo technique. Assumes knowledge of advanced probability.

Gross, Donald, and Harris, Carl M. FUNDAMENTALS OF QUEUEING THEORY. Wiley Series in Applied Probability and Mathematical Statistics. New York: Wiley, 1974. 556 p. Bibliography, pp. 453-63. Appendixes, pp. 464-546. Index, pp. 547-56.

A text for graduate and undergraduate courses in queueing theory with emphasis on both theory and applications. The presentation will appeal to both mathematicians and nonmathematicians. Topics include single-channel exponential queueing models, simple Markovian birth-death queueing models, advanced Markovian models, models with general arrival or service patterns, simulation, and applied queueing theory. The appendixes consist of tables, dictionary of symbols, and reviews of differential and difference equations, stochastic processes and Markov chains, and transforms and generating functions. Assumes knowledge of undergraduate differential and integral calculus, differential equations, and the calculus of probability. Each chapter concludes with an extensive set of problems.

Haight, Frank A. HANDBOOK OF THE POISSON DISTRIBUTION. Publications in Operations Research, no. 11. New York: Wiley, 1967. xi, 168 p. Reference List and Author Index, pp. 125-62. Subject Index, pp. 163-68.

A handbook and reference on the Poisson distribution and Poisson process with modifications and generalizations. Topics include elementary properties, models leading to the Poisson distribution, generalizations of the Poisson distribution, statistical estimation and testing, applications, tables and computer programs, and historical remarks. Every published theorem and formula is given in full, unless it involves a sequence of definitions and lemmas too

extensive for compact quotation, in which case a general statement of the results is given. Proofs are omitted, and bibliography
is virtually complete up to the date of publication.

_____. MATHEMATICAL THEORIES OF TRAFFIC FLOW. Mathematics in
Science and Engineering, vol. 7. New York: Academic Press, 1963. xiii,
242 p. Supplementary References, pp. 227-34. Author Index, pp. 235-38.
Subject Index, pp. 239-42.

A treatment of elementary probability theory and single-channel
queueing theory with applications to car flow and concentration,
the arrangement of cars on a road, the problem of traffic delays,
such as by traffic signals, and merging, multilane, and two-lane
traffic. Assumes knowledge of differential and integral calculus.
Chapters conclude with references.

Jaiswal, N.K. PRIORITY QUEUES. Mathematics in Science and Engineering,
vol. 50. New York: Academic Press, 1968. xiii, 240 p. Appendixes,
pp. 223-28. Glossary of Symbols, pp. 229-31. References, pp. 232-37.
Subject Index, pp. 238-40.

A presentation of congestion systems under priority disciplines or
similar stochastic processes. Topics include basic finite- and
infinite-source models; priority queueing models; preemptive, dis-
cretionary priority, and other disciplines. Assumes knowledge of
queueing theory and stochastic processes. Appendixes consist of
proofs of lemmas.

Khintchine, A.Y. MATHEMATICAL METHODS IN THE THEORY OF QUEUEING.
2d ed. Translated by D.M. Andrews and M.H. Quenouille. Griffin's Statistical Monographs and Courses, no. 7. New York: Hafner, 1969. 124 p.
Section Notes and References, p. 119. Bibliography, p. 120. Additional
Notes by E. Wolman, pp. 121-24.

Originally published in 1955 in Russian, this book serves to acquaint the reader with the main ideas, methods, and different
ways of thought which govern the application of the theory of
probability to questions of mass service. It is divided into three
parts as follows: the incoming stream of calls, systems with losses,
and systems allowing delay. The problems of Erland and Palm are
given special attention. Assumes knowledge of intermediate probability theory and mathematical analysis.

Kleinrock, Leonard. QUEUEING SYSTEMS. VOLUME 1: THEORY. New
York: Wiley-Interscience, 1975. xviii, 417 p. Appendix 1: Transform
Theory Refresher: Z-transform and Laplace Transform, pp. 321-62. Appendix
2: Probability Theory Refresher, pp. 363-95. Glossary of Notation, pp. 396-99.
Summary of Important Results, pp. 400-409. Index, pp. 411-17.

A text for a graduate course in the theory of queues for students
who have knowledge of probability theory and transforms, although
these subjects are reviewed in the appendixes. It is divided into

the following four parts: preliminaries, elementary queueing theory, intermediate queueing theory, and advanced queueing theory. The systems considered include birth-death, the queue M/G/1, the queue G/M/m, the queue G/G/1, and Markovian queues in equilibrium. Each chapter concludes with references and from twelve to twenty-four exercises.

Kosten, Leendert. STOCHASTIC THEORY OF SERVICE SYSTEMS. International Series of Monographs in Pure and Applied Mathematics, vol. 103. New York: Pergamon Press, 1973. xii, 168 p. Appendix: Generating Functions and Laplace Transforms, pp. 148-54. References, pp. 155-58. Author Index, pp. 159-60. Subject Index, pp. 161-65.

A monograph on waiting-line systems in the area of telecommunications with emphasis on a number of standard cases using the method and solution of "birth-and-death equations." Among the systems discussed are: M/M/c-delay, M/M/c-blocking, and M/G/1-delay. One chapter is devoted to the use of simulation in congestion theory. Assumes knowledge of elementary statistics, probability theory, and some understanding of complex variables and Laplace transforms.

Lee, Alec M. APPLIED QUEUEING THEORY. Foreword by S. Eilon. Studies in Management. New York: St. Martin's Press, 1966. xi, 244 p. Appendix 1: Notes on Sources and References, pp. 212-24. Appendix 2: List of Principal Symbols, pp. 225-26. Appendix 3: Summary of Useful Charts and Formulas, pp. 227-41. Index, pp. 242-44.

A presentation of the theory of single- and multiple-channel queueing processes with applications for the "attainment of operational solutions to queueing problems in the real world," including freight reservations, coupon sorting, airline passenger check-in, and other problems based on the author's extensive experience in the airline industry. Assumes knowledge of probability theory, mathematical statistics, and Laplace transformations.

Morse, Philip McCord. QUEUES, INVENTORIES AND MAINTENANCE: AN ANALYSIS OF OPERATIONAL SYSTEMS WITH VARIABLE SUPPLY AND DEMAND. Publications in Operations Research, no. 1. New York: Wiley-Interscience, 1958. ix, 202 p. Appendixes, pp. 175-93. Bibliography, pp. 195-97. Index, pp. 199-202.

One of the first publications to appear on queueing theory. It is primarily expository, and is not rigorous from a mathematical point of view. The applications are to waiting lines, inventory control, and maintenance of equipment. The appendixes consist of functions and tables, and a glossary of symbols. Assumes knowledge of modern probability theory.

Nair, Sreekanton S. ON CERTAIN PRIORITY QUEUES. Detroit: Management Information Services, n.d. 163 p. Bibliography, pp. 137-41. Appendix A: A Theorem on Summation of Series, pp. 142-45. Appendix B: Properties of the Taboo Probabilities, pp. 146-60. Appendix C: Proof of Lemmas in Chapter 4, pp. 161-63. No index.

A mathematical development of priority queues of the following types: single-server tandem queues with nonzero switching, single-server tandem queues with zero switching, alternating priority queues with nonzero switching, and M/G/1 queues. The presentation is rigorous with only a brief mention of applications.

Newell, Gordon Frank. APPLICATIONS OF QUEUEING THEORY. Monographs in Applied Probability and Statistics. London: Chapman and Hall, 1971. x, 148 p. Index, pp. 147-48.

A presentation of queueing theory with applications in transportation for students possessing knowledge of introductory probability theory, mathematical statistics, and elementary calculus: The emphasis is on time-dependent behavior, particularly rush hours in which the arrival rate of customers temporarily exceeds the service rate. The methods of analysis employed are "fluid approximation" employing graphical methods, and "diffusion approximation" employing some elementary properties of differential equations. No use is made of generating functions, characteristic functions, and Laplace transforms. Chapters conclude with problems.

_____. APPROXIMATE STOCHASTIC BEHAVIOR OF N-SERVER SERVICE SYSTEMS WITH LARGE N. Lecture Notes in Economics and Mathematical Systems, vol. 87. New York: Springer-Verlag, 1973. viii, 118 p. References, p. 118. No index.

Presents the G/G/n queueing system consisting of n servers in parallel with independent service times serving a general type of customer arrival process. It is limited primarily to a classification of types of behavior and the qualitative properties of these types, and methods for obtaining more quantitative results. There are no case studies or empirical applications.

Page, Eric. QUEUEING THEORY IN OPERATIONS RESEARCH. Operations Research Series. New York: Crane, Russak and Co., 1972. 187 p. Tables, pp. 152-78. Glossary, p. 179. References, pp. 181-82. Solutions to Problems, pp. 183-84. Index, pp. 185-87.

An introduction to the main areas of applied queueing theory. After a brief presentation of the theory, numerical results are given for queues which can be considered as extremes of those which occur in practice, enabling the reader to adapt the models to his own particular problems. The following queues are considered: constant service times and random arrivals of customers, the queue D/M/n, queues with more general distributions of arrival and service times, finite queues, queues with arrivals dependent on queue size, and priority queues. Assumes a level of mathematics equivalent to a first-year science degree. Chapters conclude with problems.

Panico, Joseph A. QUEUEING THEORY: A STUDY OF WAITING LINES FOR BUSINESS, ECONOMICS, AND SCIENCE. Englewood Cliffs, N.J.: Prentice-Hall, 1969. xiii, 200 p. Bibliography, pp. 196-97. Index, pp. 198-200.

A supplementary textbook built around examples drawn from the author's experience in consulting for forty-eight firms, and includes cases from aircraft, chemicals, electronics, food, glass, instrumentation, machinery, tobacco, and tool industries. It is divided into the following six parts: "Introduction," "Machine Interference: A Brief Look at Queueing Theory," "Mathematical Queueing Models," "Case Histories: Appalachia Memorial Hospital--Including Mathematical Solution," "Special Topics in Mathematical Queueing Models," and "Simulation." Assumes knowledge of calculus and probability theory, or introductory operations research. Chapters conclude with discussion and review questions.

Peck, Leslie G., and Hazelwood, R. Nichols. FINITE QUEUING TABLES. Publications in Operations Research, no. 2. New York: Wiley, 1958. xvi, 210 p. No index.

A monograph intended to provide useful tables for the solution of a variety of queueing problems. Except for a ten-page introduction, which is devoted to sample illustrations of queueing problems and an explanation of the use of the tables, the entire book consists of queueing tables arranged by increasing values of N (to 250), and X from .001 to .950. For each value of X, several values of M, the number of channels, may be found. For a given X and M, values of D and F are listed. N represents the number of customers, L is the length of queue, D is the probability of delay, X is the ratio (average servicing time)/(average servicing time plus average waiting time), and F is the ratio (average servicing time plus average idle time)/(average servicing time plus average idle time plus average waiting time).

Prabhu, Narahari Umanath. QUEUES AND INVENTORIES: A STUDY OF THEIR BASIC STOCHASTIC PROCESSES. Wiley Series in Applied Probability and Statistics. New York: Wiley, 1965. xi, 275 p. Index, pp. 273-75.

A self-contained treatise on the theory of queues requiring knowledge of modern probability theory and stochastic processes at the introductory levels. Topics include fundamental structure of stochastic models of inventories, the ruin problem of the theory of collective risk, discrete and continuous dam models, and standard models of the theory of queues. The mathematical derivations are rigorous, and the solutions of the stochastic models use combinatoric theory which reduces the stochastic variables to sums of independent variables. Chapters conclude with references and complementary results.

Riordan, John. STOCHASTIC SERVICE SYSTEMS. The SIAM Series in Applied Mathematics. New York: Wiley, 1962. x, 139 p. References, pp. 132-36. Index, pp. 137-39.

A brief account of the mathematical results on service systems in which either the demands for services or the services provided, or both, have a probabilistic or stochastic nature. It is addressed to mathematicians, and the main tools used are generating functions, Laplace and Laplace-Stieltjes transforms, and differential equations. Topics treated include the simplest traffic system, single- and many-server systems, and traffic measurements.

Ruiz-Palá, Ernesto; Ávila-Beloso, Carlos; and Hines, William W. WAITING-LINE MODELS: AN INTRODUCTION TO THEIR THEORY AND APPLICATION. Reinhold Industrial Engineering and Management Science Series. New York: Reinhold Publishing Corp., 1967. xii, 180 p. References, pp. 151-53. Appendix A: Glossary of Symbols, pp. 154-55. Appendix B: Tables, pp. 156-73. Appendix C: Notes, pp. 174-78. Index, pp. 179-80.

A text for a one-semester course in queueing theory for undergraduates in engineering, science, and management who have knowledge of elementary calculus and introductory probability, including random variables. It is also a reference for practitioners in systems engineering, traffic control, and management who have not recently been exposed to extensive formal mathematics. It emphasizes the approach to, and structure of, waiting line problems, analytical methods for solving specific problems, and simulation methods for broad classes of problems. Topics include description and study of the probability distributions of service and arrival times, single service channel with exponential distributed service times, multiple exponential channels, simulation of non-exponential distributions, and Monte Carlo simulation. Chapters conclude with problems.

Saaty, Thomas L. ELEMENTS OF QUEUEING THEORY: WITH APPLICATIONS. New York: McGraw-Hill Book Co., 1961. xv, 423 p. General Comments, Bibliography, pp. 373-413. Index, pp. 415-23.

This book is divided into four parts. The first part (three chapters) presents basic material on queueing concepts and probability theory. The second part (two chapters) discusses Poisson queues. The third part (five chapters) is concerned with non-Poisson queues, which are studied mostly for cases in which statistical equilibrium exists. The last part (five chapters) deals with the effect of various queue disciplines, special networks of queues, and a variety of queueing phenomena. The book concludes with a chapter on applications to telephone systems, car and air traffic, machine interference, inventories, dams and storage systems, hospital and medical care facility congestions, cafeteria design, coal mining, and semiconductor noise. This book is intended for graduate students who have knowledge of calculus, probability, introductory complex variables, and matrix theory. Some chapters have exercises.

Takács, Lajos. INTRODUCTION TO THE THEORY OF QUEUES. New York: Oxford University Press, 1962. x, 268 p. Appendix, pp. 219-35. Solutions of the Problems, pp. 237-66. Index, pp. 267-68.

An introduction to the probabilistic treatment of mass servicing with applications to the theory of telephone traffic, airplane traffic, road traffic, storage, operations of dams, servicing of customers, etc. It is based primarily on the author's research experience. The presentation is based mainly on the time-dependent or transient behavior of queueing processes. The appendix contains mathematical reviews of Markov chains, Poisson processes, and recurrent processes, the theorems used in the book, and a bibliography. Chapters conclude with bibliographies, and some sections of the chapters contain problems.

Wallace, Victor L. ON THE REPRESENTATION OF MARKOVIAN SYSTEMS BY NETWORK MODELS. Detroit: Management Information Services, 1970. xi, 109 p. References, pp. 109. No index.

A report on a general approach to the decomposition and recomposition of Markovian queueing networks with applications.

Chapter 14

STOCHASTIC PROCESSES

Aoki, Masanao. OPTIMIZATION OF STOCHASTIC SYSTEMS: TOPICS IN DISCRETE-TIME SYSTEMS. Mathematics in Science and Engineering, vol. 32. New York: Academic Press, 1967. xv, 354 p. Appendixes, pp. 309-38. Bibliographies, pp. 339-46. List of Symbols, pp. 347-48. Author Index, pp. 349-51. Subject Index, pp. 352-54.

A text for a graduate course in optimization of stochastic systems of the discrete type with continuous state variables taking values in some subset of Euclidean spaces. Topics include procedures for deriving optimal Bayesian control policies for discrete-time stochastic systems, estimation of problems of linear and nonlinear systems, convergence in Bayesian optimization, approximations in control and estimation problems, and stability of stochastic systems. The mathematics prerequisites include advanced calculus and advanced probability, and stochastic processes. The appendixes contain mathematical reviews of probability and statistical theory.

Athreya, Krishma, and Ney, Peter E. BRANCHING PROCESSES. New York: Springer-Verlag, 1972. xi, 287 p. Bibliography, pp. 268-81. List of Symbols, p. 282. Author Index, pp. 283-84. Subject Index, pp. 285-87.

A unified treatment of the limit theory of branching processes. All but one of the six chapters deal with single-type processes, and the Galton-Walson process is developed in detail in chapters 1 and 2. In the treatment of continuous time cases (Markov and age dependent) in chapters 3 and 4, an attempt is made to reduce problems to their Galton-Walson counterparts. Chapter 5 deals with a finite number of distinct processes, and chapter 6 consists of applications (energy distributions of nucleons undergoing binary fission, and immigration rates), and special processes (martingale methods and continuous state branching processes). The mathematics prerequisites consist of mathematical analysis and graduate-level probability. Familiarity with Markov chains, the martingale theorem, and renewal theory are also helpful. Chapters conclude with complements and problems.

Bagchi, Tapan Prasad, and Templeton, James G.C. NUMERICAL METHODS IN MARKOV CHAINS AND BULK QUEUES. Lecture Notes in Economics and Mathematical Systems, vol. 72. New York: Springer-Verlag, 1972. xi, 89 p. List of Frequently Occurring Symbols, pp. ix-xi. References, pp. 53-58. Appendix A: Program Listings, pp. 59-66. Appendix B: On the Analysis of Computational Errors, pp. 67-81. Appendix C: On Analytic Approximations, pp. 82-89. No index.

> Presents a new numerical method of computing transient state probabilities for the embedded Markov chains of queue-length processes. This method utilizes probability generating functions of random variables, employing the notion of sweeping of probabilities, and obtains exact time-dependent distributions. It presents an algorithm for computing these probabilities.

Bailey, Norman T. THE ELEMENTS OF STOCHASTIC PROCESSES WITH APPLICATIONS TO THE NATURAL SCIENCES. New York: Wiley, 1964. xi, 249 p. References, pp. 234-36. Solutions to Problems, pp. 237-41. Author Index, p. 243. Subject Index, pp. 245-49.

> A text on the theory and application of stochastic processes in discrete and continuous time. Applications are to birth-and-death processes, queues, and epidemics. Also included is a discussion of diffusion processes and an introduction to non-Markovian processes. Assumes knowledge of probability theory, linear algebra, complex variables, differential equations, and the Laplace transform. Chapters conclude with problems.

Balakrishnan, A.V. STOCHASTIC DIFFERENTIAL SYSTEMS, I. FILTERING AND CONTROL: A FUNCTION SPACE APPROACH. Lecture Notes in Economics and Mathematical Systems, vol. 84. New York: Springer-Verlag, 1973. iv, 252 p. Appendixes, pp. 223-39. References, pp. 240-42. Supplementary Notes, pp. 243-52.

> A presentation of a functional analysis approach to stochastic filtering and control problems. Topics include the linear filter theory (Bucy-Kalman), the feedback control theory with random disturbances, and stochastic differential games. The appendixes are devoted to the properties of Volterra operators and the Krein factorization theorem. The supplemental notes treat the discrete version of recursive filtering, and also likelihood ratios. Assumes substantial mathematics background, and the presentation is rigorous.

Bartholomew, D.J. STOCHASTIC MODELS FOR SOCIAL PROCESSES. 2d ed., rev. Wiley Series in Applied Probability and Statistics. New York: Wiley, 1973. xi, 411 p. Bibliography and Author Index, pp. 381-402. Subject Index, pp. 403-11.

> A contribution to the study of social phenomena by means of the theory of stochastic processes. Useful both to social scientists,

who wish to know what the theory can offer in their own fields,
and to students of mathematics and statistics, who wish to know
more about applications of stochastic processes. The models discussed
are: social and occupational mobility, Markov models for educational and manpower systems, control theory for Markov models,
continuous time models for stratified social systems, models for
duration, renewal theory models for recruitment and wastage, renewal theory models for graded social systems, and simple and
general epidemic models for diffusion of news and rumors. The
presentation is informal and nonrigorous, with the mathematical
results usually given without proof. Mathematical reviews are
omitted, and knowledge of differential and integral calculus,
linear algebra, and probability theory is required.

Bartlett, Maurice S. AN INTRODUCTION TO STOCHASTIC PROCESSES
WITH SPECIAL REFERENCE TO METHODS AND APPLICATIONS. 2d ed., rev.
New York: Cambridge University Press, 1966. xvi, 362 p. Bibliography,
pp. 342-56. Glossary of Stochastic Processes, p. 357. Index, pp. 358-62.

The second edition consists mainly of the addition of about twelve
new sections to the first edition which was published in 1955.
The mathematical results are presented without rigorous proofs,
and emphasis is on applications to such areas as population growth,
queues, theory of storage, prediction and communication problems,
statistics, and correlation analysis. The nine chapter titles are:
"General Introduction," "Random Sequences," "Processes in Continuous Time," "Miscellaneous Statistical Applications," "Limiting
Stochastic Operations," "Stationary Processes," "Prediction and
Communication Theory," "The Statistical Analysis of Stochastic
Processes," and "Correlation Analysis of Time Series." Assumes
knowledge of differential calculus and probability theory.

. STOCHASTIC POPULATION MODELS IN ECOLOGY AND EPIDE-
MIOLOGY. Methuen's Monograph on Applied Probability and Statistics.
New York: Barnes and Noble, 1960. x, 90 p. Bibliography, pp. 85-87.
Index, pp. 89-90.

An introductory monograph on stochastic models for biological
populations for biometricians with some acquaintance with stochastic processes. It is also useful to ecologists and others who wish
an introduction to the subject. Emphasis is on illustrative examples
of the models, rather than rigorous proofs. Topics include frequency distributions for single species, birth-and-death processes
and the problem of extinction, growth and interaction, competition, epidemic models, and the spatial or topological factor.

Beekman, John A. TWO STOCHASTIC PROCESSES. New York: Wiley,
1974. 192 p. Answers to Exercises, pp. 181-92. No index.

An exposition of the theory and applications of Gaussian Markov and collective risk stochastic processes. The applications are in insurance (collective risk), physics (quantum mechanics), electrical engineering, and statistics (limit laws for Kolmogorov and Kac statistics). It is written for actuarial students of risk theory, but it can also be used as a supplement in a course in probability and stochastic processes. Assumes knowledge of measure theory. Sections conclude with exercises, and chapters conclude with references.

Bharucha-Reid, A.T. ELEMENTS OF THE THEORY OF MARKOV PROCESSES AND THEIR APPLICATIONS. McGraw-Hill Series in Probability and Statistics. New York: McGraw-Hill Book Co., 1960. xi, 468 p. Appendix A: Generating Functions, pp. 439-42. Appendix B: The Laplace and Mellin Transforms, pp. 443-48. Appendix C: Monte Carlo Methods in the Study of Stochastic Processes, pp. 449-57. Name Index, pp. 459-63. Subject Index, pp. 464-68.

A text and reference on applied probability which presents a non-measure-theoretic introduction to Markov processes and a formal treatment of mathematical models based on theories which have been employed in various fields. Part 1 (on theory) consists of three chapters which are devoted to processes discrete in space and time, processes discrete in space and continuous in time, and processes continuous in space and time (diffusion processes). Part 2 (applications) consists of six chapters devoted to applications in biology, astronomy and astrophysics, chemistry, and operations research. The applications are restricted to a formal treatment of the models and no numerical results are included. The prerequisites are elementary probability theory, mathematical statistics, mathematical analysis, matrix algebra, and differential equations. Chapters conclude with exercises and references.

Bhat, U. Narayan. ELEMENTS OF APPLIED STOCHASTIC PROCESSES. Wiley Series in Applied Probability and Statistics. New York: Wiley, 1972. xvi, 414 p. Appendixes, pp. 333-405. Author Index, pp. 407-9. Subject Index, pp. 410-14.

A text for a one-semester senior or graduate course in applied stochastic processes with a prerequisite of calculus, one semester of probability theory, and linear algebra. It can be used in departments of business, economics, engineering, operations research, and computer science. The book is divided into two parts: the theory of Markov and renewal processes; and applications of the theory to queues, traffic flow problems, time series analysis, social and behavioral processes, investments, and other areas. The appendixes are devoted to mathematical and statistical reviews. Chapters conclude with references and problems.

Billingsley, Patrick. STATISTICAL INFERENCE FOR MARKOV PROCESSES. Statistical Research Monographs, vol. II. Chicago: The University of Chicago Press, 1961. 75 p. Mathematical Appendix, pp. 52-72. Bibliography, pp. 73-74. Index, p. 75.

A monograph on the theory of statistical inference on Markov processes. Part 1 deals with maximum likelihood estimates of time-
discrete Markov processes and significance tests based on the Neyman-
Pearson lemma. Proofs of the asymptotic normality of the various
statistics involved are on a new contral limit theorem for martingales. Part 2 treats the analogous problems for a Markov process
with a continuous parameter which is of the completely discontinuous or jump type. The probability theory in the monograph is
drawn mainly from Doob's STOCHASTIC PROCESSES (see below
this chapter). Additional derivations are relegated to the appendix
in order not to interrupt the statistical argument of the book, which
is aimed at applications. The appendix takes up limit theorems,
a projection theorem, and a martingale theorem.

Blumenthal, Robert McCallum, and Getoor, Ronald Kay. MARKOV PROCESSES
AND POTENTIAL THEORY. New York: Academic Press, 1968. x, 313 p.
Notes and Comments by Chapters, pp. 295-304. Bibliography, pp. 305-10.
Index of Notation, p. 311. Subject Index, pp. 312-13.

A self-contained book on potential theory for advanced graduate
students which collects within one cover most of the contents of
Hunt's fundamental papers, portions of the theory of additive
functionals, and other closely related matters. Topics include
Markov processes, excessive functions, multiplicative and additive
functionals, and dual processes. Assumes knowledge of measure
theory. The presentation is rigorous with many proofs of theorems.
Chapters conclude with problems.

Bucy, Richard S., and Joseph, Peter C. FILTERING FOR STOCHASTIC PRO-
CESSES WITH APPLICATIONS. Interscience Tracts in Pure and Applied Mathematics, no. 23. New York: Wiley-Interscience, 1968. xviii, 195 p. Notation, pp. xvii-xviii. Appendix A: Least Squares Fitting, pp. 177-79. Appendix B: Probability Review, pp. 180-88. References, pp. 189-92. Index,
pp. 191-95.

Provides a detailed derivative of the Kalmon-Bucy filter and its
asymptotic properties, and also introduces a partial solution and
analysis of the problem of nonlinear filtering. It provides applications of filtering to aerospace guidance problems. It is suitable
as a text for an advanced graduate course in filtering theory.
Assumes knowledge of probability theory and differential equations.

Chiang, Chin Long. INTRODUCTION TO STOCHASTIC PROCESSES IN BIO-
STATISTICS. Wiley Series in Applied Probability and Statistics. Huntington,
N.Y.: Krieger, 1968. xvi, 313 p. References, pp. 297-301. Author In-
dex, pp. 303-4. Subject Index, pp. 305-11.

A presentation of stochastic models describing the empirical processes of birth, death, illness, and other risks which confront
man. Emphasis is placed on specific results and explicit solutions

rather than on the general theory of stochastic processes. Part 1 is on models of population growth, including the Poisson process and time-dependent birth-death processes. The Kolmogorov differential equations are used in the discussion of the illness-death process. Part 2 is devoted to problems of survival and mortality, and consists of topics such as the construction of life tables, probability distributions of life table functions, competing risks, and medical follow-up studies. Assumes knowledge of calculus and some familiarity with differential equations and matrix algebra, and a basic knowledge of probability and statistics. Useful as a text in biostatistics or demography. Chapters conclude with problems.

Chow, Gregory C. ANALYSIS AND CONTROL OF DYNAMIC ECONOMIC SYSTEMS. A Wiley Publication in Applied Statistics. New York: Wiley, 1975. xv, 316 p. Bibliography, pp. 301-7. Index, pp. 309-16.

A presentation of a set of related techniques for analyzing the properties of dynamic stochastic models in economics and for applying these models in the determination of quantitative economic policy. It is divided into two parts: the analysis of dynamic economic systems, and control of dynamic economic systems. Emphasis is on method, and not applications. Topics include analysis of linear deterministic and stochastic systems, dynamic analysis of a sample macroeconomic model, optimal control of known linear systems using Lagrange multipliers and dynamic programming, problems of macroeconomic policy by optimal control, and control of unknown linear systems with and without learning. Useful as a text for graduate students and as a reference. Assumes knowledge of econometrics comparable to J. Johnston, ECONOMETRIC METHODS (2d ed. New York: McGraw-Hill Book Co., 1972). Chapters conclude with problems.

Chung, Kai Lai. LECTURES ON BOUNDARY THEORY FOR MARKOV CHAINS. Annals of Mathematical Studies, no. 65. Princeton, N.J.: Princeton University Press, 1970. xvi, 94 p. Basic Notations, pp. xv-xvi. Appendix: Proof of Lemma 6, chapter 3, section 3, pp. 89-91. References, pp. 93-94.

A series of four lectures given at the Universite de Strasbourg in 1967 and 1968. The four chapter titles are: "Elementary Properties of Transition Matrices," "Sample Function Behavior," "Boundary Behavior," and "Probability Interpretations and Additional Results."

_____. MARKOV CHAINS: WITH STATIONARY TRANSITION PROBABILITIES. 2d ed., rev. New York: Springer-Verlag, 1967. x, 301 p. Bibliography, pp. 292-97. Index, pp. 298-301.

A text for mathematicians on discrete and continuous Markov chains with stationary transition probabilities. Assumes knowledge of probability theory at the level of Feller's INTRODUCTION TO PROBABILITY THEORY AND ITS APPLICATIONS, vol. 1 (see below, this chapter), and the theorems of Dini, Fatou, Fubini, and Lebesgue from real function theory. Familiarity with measure theoretic concepts found in Doob's STOCHASTIC PROCESSES (see below, this chapter) is helpful. The concepts of separability and measurability are applied constantly.

Clark, A. Bruce, and Disney, Ralph L. PROBABILITY AND RANDOM PROCESSES FOR ENGINEERS AND SCIENTISTS. New York: Wiley, 1970. xv, 346 p. Bibliography, pp. 338-39. Index, pp. 341-46.

A text for a two-semester course in probability and stochastic processes (discrete-state Markov processes), with applications to engineering science and management, for students in engineering, the physical and social sciences, and management programs who possess an elementary mathematics background. Topics include sample spaces and random variables; combinatorial probability; expectation; examples of stochastic processes; finite and irreducible, and nonfinite and nonirreducible, discrete parameter Markov processes; and queueing theory. While some theorems are proved, the emphasis is on describing concepts and processes by examples. Each of the fourteen chapters concludes with a fairly lengthy set of exercises which deal with applications and mathematical derivations.

Cox, David Roxbee, and Miller, H.D. THE THEORY OF STOCHASTIC PROCESSES. Wiley Series in Applied Probability and Statistics. New York: Wiley, 1965. x, 398 p. Appendix 1: Table of Exponentially Distributed Random Quantities, p. 377. Appendix 2: Bibliography, pp. 378-85. Author Index, pp. 387-88. Subject Index, pp. 389-98.

A reference and a text on the mathematical analysis of stochastic processes, that is, systems that change in accordance with probabilistic laws, for statisticians and applied mathematicians, rather than for pure mathematicians interested in theorems. Topics include Markov chains, Markov processes and continuous state spaces in continuous time, non-Markovian processes, stationary processes, and point processes. It does not deal with the construction of models, the solution of problems by simulation, and the statistical analysis of data from stochastic processes. Assumes knowledge of matrix algebra and advanced calculus, including the theory of Laplace and Fourier transforms, and elementary probability theory.

Cramer, Harold, and Leadbetter, M.R. STATIONARY AND RELATED STOCHASTIC PROCESSES: SAMPLE FUNCTION PROPERTIES AND THEIR APPLICATIONS. Wiley Series in Probability and Mathematical Statistics. New York: Wiley, 1967. xii, 348 p. References, pp. 339-44. Index, pp. 345-48.

Presents the theory and applications of stationary stochastic processes with emphasis on processes with continuous time parameters. Complete and rigorous mathematical proofs of all results of importance from an applications point of view are given. Topics include processes with finite second-order moments, processes with orthogonal increments, normal processes, and crossing problems. The analytic properties of sample functions, such as continuity, and differentiability, are studied in some detail, and proofs of some basic ergodic theorems are given, but problems of prediction and filtering are only briefly discussed. Assumes knowledge of probability theory.

Curry, Renwick E. ESTIMATION AND CONTROL WITH QUANTIZED MEASUREMENTS. Research Monograph, no. 60. Cambridge, Mass.: MIT Press, 1970. xii, 125 p. Appendixes A to E, pp. 101-19. References, pp. 120-22. Index, pp. 123-25.

A monograph which presents the results of research on efficient estimation and control based on quantized measurements. Discrete-time problems are considered where emphasis is placed on coarsely quantized measurements and on linear and time-varying systems. Approximate and nonlinear filters are examined in detail and applied to three digital communication systems. The problem of designing linear digital compensators for closed loop, time-invariant linear systems when quantizers are present is also taken up. The optimal stochastic control of a linear system with quantized measurements subject to a quadratic cost is computed for a two-stage problem. Assumes knowledge of probability theory, random processes, and estimation theory. Presentation is not rigorous mathematically, although some theorems are proved. The appendixes are devoted to mathematical reviews and proofs of theorems.

Derman, Cyrus. FINITE STATE MARKOVIAN DECISION PROCESSES. Mathematics in Science and Engineering, vol. 67. New York: Academic Press, 1970. xiii, 159 p. Appendix A: Markovian Chains, pp. 139-42. Appendix B: Some Theorems from Analysis and Probability Theory, pp. 143-47. Appendix C: Convex Sets and Linear Programming, pp. 149-52. References, pp. 153-56. Index, pp. 157-59.

A text for a course in dynamic programming and a reference for operations researchers, statisticians, mathematicians, and engineers interested in the mathematical methods for the control of dynamic systems. The presentation is rigorous, with many proofs included, and it provides the student with the basic computational algorithms. Among the topics treated are: existence theorems; computational procedures for the discounting cost problem, the optimal first-passage problem, and for the expected average cost criterion using linear programming; state-action frequencies and problems with constraints; optimal stopping of a Markov chain; and applications to replacement, maintenance, sequential search problems, continuous sampling

plan, and stochastic traveling salesman problem. Assumes knowledge of real analysis or advanced calculus, elementary theory of
Markov chains, and the rudiments of linear programming. Chapters conclude with bibliographical remarks and problems.

Doob, Joseph L. STOCHASTIC PROCESSES. Wiley Series in Probability and
Mathematical Statistics. New York: Wiley, 1953. vii, 654 p. Supplement,
pp. 599-622. Appendix, pp. 623-39. Bibliography, pp. 641-49. Index,
pp. 651-54.

A fairly complete presentation of the theory of stochastic processes
up to 1953. Topics include Markov processes and martingales.
Assumes knowledge of probability theory, measure theory, and calculus. References to the literature and historical remarks are presented in the chapter appendixes.

Dynkin, Evgeniĭ B. MARKOV PROCESSES. Vol. 1. Translated by J. Fabius
et al. New York: Academic Press, 1965. xii, 365 p. Index, pp. 358-62.
List of Symbols, pp. 363-65.

Translated from the Russian. Volume 1 consists of chapters 1-11
divided into two parts. The first, consisting of chapters 1 to 5,
deals with the general theory of homogeneous Markov processes
with emphasis on infinitesimal and characteristic operators. The
second part is devoted to additive functionals and transformations
of processes with emphasis on Ito's theory of stochastic integrals
and stochastic integral equations.

_____. MARKOV PROCESSES. Vol. 2. Translated by J. Fabius et al.
New York: Academic Press, 1965. viii, 274 p. Appendix, pp. 201-38.
Historical-bibliographical Note, pp. 240-58. Bibliography, pp. 258-66.
Index, pp. 267-71. List of Symbols, pp. 272-74.

Translated from the Russian. Volume 2 consists of chapters 12 to
17 divided into three parts. The first part (chapters 12 and 13)
is devoted to harmonic and superharmonic functions related to a
process and probabilistic formulas for the solution of certain differential equations. The second part (chapter 14) applies the results of the previous chapters to the investigations of the n-dimen-
sional Wiener process and its transformation. In the third part
(chapters 15 to 17), continuous strong Markov processes on the
line are studied. The appendix includes definitions, statements
of theorems, references where proofs of theorems can be found,
additional results of measure theory, probability theory, and the
theory of differential equations.

_____. THE THEORY OF MARKOV PROCESSES. Edited by T. Kovary.
Translated by E.E. Brown. Englewood Cliffs, N.J.: Prentice-Hall, 1961.
ix, 210 p. Addendum A: A Theorem on the Measurability of the Instants of
First Departure, pp. 174-95. Supplementary Notes, pp. 196-201. References,

pp. 202-3. Alphabetical Index, pp. 204-6. Index of Lemmas and Theorems,
pp. 207-8. Index of Notation, pp. 209-10.

Translated from the Russian, this book is an investigation of the
logical foundations of the theory of stationary and nonstationary
Markov processes without any reference whatever to the general
theory of probability. It does, however, assume knowledge of
the elementary theory of Markov processes. Measure theory is
reviewed in chapter 1, and various Markov processes are defined
in chapter 2. The generation of subprocesses is considered in
chapter 3, and the construction of Markov processes with given
transition functions is taken up in chapter 4. Strictly Markov
processes are discussed in chapter 5, and the conditions that must
be imposed on the transition function are taken up in chapter 6.
The supplement describes some of Choquet's results concerning the
general theory of capacities.

Dynkin, Evgeniĭ B., and Yushkevich, Aleksandr A. MARKOV PROCESSES:
THEOREMS AND PROBLEMS. Translated by J.S. Wood. New York: Plenum
Books, 1969. x, 237 p. Appendix, pp. 221-30. Literature Cited, pp. 231-32.
Index, pp. 233-37.

A translation of a Russian book first published in 1967. It is a
problem introduction to the latest findings in the theory of Markov
processes and is divided into four chapters, each of which introduces a different problem. In chapter 1 it is shown that the familiar concepts of a harmonic function, potential, capacitance,
and other concepts from classical analysis have their analysis in
the simplest Markov chain, that is, a symmetric random walk on
a lattice, and may be used for the solution of purely probabilistic
problems. The probabilistic solution of differential equations is
presented in chapter 2. In chapter 3 the relationship between
Markov processes and potentials is applied in the investigation
of the optimal stopping of a Markov process. Chapter 4 is devoted to the application of the probabilistic approach to the
mathematics of Markov processes by means of specific problems,
rather than with theorems alone. Assumes knowledge of the basic
tenets of probability theory and classical analysis. Each of the
chapters concludes with a set of thirty to forty-eight problems
which supplement the text and present certain new information.

Feller, William. AN INTRODUCTION TO PROBABILITY THEORY AND ITS
APPLICATIONS. Vol. 1. 3d ed., rev. Wiley Series in Probability and
Mathematical Statistics. New York: Wiley, 1968. xviii, 509 p. Answers
to Problems, pp. 483-98. Index, pp. 499-509.

Presents probability theory as a self-contained mathematical topic
and with a variety of applications. In order to avoid measurability difficulties and to appeal to the nonmathematician, this
volume is restricted to discrete sample spaces. Topics include
laws of large numbers, Borel-Cantelli lemmas, Kolmogorov's in-

equality, branching processes, renewal theory, random walk and ruin problems, Markov chains, and stochastic processes. Useful as a text and reference on probability for more advanced topics in operations research. Chapters conclude with problems.

_____. AN INTRODUCTION TO PROBABILITY THEORY AND ITS APPLICA-TIONS. Vol. 2. Wiley Series in Probability and Mathematical Statistics. New York: Wiley, 1966. xviii, 626 p. Answers to Problems, pp. 611-14. Some Books on Cognate Subjects, pp. 615-16. Index, pp. 619-26.

Presents probability theory which is mathematically more rigorous than that contained in volume 1, but which can be read by individuals at different levels of mathematical preparation. The chapters are as self-contained as possible so that they can be read without following any given sequence. Topics include probability density and probability distribution functions, limit theorems, infinitely divisible distribution and semi-groups, the Laplace transform and its application, characteristic functions, applications of Fourier methods to random walks, and harmonic analysis. Useful as a text for graduate courses in probability, independent study, and reference.

Freedman, David. APPROXIMATING COUNTABLE MARKOV CHAINS. Holden-Day Series in Probability and Statistics. San Francisco: Holden-Day, 1971. xi, 291 p. Appendix, pp. 240-77. Bibliography, pp. 278-83. Index, pp. 284-88. Symbol Finder, pp. 289-91.

The third book completes a trilogy with the following other volumes: MARKOV CHAINS (see below), and BROWNIAN MOTION AND DIFFUSION (San Francisco: Holden-Day, 1971). These three books have a common preface and bibliography, but each book has its own index and symbol finder. It presents a general theory of countable Markov chains in discrete and continuous time which is rigorous with many theorems and proofs. The presentation does not use the notion of separability for stochastic processes, and in general avoids the uncountable axiom of choice. The appendix consists of seventeen parts dealing primarily with additional mathematical results such as martingales, the classical Lebesgue measure, metric spaces, and conditioning.

_____. MARKOV CHAINS. Holden-Day Series in Probability and Statistics. San Francisco: Holden-Day, 1971. xiv, 382 p. Appendix, pp. 329-66. Bibliography, pp. 367-72. Index, pp. 373-78. Symbol Finder, pp. 379-82.

A text for a graduate course in Markov chains which does not employ the notions of separability for stochastic processes or the uncountable axiom of choice. It is divided into two parts: discrete time, and continuous time. The appendix is devoted to a review of advanced mathematical topics, and the prerequisite is advanced probability.

Freeman, Herbert. DISCRETE-TIME SYSTEMS: AN INTRODUCTION TO THE THEORY. New York: Wiley, 1965. xiii, 241 p. Appendix A: Numerical Method for Z-transform Inversion, pp. 213-15. Appendix B: Z-transform Tables, pp. 216-20. Problems, pp. 221-36. Index, pp. 237-41.

A text for a one-semester graduate course in discrete-time systems, control systems, or general system theory for engineering students who have a background in Laplace and Fourier transforms, linear system theory, complex variables, and matrix algebra. Knowledge of difference equations, control theory, and statistics is helpful but not necessary. Topics include time-domain analysis of linear, discrete-time systems, with emphasis on state variable techniques; transformation calculus techniques for linear stationary systems; sampling a continuous function; interpolation and extrapolation of sampled data by means of approximating functions; continuous-time systems that are subjected to discrete-time inputs; and applications to systems with stochastic signals as well as to finite-state, probabilistic systems. An illustration of the design of an optimum Markov system using dynamic programming is provided. The problems at the end of the book are arranged by chapter, and the chapters conclude with references.

Gikham, Iosif Ilich, and Skorokhod, A.V. INTRODUCTION TO THE THEORY OF RANDOM PROCESSES. Translated by Scripta Technica. Saunders Mathematics Books. Philadelphia: W.B. Saunders, 1969. xiii, 516 p. Bibliographic Notes, pp. 497-502. Bibliography, pp. 503-4. Index of Symbols, p. 511. Index, pp. 513-16.

Translated from the Russian, this book is the first publication since the classic STOCHASTIC PROCESSES (see above), by J. Doob, to survey rigorously the modern results in the theory of stochastic processes. Suitable as a text for graduate students with a good background in measure-theoretic probability theory. Topics include measure theory, axiomatization of probability theory, random functions, processes with independent increments, jump Markov processes, diffusion processes, and limit theorems for random processes.

Girault, Maurice. STOCHASTIC PROCESSES. Econometrics and Operations Research, vol. 3. New York: Springer-Verlag, 1966. ix, 126 p. Answers to Problems, pp. 116-25. Index, p. 126.

A collection of original and varied stochastic models for use by research scientists in physics, chemistry, biology, medicine, population, economics, and operations research. Each class of models is presented by its simplest and most characteristic case (uniform process of Poisson, for example) after which a generalization is made disposing of certain hypotheses (uniformity, for example). The Markov processes, which are restricted to the finite case, are presented on the basis of the properties of convex-linear operations. Other classes of processes treated are: numerical processes with independent random increments, Laplace processes and second order processes, and some Markov processes on continuous-time. Assumes knowledge of elementary differential and integral calculus, and the calculus of probability. Five of the six chapters conclude with a small number of problems.

Harris, Theodore E. THE THEORY OF BRANCHING PROCESSES. Englewood Cliffs, N.J.: Prentice-Hall, 1963. xiv, 230 p. Bibliography, pp. 211-24. Index, pp. 225-30.

Emphasizes the systematic development of the mathematical theory of branching processes with important applications. Assumes knowledge of Markov chains and intermediate probability theory.

Hoel, Paul Gerhard; Port, Sidney C.; and Stone, Charles J. INTRODUCTION TO STOCHASTIC PROCESSES. The Houghton Mifflin Series in Statistics, vol. 3. Boston: Houghton Mifflin, 1972. x, 203 p. Answers to Exercises, pp. 190-97. Glossary of Notation, pp. 199-200. Index, pp. 201-3.

A text for a one-semester course in stochastic processes for graduates and undergraduates who have knowledge of the calculus of probability, mathematical statistics, and ordinary differential equations. The other volumes in the three-volume series are INTRO-DUCTION TO PROBABILITY THEORY, and the second, INTRODUC-TION TO STATISTICAL THEORY. The third volume presents an elementary account of selected topics, including Markov chains (both discrete and continuous), Poisson processes, birth-and-death processes, Gaussian and Wiener processes, Brownian motion, and processes defined in terms of Brownian motion by means of elementary stochastic differential equations. The presentation is rigorous, consisting primarily of theorems and proofs with examples. Each of the six chapters concludes with nearly two dozen exercises designed to illustrate the material covered and to extend the theoretical results.

Howard, Ronald A. DYNAMIC PROBABILISTIC SYSTEMS. Vol. 1: MARKOV MODELS. Series in Decision and Control. New York: Wiley, 1971. xvii, 577 p., I-7. Notation, pp. 543-47. Appendix A: Properties of Congruent Matrix Multiplication, pp. 549-50. Appendix B: A Paper by A.A. Markov, pp. 551-76. References, p. 577. Index, pp. I-1 to I-7.

Volume 1, chapters 1 to 9, treats the basic Markov process and its variants. Both volumes 1 and 2 can be used as texts for a one-year course in Markov processes for graduates and undergraduates who have knowledge of differential and integral calculus, matrix algebra, and a first course in probability theory. It is aimed at a mixed audience, including engineers, management science majors, economists, and so forth. Most of the mathematical results are stated without proof. Emphasis is on formulating, analyzing, and evaluating simple and advanced Markov models of systems drawn from a variety of fields: consumer purchasing, taxicab operation, inventory control, rabbit reproduction, coin tossing, gambling, family name extinction, search, car rental, machine repair, depreciation, production, action timing, reliability, reservation policy, machine maintenance and replacement, network traversal, project scheduling, space exploration, and success in business. Each chapter concludes with about two dozen problems.

_____. DYNAMIC PROBABILISTIC SYSTEMS. Vol. 2: SEMI-MARKOV DECISION PROCESSES. Series in Decision and Control. New York: Wiley, 1971. xviii, pp. 577-1109, I-12. Notation, pp. 1101-6. Appendix: Properties of Congruent Matrix Multiplication, pp. 1107-8. References, p. 1109. Index, pp. I-1 to I-12.

Volume 2, chapters 10 to 15, treats semi-Markov and decision processes. It can be used for a second semester course in Markov processes for graduates and undergraduates in a variety of areas who have knowledge of differential and integral calculus, matrix algebra, and probability theory. Emphasis is on formulating, analyzing, and evaluating simple and advanced Markov models of systems from a variety of fields ranging from genetics, to space engineering, to marketing. The chapter titles for volume 2 are "The Discrete-time Semi-Markov Process," "The Continuous-time Semi-Markov Process," "The Continuous-time Markov Processes," "Rewards," "Dynamic Programming," and "Semi-Markov Decision Processes." Each chapter concludes with about two dozen problems.

Iosifescu, Marius, and Tautu, Petre. STOCHASTIC PROCESSES AND APPLICATIONS IN BIOLOGY AND MEDICINE. Part 1: THEORY. Biomathematics, vol. 3. New York: Springer-Verlag, 1973. 331 p. Notation Index, p. 321. Subject Index, pp. 323-26. Author Index, pp. 327-31.

Volume 1 is a revised and enlarged version of chapters 1 and 2 of a book with the same title published in Romania in 1968. It introduces mathematicians and biologists with a strong mathematical and probabilistic background to the study of discrete and continuous parameter stochastic processes. Each of the two chapters concludes with a sixteen-page reference list. Topics include Markov chains, convergence of stochastic processes, processes with independent increments, diffusion processes, and renewal processes. Knowledge of probability theory, advanced calculus, and linear algebra is assumed.

_____. STOCHASTIC PROCESSES AND APPLICATIONS IN BIOLOGY AND MEDICINE. Part 2: MODELS. Biomathematics, vol. 4. New York: Springer-Verlag, 1973. 337 p. References, pp. 278-321. Notation Index, p. 323. Subject Index, pp. 325-30. Author Index, pp. 331-37.

Volume 2 is a revised and enlarged version of chapter 3 of a book with the same title published in Romania in 1968. It is both a text and reference intended to introduce mathematicians and biologists to the study of stochastic processes and their applications in biological sciences. The topics treated include population growth models, population dynamics processes, evolutionary processes, and models in physiology and pathology. Assumes knowledge of elementary calculus and probability.

Iosifescu, Marius, and Theodorescu, Radu. RANDOM PROCESSES AND LEARNING. New York: Springer-Verlag, 1969. x, 304 p. Bibliography, pp. 286-99. Notation Index, p. 300. Author and Subject Index, pp. 301-4.

A monograph which seeks to present the main results concerning the theory of random systems with complete connections, and to describe the general learning model by means of random systems with complete connections. Useful to mathematicians with a background in modern probability theory and functional analysis.

Jazwinski, Andrew H. STOCHASTIC PROCESSES AND FILTERING THEORY. Mathematics in Science and Engineering, vol. 64. New York: Academic Press, 1970. xiv, 376 p. Author Index, pp. 367-70. Subject Index, pp. 371-76.

A unified treatment of linear and nonlinear filtering theory for engineers with emphasis on applications. It is self-contained and suitable for a two-semester graduate course in linear and nonlinear filtering theory, or a one-semester course in linear filtering theory with prerequisites of advanced calculus, ordinary differential equations, and matrix algebra. It uses the probabilistic or Bayesian approach, and contains a complete treatment of mean square calculus. Chapters conclude with references.

Kappas, Demetrios A. PROBABILITY ALGEBRAS AND STOCHASTIC SPACES. Probability and Mathematical Statistics Series. New York: Academic Press, 1969. x, 267 p.

A presentation of the lattice-theoretic treatment of probability theory. Assumes knowledge of measure theory and mathematical analysis.

Karlin, Samuel, and Taylor, Howard M. A FIRST COURSE IN STOCHASTIC PROCESSES. 2d ed., rev. New York: Academic Press, 1975. xvi, 557 p. Appendix: Review of Matrix Analysis, pp. 536-52. Index, pp. 553-57.

An introduction to stochastic processes which attempts to bridge the gap between elementary probability and advanced work in the field. The three objectives are: (1) to present an introductory account of the main areas in stochastic processes; (2) to introduce students of pure mathematics to the applications of stochastic processes; and (3) to make the student who is concerned with applications aware of the mathematical structure of stochastic processes. Topics include Markov chains with discrete and continuous time processes and enumerable state space; examples and applications of Markov chains; Brownian motion; branching processes; and stationary processes. Applications are in the areas of ecology, population genetics, and queues. It is suitable as a text for a one- or two-semester course, and assumes knowledge of intermediate probability theory. Chapters conclude with elementary problems, problems, notes, and references.

Kemeny, John G., and Snell, J. Laurie. FINITE MARKOV CHAINS. The University Series in Undergraduate Mathematics. New York: D. Van Nostrand, 1960. viii, 210 p. Appendixes, pp. 207-10. No index.

An exposition of the basic concepts of finite Markov chains requiring a minimum mathematical background for students in the social and biological sciences. Chapter 1 reviews the mathematical prerequisites; chapters 2 to 6 develop the theory; and chapter 7 applies the theory in a variety of fields. It is suitable as an undergraduate text in Markov chains. Some chapters contain exercises.

Kemeny, John G.; Snell, J. Laurie; and Knapp, Anthony W. DENUMERABLE MARKOV CHAINS. The University Series in Higher Mathematics. New York: D. Van Nostrand, 1966. xi, 439 p. Notes, pp. 425-30. References, pp. 431-34. Index of Notations, pp. 435-36. Index, pp. 437-39.

A text for a course in Markov chains for students familiar with topology and measure theory of compact metric spaces. Part 1 provides background for the theory, and part 2 contains the basic results of denumerable Markov chains. Part 3 is on discrete potential theory, and part 4 examines boundary theory for both transient and recurrent chains. The presentation is rigorous, consisting largely of theorems and proofs. Chapters conclude with problems.

Kemperman, Johannes Henricus Bernardus. THE PASSAGE PROBLEM FOR A STATIONARY MARKOV CHAIN. Statistical Research Monograph, vol. 1. Chicago: University of Chicago Press, 1961. 127 p. Bibliography, pp. 122-24. Index, pp. 125-27.

Presents the procedures for studying first passage and absorption for a stationary Markov chain. Topics include the Ehrenfest model, the Bernoulli case, sequential analysis, occupancy times, queueing problems, and collective risk theory. Asymptotic formulas and derivatives of exact formulas for the probabilities or their moments are also presented. Assumes knowledge of advanced calculus or real analysis. The presentation is rigorous and consists mainly of theorems and proofs.

King, William R. PROBABILITY FOR MANAGEMENT DECISIONS. Wiley Series in Management and Administration. New York: Wiley, 1968. xix, 372 p. Appendix, pp. 357-67. Index, pp. 369-72.

Develops probability as a tool in analyzing management decisions and requires knowledge of elementary calculus for proper understanding. Topics include basic probability concepts, utility analysis, random variables and their properties, empirical and theoretical probability distributions, Bayesian analysis, and stochastic processes. Chapters conclude with references and exercises.

Leve, Gijsbert de. GENERALIZED MARKOVIAN DECISION PROCESSES. PART II: PROBABILISTIC BACKGROUND. 2d ed., rev. Mathematical Centre Tracts, 4. Amsterdam: Mathematisch Centrum, 1970. 135 p. References, p. 117. Errata and Addenda, pp. 118-29. List of Symbols, pp. 130-35.

An exposition of the probabilistic foundations of the Markovian decision processes consisting of two chapters: the fundamental stochastic processes, and the decision process. Topics include stationary strong Markov processes, stationary Markov processes and random losses, and stationary strong Markovian decision processes. The presentation is rigorous, consisting primarily of lemmas and proofs. Assumes knowledge of advanced probability theory, advanced calculus, and stochastic processes.

Leve, Gijsbert de; Tijms, H.C.; and Weeda, P.J. GENERALIZED MARKOVIAN DECISION PROCESSES. PART III: APPLICATIONS. Mathematical Centre Tracts, 5. Amsterdam: Mathematisch Centrum, 1970. ii, 108 p. References, p. 108. No index.

Presents five applications of Markovian decision processes utilizing two techniques for finding the optimal strategy: A direct approach (functional equations), and an iterative approach. The iterative approach is used to solve a single-item production problem, given demand and cost, and the (S,s) and (Q,s) strategies for continuous-time inventory models. The iterative approach is also used to solve an automobile replacement problem, and a production problem with a nondenumerable state space. The direct approach is used to solve an automobile insurance problem. Assumes knowledge of advanced calculus, stochastic processes, and mathematical analysis.

McKean, Henry P., Jr. STOCHASTIC INTEGRALS. Probability and Mathematical Statistics Series. New York: Academic Press, 1969. xiii, 140 p. List of Notations, pp. xi-xiii. References, pp. 133-38. Index, pp. 139-40.

A brief presentation of differential and integral calculus based upon Brownian motion for mathematicians and others employing probabilistic models in applied problems. The four chapter titles are: "Brownian Motion," "Stochastic Integrals and Differentials," "Stochastic Integral Equations (d=1)," and "Stochastic Integral Equations (d greater than or equal to 2)." Assumes knowledge of differential equations, functional analysis, and some familiarity of fields, independence, conditional probabilities and expectations, the Borel-Cantelli lemmas, and Brownian motion. Three of the four chapters conclude with problems and solutions.

McShane, Edward J. STOCHASTIC CALCULUS AND STOCHASTIC MODELS. Probability and Mathematical Statistics Series. New York: Academic Press, 1974. x, 239 p. References, p. 235. Subject Index, pp. 237-39.

A unified theory of stochastic calculus and stochastic models that does not require extensive knowledge of measure theory. It utilizes two kinds of stochastic integrals: A Riemann integral and an Ito integral, which are referred to as Riemann-belated and Ito-belated integrals. The purpose is to develop methods of solving differential equations involving random functions of fairly general types which utilize experimental observations. The models developed place only physically reasonable restrictions on the noise process involved. The six chapter titles are: "Introduction"; "Stochastic Integrals"; "Existence of Stochastic Integrals"; "Continuity, Chain Rule, and Substitution"; "Stochastic Differential Equations"; and "Equations in Canonical Form." Assumes knowledge of modern probability theory comparable to Feller's AN INTRODUCTION TO PROBABILITY THEORY AND ITS APPLICATIONS, volume 2 (see above), an introduction to measure theory, and differential equations.

Mandl, Petr. ANALYTICAL TREATMENT OF ONE-DIMENSIONAL MARKOV PROCESSES. New York: Springer-Verlag, 1968. xx, 192 p. List of Notations, pp. 186-87. Bibliography, pp. 188-90. Index, pp. 191-92.

A book devoted to one-dimensional, homogeneous random processes of the Markovian type with continuous-time parameters. Topics include second-order differential operators and transition functions, asymptotic behavior of transition probabilities, first passage problems, and optimal control of processes. Assumes knowledge of mathematical analysis. Chapters conclude with problems.

Martin, James John, Jr. BAYESIAN DECISION PROBLEMS AND MARKOV CHAINS. Publications in Operations Research, no. 13. New York: Wiley-Interscience, 1967. xii, 202 p. Appendixes, pp. 179-98. Bibliography, pp. 199-200. Index, pp. 201-2.

A presentation of the theoretical foundation needed for the solution of decision problems in a Markov chain with uncertain transition probabilities based on both sequential sampling and fixed-sample-size problems. The notion of a family of distributions closed under sampling is introduced, and then applied to the study of adaptive control problems. Assumes familiarity with the theory of Markov chains and Bayesian decision theory, and mathematical analysis. The appendixes contain computer programs and a glossary of symbols.

Mine, Hisashi, and Osaki, Shunji. MARKOVIAN DECISION PROCESSES.

Modern Analytic and Computational Methods in Science and Engineering, vol. 25. New York: American Elsevier, 1970. x, 142 p. Appendix: Stochastic Games, pp. 127-31. Bibliography, pp. 132-37. Author Index, pp. 139-40. Subject Index, pp. 141-42.

A presentation of the mathematical theory and algorithms of some of the most important types of Markovian decision processes, including semi-Markovian and general sequential decision processes and stochastic games. Assumes knowledge of mathematical analysis, probability theory, Markov chains, and linear programming. The presentation is rigorous, consisting of definitions, theorems, lemmas, corollaries, and proofs. Chapters conclude with references and comments.

Norman, M. Frank. MARKOV PROCESSES AND LEARNING MODELS. Mathematics in Science and Engineering, vol. 84. New York: Academic Press, 1972. xiii, 274 p. References, pp. 263-67. List of Symbols, pp. 269-70. Index, pp. 271-74.

A monograph on the probability theory which forms the basis of stochastic learning models. It is divided into three parts: distance diminishing models, slow-learning, and special models. The first part develops a theory of Markov processes that moves by a random contraction of a metric space; the second part presents an extensive theory of diffusion approximation of discrete time Markov processes that move "by small steps"; and the third part considers some special models, such as five-operator linear models, additive models, and multiresponsive models. Assumes knowledge of mathematical analysis, real analysis, and probability theory at the first-year graduate level.

Orey, Steven. LECTURE NOTES ON LIMIT THEOREMS FOR MARKOV CHAIN TRANSITION PROBABILITIES. New York: Van Nostrand Reinhold, 1971. viii, 108 p. Notes, pp. 99-102. Bibliography, pp. 103-6. Index, pp. 107-8.

A development of selected topics in the ergodic theory of discrete-time Markov chains with stationary transition probabilities and arbitrary measurable space for state space. Most of the development is concerned with asymptotic behavior of the n-step transition probabilities or the corresponding measures. The three chapter titles are: "Limit and Decomposition Theorems," "Sums of Transition Probabilities," and "Individual Ratio Limit Theorems."

Papoulis, Athansios. PROBABILITY, RANDOM VARIABLES, AND STOCHASTIC PROCESSES. McGraw-Hill Series in Systems Science. New York: McGraw-Hill Book Co., 1965. xi, 583 p. Index, pp. 577-83.

A text for a course in probability for engineers and scientists having knowledge of advanced calculus and systems theory. It is divided into the following two parts: probability and random variables, and stochastic processes. Chapters conclude with problems.

Parthasarathy, K.R. PROBABILITY MEASURES ON METRIC SPACES. Probability and Mathematical Statistics Series. New York: Academic Press, 1967. xi, 276 p. Bibliographical Notes, pp. 268–69. Bibliography, pp. 270–72. List of Symbols, p. 273. Author Index, p. 274. Subject Index, pp. 275–76.

A monograph on the theory of probability measures in abstract metric spaces, complete separable metric spaces, locally compact abelian groups, Hilbert spaces, and the spaces of continuous functions and functions with discontinuities of the first kind only. Assumes knowledge of advanced calculus and probability theory.

Parzen, Emanual. STOCHASTIC PROCESSES. Holden-Day Series in Probability and Statistics. San Francisco: Holden-Day, 1962. xi, 324 p. References, pp. 307–13. Author Index, pp. 314–15. Subject Index, pp. 316–24.

A text for a first course in stochastic processes for students with diverse interests and backgrounds who have knowledge of calculus and the elements of continuous probability theory. Topics include normal, covariance stationary, counting, Poisson, and renewal counting processes; and discrete and continuous parameter Markov chains. It stresses both theory and applications. The mathematical results are stated in one of three ways: without proof, intuitive proofs, and rigorous proofs. Numerous examples are used to illustrate the concepts. The sections conclude with complements and exercises.

Prabhu, Narahari Umanath. STOCHASTIC PROCESSES: BASIC THEORY AND ITS APPLICATIONS. New York: Macmillan, 1965. xii, 233 p. Index, pp. 229–33.

A formal treatment of some of the important classes of stochastic processes. In each chapter the main results are presented in the form of theorems, followed by examples and applications. The continuity, differentiability, and integrability properties of second-order stochastic processes are developed; and both diffusion and discontinuous Markov processes, based on the Feller-Kolmogorov equations, are considered. Finally, classical renewal theory, and the theory of recurrent events, with extensions to fluctuation theory of sums of random variables, are taken up. The applications are in the areas of random walks, branching processes, queueing theory, Brownian motion, population growth, and counter models. Assumes knowledge of advanced calculus and modern probability theory. Chapters conclude with complementary details, problems, and references.

Romanovsky, Vsevolod I. DISCRETE MARKOV CHAINS. Translated by E. Seneta. Groningen, The Netherlands: Walters-Noordhoff, 1970. xi, 409 p. Bibliography, pp. 406–8. More Recent Bibliography for Finite Discrete Markov Chains, p. 409. No index.

Translated from the Russian, this book presents a development of the matrix method of investigating the theory of Markov chains with a finite number of states and discrete time. Among the topics considered are decomposable and nondecomposable acyclic chains Cn, nondecomposable cyclic chains, characteristic functions of discrete Markov chains, chain correlations, Markov-Bruns chains, and complex chains. The applications are to areas such as the study of randomness, geophysical problems, and statistical problems. Assumes knowledge of linear and abstract algebra, and intermediate probability theory. The presentation is rigorous, with proofs of theorems provided.

Rosenblatt, Murray. MARKOV PROCESSES: STRUCTURE AND ASYMPTOTIC BEHAVIOR. New York: Springer-Verlag, 1971. xiii, 268 p. Appendixes, pp. 220-53. Bibliography, pp. 254-60. Author Index, pp. 261-62. Subject Index, pp. 263-65. Notation, pp. 267-68.

A text and reference on discrete parameter Markov processes. Topics include ergodic and prediction problems, random walks and convolution on groups and semigroups, nonlinear representations in terms of independent random variables, and mixing and the Central Limit Theorem. Of particular interest to economists is chapter 2 where applications to learning and resource flow models are taken up. The appendixes are devoted to mathematical reviews of probability theory, topological spaces, spaces and operators, and topological groups.

———. RANDOM PROCESSES. University Texts in the Mathematical Sciences. New York: Oxford University Press, 1962. x, 208 p. References, pp. 203-6. Index, pp. 207-8.

A text for a course on the theory of random processes for graduates and undergraduates with a background in advanced calculus. Previous knowledge of probability theory is not required. Topics include Markov chains, stationary processes, Markov processes, weakly stationary processes, and random harmonic analysis. Chapters conclude with notes.

Ross, Sheldon M. APPLIED PROBABILITY MODELS WITH OPTIMIZATION APPLICATIONS. Holden-Day Series in Management Science. San Francisco: Holden-Day, 1970. ix, 198 p. Appendixes, pp. 192-94. Index, pp. 195-98.

A supplement for courses in applied probability, stochastic processes, and sequential decision theory. Topics and applications include stochastic processes, Markov chains, the Poisson process, renewal theory, Brownian motion, and continuous time optimization models. Assumes knowledge of intermediate probability theory. The presentation is rigorous, consisting of theorems and propositions with proofs, but many theorems are illustrated with one or more examples. The appendixes contain a discussion of contraction mappings and counterexamples. Chapters conclude with problems and references.

Rozanov, Yu. A. STATIONARY RANDOM PROCESSES. Translated by A. Feinstein. Holden-Day Series in Time Series Analysis. San Francisco: Holden-Day, 1967. 211 p. Historical and Bibliographic References, pp. 199-204. Bibliography, pp. 205-10. Index, p. 211.

> Translated from the Russian, this book is a collection of a series of papers on new developments in the theory of random processes which are based on lectures given by the author at Moscow University in 1959-60. The paper titles (chapter titles) are: "Harmonic Analysis of Stationary Random Processes," "Linear Forecasting of Stationary Discrete-parameter Processes," "Linear Forecasting of Continuous-parameter Stationary Processes," and "Random Processes, Stationary in the Strict Sense." It is intended for the mathematically inclined reader interested in applications.

Saeks, R. RESOLUTION SPACE, OPERATORS AND SYSTEMS. Lecture Notes in Economics and Mathematical Systems, vol. 82. New York: Springer-Verlag, 1973. x, 267 p. Appendixes, pp. 204-57. References, pp. 258-67. No index.

> Presents a new operator theory which makes possible the unification of the continuous (time function) and discrete (time series) theories and simultaneously allows the formulation of a single theory which is valid for time-variable distributed and nonlinear systems. The appendixes are devoted to a review of topological group theory, operator-valued integration, spectral theory, and the representation theory for resolution and uniform spaces.

Shook, Robert C., and Highland, Harold J. PROBABILITY MODELS WITH BUSINESS APPLICATIONS. Edited by Esther H. Highland. Irwin Series in Quantitative Analysis for Business. Homewood, Ill.: Richard D. Irwin, 1969. xvii, 592 p. Appendix A: How to Read and Design a Procedure Chart, pp. 453-61. Appendix B: Recommended Readings, pp. 462-65. Appendix C: Tables, pp. 467-551. Appendix D: Answers to Problems, pp. 553-71. Appendix E: Glossary of Formulas, pp. 573-83. Index, pp. 585-92.

> An introductory text in probability designed to assist students in developing a pattern of thinking about uncertain situations. It is divided into the following three parts: probability model development from intuition about odds, random variable models, and random process models. Topics include elementary probability models, probability and decision tree diagrams, various simulation models (including inventory simulation), discrete and continuous random variables, statistical inference, random processes, generating probability functions, compound distributions and branching processes, and renewal processes. Chapters conclude with problems.

Silverstein, Martin L. SYMMETRIC MARKOV PROCESSES. Lecture Notes in Mathematics, vol. 426. New York: Springer-Verlag, 1974. 287 p. Bibliography, pp. 271-75. Comments Added in Proof, pp. 276-78. Index, pp. 279-83. Misprints and Minor Corrections, pp. 284-87.

A monograph on symmetric Markov processes and especially with Dirichlet spaces as a tool for analyzing them. Attention is given to the problem of classifying the symmetric sub-Markovian semi-groups which dominate a given one. The four chapter titles are "General Theory," "Decomposition of the Dirichlet Form," "Structure Theory," and "Examples." The presentation is rigorous, and assumes knowledge of the theory of martingales as developed by P.A. Meyer in MARTINGALES AND STOCHASTIC INTEGRALS (New York: Springer-Verlag, 1972).

Skorokhod, A.V. STUDIES IN THE THEORY OF RANDOM PROCESSES. Translated by Scripta Technica. Adiwes International Series in Mathematics. Reading, Mass.: Addison-Wesley, 1965. viii, 199 p. Supplementary Remarks, pp. 186-89. Bibliography, pp. 190-97. Index, pp. 198-99.

Originally published in Russian by Kiev University Press in 1961, this book develops probabilistic methods in stochastic differential equations and limit theorems for Markov processes. The first part of the book introduces the theory of random processes and stochastic integrals. The second part, chapters 3 to 5, is devoted to the theory of stochastic integrals. The third part, chapters 6 and 7, is devoted to various limit theorems connected with the convergence of a sequence of Markov chains to a Markov process with continuous time. Assumes knowledge of advanced calculus, mathematical analysis, and probability theory.

Spitzer, Frank. PRINCIPLES OF RANDOM WALK. The University Series in Higher Mathematics. New York: D. Van Nostrand, 1964. xi, 406 p. Bibliography, pp. 395-400. Index, pp. 401-6.

This book presents a special class of random processes, called random walk on the lattice points of ordinary Euclidean space, with emphasis on the connection between random walk and both harmonic analysis and potential theory. Topics include harmonic analysis, two-dimensional recurrent random walk, random walk on a half-line, random walk on an interval, transient random walk, and recurrent random walk. It contains almost 100 pages of examples and problems set apart in small print. Assumes knowledge of probability theory, real variables and measure theory, analytic functions, Fourier analysis, and differential and integral operators. Emphasis is on theorems and proofs.

Srinivasan, S. Kidambi. STOCHASTIC POINT PROCESSES AND THEIR APPLICATIONS. Griffin's Statistical Monographs and Courses, no. 34. New York: Hafner, 1974. xi, 174 p. References, pp. 169-74. No index.

A monograph which gives a largely heuristic account of the theory of point processes and its application in statistical physics, management science, and biology. Topics include renewal processes, stationary point processes, doubly stochastic Poisson processes, and multivariate point processes. The material in each chapter is illustrated by examples, and problems are given at the end of most of the chapters. Assumes knowledge of probability at the level of Feller's AN INTRODUCTION TO PROBABILITY THEORY AND ITS APPLICATIONS, volume 1 (New York: Wiley, 1968). Useful to graduate students specializing in probability and statistics, and research workers in the applied fields.

Stratonovich, R.L. CONDITIONAL MARKOV PROCESSES AND THEIR APPLICATION TO THE THEORY OF OPTIMAL CONTROL. Edited by R. Bellman. Translated by R.N. McDonough and N.B. McDonough. Modern Analytic and Computational Methods in Science and Mathematics, vol. 7. New York: American Elsevier, 1968. xvii, 350 p. Appendixes, pp. 306-17. Supplement, Solutions of Certain Problems of Mathematical Statistics, and Sequential Analysis, pp. 318-41. References, pp. 342-47. Index, pp. 349-50.

Translated from the Russian, this book is devoted to a discussion of the mathematics of optimal cybernetic systems processing statistical input information, and, in particular, systems of optimal filtering, optimal detection, and optimal control in the presence of noise. Part 1 is concerned with the theory of Markov processes, part 2 takes up the theory of conditional Markov processes, and part 3 applies the theory of conditional Markov processes to optimal control. Assumes knowledge of advanced probability.

Takacs, Lajos. COMBINATORIAL METHODS IN THE THEORY OF STOCHASTIC PROCESSES. Wiley Tracts on Probability and Statistics. New York: Wiley, 1967. xi, 262 p. Appendix, pp. 189-210. Solutions, pp. 211-54. Author Index, pp. 255-58. Subject Index, pp. 259-62.

An attempt to show that for a wide class of random variables and for a wide class of stochastic processes, explicit results can be obtained by using a generalization of the classical ballot theorem. Theorems are proved for the determination of the distributions of the maximum of random variables and the distributions of the supremum of stochastic processes. Then, the theory is illustrated by examples in queues, dams, storage, insurance risk, physics, engineering, games of chance, random walks, and other areas. Assumes knowledge of the elements of probability and stochastic processes. The appendix is devoted to additional theorems and mathematical developments, such as the abelian and Tauberian theorems, and interchangeable random variables.

_____. STOCHASTIC PROCESSES: PROBLEMS AND SOLUTIONS. Translated by P. Zador. Methuen's Monographs on Applied Probability and Statistics. New York: Wiley, 1960. xi, 137 p. References, pp. 131-35. Index, p. 137.

Translated from the Hungarian, this book contains a collection of

problems and their solutions and is intended for readers who are
familiar with probability theory and who wish to understand the
theory and methods of stochastic processes. Proofs of theorems
are generally omitted, or only brief outlines of proofs are provided. Topics include Markov chains and processes, stationary stochastic processes, recurrent processes, and secondary stochastic
processes. The problems are taken from the fields of natural
sciences, engineering, and industry, and their solutions are based
on mathematical models which can be applied in investigating
empirical processes in these fields.

Thompson, William Alfred, Jr. APPLIED PROBABILITY. International Series
in Decision Processes. New York: Holt, Rinehart and Winston, 1969. xiii,
175 p. Notes on Problems, pp. 169-71. Index, pp. 173-75.

A text, with extensive examples, for a course in probability for
liberal arts students with an advanced calculus background. An
attempt is made to place the various topics in their historical context, and two chapters are devoted to the application of probability to the physical and social sciences. Topics include discrete
and continuous random variables, central limit theorem, multivariate distributions, limiting extreme value distributions, and stochastic
processes. Chapters conclude with problems.

Thonstad, Tore. EDUCATION AND MANPOWER: THEORETICAL MODELS
AND EMPIRICAL APPLICATIONS. Foreword by C.A. Moser. Report no. 5.
Unit for Economic and Statistical Studies on Higher Education, the London
School of Economics and Political Science. Toronto: University of Toronto
Press, 1968. xiv, 162 p. Author Index, pp. 156-57. Subject Index, pp. 158-62.

Presents a Markov chain model of the educational system of Norway. The initial distribution of individuals by activities (grades
and branches of education) is known as well as the transition probabilities of their moving to any other activity in a specified
time period. The first model examines the implications of constant transition ratios between different school activities and
between school activities and final educations when inputs (of
pupils) into the school system are known, but there is no particular pattern of qualified manpower requirements. The second
model assumes that the stock of nonteachers is known and there
are adequate pupils to fill the places necessary in each branch
of education. Each model attempts to answer the following questions: the fraction of persons now in activity h who will be in
activity k after a given number of years; the average number of
years spent in activity k by those now in activity h; the average
number of years of schooling left for persons entering activity h;
the fraction of pupils now at the beginning of activity h who will
graduate with education e after a given number of years; the

fraction of pupils in activity h who will have completed education
e within a given number of years; and the final educational distribution of persons entering a given school activity. Assumes
some knowledge of mathematical statistics, linear algebra, and
stochastic processes.

Tintner, Gerhard, and Sengupta, Jati K. STOCHASTIC ECONOMICS: STO-
CHASTIC PROCESSES, CONTROL, AND PROGRAMMING. New York: Aca-
demic Press, 1972. xi, 315 p. References and Bibliography, pp. 269-96.
Author Index, pp. 297-304. Subject Index, pp. 305-15.

An application of stochastic processes to the theory of economic
development, stochastic control theory, and various aspects of
stochastic programming. It is based on research results and concepts arising from the authors' own research in growth, resource
allocation, and quantitative economic policy. It is useful to
economists, mathematicians, operations research analysts, and
system engineers. Assumes knowledge of advanced probability
and a course in operations research.

Tucker, Howard G. A GRADUATE COURSE IN PROBABILITY. Probability
and Mathematical Statistics Series. New York: Academic Press, 1967. xiii,
273 p. Suggested Reading, p. 270. Index, pp. 271-73.

A text for a one-year course in probability with prerequisites of
real analysis or measure theory. While no previous knowledge of
probability theory is assumed, the reader should be familiar with
the Lebesgue integral over abstract spaces.

Wong, Eugene. STOCHASTIC PROCESSES IN INFORMATION AND DYNAMI-
CAL SYSTEMS. McGraw-Hill Series in Systems Science. New York: McGraw-
Hill Book Co., 1971. xii, 308 p. References, pp. 261-64. Solutions to
Exercises, pp. 265-302. Index, pp. 303-8.

A graduate-level text in stochastic processes for students whose
primary interest is applications. It consists of a rigorous, yet
readily accessible, treatment of those topics in the theory of
continuous-parameter stochastic processes that are important in
the analysis of information processing and dynamical systems.
The two main topics are: second-order properties of stochastic
processes, requiring a good undergraduate background in probability theory and linear system analysis; and stochastic integrals
with applications to white noise in dynamical systems, requiring
familiarity with measure theory. Proofs of theorems are provided.
Chapters conclude with exercises.

Chapter 15

SYSTEMS AND SIMULATION

Ackoff, Russell L., and Emery, Fred E. ON PURPOSEFUL SYSTEMS. Chicago: Aldine-Atherton, 1972. xii, 288 p. Appendix 1: Errors of Observations, pp. 265-67. Appendix 2: The Form of Perceptions and Observations, pp. 268-73. Appendix 3: On Newcomb's and Rapoport's Hypothesis, pp. 274-80. Author Index, pp. 281-83. Subject Index, pp. 284-88.

A presentation of human behavior as systems of purposeful (teleological) events. It attempts to synthesize findings in the many disciplines of science. Part 1 lays the foundations on which the conceptual system is built, part 2 takes up the process that makes purposeful behavior possible, part 3 is devoted to interactions of purposeful systems, and part 4 is on social systems whose components are purposeful. Assumes knowledge of differential calculus. Chapters conclude with references.

Barton, Richard F. A PRIMER ON SIMULATION AND GAMING. Englewood Cliffs, N.J.: Prentice-Hall, 1970. x, 239 p. Bibliography, pp. 226-34. Index, pp. 235-39.

A nontechnical introduction to simulation and gaming for the administrative professions, the behavioral sciences, and education. It discusses four techniques: analysis, man-model simulation, man-computer simulation, and all-computer simulation. Flow diagrams for computer programs are explained, and several simulation languages are discussed.

Beazer, William F.; Cox, William A.; and Harvey, Curtis A.; with the assistance of Nancy Watkins. U.S. SHIPBUILDING IN THE 1970S. Studies in Business, Technology, and Economics. Lexington, Mass.: D.C. Heath, 1972. xviii, 180 p. Appendix A: A Linear Programming Model of the U.S. Shipbuilding Industry, pp. 109-20. Appendix B: Costs, Labor Use, and Ways Requirements Data, pp. 121-24. Appendix C: The Processing of Yard Employment Data, pp. 125-26. Appendix D: Production Functions, Labor Supply, and the Costs of Output Expansion, pp. 127-33. Appendix E: Economics of Splitting Ship Runs among Yards, pp. 135-39. Appendix F: Yard Technology and Investment, pp. 141-46. Appendix G: Conversion of Lump-sum Investment

Costs to Periodic Rental Charges, pp. 147–48. Appendix H: Calculation of the Savings from Investment, pp. 149–54. Appendix I: Derivation of Capacity-Utilization Quotas for Work-dispersion Policies, pp. 155–56. Appendix J: Estimating Steel Throughput Capacity, pp. 157–58. Bibliography, pp. 161–67. Notes, pp. 169–77. Index, pp. 179–80.

> A study of U.S. private shipbuilding programs (excluding naval shipyards) with emphasis on the effects of alternative shipbuilding programs and government policies on the size and location of the industry and the costs of ships. It is divided into the following three parts: historical perspectives, method of analysis and processing of data, and simulation results. The study estimates the variable costs of building ships in each of fifteen shipyards both within present facilities and manpower and within improved facilities. A linear programming model is used to minimize construction costs of ships to be built between 1969 and 1980. The model is also used for testing several alternative shipbuilding programs by simulation.

Bonini, Charles P. SIMULATION OF INFORMATION AND DECISION SYSTEMS IN THE FIRM. Foreword by C.H. Faust. The Ford Foundation Doctoral Dissertation Series. Englewood Cliffs, N.J.: Prentice-Hall, 1963. xvi, 160 p. Bibliography, pp. 153–60. No index.

> A report on a simulation model of a hypothetical business firm. The model is a synthesis of theories from several disciplines of economics, accounting, organization theory, and behavioral science, and its purpose is to study the effects of informational, organizational, and environmental factors upon the decisions of a business firm. The analysis is accomplished by making alterations in the model and by observing the effects of these alterations upon the firm's behavior. Familiarity with computer programming is helpful.

Brockett, Roger W. FINITE DIMENSIONAL LINEAR SYSTEMS. Series in Decision and Control. New York: Wiley, 1970. xii, 244 p. References, pp. 228–33. Glossary of Notation, pp. 235–37. Index, pp. 239–44.

> A text for a one-semester course on dynamical systems for engineers, economists, and mathematicians. Many of the standard topics of applied mathematics are treated, including the solubility of linear systems of equations, ordinary differential equations, calculus of variations, and the basic ideas of vector analysis. The presentation is rigorous, consisting of theorems and proofs, followed by examples, but it can be followed by anyone having a linear algebra background. In order to make the book more widely accessible, considerable background material on linear algebra is included in summary form. Sections conclude with exercises, and the chapters conclude with notes and references. The four chapter titles are: "Linear Differential Equations," "Linear Systems," "Least Squares Theory," and "Stability."

Carlson, John G.H., and Misshauk, Michael J. INTRODUCTION TO GAMING: MANAGEMENT DECISION SIMULATIONS. The Wiley Series in Management and Administration. New York: Wiley, 1972. viii, 184 p. Bibliography, pp. 6-7. Appendixes, pp. 175-84. No index.

A text for undergraduates in (1) introductory simulation and gaming courses, (2) management decision gaming techniques, (3) model design and decision processing, and (4) management laboratories. More than half of the book consists of fifteen illustrative game cases, each with tables and exhibits. Other topics include concepts in game design, how to design a business game, behavior in gaming situations, and briefing material. The appendixes contain interest tables and random number tables, and a management simulation game in time-sharing BASIC.

Carter, L.R., and Huzan, E. A PRACTICAL APPROACH TO COMPUTER SIMULATION IN BUSINESS. Unwin Professional Management Library. London: George Allen and Unwin, 1973. 298 p. References and Bibliography, pp. 221-22. Appendixes, pp. 223-93. Index, pp. 295-98.

A self-contained introduction to the general principles of simulation in business for managers, systems analysts, industrial engineers, and operational research workers who have little mathematical backgrounds. The simulation programs are written in FORTRAN, BASIC, and CSL (a special purpose simulation language). One chapter of the book is devoted to basic statistics. Other topics include building simulation models, the use of computer languages for computer simulation studies, maintenance policies, and production control. Fourteen of the twenty-two appendixes present the computer programs, seven consist of statistical tables, and the remaining appendix is devoted to queueing formulas.

Chen, Chi-Tsong. INTRODUCTION TO LINEAR SYSTEM THEORY. Holt, Rinehart and Winston Series in Electrical Engineering, Electronics, and Systems. New York: Holt, Rinehart and Winston, 1970. xvi, 431 p. Glossary of Symbols, pp. xv-xvi. Appendixes, pp. 399-416. References, pp. 417-23. Index, pp. 425-31.

A text for undergraduate and graduate courses in the analysis and design of linear systems and a reference for engineers and applied mathematicians. Assumes knowledge of matrix algebra and differential equations. The presentation is based on the use of transfer functions. The appendixes contain additional mathematical results.

Chorafas, Dimitris N. CONTROL SYSTEMS FUNCTIONS AND PROGRAMMING APPROACHES. Vol. A: THEORY. Foreword by H.B. Fancher. Mathematics in Science and Engineering, vol. 27A. New York: Academic Press, 1966. xxvi, 395 p. Index, pp. 379-95.

A nonmathematical introduction to the role of computer programming in systems analysis and control. It is divided into the following five parts: the dynamics of digital automation; data collection and teletransmission; numerical, logical, and stochastic processes; mathematics for systems control; and programming for real-time duty. Knowledge of college algebra and computer programming is helpful.

_____. CONTROL SYSTEMS FUNCTIONS AND PROGRAMMING APPROACHES. Vol. B: APPLICATIONS. Foreword by O.W. Rechard. Mathematics in Science and Engineering, vol. 27B. New York: Academic Press, 1966. xx, 276 p. Index, pp. 249-76.

Volume B consists of part 6, process type cases and data control; part 7, applications in the metals industry; and part 8, guidance for discrete particles. It is nonmathematical, consisting of such topics as the nature of information that can be obtained about the system, where and how it can be obtained and how it can be transmitted to a digital computer, the transformations of the input information that are required, and the timing requirements of the input information.

_____. SYSTEMS AND SIMULATION. Foreword by R. Bellman. Mathematics in Science and Engineering, vol. 14. New York: Academic Press, 1965. xvi, 503 p. Index, pp. 489-503.

A reference for engineers and practitioners on mathematical systems consisting of seven parts: basic notions (three chapters), the mathematics of simulation (six chapters), evaluating industrial systems (four chapters), applications with stochastic processes (four chapters), research on traffic and cargo problems (three chapters), hydrological applications (three chapters), and simulation by analog means (two chapters). Prerequisites include a course in calculus and some knowledge of operations research. The presentation is not rigorous mathematically, and applications are to inventories, traffic control and cargo handling, military problems, economics, industrial location, and so forth. No computer programs are included.

Churchman, Charles West. THE SYSTEMS APPROACH. New York: Dell Publishing Co., 1968. xi, 243 p. Supplement 1: Some Exercises in Systems Thinking, pp. 235-38. Supplement 2: Suggested Readings, pp. 239-43. No index.

A nontechnical examination of the systems approach from the point of view of its validity, particularly in government planning. It is divided into the following four parts: what is a system, with input-output as an illustration; applications of systems thinking; systems approach to the future; and the systems approach and the human being.

Cleland, David I., and King, William R. SYSTEMS ANALYSIS AND PROJECT MANAGEMENT. McGraw-Hill Series in Management. New York: McGraw-Hill Book Co., 1968. xvi, 315 p. Appendix 1: Specifications of the Project Plan, pp. 289-97. Appendix 2: The Project Manual, pp. 298-300. Appendix 3: Project Management Checklist: Authority and Responsibility of the Project Manager, pp. 301-8. Index, pp. 311-15.

A nontechnical presentation of systems analysis and project management to demonstrate their unity and applicability in a wide variety of industrial and governmental management environments. It provides a framework for the integration of traditional management theory with the newer concepts of operations research and project or systems theory. The book is divided into the following three parts: basic systems concepts, systems analysis for strategic decisions, and project management in executing decisions. Useful as a supplementary text in undergraduate and graduate courses in management. The chapters conclude with recommended references, and there are no prerequisites except the rudiments of management theory.

Colella, A.M.; O'Sullivan, Michael J.; and Carlino, D.J. SYSTEMS SIMULATION: METHODS AND APPLICATIONS. Foreword by M.E. Connelly. Lexington, Mass.: D.C. Heath, 1974. xxv, 293 p. Appendix A: MAPTACS: A Computerized Municipal Management Information and Planning System for Systems Simulation, pp. 277-87. Index, pp. 287-92. About the Author, p. 293.

An application-oriented book that can serve as a simulation handbook for engineers, mathematicians, computer programmers and other technical specialists, and for those involved in engineering management and administration. Topics include systems modeling, computational requirements and their computer representations, computer aspects of simulation, and applications of the simulation process to: adaptive control, orbit simulation, aircraft landing systems, lunar landing, regression and parametric estimation, automated highway systems, and simulation of a total digital computer system. Assumes knowledge of differential and integral calculus, basic statistics, and computer systems.

Cone, Paul R.; Basil, Douglas C.; Burak, Marshall J.; and Megley, John E. III. EXECUTIVE DECISION MAKING THROUGH SIMULATION. 2d ed. Columbus, Ohio: Charles E. Merrill, 1971. viii, 264 p. Index, pp. 261-64.

Consists of a complete package, combining industry and company case studies, a computer-based simulation management game, and a behavioral lab. It is divided into the following six parts: elements of strategy, learning through case study and simulation, the industry and its environment, company cases, the simulation, and decision forms. Decisions are of three major types: (1) repetitive year-by-year decisions involving both long-range and short-run planning, (2) repetitive quarter-by-quarter decisions involving the planning of the next quarter's operation, and (3) nonrepetitive decisions to meet unusual situations such as strikes, fire or accident, antitrust actions, recall of tires, pollution control, and so on. The cases consist of the five major tire and rubber manufacturers, including Firestone and Goodrich. Useful as a supplementary and complementary text for management and business policy (corporate strategy) courses.

Couger, J. Daniel, and Knapp, Robert W., eds. SYSTEM ANALYSIS TECHNIQUES. New York: Wiley, 1974. xii, 509 p. No index.

A text for an introductory course in information systems assuming no mathematical prerequisites. It consists of twenty-eight contributions (by different authors) grouped under the following headings: the system perspective (two contributions), techniques for analyzing systems (twenty contributions), and cost-effectiveness analysis (six contributions). The contributions are supplemented by additional material to enable the book to serve as a textbook. Among the techniques used are flowcharting and gridcharting, decision tables, and information algebra. The contributors include J.D. Couger, J. Emery, L. Fried, C.B.B. Grindley, H.J. Lynch, R.G. Murdick, J. Rhodes, Wm. Sharpe, and D. Teichroew.

Deutsch, Ralph. SYSTEMS ANALYSIS TECHNIQUES. Prentice-Hall Network Series. Englewood Cliffs, N.J.: Prentice-Hall, 1969. xvi, 472 p. Bibliography, pp. 457-64. Index, pp. 465-72.

A text for a course in systems analysis for engineers and others working in operations research and related fields, and a reference for the practitioner in systems design. It lays the foundation for the mathematical techniques, and includes topics such as optimization based on loss and risk functions; signal theory; linear and nonlinear systems; estimation, including least squares and maximum likelihood methods; detection and decisions; approximations, representations, and perturbations; dynamic and linear programming; simulation; and information theory. The mathematical results are stated without proof, but some derivations are included, and considerable descriptive material is provided. Assumes knowledge of advanced engineering mathematics, statistics, and probability. Each of the eleven chapters concludes with problems.

Emshoff, James R., and Sisson, Roger L. DESIGN AND USE OF COMPUTER SIMULATION MODELS. New York: Macmillan, 1970. xvii, 302 p. Appendix: An Example, pp. 273-96. Index, pp. 297-302.

A text for a course in simulation for students in industrial engineering, management science, and operations research who possess some knowledge of FORTRAN, statistics, and calculus. The simulation languages employed are: FORTRAN, GPSS (general purpose simulation system) IBM, SIMSCRIPT, GASP, CSMP, DYNAMO, and JOB SHOP SIMULATOR. Topics include: simulation methodology, model building and use, developing the simulation model and program, model design, analyzing a simulation run, experimental optimization, and simulation models of human behavior and of a computer center's operations. A large number of programs, printouts of simulation exercises, and flow charts are used to illustrate the concepts discussed. Chapters conclude with references and problems.

Engwall, Lars. MODELS OF INDUSTRIAL STRUCTURE. Lexington, Mass.: Lexington Books, 1973. xiii, 179 p. Appendix A: The Computer Program Used for the Discrete Simulation Model, pp. 145-50. Appendix B: The Computer Program Used for the Continuous Simulation Model, pp. 151-55. Notes, pp. 157-63. List of Designations, pp. 165-67. Bibliography, pp. 169-76. Author Index, pp. 177-78. Subject Index, p. 179.

Based on the author's dissertation, this monograph presents both analytical and simulation models for measuring industry concentration using data for the Swedish and American economies primarily. It is divided into the following four parts: the topics and related concepts, analytical models, simulation models, and conclusions. Two theoretical distributions for measuring concentration are investigated: the Pareto distribution and the lognormal distribution. The discrete models considered are based on Markov chains. This monograph deals primarily with the empirical testing of the models and the measuring of concentration ratios; it assumes knowledge of intermediate econometric techniques.

Enrick, Norbert Lloyd. MANAGEMENT PLANNING: A SYSTEMS APPROACH. New York: McGraw-Hill Book Co., 1967. xiii, 217 p. Appendix: Glossary of Programming Terms, pp. 209-13. Index, pp. 215-17.

A use-oriented introduction to mathematical programming for operations planning for the manager and student. The first half of the book deals with theory and methods, and the second half applies the theory to fifteen case studies. The book attempts to answer questions such as: Where is mathematical programming needed and likely to be of value? What types of data are needed? What functions must be analyzed? and, How are the results interpreted?

Everling, Wolfgang. EXERCISES IN COMPUTER SYSTEMS ANALYSIS. Lecture Notes in Economics and Mathematical Systems, vol. 65. New York: Springer-Verlag, 1972. viii, 184 p. References, pp. vii-viii. Appendix, pp. 167-84.

Presents system analysis as a set of concepts and computations which establish whether or not a given system satisfied given design criteria. Topics include communications network design computer center analysis, and queueing applications. Considerable space is devoted to exercises and their solutions using numerical tables and computer programs. Assumes knowledge of calculus, probability, and programming concepts.

Formby, John. AN INTRODUCTION TO THE MATHEMATICAL FORMULATION OF SELF-ORGANIZING SYSTEMS. New York: Van Nostrand Reirhold,

1965. viii, 200 p. Appendix 1: Biographical Notes, pp. 181-91. Appendix 2: Selected References, pp. 192-94. Index, pp. 195-200.

A monograph dealing with the roles played by analytical thought and the precise formulation of problems and concepts in self-organizing systems (self-regulating, self-adjusting, self-optimizing, and adaptive systems). It is useful to industrialists and others who have a formal mathematics background and who have an interest in such systems. The mathematics used include trigonometry, calculus (including partial derivatives), and analytical geometry. Topics include deterministic and probabilistic models, algebraic structures, boundary value theory, theories of optimization, operational techniques, nonlinear systems, simulation, digital automative systems, self-adjusting and self-optimizing systems, error actuated devices, and digital automative systems. The aim of the book is to convince the reader of the wide applicability of mathematics in systems analysis.

Gibbs, G. Ian, ed. HANDBOOK OF GAMES AND SIMULATION EXERCISES. Beverly Hills, Calif.: Sage Publications, 1974. x, 226 p. Appendix, pp. 221-26. No index.

Contains sources of information on gaming and simulation for reference. For each game, an attempt was made to include the title, year of publication, country of origin, author, publisher, target population, number of players, approximate duration, and major aims and objectives. The first four chapters of fifty-three pages are devoted to the introduction, vocabulary of gaming and simulation, bibliography, synthesis, and prognosis. Chapter 5, pages 54-221 is entitled "Register of Games and Simulation." The appendix contains an alphabetical listing of games on which records are incomplete.

Goode, Harry H., and Machol, Robert E. SYSTEMS ENGINEERING: AN INTRODUCTION TO THE DESIGN OF LARGE-SCALE SYSTEMS. McGraw-Hill Series in Control Systems Engineering. New York: McGraw-Hill Book Co., 1957. xii, 551 p. References, pp. 521-26. Solutions to Problems, pp. 527-35. Index, pp. 537-51.

A text for a course in systems engineering for students of engineering and related fields with emphasis on applications. It develops no general theory and can be followed by anyone with knowledge of basic calculus. It is divided into six parts as follows: introduction; probability--the basic tools of exterior system design; exterior system design; computers--the basic tool of interior system design; interior system design in which queueing theory, game theory, linear programming, cybernetics, simulation, information theory, servomechanism theory, and human engineering are treated; and epilogue. Some chapters conclude with bibliographies and problems.

Greenlaw, Paul S.; Herron, Lowell W.; and Rawdon, Richard H. BUSINESS SIMULATION IN INDUSTRIAL AND UNIVERSITY EDUCATION. Prentice-Hall International Series in Management. Englewood Cliffs, N.J.: Prentice-Hall, 1962. xii, 356 p. Appendix 1: Marketing Decision Simulation, pp. 257-69. Appendix 2: Summary of Business Simulations, pp. 270-341. Bibliography, pp. 342-49. Index of Business Simulations, pp. 351-53. General Index, pp. 354-56.

> A guide and a comprehensive source of information on the design, administration, and educational uses of business simulation, both in university curricula and in the management training programs of industry. Topics include game construction, an example of a game designed to illustrate the use of simulation in presenting a new concept (that of systems performance), game administration and approaches for operating gaming services, and problems involved in the use of gaming. There are no mathematics or computer programming prerequisites. Chapters conclude with summaries.

Guetzkow, Harold; Kotler, Philip; and Schultz, Randall L., eds. SIMULATION IN SOCIAL AND ADMINISTRATIVE SCIENCES: OVERVIEWS AND CASE EXAMPLES. Englewood Cliffs, N.J.: Prentice-Hall, 1972. xiv, 768 p.

> A collection of overviews, case examples, and readings designed to give the reader a comprehensive understanding of the underlying theory and techniques of simulation in the social and administrative sciences. The cases and readings are reprints of articles published elsewhere, but the overviews were not previously published. The presentation is grouped under the following headings: introduction, simulation in social sciences, simulations in administrative science, and methodology and theory. Among the contributors are: I. Adelman, P. Kotter, D.F. Marble, Wm. T. Newell, T.H. Naylor, R.L. Schultz, and E.M. Sullivan.

Hammersley, John Michael, and Handscomb, D.C. MONTE CARLO METHODS. Methuen's Monographs on Applied Probability and Statistics. London: Methuen, 1964. vii, 178 p. References, pp. 150-68. Index, pp. 169-78.

> A fairly complete survey of the history of Monte Carlo methods prior to 1964. Chapters 1 to 3 are devoted to a short resume of statistical terms and random, pseudorandom, and quasirandom numbers. Chapter 4 takes up direct simulation used in industrial and operational research, chapters 5 and 6 present the main theoretical results on Monte Carlo techniques, and chapters 7 to 12 deal with applications in the solution of linear operator equations, radiation shielding and reactor criticality, statistical mechanics, long polymer molecules, percolation processes, and multivariate problems. The mathematical results are presented without proof, although some derivations are included, and require knowledge of differential and integral calculus and introductory normed linear spaces.

Henshaw, Richard C., and Jackson, James R. THE EXECUTIVE GAME. Rev. ed. Homewood, Ill.: Richard D. Irwin, 1972. viii, 161 p. Index, pp. 157-61.

Simulates a small industry (oligopoly) by an electronic computer using two models, one providing management with more options than the other. The participants of the game are organized into teams (ideally from three to five members) which operate their hypothetical companies in competition with one another. The purpose of the game is to maximize the expected rate of return on the capital invested in the company. Each team manages the company by making the following quarterly decisions: price of product, marketing budget, research and development budget, maintenance budget, production volume scheduled, investment in plant and equipment, purchase of materials, and dividends declared. There are no prerequisites for participating in the game. Useful as a supplement in beginning courses in management and for seminars for practicing managers.

Johnson, Richard A.; Kast, Fremont E.; and Rosenzweig, James E. THE THEORY AND MANAGEMENT OF SYSTEMS. 2d ed., rev. McGraw-Hill Series in Management. New York: McGraw-Hill Book Co., 1967. xiv, 513 p. Bibliography, pp. 489-502. Index, pp. 503-13.

Primarily a conceptual treatise, rather than a text, although questions and case illustrations are included in the second edition to increase its appeal for classroom use. It is a nonmathematical treatment of the role of systems in the management of business with emphasis on the application of psychological principles. It is divided into four parts as follows: systems concepts and management, applications, implementation, and the future. About sixty pages of the last chapter contain a detailed description of the Weyerhaeuser Company as an illustration of systems in a large, complex corporation.

Kleijnen, Jack P.C. STATISTICAL TECHNIQUES IN SIMULATION (IN TWO PARTS). Part I. Statistics: Textbooks and Monographs, vol. 9. New York: Marcel Dekker, 1974. xv, 285 p. No index.

A monograph dealing with the statistical design and analysis of simulation and Monte Carlo experiments, with emphasis on digital simulation with models of management and economic systems. Part I consists of three chapters, each chapter ending with exercises, notes, and a long list of references.

Topics include Monte Carlo simulation models; Lewis-Learmonth random number generator; sampling two correlated variables; statistical aspects of simulation, including input data analysis, starting conditions, validation, and analysis of steady-state output and runlength; and variance reduction techniques, including stratification, selective sampling, control variables, importance sampling, antithetic variates, and common random variables. A few flow charts and a computer program for a random number generator are presented. This book provides a working knowledge of the statistical techniques of simulation, and knowledge of mathematical statistics is assumed.

———. STATISTICAL TECHNIQUES IN SIMULATION (IN TWO PARTS). Part II. Statistics: Textbooks and Monographs, vol. 9. New York: Marcel Dekker, 1975. xv, pp. 287-775. Solutions to Exercises, pp. 747-53. Author Index, pp. 755-66. Subject Index, pp. 767-75.

A continuation of part I. Part II consists of three chapters, each ending with exercises, notes, and references. Topics include the design and analysis of experiments, including factorial and screening designs, and an example of Box's formula for augmented designs; the effect of sample size on the reliability of statements based on a sample, including evaluation of a single population and the comparison of two populations, multiple comparison procedures, and multiple ranking procedures; and a case study of a Monte Carlo experiment to investigate the robustness of the Bechhofer and Blumenthal's multiple ranking procedure. The index covers both parts.

Kochenburger, Ralph J. COMPUTER SIMULATION OF DYNAMIC SYSTEMS. Englewood Cliffs, N.J.: Prentice-Hall, 1972. xii, 530 p. Index, pp. 525-30.

A reference for engineers engaged in research and development involving simulations in industrial, governmental, and educational institutions; and a text for a one-semester course at the graduate or undergraduate levels. Topics include principles of analogue computers, linear and nonlinear simulation operations with digital computers, boundary value problems, and optimization of simulated systems. Assumes knowledge of digital-computer programming and Laplace transforms. Chapters conclude with problems and references.

Kornai, János. ANTI-EQUILIBRIUM: ON ECONOMIC SYSTEMS THEORY AND THE TASKS OF RESEARCH. New York: American Elsevier, 1971. xx, 402 p. Principle Notation, pp. xvii-xix. References, pp. 377-90. Author Index, pp. 391-94. Subject Index, pp. 395-402.

An examination of the interactions of a series of partial systems— firms, households, government offices, social institutions, and so on. Emphasis is on the control and the flow of information. It is divided into the following four parts: starting points, concepts and questions, pressure and suction in the market, and retrospection and a look forward. Topics include basic concepts of general equilibrium theory; a general model of the economic system; information structure; structure of the decision process; survey and critique of preference and utility; classification and aggregation; and market equilibrium. Only a few mathematical equations are used in the presentation, and elementary set theory is the main prerequisite.

Kupperman, Robert H., and Smith, Harvey A. MATHEMATICAL FOUNDA-
TIONS OF SYSTEM ANALYSIS. Vol. 1. Reading, Mass.: Addison-Wesley,
1969. vi, 214 p. References, pp. 182-83. List of Symbols, pp. 184-85.
Answers and Hints to Selected Problems and Exercises, pp. 189-206. Index,
pp. 209-14.

> A text for a one-semester course in system analysis for graduates
> and well-prepared undergraduates in business, economics, engi-
> neering, and systems analysis. It is the first of a two-part series
> on the theory of constrained optimization developed from lecture
> notes prepared for the Defense Education Program by the Institute
> of Defense Analysis and the University of Maryland. Topics in-
> clude linear algebra, linear programming, the calculus of a func-
> tion of a single variable and functions of several variables, and
> nonlinear programming. Assumes knowledge of calculus and ele-
> mentary probability theory. Chapters contain trivial exercises,
> problems, helpful hints, and answers.

Langholm, Odd. FULL COST AND OPTIMAL PRICE: A STUDY IN THE DY-
NAMICS OF MULTIPLE PRODUCTION. Scandinavian University Books. Oslo:
Universitetsforlaget, 1969. vii, 87 p. Mathematical Appendixes, pp. 51-62.
FORTRAN Programs, pp. 63-67. Figures, pp. 68-79. Tables, pp. 80-86.
References, p. 87.

> A report on a research project in which computer simulation is
> used to test the rationality of alternative methods of price calcu-
> lation in industrial firms. The results of the simulation experi-
> ment are close to those obtained from full cost pricing. Assumes
> knowledge of calculus and computer programming.

Lee, Alec M. SYSTEMS ANALYSIS FRAMEWORK. Studies in Management.
New York: Wiley, 1970. xi, 283 p. Notes and References, pp. 269-78.
Index, pp. 279-83.

> A presentation of systems analysis in broad, general principles and
> concepts rather than in terms of specific techniques. The chapters
> are grouped under three headings: definition of systems analysis
> and introduction of basic concepts, approaches to solution of sys-
> tems problems, and potential areas for systematization in a gener-
> alized organization. While equations are not used, an under-
> standing of the mathematical and statistical concepts of operations
> research is helpful.

McKenney, James L. SIMULATION GAMING FOR MANAGEMENT DEVELOP-
MENT. Foreword by B. Fox. Boston: Graduate School of Business Admini-
stration, Harvard University, 1969. xvi, 189 p. Appendix A: Harvard Busi-
ness School Management Simulation Instructions to Participants, pp. 137-82.
Appendix B: Management Control Methods--the Cash Forecast, pp. 183-86.
Bibliography, pp. 187-89. No index.

> Presents the results of a research project aimed at developing a
> simulation model which would be suitable for a management game.

McMillan, Claude; Gonzales, Richard F.; and Schriber, Thomas J. SYSTEMS ANALYSIS: A COMPUTER APPROACH TO DECISION MODELS. 3d ed., rev. Irwin Series in Quantitative Analysis for Business. Homewood, Ill.: Richard D. Irwin, 1973. xiv, 610 p. Appendixes, pp. 581-607. Index, pp. 609-10.

A text on computer simulation models in operations research using FORTRAN with BASIC language counterpart programs for the major FORTRAN programs. Models of inventory and queueing systems, network analysis, linear programming, and other management planning models are solved by these programs. Numerous flow charts and FORTRAN statements are used to illustrate concepts. Chapters conclude with exercises and references.

Meier, Robert C.; Newell, William T.; and Pazer, Harold J. SIMULATION IN BUSINESS AND ECONOMICS. Englewood Cliffs, N.J.: Prentice-Hall, 1969. x, 369 p. Appendix A: Tables of Random Variates, pp. 333-36. Appendix B: Inventory Simulation Programs, pp. 337-53. Appendix C: Industrial Dynamics Model, p. 355. Appendix D: Flow Charts for Heuristic Program, pp. 356-61. Index, pp. 363-69.

A text for an introductory course in simulation for students who have some knowledge of algebra, statistics, and optimizing techniques. It presents the basic concepts involved in simulation; describes applications of simulation to business and economic analysis (including PERT, waiting lines, job shop scheduling, forecasting, quality control, econometric models, market processes, and games and gaming); and discusses technical problems associated with the use of simulation. Emphasis is on the design and operations of computer models, and some of the exercises at the end of each chapter are for hand and machine computation, while others are appropriate for computer solution. Any general programming language such as FORTRAN can be used for writing programs to work the exercises. Heuristic methods are also presented. Chapters conclude with bibliographies as well as exercises.

Mihram, G. Arthur. SIMULATION: STATISTICAL FOUNDATIONS AND METHODOLOGY. Mathematics in Science and Engineering, vol. 92. New York: Academic Press, 1972. xv, 526 p. Appendix: Statistical Tables, Glossary, Bibliography, pp. 497-518. Bibliographic Sources, pp. 519-20. Subject Index, pp. 521-26.

A text for a course in systems theory which consists mainly of a presentation of dynamic, stochastic, and simulation models with emphasis on techniques for the generation of random numbers and for the generation therefrom of random variables of many distributional families. Mention is made of diverse applications in ecology, sociology, economics, and other disciplines. Prerequisites include probability calculus, differential and integral calculus, vector and matrix operations, and some knowledge of computer programming. Exercises follow the sections of the book.

Mize, Joe H., and Cox, J. Grady. ESSENTIALS OF SIMULATION. Prentice-Hall International Series in Industrial Engineering and Management Science. Englewood Cliffs, N.J.: Prentice-Hall, 1968. xiii, 234 p. Appendix 1: Tables, pp. 201-19. Appendix 2: FORTRAN Program Listings, pp. 221-30. Index, pp. 231-34.

An introduction to the concepts of simulation methodology as they are applied to the analysis and design of systems. It is for graduates and undergraduates, and it avoids mathematical rigor, but it still requires knowledge of calculus, probability theory and FORTRAN programming. Topics include Monte Carlo simulation from a variety of probability models, and applications in industry, engineering, and government, particularly the military. Chapters conclude with references.

Morton, Michael S. Scott. MANAGEMENT DECISION SYSTEMS: COMPUTER-BASED SUPPORT FOR DECISION MAKING. Foreword by J.L. McKenney. Boston: Graduate School of Business Administration, Harvard University, 1971. xv, 216 p. Appendix A: Introduction, pp. 157-71. Appendix B: Software Goals, pp. 172-79. Appendix C: Software Architecture, pp. 180-92. Appendix D: Detailed Observations--Decision Making with the Management Decision System, pp. 193-210. Bibliography, pp. 211-16. No index.

A presentation of a research project called a Management Decision System aimed at improving managerial decisions. It discusses the kinds of problems, the type of technology, and types of analyses that form the basis for the system, and it utilizes interactive terminals in the decision-making process. The results are based on two years of research on the behavior of a decision-making team involved in the planning and control of a production-marketing system at the Harvard Business School.

Mosbaek, Ernest J., and Wold, Herman O.A., et al. INTERDEPENDENT SYSTEMS: STRUCTURE AND ESTIMATION. New York: American Elsevier, 1970. xxii, 542 p. Appendix 1: Symbols, Abbreviations, and Formats for References, pp. 497-502. Appendix 2: Normalizing Variables, pp. 503-6. Notes, pp. 507-21. References, pp. 522-31. Subject Index, pp. 532-40. Author Index, pp. 541-42.

A monograph on the results of research designed to compare the new fix-point method with other familiar methods, such as maximum likelihood, for estimating the parameters of simultaneous equation models. Forty-six models are presented, and no model has more than seven equations. The Monte Carlo method is used to simulate the data for analysis, and the properties of estimators evaluated are predictive power, convergence, consistency, bias, and dispersion. Chapter 10 (about 100 pages) compares the estimates of the parameters with the true values of the parameters. Assumes knowledge of econometric techniques. Useful as a reference for the applied econometrician.

Naylor, Thomas H. COMPUTER SIMULATION EXPERIMENTS WITH MODELS OF ECONOMIC SYSTEMS. New York: Wiley, 1971. xviii, 502 p. Appendix A: Pseudorandom Number Generators, pp. 381-95. Appendix B: Random Variable Generators, pp. 396-405. Appendix C: Simulation Languages, by P.J. Kiviat, pp. 406-89. Author Index, pp. 491-96. Subject Index, pp. 497-502.

A sequel to an earlier book, COMPUTER SIMULATION TECHNIQUES, by T.H. Naylor, J.L. Balintfy, D.S. Burdick, and K. Chu (see below). It describes a six-step procedure for designing computer simulation experiments: formulation of a problem, formulation of a mathematical model, formulation of a computer program, validation, experimental design, and data analysis. It reviews several management science and economic models, such as Markov chain models, queueing models, and models of the firm and industry, and illustrates validation and design using analysis of variance, sequential sampling, and spectral analysis. Special attention is given to three methodological problems: variance reduction, stopping rules, and simulation versus analytical solutions. The book concludes with three simulation experiments: the tobacco industry, the effects of alternative governmental resource allocation policies, and the monetary sector of the U.S. economy. References are included at the ends of the chapters and in the appendixes. Assumes knowledge of calculus, mathematical statistics, and computer programming. The chapters were written by J.M. Boughton and others.

Naylor, Thomas H.; Balintfy, Joseph L.; Burdick, Donald S.; and Chu, Kong. COMPUTER SIMULATION TECHNIQUES. Foreword by C.J. Grayson. New York: Wiley, 1966. xiii, 352 p. Author Index, pp. 345-48. Subject Index, pp. 349-52.

A text for a course in computer simulation techniques for students who have knowledge of calculus, mathematical statistics, and probability theory, and who have some familiarity with computer programming. Chapter 2 outlines a nine-step procedure for planning, designing, and carrying out simulation experiments, and the remaining chapters (3 through 9) elaborate on the steps. The presentation includes techniques, flow charts, FORTRAN subroutines for generating stochastic variates from probability distributions, computer models for simulating business and economic systems, and models of the firm and industry. Queueing, inventory, and scheduling systems provide applications of the models. One chapter is devoted to simulation languages, such as GPSS, SIMSCRIPT, GASP, DYNAMO, SIMULATE, and others. Chapters conclude with references and bibliographies.

Optner, Stanford L. SYSTEMS ANALYSIS FOR BUSINESS AND INDUSTRIAL PROBLEM SOLVING. Prentice-Hall International Series in Industrial and Management Science. Englewood Cliffs, N.J.: Prentice-Hall, 1965. xi, 116 p. References, p. 109. Index, pp. 111-16.

A nontechnical introduction to the description of business systems.

_____. SYSTEMS ANALYSIS FOR BUSINESS AND MANAGEMENT. 2d ed., rev. Englewood Cliffs, N.J.: Prentice-Hall, 1968. x, 277 p. Bibliography, pp. 263-72. Index, pp. 273-77.

An introductory text which explores a wide range of data processing and management science topics. It is divided into two parts: systems analysis, and business management case studies. Topics in the first part include fundamentals of integrated systems design, electronic data processing in business, characteristics of computer equipment, constructing the systems costs, and applying statistical-mathematical techniques. Chapters conclude with problems.

Orcutt, Guy H.; Greenberger, Martin; Korbel, John; and Rivlin, Alice M. MICROANALYSIS OF SOCIOECONOMIC SYSTEMS: A SIMULATION STUDY. New York: Harper and Brothers, 1961. xviii, 425 p. List of Tables, pp. vii-x. Figures, pp. xi-xiii. Index, pp. 411-25.

An exploration of the possibility of constructing working models of socioeconomic systems based on computer simulation, sample surveys, and multivariate analysis. It is divided into the following five parts: the problem and an approach to its solution, a demographic model of the U.S. household sector, extending the model of the household sector: exploratory studies, computer simulation of the demographic model, and analysis and prospects. Although the presentation is nonmathematical, knowledge of econometric techniques is helpful. Useful to economists, demographers, operations research analysts, and others. References are given in footnotes throughout the book.

Padulo, Louis, and Arbib, Michael A. SYSTEM THEORY: A UNIFIED STATE-SPACE APPROACH TO CONTINUOUS AND DISCRETE SYSTEMS. Philadelphia: W.B. Saunders, 1974. xvii, 779 p. Appendix A-1: The Laplace Transform, pp. 739-40. Appendix A-2: The Z-transform, pp. 741-42. Appendix A-3: An Outline of Optimal Control Theory, pp. 743-67. List of Symbols, pp. 769-71. Index, pp. 773-79.

A text for a one-year course in systems theory for graduates and undergraduates in a variety of fields, such as computer science, operations research, mathematics, engineering, and systems science. It attempts to unify continuous-time and discrete-time techniques, and includes both classical linear systems and automata. The eight chapter titles are: "Systems and the Concept of State"; "System Dynamics and Local Transition Functions"; "Global Behavior of Continuous-time Systems"; "Constant Linear Systems"; "Controllability, Observability, and Stability"; "Linear Systems and Linearization"; "Reachability and Observability"; and "Algebraic Approaches to System Realization." Extensive use is made of examples from circuit theory and mechanics (or freshman physics). Assumes knowledge of differential and integral calculus,

and differential equations. The sections conclude with problems
ranging from drill on the material covered to extensions for the
interested reader.

Peters, P.J.L.M. INTERRELATED MACRO-ECONOMIC SYSTEMS. Foreword
by D.B. Schoute. Tilburg Studies on Economics, no. 9. Rotterdam: Tilburg
University Press, 1974. xv, 170 p. Appendixes, pp. 137-55. Summary,
pp. 156-62. References, pp. 163-65. List of Symbols and Some Abbreviations, pp. 166-70. No index.

A study on the problems of disaggregating previously developed
macroeconomic models. It presents the idea that it is impossible
to understand the functioning of an economy by using statistical
observations only. The study is based on a multiregion model in
which the cyclical and structural models are disaggregated in
regions. Other models are interpreted as multisector models.
This study can be regarded as a contribution towards the integration of the multisector growth theory and the short-term business
cycle theory. It is confined with mathematical developments,
and not applications, and assumes knowledge of linear difference
equations.

Porter, William A. MODERN FOUNDATIONS OF SYSTEMS ENGINEERING.
New York: Macmillan, 1966. xiii, 493 p. Appendixes, pp. 355-72. References, pp. 473-86. Index, pp. 487-93.

A graduate text in systems engineering with prerequisites of advanced calculus, differential equations, Laplace transforms, linear
circuits, and matrices. About one-quarter of the book deals with
geometric methods in optimal control. The appendixes are devoted to mathematical reviews. Sections conclude with exercises,
and the chapters conclude with references.

Reitman, Julian. COMPUTER SIMULATION APPLICATIONS: DISCRETE-EVENT
SIMULATION FOR SYNTHESIS AND ANALYSIS OF COMPLEX SYSTEMS.
Wiley Series on Systems Engineering and Analysis. New York: Wiley, 1971.
xiv, 422 p. Index, pp. 415-22.

An introduction to discrete-event simulation for systems in which
relationships between variables cannot be expressed analytically
or the major system attributes are characterized by stochastic processes. A unique feature is that it shows how to solve complex
problems by computers with minimum knowledge of mathematics
and computer programming by using the simulation languages GPSS
and SIMSCRIPT. Part 1 is devoted to the simulation languages,
part 2 treats five different applications, and part 3 relates the
applications to each other. The applications are in traffic problems, production control, weapons systems analysis, and performance of a computer system. Useful to systems designers, engineers, and operations research analysts. Chapters conclude with
problems and references.

413

Schmidt, Joseph William, and Taylor, Robert Edward. SIMULATION AND ANALYSIS OF INDUSTRIAL SYSTEMS. Irwin Series in Quantitative Analysis for Business. Homewood, III.: Richard D. Irwin, 1970. xiii, 644 p. Appendix A: Statistical Tables, pp. 611-24. Appendix B: Sample Output from General Inventory Simulation Programs, pp. 625-29. Appendix C: Search Programs, pp. 630-38. Index, pp. 639-44.

A text for a course in simulation with applications to queueing, inventories, reliability, maintenance, and quality control. It may also be used for an introductory course in operations research. It is divided into the following four sections: probability theory and mathematical modeling, simulation modeling, model validation and analysis of results, and simulation languages, namely GPSS/360, SIMSCRIPT, and SIMSCRIPT II. All programs are written in FORTRAN IV. Assumes knowledge of calculus, elementary statistics, and FORTRAN programming. Chapters conclude with references.

Seiler, Karl III. INTRODUCTION TO SYSTEMS COST-EFFECTIVENESS. Publications in Operations Research, no. 17. New York: Wiley-Interscience, 1969. xii, 108 p. Other References, p. 103. Index, pp. 105-8.

A condensed text and reference on the cost-effectiveness approach to systems planning. Topics include system cost models, such as fixed and variable costs and differential cost models; systems effectiveness models, such as probability product models and reliability; and system cost-effectiveness models, such as mathematical programming and games. The presentation is not rigorous from a mathematical viewpoint, and the models and other formal results are explained in great detail. Assumes knowledge of differential calculus and probability. Chapters conclude with references.

Shreĭder, Iuliĭ Anatolevich, ed. THE MONTE CARLO METHOD: THE METHOD OF STATISTICAL TRIALS. Translated by G.J. Tee. Edited by D.M. Parkyn. International Series of Monographs in Pure and Applied Mathematics, vol. 87. New York: Pergamon Press, 1966. xii, 381 p. Appendix 1: Table of Random Digits, pp. 349-51. Appendix 2: Table of Normal Variables, pp. 352-55. Bibliography, pp. 356-70. Index, pp. 371-78.

Originally published in Russian in 1962, this book consists of contributions by the following individuals: N.P. Buslenko, D.I. Golenko, Yu. A. Schreider, T.M. Sobol, and V.G. Sragovich. In seven chapters it describes concepts and techniques of the Monte Carlo method and presents a wide range of applications, primarily in the fields of neutron physics, servicing processes, communications theory, the investigation of learning machines, and the reliability of complicated electronic apparatus. One chapter contains a detailed study of the computation of multidimensional integrals. It is designed for a varied audience, and assumes knowledge of mathematical analysis and probability theory, including Lyapunov's theorem. Knowledge of mathematical statistics is also helpful.

Smith, John. COMPUTER SIMULATION MODELS: TECHNIQUES--CASE STUDIES. New York: Hafner, 1968. xi, 112 p. Appendix: Programming Languages, pp. 105-9. Index, pp. 111-12.

An introduction to computer simulation techniques which provides a grounding in statistical theory, describes the models that can be constructed, and gives some indication of application areas. It is divided into two parts: simulation techniques, and case studies. The simulation models include physical systems, firms and industries, nuclear reactor calculations, queueing and storage, and models of combat. The case studies are in air traffic networks, standby aircraft, plant utilization and storage, and handling operations at a container-ship berth. There are no mathematical prerequisites.

Smith, V. Kerry. MONTE CARLO METHODS: THEIR ROLE FOR ECONOMETRICS. Lexington, Mass.: D.C. Heath, 1973. xiv, 153 p. Appendix A: Random Number Generation, pp. 125-28. Appendix B: Test Statistics: Parametric and Nonparametric, pp. 129-32. Appendix C: A Brief Review of Econometric Estimators, pp. 133-38. References, pp. 139-49. Index, pp. 151-53.

A description of the Monte Carlo method as a research tool. It deals with the use of sampling experiments for the evaluation of the properties of one or more estimators. The method is illustrated with two examples: estimators for population variance, and estimation with a two-equation linear simultaneous model. The analytical and Monte Carlo methods are compared with emphasis given to the strengths and weaknesses of each, followed by a review of the literature on Monte Carlo simulation for both single equation and simultaneous equation models. The sampling experiments based on both single equation and simultaneous equation models are presented, and the future of Monte Carlo simulation in econometrics is discussed. Assumes knowledge of introductory mathematical statistics or econometric theory. Chapters conclude with notes and references.

Tocher, K.D. THE ART OF SIMULATION. Electrical Engineering Series. London: English Universities Press, 1963. viii, 184 p. References, pp. 180-82. Index, pp. 183-84.

An introduction to the techniques of simulation, including sampling methods, random number tables, random number generators, pseudo-random numbers, and elementary sampling experiments. The main applications are simulations based on queueing theory. Assumes knowledge of basic statistics and elementary calculus.

Tustin, Arnold. THE MECHANISM OF ECONOMIC SYSTEMS: AN APPROACH TO THE PROBLEM OF ECONOMIC STABILISATION FROM THE POINT OF VIEW OF CONTROL SYSTEM ENGINEERING. London: Heinemann, 1953. 161 p. Classified References, pp. 155-58. Index, pp. 159-61.

A development of the analogy that exists between economic systems and certain physical systems. It discusses the usefulness of concepts in the theory of control to the problems of economic fluctuations and economic regulation. Assumes knowledge of differential and integral calculus.

Van Gigch, John P. APPLIED GENERAL SYSTEMS THEORY. Foreword by C.W. Churchman. New York: Harper and Row, 1974. viii, 439 p. Review Questions and Term Project, pp. 417-32. Index, pp. 433-39.

A presentation of the conceptual, rather than the quantitative, approach to systems theory for a diverse group--graduates and undergraduates, as well as managers in business, industry, and government. Topics include: a discussion of the systems approach, decision making, value theory, quantification and measurement, optimization, complexity of systems, control, and program planning and budgeting. Little mathematics is employed, and only a few quantitative examples are included, one being a simple linear program. Chapters conclude with references.

White, Harry J., and Tauber, Selmo. SYSTEMS ANALYSIS. Philadelphia: W.B. Saunders, 1969. x, 499 p. Problems, pp. 461-92. Index, pp. 493-99.

A text for a course in discrete systems analysis for seniors and graduate students in applied science, engineering, and mathematics, as well as a reference for practicing engineers. Part 1 (four chapters) takes up mathematical concepts including linear algebra, extrema of functions, calculus of variations, and systems of differential equations. Part 2 (two chapters) presents a generalized theory of mechanics based on Hamilton's principle, and the essential aspects of electromagnetic fields and energy relations based on Maxwell's equations and the Lorentz law of force. Part 3 (six chapters) is devoted to applications of linear systems to electric networks, electromechanical systems, structural mechanics, the application of nonlinear systems to satellite and orbit problems of space science, the generalized energy converters, the theory of vibrations, and feedback systems. The book concludes with a chapter on the use of systems methods in the physical and engineering sciences, biology, economics, industry, and operations research. Assumes a level of mathematical knowledge comparable to that possessed by senior engineering students. Chapters conclude with bibliographies.

Windeknecht, Thomas G. GENERAL DYNAMICAL PROCESSES: A MATHEMATICAL INTRODUCTION. Mathematics in Science and Engineering, vol. 78. New York: Academic Press, 1971. xi, 179 p. References, pp. 173-76. Index, pp. 177-79.

A text for a one-semester course in general systems theory for engineers and applied mathematicians with a good background in axiomatic set theory or modern algebra and an undergraduate background in engineering controls and computers. The treatment is rigorous, and complete proofs are given throughout. The five

chapter titles are: "General Processes," "Basic Interconnections,"
"Time-evolution," "Strong Types of Causality," and "State De-
compositions." Chapters conclude with a large number of exercises.

Wymore, A. Wayne. A MATHEMATICAL THEORY OF SYSTEMS ENGINEERING--
THE ELEMENTS. New York: Wiley, 1967. xii, 353 p. Table of Symbols,
pp. 343-46. Bibliography, pp. 347-49. Index, pp. 351-53.

A contribution to two lines of thought: (1) a definition of systems engineering and the mathematical tools for the practice of
systems engineering, and (2) the development of a general theory
of systems. Useful for a mixed audience, including mathematicians, engineers, economists, operations researchers, sociologists,
and others. Mathematical rigor in the form of theorems and
proofs is maintained, but many discussion sections are devoted to
a full explanation of the applications of the theorems.

Chapter 16

COLLECTIONS OF ARTICLES, PAPERS, ESSAYS,

AND OTHER CONTRIBUTIONS

Abadie, J., ed. INTEGER AND NONLINEAR PROGRAMMING. New York: American Elsevier, 1970. x, 544 p. Appendixes, pp. 511-36. Author Index, pp. 537-40. Subject Index, pp. 541-44.

A collection of twenty-six papers by noted authors presented at a NATO summer school in Bandol, France. The papers deal with the theory, computation, and application of integer and nonlinear programming. Among the topics treated are convergence theories and properties on nonlinear programming; algorithms for quadratic and integer programming, and optimal control problems; methods of constrained and unconstrained optimization; and duality theorems and properties of linear and nonlinear programming. The appendixes contain papers on a product-form algorithm using contracted transformation vectors, description of the variable-metric method, and numerical experiments with the GRG method. Among the contributors are E. Balas, E.M.L. Beale, M.J. Beckmann, G.B. Dantzig, S. Vajda, and P. Wolfe.

_____. NONLINEAR PROGRAMMING. New York: American Elsevier, 1967. xxii, 316 p. Appendix: The Product Form of the Simplex Method, pp. 305-9. Author Index, pp. 311-13. Subject Index, pp. 314-16.

A collection of fourteen papers presented at the NATO summer school on nonlinear programming in Menton, France, in 1964. the papers are concerned with the following three topics: the Kuhn-Tucker and duality theorems; numerical methods; and stochastic nonlinear programming, hydrodynamics, and optimal control theory. A background in linear algebra is sufficient for understanding most of the papers. The contributors are J. Abadie, M.L. Balinski, E.M.L. Beale, R.W. Cottle, G.B. Dantzig, P. Huard, H.W. Kuhn, B.T. Lieu, J.J. Moreau, J.B. Rosen, S. Vajda, A. Whinston, and P. Wolfe.

Anderssen, R.S.; Jennings, L.S.; and Ryan, D.M., eds. OPTIMIZATION. St. Lucia, Queensland, Australia: University of Queensland Press, 1972. vi, 237 p. No index.

Seventeen papers presented at a seminar on optimization held at the Australian National University on December 8, 1971. The papers dealt with topics in constrained and unconstrained optimization. The former included contributions on projection methods, methods for nonlinear programming, integer programming, dynamic programming, and optimal control theory. The latter included contributions on function minimization, global optimization, and statistical optimization. In addition to the editors, the contributors include A.J. Bayes, R.P. Brent, D.L. Jupp, J.H.T. Morgan, M.R. Osborne, and D.M. Ryan.

Antosiewicz, H.A., ed. SECOND SYMPOSIUM IN LINEAR PROGRAMMING. 2 vols. Washington, D.C.: National Bureau of Standards and Directorate of Management Analysis, U.S. Air Force, 1955. vi, 685 p.

Thirty-three papers presented at the second symposium on linear programming. Topics include the dynamics of production, linear programming under uncertainty, dynamic programming, existence theorems, and applications in economics and business. Among the contributors are M.J. Beckmann, A. Charnes, G.B. Dantzig, T. Marschak, and P.A. Samuelson.

Arrow, Kenneth J.; Hurwicz, Leonid; and Uzawa, Hirofumi, eds. STUDIES IN LINEAR AND NON-LINEAR PROGRAMMING. Stanford Mathematical Studies in the Social Sciences, II. Stanford, Calif.: Stanford University Press, 1958. x, 229 p. No index.

Fourteen heretofore unpublished essays grouped under the following three headings: existence theorems, including their extensions to infinite-dimensional spaces as well as extensions and simplifications of earlier theorems for finite-dimensional spaces; gradient methods for solving concave programming and linear programming models; and methods of linear and quadratic programming. Topics in the latter heading include an elementary method for solving linear programs based on a theorem on extreme points of convex polyhedral cones, formulation of price speculation in a commodity as a programming problem, a feasibility algorithm for one-way substitution in process analysis, and nonlinear programming in economic development. Knowledge of advanced mathematics, including set theory and mathematical analysis, and linear programming is assumed. The authors are: K.J. Arrow, H.B. Chenery, L. Hurwicz, S. Karlin, S.M. Johnson, T. Marschak, R.M. Solow, and H. Uzawa.

Arrow, Kenneth J.; Karlin, Samuel; and Scarf, Herbert, eds. STUDIES IN APPLIED PROBABILITY AND MANAGEMENT SCIENCE. Stanford Mathematical Studies in the Social Sciences, 8. Stanford, Calif.: Stanford University Press, 1962. 287 p. No index.

A collection of fifteen individual research papers on probability

and management science with applications to inventory theory and
policy, queueing and dam theory, replacement and maintenance
problems, reliability structures, and capital adjustment policies.
Some of the papers are portions of Ph.D. dissertations submitted
to Stanford University. Several papers represent contributions pre-
sented in STUDIES IN THE MATHEMATICAL THEORY OF INVEN-
TORY AND PRODUCTION, by the same editors (Stanford, Calif.:
Stanford University Press, 1958). The contributors are K.J. Arrow,
R.E. Barlow, C.R. Carr, A.J. Clark, J. Gani, D.P. Gaver, Jr.,
D. Iglehart, D.W. Jorgenson, S. Karlin, R.G. Miller, Jr., R.
Pyke, R. Radner, D.M. Roberts, H. Scarf, R.C. Singleton, and
P.W. Zehna.

Attinger, E.O., ed. GLOBAL SYSTEMS DYNAMICS. New York: Wiley-
Interscience, 1970. xiii, 353 p. List of Participants, pp. viii-ix. Subject
Index, pp. 337-50. Name Index, pp. 351-53.

A collection of seventeen papers presented at the international
symposium on global systems dynamics at the University of Virginia
in Charlottesville, 1969. The papers are grouped under the following headings: methodology, human factors in systems planning,
integration of large complex systems, and the analysis of specific
social systems. No formal mathematics prerequisites are required.

Avi-Itzhak, Benjamin, ed. DEVELOPMENTS IN OPERATIONS RESEARCH.
2 vols. Foreword by P. Naor. New York: Science and Breach, 1971.
Vol. 1; xvi, 292 p. Vol. 2; xvi, pp. 295-622. Author Index, p. 619.
Subject Index, pp. 620-22.

A collection of thirty-seven papers representing a major portion of
those presented at the third annual Israel conference on operations
research held in Tel-Aviv on July 1-4, 1969. The papers are
grouped under the following nine headings: mathematical programming, application of optimization methods, stochastic systems, applications of probability and statistics, planning, traffic and transportation, public services, game theory, and military applications.
The emphasis was on applications, and the papers dealt with such
topics as allocation of cotton to mills, queues, estimating losses
from bank installment loans, drainage projects, capital investment,
assembly line balancing, and design of a regional flood management system. Among the contributors are D. Armon, A. Ben-
Israel, H. Eisner, D. Goldfarb, P.L. Hammer, B. Ivi-Itzhak,
G.J. Kelleher, G.S. Lasdon, O. Levin, P. Naor, C.C. Pegels,
and H.B. Wolfe.

Balakrishnan, A.V., ed. COMPUTATIONAL METHODS IN OPTIMIZATION
PROBLEMS. Lecture Notes in Operations Research and Mathematical Economics,
vol. 14. New York: Springer-Verlag, 1969. 191 p. No index.

A collection of nineteen papers presented at the second International Conference on Computing Methods and Optimization Problems, San Remo, Italy, September 9-13, 1968. The conference focused on recent advances in computational methods for optimization problems in diverse areas, including computational aspects of optimal control and trajectory problems; computational techniques in mathematical programming and in optimization problems in economics, meteorology, biomedicine, and related areas; identification and inverse problems; computational aspects of decoding and information retrieval problems; and pattern recognition problems. The authors represent Italy, France, United States, and the USSR.

_____. CONTROL THEORY AND THE CALCULUS OF VARIATIONS. New York: Academic Press, 1969. xiii, 422 p. No index.

A collection of twelve papers by authors from engineering and mathematics departments which were presented at the workshop on calculus of variations and control theory, University of California, Los Angeles, in July 1968. The topics treated include quadratic variational theory, theorems for elliptic systems, convex sets, estimation and control problems, filter problems, Lagrange multipliers, and an existence theorem for piecewise continuous control functions. The presentations are rigorous and assume knowledge of calculus of variations. The contributors are J.P. Aubin, A.V. Balakrishnan, L.D. Berkowitz, W.H. Fleming, H. Halkin, M.R. Hestenes, V. Klee, E.J. McShane, C.B. Morrey, Jr., L.W. Neustadt, W.T. Reid, and L.C. Young.

_____. SYMPOSIUM ON OPTIMIZATION. Lecture Notes in Mathematics, vol. 132. New York: Springer-Verlag, 1970. iv, 350 p. No index.

A collection of twenty-three papers presented at the symposium on optimization held in Nice, June 29-July 5, 1969. The topics dealt mainly with the mathematical theory of optimization. Among the contributors are C. Aumasson, L.D. Berkovitz, P. Faurre, P.J. Laurent, J.L. Lions, L.W. Neustadt, L.S. Pontryagin, L.W. Taylor, and P.K.C. Wang.

_____. TECHNIQUES OF OPTIMIZATION. New York: Academic Press, 1972. ix, 509 p. No index.

Thirty-six papers presented at the 4th IFIP Colloquium on Optimization Techniques held in Los Angeles on October 19-22, 1971. The colloquium was devoted to computational aspects of optimization problems arising in a variety of disciplines, from the more traditional areas such as aerospace to the more recent areas such as public systems. The papers are grouped under the following nine headings: pattern classification-identification (four papers), numerical methods (five), distributed systems (four), optimal control (five), differential games (three), stochastic control (three), economics (four), linear and nonlinear programming (four), and public systems (four). The four papers in economics are: "Optimal Systems for Information and Decision," by J. Marschak; "Shadow Prices for Infinite Growth Programs: the Functional Analysis Approach," by M. Majumdar and R. Radner, "Price Decentralization in the Case of Interrelated Payoffs," by A. Bensousson; and "A One-sector Model of Economic Growth with Uncertain Technology: An Example of Steady State Analysis in a Stochastic Optimal Control Problem," by W.A. Brock and L.J. Mirman.

Balakrishnan, A.V., and Neustadt, Lucien W., eds. COMPUTING METHODS IN OPTIMIZATION PROBLEMS. New York: Academic Press, 1964. x, 327 p. No index.

A collection of fourteen papers presented at a conference on computing methods on optimization held at the University of California, Los Angeles, in 1964. The papers dealt with such topics as variational theory, trajectory optimization techniques, computing optimum paths in the problem of Bolza, dynamic optimization, gradient methods, and minimizing functionals on Hilbert spaces. The contributors are G.A. Bekey, R.E. Bellman, W. Brunner, E.J. Fadden, F.D. Faulkner, E.G. Gilbert, A.A. Goldstein, P. Halbert, H. Halkin, M.R. Hestenes, R.H. Hillsley, H.C. Hsieh, R.E. Kalaba, R.E. Kopp, R.B. McGhee, R. McGill, H. Moyer, B. Paiewonsky, R. Perret, G. Pinkham, H.M. Robbins, H.H. Rosenbrock, R. Rouxel, C. Storey, and P. Woodrow.

_____. MATHEMATICAL THEORY OF CONTROL. Foreword by C. Starr. New York: Academic Press, 1967. xv, 459 p.

A collection of forty-two papers presented at a conference on the mathematical theory of control held at the University of California, January 30-February 1, 1967. Topics include optimal control theory, control theory and partial differential equations, differential games, stochastic control, and stability theory. The conference was international in character and almost every country active in control theory and its applications was represented. Among the contributors are V.G. Boltyanskii, A. Blaquière, R. Conti, V.F. Dem'yanov, H. Hermes, O.L. Mangasarian, L.S. Pontryagin, and J.B. Rosen.

Balinski, M.L. PIVOTING AND EXTENSIONS: IN HONOR OF A.W. TUCKER. Mathematical Programming Studies 1. New York: American Elsevier, 1974. vii, 205 p. No index.

Thirteen papers, many of which were presented at the 8th International Symposium on Mathematical Programming held at Stanford University in August 1973, in honor of Professor A.W. Tucker's contributions to the computational and theoretical tool in mathematical programming called pivoting. Topics include the structure of convex polytopes, complementary pivoting, a computational algorithm for a class of parametric linear programs, balanced matrices, derivation of a bound for error-correcting codes using pivoting techniques, pivotal theory of determinants, the Lemke-Howson method, and an algorithm for a least-distance programming problem. The contributors include I. Adler, M.L. Balinski, R.W. Cottle, G.B. Dantzig, R.J. Duffin, B.C. Eaves, A.J. Hoffman, L.S. Shapley, and P. Wolfe.

Barish, Norman N., and Verhulst, Michael, ed. MANAGEMENT SCIENCES IN THE EMERGING COUNTRIES: NEW TOOLS FOR ECONOMIC DEVELOPMENT. Elmsford, N.Y.: Pergamon, 1965. x, 261 p. No index.

Eleven papers presented at a symposium sponsored by the College on Management Economics of the Institute of Management Sciences, Brussels, in 1961. Three additional papers are also included. The papers are grouped under the following headings: planning for development (five papers), resource allocation decisions (four papers), and case experiments (five papers). Among the topics treated in the papers are: data problems in emerging countries, industrialization incentives, allocation of medical resources to the control of leprosy, operations research in China, and capital/labor ratios and the industrialization of West Africa. There are no mathematics prerequisites.

Bass, Frank M.; King, Charles W.; and Pessemier, Edgar A., eds. APPLICATIONS OF THE SCIENCES IN MARKETING MANAGEMENT. The Wiley Management Series. New York: Wiley, 1968. xiv, 456 p. Index, pp. 455-56.

Fifteen papers presented at a symposium held at Purdue University in July 1966. The papers are grouped under the following headings: consumer behavior and normative models (five papers), behavioral theories of consumer behavior (four papers), and experimental methods and simulation models in marketing management (six papers). Among the topics treated are stochastic models of consumer behavior, simulation of consumer behavior, new-product introduction, the role of psychology, market structure studies, and experimental research in marketing. An introductory paper, "The Interdisciplinary Approach to Marketing—a Management Overview," by Wm. Lazer is also included. There are no mathematical prerequisites.

Bauer, Friedrich L., ed. ADVANCED COURSE IN SOFTWARE ENGINEERING. Lecture Notes in Economics and Mathematical Systems, vol. 81. New York: Springer-Verlag, 1973. xii, 545 p. Appendix: Software Engineering, by F.L. Bauer, pp. 522-45.

Eighteen papers presented at an advanced course in software engineering held at the University of Munich, February 21 to March 3, 1972, under sponsorship of the Mathematical Institute of the Technical University of Munich and the Leibniz Computing Center of the Bavarian Academy of Sciences. The topics treated include reliability software, portability and adaptability, modularity, debugging and testing, documentation, project management, and performance measurement. The contributors are F.L. Bauer, J.B. Dennis, R.M. Graham, M. Griffiths, G. Goos, C.C. Gotlieb, H.J. Helms, K.W. Morton, P.C. Poole, D. Tsichritzis, and W.M. Waite.

Beale, Evelyn Martin Landsdowne, ed. APPLICATIONS OF MATHEMATICAL PROGRAMMING TECHNIQUES. Foreword by E.M.L. Beale. New York: American Elsevier, 1970. ix, 451 p. No index.

Twenty-nine papers presented at a conference on mathematical programming techniques held at the University of Cambridge, June 24-28, 1968. This conference was sponsored by the NATO Science Committee and was attended by about 140 individuals from fourteen countries. The papers are grouped under the following headings: a survey paper by G.B. Dantzig, linear programming, economic applications, matrix generators and output analyzers, project scheduling, nonconvex programming methods, integer programming applications, geometrical programming, strategic deployment problems, and nonlinear programming. This book can be used as a supplement in a course in mathematical programming.

Bell, D., and Griffin, A.W.J., eds. MODERN CONTROL THEORY AND COMPUTING. New York: McGraw-Hill Book Co., 1969. xii, 221 p. References, pp. 217-18. Further Reading, p. 218. Index, pp. 219-21.

A collection of nine contributions on control theory and computing techniques. Topics include a review of classical techniques; Liapunov's direct method in nonlinear control systems; digital, analogue, and hybrid computing; random processes; adaptive control systems; and multivariable systems. Optimal controls are derived by using the techniques of dynamic programming and the maximum principle. The contributors are D. Bell, W. Fishwick, A.W.J. Griffin, M.T.G. Hughes, O.L.R. Jacobs, and J.K.M. MacCormac. This book is suitable for an introductory text for seniors and graduate students and as a reference for practicing engineers. The chapters conclude with references, and several chapters have appendixes containing computer programs and mathematical derivations.

Bell, David John, ed. RECENT MATHEMATICAL DEVELOPMENTS IN CONTROL. New York: Academic Press, 1973. xiv, 446 p. Author Index, pp. 437-42. Subject Index, pp. 443-46.

A collection of twenty-nine papers representing the proceedings of a conference held September 5-7, 1972, at the University of Bath, Somerset, England. The purpose of the conference was to make known new developments in the theory and practice of automatic control in the following five areas: stability of nonlinear systems, optimal control, filtering theory, control of systems governed

by partial differential equations, and algebraic systems. The papers
dealt with such topics as optimal control via functional analysis,
stability of nonlinear feedback systems, linear time-optimal control problems, domains of controllability, conditions for compactness of the time-phase space in control problems, filters for time
lag systems, the infinite dimensional Riccati equations, multivariable circle theorems, and the theory of multivariable infinite matrices. Among the contributors are J.C. Allwright, S. Barnett,
M.H.A. Davis, M.J. Denham, P.J. Fleming, C.J. Harris, T.S.
Rao, R.W.H. Sargent, C. Storey, and J.H. Westcott.

Bellman, Richard E., ed. MATHEMATICAL OPTIMIZATION TECHNIQUES.
Berkeley: University of California Press, 1963. xii, 346 p. Name Index,
pp. 341-43. Subject Index, pp. 344-46.

A collection of seventeen papers presented at the Symposium on
Mathematical Optimization Techniques held in Santa Barbara,
October 18-20, 1960. The papers are grouped under the following
headings: aircraft; rockets and guidance; communication, prediction, and decision; programming, combinatorics, and design; and
models, automation, and control. This book is useful mainly as
a source of applications of optimization techniques, including the
calculus of variations, linear and convex programming, dynamic
programming, game theory, decision theory, and the techniques
used in prediction and estimation. Among the contributors are
R.E. Bellman, C. Derman, S.E. Dreyfus, R.P. Ten Dyke, E.
Elfving, Wm. J. Hall, T. Kailath, R.E. Kalman, J.B. Kruskal,
J.P. LaSalle, D. Middleton, A. Miele, E. Parzen, A.W. Tucker,
and P. Wolfe.

Bellman, Richard E., and Denman, E.D., eds. INVARIANT IMBEDDING.
Lectures in Operations Research and Mathematical Systems, vol. 52. New
York: Springer-Verlag, 1971. iv, 148 p. No index.

A collection of nine papers presented at a summer workshop on invariant embedding held at the University of California from June to
August, 1970. Topics and applications include linear differential
equations subject to both initial value and two-point boundary
value conditions; Cauchy problems for ordinary differential, difference, and integral equations; a simple algorithm for the solution of the potential equation over a rectangle in partial differential equations, reduction of matrix integral equations, optimal control, distributed control problems, wave propagation through longi-
tudinally and transversally inhomogeneous slabs-I, neutron transport
theory, and dynamic programming and invariant embedding in structural mechanics. This book is useful to anyone interested in applying invariant embedding in economics, engineering, biology, and
physics.

Bensoussan, Alain, and Lions, J., eds. CONTROL THEORY, NUMERICAL
METHODS AND COMPUTER SYSTEMS MODELING. Lecture Notes in Economics and Mathematical Systems, vol. 107. New York: Springer-Verlag, 1975.
viii, 757 p. No index.

A collection of fifty papers presented at the International Conference on Control Theory, Numerical Methods, and Computer Systems Modeling held in Rocquencourt, France, June 17-21, 1974.
Two hundred specialists representing twelve countries participated
in the conference. The papers are grouped under the following
headings: filtering theory (four papers), game theory (six papers),
control of stochastic distributed parameter systems and random
fields (six papers), stochastic control (six papers), control of distributed parameter systems (four papers), discrete time problems
and numerical methods (six papers), control of jump processes
and computer model applications (five papers), free boundary value
problems and control theory (six papers), and applications of control theory in economics, process control, and pattern recognition
(six papers). Twenty papers are written in French and the remainder in English. Most of the papers assume knowledge of advanced
mathematics, such as the calculus of variations and advanced probability. Among the contributors are J.P. Aubin, A.V. Balakrishnan, V.E. Benes, J. Casti, E. Diday, W.H. Fleming, A.
Friedman, F. Levieux, H. Moulin, R.T. Rockafellar, P. Varaiya,
and E. Wong.

Bharucha-Reid, A.T., ed. PROBABILISTIC METHODS IN APPLIED MATHE-
MATICS. Vol. 1. New York: Academic Press, 1968. x, 291 p. Author
Index, pp. 283-87. Subject Index, pp. 288-91.

Volume 1 is the first in a series devoted to the applications of
stochastic processes. It contains the following contributions: "Random Eigenvalue Problems," by Wm. E. Boyce; "Wave Propagation
in Random Media," by V. Frisch; and "Branching Processes in
Neutron Transport Theory," by T.W. Mullikin. Each contribution
is self contained and fully referenced, and assumes knowledge of
measure-theoretic probability and the elements of stochastic processes and functional analysis.

_____. PROBABILISTIC METHODS IN APPLIED MATHEMATICS. Vol. 2.
New York: Academic Press, 1970. x, 220 p. Author Index, pp. 213-16.
Subject Index, pp. 217-20.

Volume 2 contains the following contributions: "Random Algebraic
Equations," by A.T. Bharacha-Reid; "Axiomatic Quantum Mechanics and Generalized Probability Theory," by S. Gudder; and
"Random Differential Equations in Control Theory," by W.M.
Wonham. Each contribution is self-contained and fully referenced,
and assumes knowledge of measure-theoretic probability, stochastic processes, and functional analysis.

Blaquière, Austin, ed. TOPICS IN DIFFERENTIAL GAMES. New York: American Elsevier, 1973. ix, 450 p. Author Index, pp. 449–50.

Ten contributions grouped under the following two headings: zero-sum differential games, and non-zero-sum differential games. The first part emphasizes the theory, while the second part deals with properties of non-zero-sum games and their applications to economics. The authors are A. Blaquière, J. Case, P. Caussin, M.D. Ciletti, R. Isaacs, L. Juricek, G. Leitmann, B. Pchenitchny, M. Shubik, H. Stalford, W. Whitt, and K.E. Wiese.

Bock, Robert H., and Holstein, William K., eds. PRODUCTION PLANNING AND CONTROL: TEXT AND READINGS. Columbus, Ohio: Charles E. Merrill, 1963. ix, 417 p. Appendix A: An Introduction to Linear Programming, pp. 372–91. Appendix B: The Fundamentals of Calculus, pp. 392–407. Index, pp. 409–17.

A collection of twenty reprints of articles on production management which are oriented toward management science. The articles are grouped under the following headings: planning and controlling production levels, with emphasis on linear programming and PERT applications; inventory control; and facilities planning, with emphasis on game theory and capital budgeting. Each part contains an introduction and appendixes, which consist of reviews of the necessary mathematics. This book is useful as a supplemental reading for a course in production management. The contributors include M. Anshen, D.G. Boulanger, K.J. Cohen, J.W. Dudley, J.C. Hetrick, E. Koenigsberg, J. Muth, H. Simon, N.V. Reinfeld, R. Schlaifer, J.B. Stewart, and L. Wester.

Brennan, James, ed. APPLICATIONS OF CRITICAL PATH TECHNIQUES. New York: American Elsevier, 1968. 447 p. No index.

A collection of twenty-three papers presented at an international conference held in Brussels on July 31–August 4, 1967, under the aegis of the NATO Scientific Affairs Committee. The papers are concerned with applications of critical path techniques in business and government in five NATO countries, excluding the United States. The application areas are: construction, nuclear power, petro-chemicals and petroleum, blast furnaces, post office administration, aircraft, shipbuilding, railroads, research and development, and investment planning.

_____. OPERATIONAL RESEARCH IN INDUSTRIAL SYSTEMS. New York: American Elsevier, 1972. vi, 325 p. No index.

A collection of twenty-one papers presented at a conference held in St. Louis, France, in July 1970, under the aegis of the NATO Scientific Affairs Committee. Seven papers are written in French. The papers deal with the applications of operations research techniques in such areas as corporate planning, market research, forecasting, scheduling sections in a shipyard, factory scheduling, purchase of appliances, inventory control, and plant location. Among the authors are O. Hart , L. Kaufman, and C. Kintz.

Buffa, Elwood Spencer, ed. READINGS IN PRODUCTION AND OPERATIONS MANAGEMENT. New York: Wiley, 1966. xi, 608 p. No index.

A collection of thirty-five papers originally published elsewhere, arranged under the following headings: introductory materials, analytical methods in production and operations management, design of information and production systems, operation and control of production systems, and a forward look. The papers deal with the techniques of linear programming, queueing theory, inventory management, PERT, employment scheduling, and simulation. They are concerned with both methodology and applications. The authors include M. Anshen, A.L. Arcus, E.H. Bowman, E.S. Buffa, F. Hanssmann, F.S. Hillier, A. Reisman, M.K. Starr, H.W. Steinhoff, and D.B. Thompson.

Caianiello, Eduardo R., ed. FUNCTIONAL ANALYSIS AND OPTIMIZATION. New York: Academic Press, 1966. xiii, 225 p. Author Index, pp. 221-23. Subject Index, pp. 224-25.

A collection of sixteen contributions presented at the seventh International School of Ravello in June 1965, on optimization using functional analysis. The conference was sponsored by NATO. The papers dealt with such topics as the state-space theory of nonlinear systems, linear controllability, the bang-bang principle, convexity, optimal control, optimization problems for linear parabolic equations, and the method of quasi reversibility. The contributors are H.A. Antosiewicz, J.P. Aubin, A.V. Balakrishnan, E.R. Caianiello, C. Castaing, R. Conti, W. De Backer, J. Douglas, Jr., W.H. Fleming, H. Halkin, L. Lattes, J.H. Lions, L. Markus, J.J. Moreau, C. Muses, and A. Straszak.

Chedzey, Clifford Stanley, ed. SCIENCE IN MANAGEMENT: SOME APPLICATIONS OF OPERATIONAL RESEARCH AND COMPUTER SCIENCE. The British Library of Business Studies. London: Routledge and Kegan Paul, 1970. xv, 357 p. List of Contributors, p. xv. Index, pp. 351-57.

Twenty-nine contributions (essentially case histories) on applications of operational research and computer science in management grouped under the following five headings: ideas and means; methods and models; first-level applications: tactical projects; second-level applications: integrating activities; and third-level applications: strategic studies. The applications are based on the experiences of one consulting company, Management Sciences, Ltd., in actual work undertaken. Among the applications are post office queues, forecasting, product blending, production planning in the canning industry, controlling yields in paper and glass manufacture, and distribution of high-fashion footwear. Assumes familiarity with operations research techniques.

Chen, Gordon K.C., and Kaczka, Eugene E., eds. OPERATIONS AND SYSTEMS ANALYSIS: A SIMULATION APPROACH. Boston: Allyn and Bacon, 1974. x, 452 p. No index.

Thirty-one articles previously published in journals and elsewhere arranged in the following parts: introduction (three articles), simulation methodology and model building (eleven articles), models and applications: processes in operations management (eight articles), models and applications: functional problems (seven articles), and models and applications: large systems (two articles). The articles will aid those who wish to construct simulation models in understanding the technical details of the approach, and also those who wish to use simulation in evaluating its worth. Topics include computational procedures for generating and testing random numbers, problems in digital simulation, verification of computer simulation models, validation of simulation results, inventory levels, simulation of a radio-dispatched truck fleet, simulation of manpower development, and a sugar refinery simulation model. Each part begins with an introduction and each of the twelve chapters begins with an abstract of the articles in that chapter. Useful either as a textbook or as a readings book. Knowledge of basic statistics and the techniques of operations research is helpful for some of the articles. Among the authors are G.C. Armour, E.S. Buffa, R.M. Brooks, J. Cohen, J.E. Cooke, G.B. Davis, P.S. Greenlaw, J.M. Hoff, H. Kidera, P. Kotler, A.S. Manne, T.H. Naylor, S.K. Sen, K. Wertz, and T.H. Wonnacott.

Chover, Joshua, ed. MARKOV PROCESSES AND POTENTIAL THEORY. Publication No. 19 of the Mathematics Research Center, U.S. Army, University of Wisconsin. New York: Wiley, 1967. x, 235 p. Index, pp. 233-35.

A collection of thirteen papers presented at a symposium on Markov processes and potential theory held in Madison, Wisconsin, on May 1-3, 1967, and sponsored by the U.S. Army and the University of Wisconsin. Representatives of the following countries attended: United States, Poland, West Germany, France, and England. Topics treated by the papers include Markov processes with infinities, limit theorems for branching processes, random walks, Markov processes in axiomatic potential theory, and multiplicative decomposition of positive super-martingales. The authors are: H. Bauer, R.M. Blumenthal, G.E. Denzel, J.G. Kemeny, J. Lamperti, P.A. Meyer, S. Orey, D.S. Ornstein, S.C. Port, J.L. Snell, C.J. Stone, D. Stroock, R.G. Getoor, and S. Watanabe.

Churchman, Charles West, and Verhulst, Michael, eds. MANAGEMENT SCIENCES: MODELS AND TECHNIQUES. Vol 1. Elmsford, N.Y.: Pergamon Press, 1960. xxix, 602 p. Index, pp. 595-602.

A collection of thirty-nine papers of the seventy-eight papers presented at the sixth international meeting of the Institute of Manage-

ment Sciences held in Paris, September 7-11, 1959. Most of
the major nations of the world were represented at the conference.
The papers in volume 1 are grouped under the following headings:
management economics, simulation, management games, decision
processes, fundamentals in management education, production and
inventory management, and computer simulation. Among the contributors are W.W. Cooper, S. Elmaghraby, B.H.P. Rivett, A.
Vazsonyi, and H.M. Wagner.

_____. MANAGEMENT SCIENCES: MODELS AND TECHNIQUES. Vol. 2.
Elmsford, N.Y.: Pergamon Press, 1960. xiv, 509 p. Index, pp. 505-9.

A collection of thirty-nine papers presented at the sixth international conference of the Institute of Management Science. Among
the topics covered by the papers are measurement problems in
management, such as measuring executive performance; the relationship of management science to sociology; management communications; linear programming and queueing theory; the dynamics of
research and development projects; long range planning; case histories in hospital care, machine tool operators, natural gas, steel,
and farming; and methodology. Most of the papers are nonmathematical.

Clarke, A. Bruce, ed. MATHEMATICAL METHODS IN QUEUEING THEORY.
Lecture Notes in Economics and Mathematical Systems, vol. 98. New York:
Springer-Verlag, 1974. 374 p. List of Participants, pp. 373-74. No index.

Eighteen papers presented at a conference on mathematical methods
in graph theory held at Western Michigan University on May 10-12,
1973. The papers dealt with queueing theory and applications
and considered such topics as combinatorial methods in the theory
of queues, algebraic techniques for numerical solutions of queueing
networks, occupational time problems in the theory of queues, the
Wiener-Hopf technique, multiple-server systems, optimal control
of queueing systems, and heavy traffic limit theorems for queues.
Among the contributors are R.L. Disney and W.P. Cherry, C.M.
Harris, J. Keilson, A. Kuczura, R. Loulou, M.F. Neuts, G.F.
Newell, N.U. Prabhu, P. Purdue, M.J. Sobel, L. Takács, and
V.L. Wallace.

Clarkson, Geoffrey P.E., ed. MANAGERIAL ECONOMICS: SELECTED READ-
INGS. Baltimore: Penguin Books, 1968. 429 p. Further Readings, pp. 415-
16. Acknowledgments, p. 417. Author Index, pp. 419-23. Subject Index,
pp. 425-29.

A collection of eleven articles by different authors, all previously
published in journals and elsewhere, arranged under the following
headings: economic theory and business behavior (two articles),
some managerial decision models (three articles), financial decision making (two articles), production and inventory systems (two
articles), and mathematical programming (two articles).

COMPUTING METHODS IN OPTIMIZATION PROBLEMS. Lecture Notes in Operations Research and Mathematical Economics, vol. 14. New York: Springer-Verlag, 1969. vi, 191 p. No index.

A collection of nineteen papers presented at the second international conference on computing methods in optimization held in San Remo, Italy, September 9-13, 1968. The conference was sponsored by the Society of Industrial and Applied Mathematics (SIAM), with the cooperation of the University of California and the University of Southern California. It focused on recent advances in computational methods for optimization problems in diverse areas including optimal control and trajectory problems; mathemathical programming; optimization problems in economics, meteorology, biomedicine, and related areas; identification and inverse problems; decoding and information retrieval, and pattern recognition. Among the authors are M. Auslender, P.A. Clavier, H.O. Fattorine, F.C. Ghelli, H.J. Kelley, A. Miele, R. Petrović, and R.G. Stefanek.

Conti, Roberto, and Ruberti, Antonio, eds. 5TH CONFERENCE ON OPTIMIZATION TECHNIQUES. PART I. Lecture Notes in Computer Science, vol. 3. New York: Springer-Verlag, 1973. xiii, 565 p. No index.

A collection of fifty-seven papers presented at the fifth IFIP colloquium on optimization held in Rome, May 7-11, 1973. The emphasis was on applications, and the papers are grouped under the following headings: system modeling and identification (eleven papers), distributed systems (six papers), game theory (four papers), pattern recognition (five papers), stochastic control (four papers), optimal control (eleven papers), mathematical programming (ten papers), and numerical methods (six papers). Among the authors are A.V. Balakrishnan, R.R. Mohler, K. Malanowski, C. Marchal, Y.A. Rosanov, and A. Szymanski.

_____. 5TH CONFERENCE ON OPTIMIZATION TECHNIQUES. PART II. Lecture Notes in Computer Science, vol. 4. New York: Springer-Verlag, 1973. xiii, 389 p. No index.

A collection of thirty-six papers presented at the 5th IFIP conference on optimization techniques held in Rome, May 7-11, 1973. The papers are grouped under the following headings: urban and society systems (nine papers), computer and communication networks (five papers), environmental systems (nine papers), economic models (seven papers), and biological systems (six papers). Among the authors are M. Aoki, C. Badi, E.J. Beltrami, G. Burgess, A. Cerutti, F. Maffioli.

Cooper, William Wager; Leavitt, H.J.; and Shelly, M[aynard].W. II, eds. NEW PERSPECTIVES IN ORGANIZATION RESEARCH. New York: Wiley, 1964. xxii, 606 p. Bibliography, pp. 567-97. Alternative Author Designations, pp. 599-606.

Twenty-nine papers taken from the following two sources: Office of Naval Research Conference on Research in Organizations held at Carnegie Institute of Technology, June 22-24, 1962, and the Ford Foundation Seminar on the Social Science of Organizations held at the University of Pittsburgh, June 10-23, 1962. The conference participants were individuals active in either management science or the behavioral sciences, including economics. The papers are grouped under the following five parts: general perspectives, behavioral science perspectives, interdisciplinary perspectives, management science perspectives, and perspectives for further research.

Dantzig, George B., and Eaves, B.C., eds. STUDIES IN OPTIMIZATION. Studies in Mathematics, vol. 10. Washington: Mathematical Association of America, 1974. viii, 180 p. Index, pp. 175-80.

A collection of eight papers (five previously published) in mathematical optimization in the following topical areas: mathematical structure of optimization models, the existence and attributes of optimal (or near optimal) solutions, and the design and use of algorithms for computation of solutions. The contributors are R.W. Cottle, G.B. Dantzig, B.C. Eaves, J. Edmonds, D.R. Fulkerson, H.W. Kuhn, H. Scarf, L. Shapley, A.W. Tucker, and A.F. Veinott, Jr.

Dantzig, George B., and Veinott, Arthur F., Jr., eds. MATHEMATICS OF DECISION SCIENCES. PART 1. Lectures in Applied Mathematics, vol. 11. Providence, R.I.: American Mathematical Society, 1968. vii, 443 p. Subject Index, pp. 431-35. Author Index, pp. 437-43.

Twenty papers presented at the Fifth Summer Seminar on the mathematics of the decision processes sponsored by the American Mathematical Society and held at Stanford University from July 10 to August 11, 1967. The papers are grouped under the following headings: control theory, mathematical economics, dynamic programming, applied probability and statistics, mathematical psychology and linguistics, and computer science. The authors are K.J. Arrow, R.E. Barlow, H. Chernoff, G.B. Dantzig, C. Derman, A.J. Ehrenfeucht, D. Gale, D.L. Iglehart, S. Karlin, D. Krantz, H.W. Kuhn, W.F. Miller, L.W. Neustadt, M.F. Norman, S. Peters, E. Polak, H. Robbins, J.B. Rosen, D. Siegmund, W.R. Sutherland, A.H. Taub, and A.F. Veinott, Jr.

Davar, Ruston S., ed. EXECUTIVE DECISION MAKING: MODERN CONCEPTS AND TECHNIQUES. Bombay: Progressive Corp., 1966. xv, 319 p. Author Index, p. 315. Subject Index, pp. 317-19.

A collection of twenty-six reprints of articles previously published, grouped under the following headings: decision making--theory and process, operations research and its weapon, managerial control

and analysis, and specific applications. The purpose of this collection is to introduce modern decision-making techniques to managers. The articles treat such topics as linear programming, queueing models, break-even analysis, PERT, and inventories. There are
no mathematical prerequisites.

Dean, Burton Victor, ed. OPERATIONS RESEARCH IN RESEARCH AND DE-
VELOPMENT: PROCEEDINGS OF A CONFERENCE AT CASE INSTITUTE OF
TECHNOLOGY. New York: Wiley, 1963. xii, 289 p. Index, pp. 287-89.

Twelve papers presented at a conference on the applications of
the methodology of operations research to the solution of research
and development management problems. The papers deal with such
topics as the measurement of value of scientific research, network
planning, selection of projects, and case studies. Knowledge of
calculus is required for three articles. The contributors are Wm.
F. Ashley, M.T. Austin, P.G. Carlson, R.J. Freeman, D.B.
Hertz, E.A. Johnson, D.G. Malcolm, T.A. Marschak, M.W.
Martin, Jr., P.V. Norden, J. Perlman, A.H. Rubenstein, H.A.
Shepard, and H.K. Weiss.

Elmaghraby, Salah Eldin, ed. SYMPOSIUM ON THE THEORY OF SCHEDUL-
ING AND ITS APPLICATIONS. Lecture Notes in Economics and Mathematical Systems, vol. 86. New York: Springer-Verlag, 1973. viii, 473 p. No
index.

Twenty-six papers presented at a symposium held at North Carolina
State University at Raleigh on May 15-17, 1972, under the sponsorship of the Office of Naval Research. The papers are grouped
under the following headings: survey papers (two papers), applications (seven papers), theory (three papers), and models of processes (fourteen papers). Topics include the engine scheduling
problem in a railway network, applications in the chemical industry, optimization techniques for functions of permutations,
scheduling of a multiproduct facility, scheduling problems using
Lagrange multipliers, and scheduling with sequence independent
change-over costs. Assumes knowledge of mathematical programming. Among the contributors are G.E. Bennington, G.H. Bradley, M.L. Fisher, L.F. McGinnis, T. Prabhakar, J.G. Rau, M.S.
Salvador, H.W. Steinhoff, Jr., J.D. Wiest, and V.A. Zaloom.

Flagle, Charles D.; Huggins, William H.; and Roy, Robert H., eds. OPERA-
TIONS RESEARCH AND SYSTEMS ENGINEERING. Baltimore: Johns Hopkins
Press, 1960. x, 889 p. Index, pp. 881-89.

A collection of twenty-seven papers by twenty authors presented
at a two-week course in management at the Johns Hopkins University. The disciplines represented are physics, economics, statistics, psychology, engineering, and mathematics. The topics
treated include statistical quality control, electronic digital com-

puters, inventory systems, queueing theory, simulation techniques, game theory, information theory, and the design of experiments. Most of the papers can be understood without knowledge of higher mathematics.

Fletcher, Roger, ed. OPTIMIZATION. New York: Academic Press, 1969. xviii, 354 p. Author Index, pp. 349-52. Subject Index, pp. 353-54.

A collection of twenty-two papers on optimization techniques and nonlinear programming presented at the symposium of the Institute of Mathematics and Its Applications held at the University of Keele, England, on March 25-28, 1969. The papers are published in the order of presentation and treat theory, solution procedures, and applications. Among the topics treated are unconstrained and constrained optimization by hill-climbing methods, constrained optimization by simplex-like methods, generalized Lagrangian functions, and acceleration techniques for nonlinear programming. The conference contrasts the difference in American work in operations research problems, characterized by many variables for which nonnegativity is an important constraint, and the British work, which has fewer variables with nonnegativity being relatively unimportant. The conference also makes a transition from penalty function methods to more direct methods for solving nonlinear programs by hill-climbing methods. Among the authors are E.M. Beale, R. Fletcher, D. Goldfarb, P. Huard, T.O.M. Kronsjo, A.P. McCann, R.W.H. Sargent, and A.W. Tucker.

Fox, Karl A., ed. ECONOMIC ANALYSIS FOR EDUCATION PLANNING: RESOURCE ALLOCATION IN NONMARKET SYSTEMS. Baltimore: Johns Hopkins University Press, 1972. xii, 376 p. References, pp. 347-66. Author Index, pp. 367-70. Subject Index, pp. 371-76.

A study of resource allocation where there is no market pricing mechanism (a nonmarket system) by using those aspects of linear, nonlinear, and dynamic programming that might have relevance for educational planning. This book consists of ten chapters by various authors describing a research project undertaken at Iowa State University during the 1960s to study resource allocation within a university. The contributors are K.A. Fox, T.K. Kumar, B.C. Sanyal, and J.K. Sengupta. This book is useful to anyone interested in the application of economic theory in educational planning.

Geoffrion, Arthur M., ed. PERSPECTIVES ON OPTIMIZATION: A COLLECTION OF EXPOSITORY ARTICLES. Reading, Mass.: Addison-Wesley, 1972. xiii, 238 p. No index.

Eight surveys on unconstrained optimization, mathematical programming, integer programming, and optimization in graphs and networks. The emphasis is on optimization in a static rather than dynamic setting, and on useful theory and computationally effective algorithms, rather than on applications. It is useful as a supplementary text for courses in optimization. Seven of the surveys have been published in journals, and the eighth is based on portions of a book. Assumes knowledge of undergraduate mathematics. The contributors are S.E. Dreyfus, D.R. Fulkerson, R. Garfinkel. A.M. Geoffrion, D.G. Luenberger, R.E. Marsten, G.L. Nemhauser, and M.J.D. Powell.

Gill, P.E., and Murray, W., eds. NUMERICAL METHODS FOR CONSTRAINED OPTIMIZATION. New York: Academic Press, 1974. xiv, 283 p. References, pp. 261-68. Author Index, pp. 269-71. Subject Index, pp. 273-83.

Nine contributions based on the proceedings of a symposium on numerical methods for constrained optimization held at the National Physical Laboratory on January 10-11, 1974. The symposium was sponsored jointly by the Institute of Mathematics and Its Applications and by the National Physical Laboratory. The contributions are expository and emphasis is on methods which can be used as practical tools. Topics include Newton and quasi-Newton methods for linearly constrained optimization, methods for large-scale linearly constrained problems, reduced-gradient and projection methods for nonlinear programming, penalty and barrier functions, direct search methods in constrained optimization, and methods related to Lagrangian functions. Familiarity with the rudiments of numerical linear algebra and unconstrained-optimization is helpful, but much of the book can be understood by the nonspecialist. The contributors are R. Fletcher, P.E. Gill, W. Murray, M.J.D. Powell, D.M. Ryan, R.W.H. Sargent, and W.H. Swann.

Graves, Robert L., and Wolfe, Philip, eds. RECENT ADVANCES IN MATHEMATICAL PROGRAMMING. New York: McGraw-Hill Book Co., 1963. ix, 347 p. Index, pp. 345-47.

A collection of twenty-three of the forty-three papers presented at the fourth symposium on mathematical programming held in Chicago, June 18-22, 1962. The following four survey papers are included: "Combinatorial Theory Underlying Linear Programming," by A.W. Tucker; "Methods of Nonlinear Programming," by P. Wolfe; "Linear Programming under Uncertainty," by A. Madansky; and "Flows in Networks," by D.R. Fulkerson. Among the other authors are E.M.L. Beale, G.B. Dantzig, J.B. Rosen, and A.C. Williams. Abstracts of the twenty remaining papers are also presented.

Greenwood, William T., ed. DECISION THEORY AND INFORMATION SYSTEMS: AN INTRODUCTION TO MANAGEMENT DECISION MAKING. Cincinnati, Ohio: South-Western, 1969. xiii, 818 p. Author Index, pp. 805-6. Subject Index, pp. 807-18.

A collection of forty-seven reprints of articles previously published
in journals and elsewhere, grouped under the following headings:
decision theory, practice, and structure; business-management problem solving; decision information systems; information decision
models using simulation, dynamic programming, linear programming,
queueing models, and game theory; decision environment systems;
organizational decision behavior; and computer information control systems. The articles discuss topics from a general point of
view, but do not treat them rigorously from a mathematical point
of view.

Groff, Gene K., and Muth, John F., eds. OPERATIONS MANAGEMENT:
SELECTED READINGS. Irwin Series in Quantitative Analysis for Business.
Homewood, Ill.: Richard D. Irwin, 1969. vi, 440 p. No index.

A collection of thirty-three previously published articles, grouped
under the following headings: introduction (two articles), managerial decision making (eight articles), design for production
(seven articles), operations scheduling and control (one article),
and controlling performance (five articles). The papers treat such
topics as heuristic programming, queueing problems, computer simulation, control theory, inventories, network models, and linear
programming. Among the authors are R.E. Bellman, E.S. Buffa,
E.W. Davis, P.C. Fishburn, J.E. Magee, and M.K. Starr.

Grouchko, Daniel, ed. OPERATIONS RESEARCH AND RELIABILITY. New
York: Gordon and Breach, 1971. xvi, 625 p. No index.

Twenty-seven papers presented at the conference on the theory and
practice of reliability sponsored by the Scientific Committee of
NATO and held in Turin, Italy, from June 30 to July 4, 1969,
with participation of specialists from fifteen countries. The papers
dealt with applications in the following areas: military telecom-
munication satellite systems, aircraft, digital systems, production
systems subject to breakdown, prediction in cost effectiveness analysis, and statistical test of reliability. Some papers also dealt
with problems of gathering reliability data, while others were concerned with the theory of reliability. Among the contributors are
R.E. Barlow, C.F. Bell, R. Chaplin, F. Ferrazzano, S.S. Gupta,
G. Mariani, J.M. Miller, K. Niemeyer, R.W. Watts, and A.K.
Weaver. Three papers are printed in French.

Gurland, John, ed. STOCHASTIC MODELS IN MEDICINE AND BIOLOGY.
Madison: University of Wisconsin Press, 1964. xvi, 393 p. Index, pp. 387-93.

Thirteen papers presented at a symposium on stochastic models held
at the University of Wisconsin on June 12-14, 1963. The applications include epidemics, queue processes, genetics, evolution,
computing risks in illness and death, sclerosis, and carcinogenesis.

Among the contributors are W.G. Cochran, H. Lucas, J. Neyman, A. Rapoport, and N. Wiener.

Hammer, Preston C., ed. ADVANCES IN MATHEMATICAL SYSTEMS THEORY. University Park: Pennsylvania State University Press, 1969. 174 p. No index.

Eight heretofore unpublished essays dealing with systems theory by the following scholars: R.B. Banerji, P.C. Hammer, M.D. Mesarović, A.J. Perlis, and A.W. Wymore. The topics include filters, approximation spaces, mathematical theory of general systems, continuity, and programming systems. Assumes knowledge of calculus, set theory, and real analysis.

Hertz, David B., and Eddison, Roger T., eds. PROGRESS IN OPERATIONS RESEARCH. Vol. 2. Publications in Operations Research, no. 9. New York: Wiley, 1964. xii, 455 p. Author Index, pp. 439-49. Subject Index, pp. 451-55.

Seventeen heretofore unpublished contributions on applications of operations research in such areas as capital budgeting, government operations, transportation, agriculture, petroleum, textiles, mining, space programs, and steel. Some contributions are exploratory and require little formal mathematics for understanding, while others assume knowledge of algebra and probability. Among the contributors are N.T.J. Bailey, S. Beer, B. Bernholtz, T. Fabian, D.S. Leckie, A. Muir, E. Ritchie, B.H.P. Rivett, and M. Van Buren.

Hertz, David B., and Melese, Jacques, eds. PROCEEDINGS OF THE FOURTH INTERNATIONAL CONFERENCE ON OPERATIONAL RESEARCH. Publications in Operations Research, no. 15. New York: Wiley, 1966. xxxvi, 1092 p. Index, pp. 1085-92.

A collection of ninety-one papers (in addition to a large number of abstracts of other papers) presented at the fourth international conference on operational research held at the Massachusetts Institute of Technology from August 29 to September 2, 1966. The conference was organized into the following three main sessions: recent advances in techniques of mathematical programming (five papers), progress in techniques of decision theory (six papers), and advances in techniques of modeling (five papers). Fifty-two papers were presented in the following specific sessions: the theory of graphs (three papers), marketing (six papers), transportation (thirteen papers), urban planning (three papers), investment policy analysis (five papers), scheduling problems (seven papers), simulation (six papers), natural resources (four papers), and distribution systems (five papers). Four additional sessions were organized to accommodate eighteen papers on stochastic models, business applications, new applications, and mathematical programming. This book concludes with five papers on informal meetings and conference summaries. The table of contents appears in both French and

English, and some of the papers are written in French.

Hertz, David B., and Rubenstein, Albert H., eds. RESEARCH OPERATIONS IN INDUSTRY. New York: Columbia University Press, 1953. xiv, 444 p. No index.

Twenty-nine papers presented at the third annual conference on industrial research held in June 1952, at Columbia University, and selected papers from the first and second conferences. The papers are grouped under the following headings: philosophy and management's appraisal of research (four papers), economics, costs, and budgeting (two papers), personnel in industrial research (six papers), the planning of research facilities (two papers), research methodology and design of experiments (four papers), operations research (four papers), and communications and technical information services (four papers). Among the authors are R.A. Ackoff, R.T. Eddison, D.B. Hertz, W.H. Kliever, P.M. Morse, C.W. Churchman, and J.W. Tukey.

Hu, Te Chiang, and Robinson, Stephen M., eds. MATHEMATICAL PROGRAM-MING. Publication No. 30 of the Mathematics Research Center, University of Wisconsin. New York: Academic Press, 1973. x, 295 p. Index, pp. 291-95.

Proceedings of an advanced seminar on mathematical programming held in Madison, September 11-13, 1972, under the auspices of the Mathematics Research Center with financial support from the U.S. Army. Of the ten lectures, four were on integer programming, two were on game theory, and one each on large-scale systems, nonlinear programming, dynamic programming, and combinatorial equivalence. The contributors are R.W. Cottle, H.P. Crowder, G.B. Dantzig, E.V. Denardo, B.C. Eaves, D.R. Fulkerson, C.B. Garcia, R.S. Garfinkel, G.H. Golub, R.E. Gomory, E.L. Johnson, C.E. Lemke, H. Luethi, A.S. Manne, G.L. Nemhauser, L.S. Shapley, R.B. Wilson, and D.J. Wilde.

Karreman, Herman F., ed. STOCHASTIC OPTIMIZATION AND CONTROL. Publication No. 20 of the Mathematics Research Center, University of Wisconsin. New York: Wiley, 1968. xii, 217 p. Index, pp. 211-17.

Nine papers presented at an advanced seminar held at the University of Wisconsin, October 2-4, 1967. Four papers dealt with the theory of stochastic optimization; one paper was on the theoretical aspects of system identification techniques; two papers dealt with applications in space exploration; and the last two papers were concerned with military applications. The authors are A.V. Balakrishnan, J.V. Breakwell, T.A. Brown, S.E. Dreyfus, W.H. Fleming, A. Klinger, H.J. Kushner, Wm. C. Lindsey, W.L. Nelson, R.M. Van Slyke, C.L. Weber, and R.J.B. Wets.

Kelleher, Grace J., ed. THE CHALLENGE TO SYSTEMS ANALYSIS: PUB-
LIC POLICY AND SOCIAL CHANGE. Publications in Operations Research,
no. 20. New York: Wiley, 1970. viii, 150 p. Index, pp. 149-50.

Fifteen papers on systems analysis, some of which were initially
presented at a forum on System Analysis and Social Change held
in Washington, D.C., in March 1968 under the sponsorship of the
American Institute of Aeronautics and Astronautics and the Operations Research Society of America. Topics include the evolution
of operations research and systems analysis, challenges in urban
and suburban areas, pollution control, safety, communications,
education, population, and food supply. The contributors include
J.H. Engel, W. Holst, G.J. Kelleher, H.W. Maier, P. Morse,
and O.E. Teague.

Kuhn, Harold W[illiam]., ed. PROCEEDINGS OF THE PRINCETON SYMPOSIUM
ON MATHEMATICAL PROGRAMMING. Princeton, N.J.: Princeton University
Press, 1970. vi, 620 p. No index.

A collection of thirty-three contributions presented at the symposium on mathematical programming held at Princeton University on
August 14-18, 1967. The papers are grouped under the following
headings: large-scale systems (four papers), programming under
certainty (five papers), integer programming (six papers), algorithms
(three papers), applications (four papers), theory (three papers),
nonlinear programming (five papers), pivotal methods (three papers),
and abstracts. Among the contributors are E.M.L. Beale, A.
Charnes, P. Huard, R.T. Rockafellar, A.W. Tucker, and G.
Zoutendijk.

Kuhn, Harold William, and Szegö, Giorgia P., eds. DIFFERENTIAL GAMES
AND RELATED TOPICS. New York: American Elsevier, 1971. x, 489 p.
Author Index, pp. 487-89.

Twenty-five highly mathematical papers presented at the International Summer School in game theory held during June 15-27,
1970, in Varenna, Italy. The papers are grouped under the following four headings: theory of differential games (five papers),
special mathematical problems (six papers), computational problems (three papers), and applications (eleven papers). The application papers are devoted to the solutions to economic problems,
and treat topics such as noncooperative games, economic diffusion
processes, the generalized linear-quadratic-gaussian problem, open-
loop Nash equilibrium strategies for an N-person game, decomposition and competition in multilevel environment control systems,
noncooperative equilibria and strategy spaces in an oligopolistic
market, and concepts of preference in economics and biology.
The contributors include M.J. Beckmann, L. Berkovitz, U. Bertele,
A. Blaquiere, V.N. Burkov, A. Cellina, K.C. Chu, B. Grodal,
R.E. Levitan, E. Polak, R.T. Rockafellar, D. Schmeidler, G.P.
Szegö, J.E. Turner, and P. Varaiya.

_____. MATHEMATICAL SYSTEMS THEORY AND ECONOMICS, I. Lecture Notes in Operations Research and Mathematical Economics, vol. 11. New York: Springer-Verlag, 1969. viii, 292 p. No index.

Volume I of a two-volume series. It consists of twelve papers presented at the International Summer School on mathematical systems and economics, Varenna, Italy, on June 1-2, 1967. The papers are grouped under the following headings: basic theories (seven papers), and optimal control of economic systems (five papers). The seven papers on basic theories consist of a representative sample of the fundamentals of general systems theory, of the theory of dynamical systems, and the theory of control; the five papers on optimal control deal with applications to investment policy and the control and identification of dynamic Keynesian economic systems, and the application of Pontryagin's maximum principle to economics. Among the authors are: N.P. Bhatia, G. Debreu, A.R. Dobell, C. Castaing, H.W. Kuhn, M. Kurz, E.B. Lee, L. Markus, J. Nagy, and J.A. Yorke.

_____. MATHEMATICAL SYSTEMS THEORY AND ECONOMICS, II. Lecture Notes in Operations Research and Mathematical Economics, vol. 12. New York: Springer-Verlag, 1969. ii, pp. 293-486. No index.

Volume II of a two-volume series. It consists of fifteen papers presented at the International Summer School on mathematical systems theory and economics, Varenna, Italy, on June 1-2, 1967. The papers are grouped under the following two headings: special mathematical problems (nine papers), and special applications (six papers). The special problems include semidynamical systems, controllability of linear difference-differential systems, the core and competitive equilibria, stability of sets with respect to abstract processes, and the invariance of contingent equations. Special applications include production processes with tree structure, feedback and the dynamics of market stability, optimal accumulation in a Listian model, and testing econometric models by means of time series analysis. Among the authors are: F. Albrecht, A.J. Blikle, M. Beckmann, A. Halanay, R. Kulikowski, E.B. Lee, and A. Straszak.

Kuhn, Harold W[illiam]., and Tucker, A.W., eds. CONTRIBUTIONS TO THE THEORY OF GAMES. Vol. 1. Annals of Mathematics Studies, no. 24. Princeton, N.J.: Princeton University Press, 1950. xv, 201 p. Bibliography, pp. 193-201.

A collection of thirteen papers on zero-sum two-person games, one paper on the three-person game, and one paper on isomorphism of games and strategic equivalence. The papers are grouped into finite games (eleven papers) and infinite games (four papers). The contributors are H.F. Bohnenblust, G.W. Brown, M. Dresher, D. Gale, S. Karlin, H.W. Kuhn, J.C.C. McKinsey, J.F. Nash, J. von Neumann, L.S. Shapley, S. Sherman, R.N. Snow, A.W. Tucker, and H. Weyl.

_____. CONTRIBUTIONS TO THE THEORY OF GAMES. Vol. 2. Annals of Mathematics Studies, no. 28. Princeton, N.J.: Princeton University Press, 1953. viii, 395 p. Bibliography, pp. 389-95.

Twenty-one papers grouped under the following headings: finite zero-sum two-person games (five papers), infinite zero-sum two-person games (five papers), games in extensive form (six papers), and general n-person games (five papers). The contributors are K.J. Arrow, E.W. Barankin, D. Blackwell, R. Bott, N. Dalkey, M. Dresher, D. Gale, D.B. Gillies, I. Glicksberg, O. Gross, S. Karlin, H.W. Kuhn, J.P. Mayberry, J.W. Milnor, T.S. Motzkin, J. von Neumann, H. Raiffa, L.S. Shapley, M. Shiffman, F.M. Stewart, G.L. Thompson, and R.M. Thrall.

Langer, Rudolph E., ed. NONLINEAR PROBLEMS. Publication No. 8 of the Mathematics Research Center, University of Wisconsin. Madison: University of Wisconsin Press, 1963. xiii, 321 p. Abstracts, pp. 277-312. Index, pp. 313-21.

A collection of fifteen papers and twenty-six abstracts of papers presented at the sixth symposium in nonlinear methods held at the University of Wisconsin from April 30 to May 2, 1962. The papers dealt with such topics as the Stokes paradox, Lyapunov's direct method, nonlinear "eigenvalue" problems, boundary value problems for nonlinear elliptic equations in n variables, the initial value problem for the Navier-Stokes equations, and uniqueness and differentiability of solutions of ordinary differential equations. Assumes knowledge of mathematical analysis, advanced calculus, and differential equations. The contributors are M.L. Cartwright, C.L. Dolph, R. Finn, D. Gilbarg, W. Hahn, P. Hartman, W.T. Koiter, P.D. Lax, J.E. Littlewood, J. Moser, L. Nirenberg, S.I. Pai, E.H. Rothe, H.H. Schaefer, and J. Serrin.

Lavi, Abraham, and Vogl, Thomas P., eds. RECENT ADVANCES IN OPTIMIZATION TECHNIQUES. New York: Wiley, 1966. xiii, 656 p. No index.

A collection of thirty-one papers presented at a symposium on recent advances in optimization held at the Carnegie Institute of Technology on April 21-23, 1965. The papers represent primarily a wide range of applications to real life problems and are grouped under the following headings: optimization of static systems, essentially nonlinear programming, integer programming and search techniques; and trajectory optimization, controller synthesis, and performance optimization of dynamic systems with deterministic or stochastic inputs. Among the authors are A. Charnes, D.A. Gall, R.A. Howard, F.H. Kishi, and G.L. Thompson.

Lawrence, J.R., ed. OPERATIONAL RESEARCH AND THE SOCIAL SCIENCES. New York: Tavistock Publications, 1966. xxxiv, 669 p. Combined Bibliography by C. Forrester, pp. 641-59. Name Index, pp. 661-69.

A collection of forty-two papers presented at an international conference held at Gonville and Caius College, Cambridge, England, on September 14-18, 1964. The papers are grouped under the following headings: organization and control (seventeen papers); social effects of policies and their measurement (six papers); conflict resolution and control (four papers); and models, decisions, and operational research (seven papers). There are no mathematics prerequisites.

Lawrence, John, ed. OR 69. PROCEEDINGS OF THE FIFTH INTERNATIONAL CONFERENCE ON OPERATIONAL RESEARCH. VENICE, 1969. New York: Tavistock Publications, 1970. cxxvi, 955 p. Abstracts (English), pp. xvii-lxxiv. Resumes (French), pp. lxxv-cxxvi. No index.

A collection of seventy-eight papers presented at the fifth international conference on operational research held in Venice on June 22-27, 1969. The papers are grouped under the following headings: prologue (two papers), corporate planning and corporate objectives (three papers), social and political science (six papers), stochastic processes (six papers), transport and traffic (seven papers), market strategy and marketing (six papers), mathematical methods of optimization (seven papers), informatics (six papers), network flows and graph theory (six papers), operational research in the public sector (five papers), decision analysis (five papers), general interest papers (fourteen papers), working group reports (four papers), and epilogue (one paper). About 450 delegates representing twenty-three countries attended, and some of the papers are printed in French. Among the contributors are R.L. Ackoff, J.P. Bansard, E.M.L. Beale, A. Charnes, R.T. Eddison, S. Enke, S.K. Gupta, R.A. Howard, K. Kayukawa, H.W. Kuhn, J.E. Matheson, T. Matsuda, M.K. Starr, R.C. Tomlinson, and P.M. Tullier.

Leitmann, George, ed. TOPICS IN OPTIMIZATION. Mathematics in Science and Engineering, vol. 31. New York: Academic Press, 1967. xv, 469 p. Author Index, pp. 463-65. Subject Index, pp. 466-69.

A collection of ten contributions to the field of optimization of dynamical systems heretofore unpublished and grouped into the following two parts: variational techniques which constitute essentially extensions of the classical calculus of variations, and optimal control theory and its applications. The contributors are A. Blaquière, E.K. Blum, S.P. Diliberto, B. Garfinkel, H. Halkin, H.J. Kelley, R.E. Kopp, G. Leitmann, A.I. Lurie, K.A. Lurie, H.G. Moyer, and B. Paiewonsky.

Lewis, Peter A., ed. STOCHASTIC POINT PROCESSES: STATISTICAL ANALYSIS, THEORY, AND APPLICATIONS. Series in Applied Probability and

Statistics. New York: Wiley-Interscience, 1972. xii, 894 p. Author Index, pp. 887-94.

A collection of thirty-five papers presented at a conference held at the IBM Research Center, Yorktown Heights, New York, on August 2-6, 1971. An introductory paper by Lewis is also included. The papers are arranged under the following headings: statistical analysis (nine papers), models (five papers), theory (twelve papers), and applications (nine papers). Among the topics treated are: univariate and multivariate point processes, nonhomogeneous Poisson processes, statistical properties of traffic counts, multidimensional and infinitely divisible point processes, and hyperplane processes. The applications are in the areas of epidemiology, reliability, photographic science, forestry, neurophipiology, and lunar and planetary surfaces. Among the authors are M.S. Bartlett, D.R. Cox, D.P. Gaver, M.R. Leadbetter, J. Neyman, G.P. Patil, A. Ramakushnan, M.W. Sachs, and W.G. Warren.

Livingstone, John Leslie, ed. MANAGEMENT PLANNING AND CONTROL: MATHEMATICAL MODELS. McGraw-Hill Accounting Series. New York: McGraw-Hill, 1970. xviii, 616 p. No index.

A collection of twenty-three carefully selected journal articles from the ACCOUNTING REVIEW, the JOURNAL OF ACCOUNTING RESEARCH, MANAGEMENT SCIENCE, and MANAGEMENT ACCOUNTING. The papers are grouped under the following headings: mathematical representation of cost information systems (seven articles), cost standards and control (five articles), cost-profit-volume analysis (six articles), and planning and budgeting future operations (five articles). The tools of analysis include input-output analysis, PERT, breakeven analysis, linear and integer programming, and sensitivity analysis. Among the authors are N. Churchill, W.W. Cooper, W. Frank, Y. Ijiri, R.K. Jaedicke, R.E. Jensen, F.K. Levy, R.P. Manes, W.R. Ross, G.L. Salamon, J.D. Wiest, and G.A. Zeisel.

Lock, Dennis, ed. A GUIDE TO MANAGEMENT TECHNIQUES WITH GLOSSARY OF MANAGEMENT TERMS. Foreword by Sir Richard Powell. New York: Wiley, 1972. xvi, 482 p. No index.

A collection of essays by different authors which explain the use and procedures of thirty-two topics and a large number of management terms and techniques. It is divided into the following sections: general management and organization (six papers); financial management (six papers); personnel management, industrial relations, and training (six papers); marketing, including marketing research by input-output models, product profitability analysis, and customer cost analysis (five papers); production and distribution, including scheduling of resources, computers in production control, and maintenance programming (six papers); and decision making in management (three papers). Included are an explanation of the techniques, the procedures for implementation, and the benefits to be derived from the techniques.

Lombaers, H.J.M., ed. PROJECT PLANNING BY NETWORK ANALYSIS. New York: American Elsevier, 1969. xii, 457 p. Author Index, p. 457.

Fifty-five papers presented at an international conference for planning and network analysis held in Amsterdam on October 6-10, 1969. The papers are grouped under the following headings: organization management and training (two papers), analysis of structure and stochastic aspects (fourteen papers), allocation of resources and scheduling (ten papers), new developments in computer applications (seven papers), and applications (twelve papers). The papers give a reasonably complete view of the present state of project planning by means of network analysis.

Lootsma, F.A., ed. NUMERICAL METHODS FOR NON-LINEAR OPTIMIZATION. New York: Academic Press, 1972. xiv, 440 p. Author Index, pp. 429-33. Subject Index, pp. 435-39.

A collection of twenty-nine papers presented at the conference on numerical methods for nonlinear optimization held at the University of Dundee (Scotland) from June 28 to July 1, 1971. The papers were confined to topics in nonlinear unconstrained and constrained optimization, with emphasis on the numerical aspects. Methods for integer linear programming are omitted. Among the contributors are E.M.L. Beale, M.C. Biggs, L.C.W. Dixon, U. Eckhardt, D. Goldfarb, D.M. Himmelblau, S.T. Loney, F.A. Lootsma, M.R. Osborne, and M.J.D. Powell.

McCloskey, Joseph F., and Coppinger, John M., eds. OPERATIONS RESEARCH FOR MANAGEMENT. Vol. 2: CASE HISTORIES, METHODS, INFORMATION HANDLING. Introduction by the Earl of Halsbury. Baltimore: Johns Hopkins Press, 1956. xxxvi, 563 p. Authors, p. xiii. Appendix A: Bibliography on Queueing Theory, by V. Riley, pp. 539-56. Index, pp. 557-63.

Twenty-eight papers presented before the Johns Hopkins University informal seminar on operations research during 1953-54 and 1954-55. The papers are grouped under the following headings: case histories (thirteen papers), methods (nine papers), and information handling (six papers). In addition, an introductory paper, "From Plato to the Linear Program," by the Earl of Halsburg, is also included. The contributors include M. Astrachan, G.D. Camp, L.S. Christie, J.W. Dunlap, M.M. Flood, W.H. Glanville, H.H. Jacobs, J. Macy, Jr., B.H.P. Rivett, Wm. S. Vickrey, and W.J. Youden.

McCloskey, Joseph F., and Trefethen, Florence N., eds. OPERATIONS RESEARCH FOR MANAGEMENT. Vol. 1. Baltimore: Johns Hopkins Press, 1954. xxiv, 407 p. Selected Bibliography, pp. 381-401. Index, pp. 403-7.

A collection of twenty-three papers presented at a seminar on operations research held at the Johns Hopkins University in 1952. Volume I is addressed primarily to management and the papers are grouped under the following headings: general (five papers), methodology (nine papers), and case histories (eight papers). An introductory paper, "The Executive, the Organization, and Operations Research," by E.A. Johnson, is also included. Among the contributors are R.L. Ackoff, D.H. Blackwell, W.E. Cushen, A.H. Hausrath, L.J. Henderson, C. Hitch, H.C. Levinson, B.O. Marshall, Jr., P.M. Morse, G.S. Pettee, C.W. Thornthwaite, and F.N. Trefethen.

Mansfield, Edwin, ed. MANAGERIAL ECONOMICS AND OPERATIONS RESEARCH: TECHNIQUES, APPLICATIONS, CASES. 3d ed. New York: W.W. Norton, 1975. xiv, 602 p. No index.

A collection of forty-eight reprints of articles dealing with many facets of managerial economics and operations research. It is aimed primarily at students in business and economics who have only modest training in mathematics. The articles are grouped under the following headings: the decision-making process (four articles), costs and production (four articles), profits, demand, and pricing (six articles), capital budgeting and investment (four articles), business and economic forecasting (five articles), linear programming (four articles), decision theory and scheduling techniques (five articles), game theory, inventory policy, and queueing analysis (five articles); the role of the computer in industrial management (four articles), and economic analysis and the public sector (seven articles). Assumes knowledge of introductory operations research and one year of basic statistics. Among the contributors are Wm. J. Baumol, J. Dean, R. Dorfman, D.B. Hertz, C.J. Hitch, E. Mansfield, H. Raiffa, and H.A. Simon.

Mansour, M., and Schaufelberger, W., eds. 4TH IFAC/IFIP INTERNATIONAL CONFERENCE ON DIGITAL COMPUTER APPLICATIONS TO PROCESS CONTROL. Part I. Lecture Notes in Economics and Mathematical Systems, vol. 93. New York: Springer-Verlag, 1974. xvii, 544 p. No index.

Forty-three papers presented at the 4th IFAC/IFIP International Conference on Digital Computer Applications to Process Control held in Zurich, Switzerland, March 19-22, 1974. The papers in this volume are grouped under the following headings: digital computer algorithms (fourteen papers), new developments of process computers hardware and software (eight papers), digital control in chemical and oil industries (nine papers), digital control in cement industry (six papers), and digital control in pulp and paper industries (six papers). Topics include computer algorithms for automatic design of linear control systems, suboptimal adaptive control, learning control, and the control of heat exchangers; production and test of system software for minicomputers; structure and hierarchial organization of process control; the automatic control system of the ammoniac nitrate production; rotary cement kiln control combined algorithm; control of the homogenisation of the cement raw meal; and dynamics and control of multiple-effect evaporator and drum washer plants. Among the contributors are K.J. Astrom, Z. Binder, J. Boyd, T. Cegrell, D.G. Fisher, H. Gran, H. Hammer, M.J. Hessen, K. Kloster, C. Marin, S. Narita, A. Niemi, A. Ramaz, D.E. Seborg, P. Verebely, and B. Wittenmark.

_____. 4TH IFAC/IFIP INTERNATIONAL CONFERENCE ON DIGITAL COMPUTER APPLICATIONS TO PROCESS CONTROL. Part II. Lecture Notes in Economics and Mathematical Systems, vol. 94. New York: Springer-Verlag, 1974. xviii, 546 p. No index.

Forty-four papers presented at the 4th IFAC/IFIP International Conference on Digital Computer Applications to Process Control held in Zurich, Switzerland, on March 19-22, 1974. The papers in part II are grouped under the following headings: digital control in metallurgical processes (ten papers); digital control in power systems (twenty-three papers); digital control in other processes: material handling and storage, etc. (seven papers); reliability of digital control (two papers); and economics of digital control (two papers). Some of the papers are written in German, and many papers describe applications verbally and with graphs and do not require formal mathematics knowledge. The contributors include P. Andersen, D.R. Bjork, J. Burley, C.E. Carter, D.R. Hirst, D.W. Huber, A.J. Kisiel, H.A. Kuhr, J. Lenschow, A.G. Longmuir, O.P. Malik, J. Marney, P. Petrov, B. Qvarnstrom, and T. Yasui.

Mayne, David Q., and Brockett, Roger W., ed. GEOMETRIC METHODS IN SYSTEM THEORY. NATO Advanced Study Institutes Series. Boston: Reidel, 1973. vi, 314 p. No index.

A collection of twenty-one papers presented at the NATO Advanced Study Institute held in London from August 27 to September 7, 1973. The papers deal with the use of differential geometry in the study of control theory. The first five papers are mainly expository and are intended as a combination textbook and guide to the literature, while the remaining papers explore in some depth the theory. The use of differential geometry serves to overcome the limitations of linear theory and enables one to analyze large classes of nonlinear problems without difficulty. Among the contributors are J.M.C. Clark, D.L. Elliott, E. Fornasini, J. Grate, R. Hermann, R.M. Hirschorn, C. Lobry, L. Markus, A. Ruberti, H.J. Sussman, and A.S. Willsky.

Menges, Gunter, ed. INFORMATION, INFERENCE AND DECISION. Theory and Decision Library. Boston: Reidel, 1974. viii, 195 p. Index of Names, pp. 189-91. Index of Subjects, pp. 192-95.

Ten heretofore unpublished papers on issues from statistical inference, philosphy, and epistemology written by statisticians and decision theorists who belong to the former Saarbrucken school of statistical decision theory. The papers are grouped under the following four headings: objective theory of inductive behavior (three papers); problems of inference (two papers); probability, information and utility (three papers); and semantic information (two papers). The authors are M.J. Beckmann, M. Behara, D.A.S. Fraser, J. Kalbfleisch, B. Leiner, J. Marschak, H. Schneeweiss, H.J. Skala, and D.A. Sprott.

Mensch, A., ed. THEORY OF GAMES: TECHNIQUES AND APPLICATIONS. Foreword by E. Torchet. New York: American Elsevier, 1966. 490 p. No index.

A collection of thirty-four papers presented at a conference on the theory of games in Toulon, France, from June 29 to July 3, 1966, under the aegis of the NATO Scientific Affairs Committee. The papers are divided into the following three parts: mathematical methods (thirteen papers), military applications (seventeen papers), and discussions and synthesis (four papers). The papers treat such topics as matrix games, infinite games, n-person games, missile penetration, resource allocation models for tactical air war, role of differential games in warfare, and non-zero-sum theory. Among the contributors are E.M.L. Beale, B. Contini, J.M. Danskin, M. Dresher, A. Ghouila-Houri, R. Isaacs, O. Morgenstern, T. Schelling, R.W. Shephard, and M. Shubik.

Mesarović, Mihajlo D., and Reisman, Arnold, eds. SYSTEMS APPROACH AND THE CITY. New York: American Elsevier, 1972. xx, 481 p. Index, pp. 461-81.

Twenty contributions presented at the Fifth Systems Symposium held at Case Western Reserve University on November 9-11, 1970. With the exception of one contribution, no knowledge of mathematics nor of systems science is assumed. Intended for students and practitioners concerned with urban problems who are interested in what the systems approach can do in the urban setting and for systems scientists who are interested in applications to urban problem solving. Topics include systems analysis of urban air pollution, health systems, systems analysis and social welfare planning, systems analysis of crime control, models of a total criminal justice system, design of a freeway control system, and the systems approach to urban planning. Among the authors are A. Blumstein, C. Flagle, S. Goldstone, S.J. Mantel, Jr., R.L. Meier, M.D. Mesarovic, and E.S. Savas.

Meyer, Herbert A., ed. SYMPOSIUM ON MONTE CARLO METHODS. A Wiley Publication in Applied Statistics. New York: Wiley, 1956. xvi, 382 p. Bibliography, pp. 283-370. Name Index, pp. 371-76. Source Index, pp. 377-79. Subject Index, pp. 381-82.

A collection of twenty papers presented at a symposium at the University of Florida on March 16-17, 1954, under the sponsorship of Wright Air Development Center of the Air Research and Development Command. The papers are arranged in the order of presentation and cover both theory and application. Among the contributors are G.E. Albert, H. Kahn, A.W. Marshall, J.W. Tukey, J.E. Walsh, and A. Walther. The applications treated by the papers include tactical games, gamma ray diffusion, multiple stage sampling procedures, generation of numbers, confidence interval estimates for insurance mortality rates, and determinants of eigenvalues and dynamic influence coefficients for complex structures, such as airplanes.

Meyer, John R., ed. TECHNIQUES OF TRANSPORT PLANNING. Vol. 2. Foreword by K. Gordon. Washington, D.C.: Brookings Institution, 1971. xiv, 228 p. Appendixes A to C, pp. 161-223. Index, pp. 225-28.

Volume 1 of TECHNIQUES OF TRANSPORT PLANNING surveyed the underlying principles and synthesized the literature of conventional project evaluation, or cost-benefit analysis, as a tool for making transport investment decisions. In volume 2 the authors apply systems analysis to the transport network of a developing country, Columbia, using a series of interacting models to simulate a network and the economy of Columbia as a whole. Part 1 of volume 2 contains the following contributions: "The Macroeconomic Model," by D.T. Kresge, P.O. Roberts, J.R. Meyer, H. Luft, D.N. Dewes, and S.R. Ginn; "Specification and Evaluation of Alternative Transport Plans," by D.T. Kresge, J.R. Meyer, and P.O. Roberts; and "The System Approach: Summary and Conclusion," by J.R. Meyer. The appendixes contain three additional models as follows: highway cost-performance model, the railway cost-performance model, and the transfer model. Assumes knowledge of econometric techniques.

Milsum, John H., ed. POSITIVE FEEDBACK: A GENERAL SYSTEMS APPROACH TO POSITIVE/NEGATIVE FEEDBACK AND MUTUAL CAUSALITY. Elmsford, N.Y.: Pergamon Press, 1968. x, 169 p. Contributing Authors, pp. 164-66. Index, pp. 167-69.

A collection of nine essays on positive feedback systems requiring no formal mathematics background for comprehension. The contributors are K.E. Boulding, C.R. Dechert, E. Kramer, J.H. Milsum, H.M. Paynter, A. Rapoport, M.D. Rubin, G.B. Slobodkin.

MIT Operations Research Center. NOTES ON OPERATIONS RESEARCH. Cambridge, Mass.: MIT Press, 1959. viii, 256 p. Bibliography, pp. 251-54. Index, pp. 255-56.

A collection of eleven papers presented at a seminar in operations research in Brussels in August 1959 by representatives from NATO countries. The papers dealt with such topics as search techniques, Markov processes, queueing systems, control processes, sequential decision processes, reliability and maintenance, production scheduling, information theory, and simulation. The contributors are H.P. Galliher, R.A. Howard, G.E. Kimball, B.O. Koopmans, P.M. Morse, and G.P. Wadsworth.

Moiseev, Nikita N., ed. COLLOQUIUM ON METHODS OF OPTIMIZATION. Lecture Notes in Mathematics, vol. 112. New York: Springer-Verlag, 1970. 293 p. No index.

Eighteen papers presented at a colloquium on methods of optimization held in Novosibirsk, USSR, in June 1968. The papers dealt with such topics as game theory; stochastic processes; dynamic multibranch industrial models; optimal control; and computer algorithms for finding solutions to optimization problems.

Moore, Peter Gerald, and Hodges, Stewart Dimont, eds. PROGRAMMING FOR OPTIMAL DECISIONS: SELECTED READINGS IN MATHEMATICAL PROGRAMMING TECHNIQUES FOR MANAGEMENT PROBLEMS. Baltimore: Penguin Books, 1970. 360 p. Further Reading, pp. 343-47. Acknowledgments, p. 349. Author Index, pp. 353-56. Subject Index, pp. 359-60.

Seventeen papers previously published in journals and elsewhere grouped under the following headings: applications of linear programming (four papers), applications of other programming techniques (seven papers), and theoretical developments (six papers). An introductory paper, "Basic Optimization Techniques--a Brief Survey," by J.E. Mulligan, is also included. The applications of linear programming are in forest management, transistor production, agricultural production, iron foundries, portfolio selection, separable programming applied to ore purchasing, open-pit mining, models for hospital menu planning, job shop scheduling, and warehouse location. The theoretical developments include nonlinear programming, linear programming under uncertainty, integer programming, and optimum seeking with branch and bound techniques. Among the contributors are E.M.L. Beale, A. Charnes, W.W. Cooper, S.E. Elmaghraby, R.L. Gue, K.B. Haley, and P. Wolfe.

Morse, Philip McCord, ed. OPERATIONS RESEARCH FOR PUBLIC SYSTEMS. Cambridge, Mass.: MIT Press, 1967. ix, 212 p. List of Lectures, Special Summer Program, pp. 207-8. Index, pp. 209-12.

Nine papers presented at a summer program on operations research and public affairs held at MIT on September 6-10, 1966. The applications dealt with in the papers include operations research in local government, simulation models and urban planning, traffic, transportation networks, medical and hospital practice, crime and criminal justice, mathematical techniques: probabilistic models, and mathematical techniques: mathematical programming. Few formulas are used in the presentation.

Murray, William Allan, ed. NUMERICAL METHODS FOR UNCONSTRAINED OPTIMIZATION. New York: Academic Press, 1972. xi, 144 p. A Glossary of Symbols, pp. ix-x. Appendix: Some Aspects of Linear Algebra Relevant to Optimization, pp. 131-34. References, pp. 135-40. Author Index, pp. 141-42. Subject Index, pp. 143-44.

Eight contributions based on papers presented at the joint IMA/NPL conference held at the National Physical Laboratory on January 7-8, 1971. It provides a comprehensive and detailed survey of numerical methods available for unconstrained optimization. Among the methods discussed are: direct search, second derivative, conjugate direction, and quasi-Newton. Other topics include causes and cures of failure in finding an optimization algorithm, and a survey of algorithms for unconstrained optimization. Assumes knowledge of calculus and numerical methods. The contributors are C.G. Broyden, R. Fletcher, W. Murray, M.J.D. Powell, and W.H. Swann.

Muth, John F., and Thompson, Gerald L., eds. INDUSTRIAL SCHEDULING. Prentice-Hall Series in Management. Englewood Cliffs, N.J.: Prentice-Hall, 1963. xviii, 387 p. Bibliography, pp. 379-87. No index.

A collection of twenty-two papers on factory scheduling, fourteen of which were presented at a conference held at the Graduate School of Industrial Administration, Carnegie Institute of Technology on May 10-12, 1961. The papers are grouped under the following headings: structure of scheduling problems, scheduling problems in inventory control, integer programming models in scheduling, simulation of scheduling procedures and scheduling performance, and scheduling complex activities. The techniques applied include critical path methods, network analysis, integer programming, lot-size programming, average cost method of scheduling, and Monte Carlo algorithms for solving production scheduling problems. Some papers require knowledge of differential calculus. Among the contributors are M. Allen, R.W. Conway, R.E. Gomory, S.M. Johnson, F.K. Levy, A.S. Manne, J.F. Muth, V. Van Ness, G.L. Thompson, and H.M. Wagner.

Naylor, Thomas H., ed. THE DESIGN OF COMPUTER SIMULATION EXPERIMENTS. Durham, N.C.: Duke University Press, 1969. x, 417 p. No index.

Twenty-five papers presented at a symposium sponsored by the College of Simulation and Gaming of The Institute of Management Science and Duke University and held at Duke University on October 14-16, 1968. The papers are grouped under the following five headings: introduction (one), experimental designs (four), data analysis (five), methodological problems (eight), and applications (seven). Topics include factor selection, response surface designs, sequential designs, regression analysis and analysis of variance, ranking procedures, time series analysis, simulation versus analytical solutions, theoretical and practical Monte Carlo techniques, simulation languages, and distributions of blocks of signs. The applications are in economic policy, nonlinear econometric models, life insurance models, risk theory models, multidimensional verification, and simulation techniques. Most of the papers assume knowledge of mathematical statistics. The authors include H. Chernoff, N.R. Draper, M.K. Evans, S.S. Gupta, R.H. Hayes, J.S. Hunter, H.S. Krasnow, T.H. Naylor, M. Pfaff, J.S. Ramberg, and D. Watts.

Nemchinov, Vasiliĭ S., ed. THE USE OF MATHEMATICS IN ECONOMICS. Translated by Oliver and Boyd. English ed. edited with an introduction by A. Nove. Cambridge, Mass.: MIT Press, 1965. xxi, 377 p. Short Bibliography on Linear Programming and Related Problems, by A.A. Korbut, pp. 357-67. Postscript by V.S. Nemchinov, pp. 369-77.

Contains six contributions from the original Russian publication of 1959. The longest contribution is "Cost-Benefit Comparisons in a Socialist Economy," by V.V. Novozhilov. Topics treated by the remaining contributions include input-output analysis; mathematical methods of production planning and organization; the use of mathematical models, including linear programming, in economic planning; and methods of establishing the shortest running distances for freights when setting up transportation systems. Assumes knowledge of linear programming and input-output techniques. The contributors are O.R. Lange, A.L. Lur'e, L.V. Kantorovich, V.S. Nemchinov, and V.V. Novozhilov.

Oettli, W[erner]., and Ritter, K., eds. OPTIMIZATION AND OPERATIONS RESEARCH. Lecture Notes in Economics and Mathematical Systems, vol. 117. New York: Springer-Verlag, 1976. iv, 316 p. No index.

Twenty-seven papers presented at a conference held at Oberwolfuch from July 27 to August 2, 1975. The conference was devoted to optimization problems in operations research and the following topics were included in the papers: optimization of elastic structures by mathematical programming; minimization under linear equality constraints; generalized Stirling-Newton methods; a method for computing pseudoinverses; approximations to stochastic optimization problems; dual methods in convex control problems; preference optimality; and decomposition procedures for convex programs. Assumes knowledge of higher mathematics, including set theory, mathematical analysis, and Hilbert space theory. The contributors include A. Bachem, L.C.W. Dixon, W. Gaul, J. Hartung, M. Köhler, V. Kovacevic, S.M. Robinson, K. Schumacher, and J. Stahl.

Ogander, Mats, ed. THE PRACTICAL APPLICATION OF PROJECT PLANNING BY NETWORK TECHNIQUES. 3 vols. New York: Wiley, Halsted Press, 1972. Vol. I: 411 p.; Vol. II: 658 p.; Vol. III: 596 p.

A collection of 109 papers presented at the Third International Congress on Project Planning by Network Techniques in Stockholm in May 1972. All papers have been typed by the authors on original sheets according to a typing instruction for contributors, and the papers have been printed via offset printing. Volume 1 contains twenty-five papers arranged in Block A which is concerned with network planning in action. Volume 2 contains forty-three papers arranged in Block B on the practical applications of network planning. Volume 3 contains forty-one papers arranged in Blocks C and D under the headings, respectively, of: computer programs for integrated network planning-time, cost, resource planning and techniques; and project management and information systems using network techniques. One paper, "On the Integration of Project Management and Informations Systems," by P. Norden, is printed separately. The papers deal mainly with concepts, applications, and interpretation of results, and very few equations are employed. Each volume contains an index of articles.

Palda, Kristian S., ed. READINGS IN MANAGERIAL ECONOMICS. Englewood Cliffs, N.J.: Prentice-Hall, 1973. x, 320 p. No index.

Twenty-nine articles and essays, previously published in journals, books, and elsewhere, arranged under the following headings: the goals of the firm (five papers), the production function submodel of the firm (eight papers), the R & D submodel of the firm (three papers), the marketing submodel of the firm (four papers), the financial submodel of the firm (six papers), and the overall model of the firm (three papers). Assumes knowledge of microeconomic theory, elementary calculus, and regression analysis. Each group of readings is preceded by a brief introduction by the editor pinpointing the highlights of each selection. Techniques and methodology in management science are, for the most part, omitted. For example, only one paper deals with linear programming. Econometrics is stressed in many of the papers and the empirical verification of the models' parameters are provided. Topics include multidimensional utility, production and factor shares in the halibut fishing industry, multiple regression analysis of cost behavior, determinants of industrial research, a simultaneous equation regression study of advertising and sales of cigarettes, a portfolio analysis of conglomerate diversification, and an econometric model of a Japanese pharmaceutical company. Among the authors are Wm. J. Baumol, G.J. Benston, S. Danø, P. Kotler, D.C. Mueller, K.S. Palda, K.V. Smith, H. Tsurumi, and Y. Tsurumi.

Patil, Ganapati P., ed. VOLUME 1. RANDOM COUNTS IN SCIENTIFIC WORK: RANDOM COUNTS IN MODELS AND STRUCTURES. The Penn State Statistics Series. University Park: Pennsylvania State University Press, 1970. 268 p. No index.

Fourteen papers out of forty-two papers which were either presented

at the Biometric Society symposium in Dallas in December 1968,
or invited from specialists who were not present. The symposium
dealt with the formulation of discrete models and methods to solve
problems that generate or require counted data. Volume 1 is devoted to the analysis of random numbers generated by theoretical
distributions such as the binomial and negative binomial distributions, the Poisson and inverse Poisson distributions, classes of discrete distributions, and discrete Markov chains. The contributors
are M.T. Boswell, G.P. Patil, V.R. Rao Uppuluri, W.J. Blot,
B. Hoadley, K.G. Janardan, R.T. Leslie, Z. Govindarajulu, H.
Makabe, B.M. Bennett, E. Nakamura, D.O. Bowman, L.R. Shenton,
M. Sandelius, J.J. Gart, H. Grimm, D.A. Sprott, S.E. Fienberg,
P.W. Holland, C.K. Tsao, and G.P. Patil.

———. VOLUME 2. RANDOM COUNTS IN SCIENTIFIC WORK: RANDOM
COUNTS IN BIOMEDICAL AND SOCIAL SCIENCES. The Penn State Statistics
Series. University Park: Pennsylvania State University Press, 1970. 267 p.
No index.

Fifteen out of forty-two papers presented at the symposium of the
Biometric Society in Dallas in December 1968, or invited from
specialists. Volume 2 deals with the applications of some theoretical distributions to various problems in health and with further
considerations of theoretical distributions, such as the minimum
Chi-square estimation for the log-zero-Poisson distribution and
normal approximations to the Poisson distribution. The contributors are V.P. Bhapkar, P. Froggatt, S. Iwao, S.W. Joshi, A.W.
and C.D. Kemp, S.K. Katti, H.O. Lancaster, F.M. Lord, P.T.
Ma, S.K. Mitra, W. Molenaar, J.K. Ord, G.P. Patil A.V.
Rao, C. Särndal, J. Stene, and K. Subrahmaniam.

———. VOLUME 3. RANDOM COUNTS IN SCIENTIFIC WORK: RANDOM
COUNTS IN PHYSICAL SCIENCES, GEOSCIENCE, AND BUSINESS. The Penn
State Statistics Series. University Park: Pennsylvania State University Press,
1970. 232 p. No index.

Thirteen out of forty-two papers presented at a symposium in Dallas
in 1968. Volume 3 deals with applications to such topics as
models of pollen studies, regularity in spatial distributions, statistical aspects of amounts and duration of rainfall, market research, multiple security analysis, and the classical occupancy
problem. Among the contributors are A.C. Cohen, M.F. Dacey,
G.J. Kelleher, J.E. Mosimann, G.P. Patil, S.J. Press, L.R.
Shenton, and J.E. Walsh.

Prékopa, András, ed. INVENTORY CONTROL AND WATER STORAGE. Amsterdam: North-Holland, 1973. 382 p. No index.

Twenty-seven papers presented at a conference organized by the
Bolyai Janos Mathematical Society and held in Gyor, Hungary,
on September 11-17, 1971. The papers are arranged alphabeti-

cally by the authors' names and deal with two subjects which are
similar with respect to mathematical models employed. These subjects are inventory control and water storage, and the topics and
applications considered are inventory control in chemical companies, stochastic optimization of water resource allocation, multistage stochastic inventory models, limit theorems for storage models,
mass service theory, and first emptiness problems for storage. Among
the authors are A. Bárász, H.J. Girlich, D.P. Kennedy, Q. Mann,
S. Opricovic, I.A. Pappas, N.U. Prabhu, A. Prékopa, J. Stahl,
J. Wessels, and T.M. Whitin.

PROCEEDINGS OF THE SECOND SYMPOSIUM, LINEAR PROGRAMMING.
Vol. 1. Washington, D.C.: National Bureau of Standards, 1955. vi, 396 p.
No index.

A collection of thirty-three papers presented at the second symposium on linear programming held in Washington, D.C., during January
27-29, 1955, under sponsorship of the Office of Scientific Research
of the Air Research and Development Command and the National
Bureau of Standards. The papers are grouped under the following
headings: applications (twelve papers), economic theory (six
papers), computations (eight papers), theory of linear inequalities
(six papers), and developments in linear programming (one paper).
Among the contributors are G.B. Dantzig, L.W. McKenzie, R.
Radner, P.A. Samuelson, S. Vajda, and A. Vazsonyi.

Quade, Edward S., ed. ANALYSIS FOR MILITARY DECISIONS. New York:
American Elsevier, for the Rand Corporation, 1970. vii, 382 p. Appendixes,
pp. 331-61. Bibliography, pp. 362-63. Index, pp. 365-82.

A series of nineteen lectures (some extensively revised) which
formed a course on the analysis for military decisions given to
military officers and civilians associated with the armed forces.
The lectures are grouped under the following headings: orientation, elements and methods, special aspects, and summary. The
lectures are nontechnical and of value to anyone interested in a
critical evaluation of the military aspects of national security.
The appendixes are devoted to the lunar base problem and a
missile comparison.

Rao, H.S. Subba; Jaiswal, N.K.; and Ghosal, A[mitava]., eds. ADVANCING
FRONTIERS IN OPERATIONAL RESEARCH. Foreword by J.E. Walsh. Inter-
national Monographs on Advanced Mathematics and Physics. Delhi-7, India:
Hinduston Publishing Corp., 1969. xxv, 463 p. No index.

A collection of forty-one papers presented at the International Seminar in New Delhi on August 7-10, 1967 (Second Operations Research Around-the-World Meeting) sponsored by the Operational
Research Society of India and International Federations of Operational Research Societies. The papers are grouped under the fol-

lowing headings: business and industrial uses of operations research; queueing, scheduling, and stochastic processes; programming; reliability, replacement, and maintenance; computer, statistics, models, and measures; operational research in planning; comparative operations research; inventory and production control; simulation; and control and reduction of military conflicts. Topics include a telephone installation problem, inventory reduction in India, application of linear programming to production planning in a textile mill, lexicographic search algorithm for scheduling, sequential competitive budding, systems effectiveness with preemptive resume repair policy, organization of health care in a developing country, operations research in weather modification programs, and cross-cultural comparisons of operations research. Among the authors are P.C. Bagga, P.N. Chowdhury, A. Ghosal, A.V.K. Iyengar, M. Knayer, N.K. Malhotra, R. Natarajan, Wm. A. Reinke, L.H. Smith, and V.R. Rao Uppuluri.

Rapoport, Anatol, ed. GAME THEORY AS A THEORY OF CONFLICT RESOLUTION. Theory and Decision Library. Boston: Reidel, 1974. 283 p. No index.

Eleven heretofore unpublished papers on game theory arranged under the following two headings: two-person games (five papers), and n-person games (six papers). Among the papers on two-person games, one presents a brief overview of the experimental work with the prisoner's dilemma, three deal with theory, and one paper is on a cooperative two-person game which is not a prisoner's dilemma. Among the papers on n-person games, two papers deal with empirical tests of solutions of n-person games, three papers are purely theoretical, and one paper is on metagame theory. The contributors are T. Burns, M. Freimer, A.D. Horowitz, J.P. Kahon, D.M. Kilgour, J.D. Laing, L.D. Meeker, R.J. Morrison, J. Perner, A. Rapoport, C.S. Thomas, and P.L. Yu.

Rappaport, Alfred, ed. INFORMATION FOR DECISION MAKING: QUANTITATIVE AND BEHAVIORAL DIMENSIONS. Englewood Cliffs, N.J.: Prentice-Hall, 1970. xi, 398 p. No index.

An anthology of thirty-three selections arranged in ten chapters, with editorial comment for each chapter. The selections, previously published in journals, are grouped under the following headings: an overview of information and decisions (four selections); budgeting and financial models (four selections); cost-volume-profit analysis (three selections); capital budgeting (three selections); costs: estimation, standard, and allocation (five selections); decentralization and performance evaluation (five selections); decentralization and transfer pricing (four selections); and behavioral aspects of information (five selections). There are no mathematical prerequisites.

Rima, Ingrid H., ed. A FORUM ON SYSTEMS MANAGEMENT. Foreword by S.L. Wolfbein. Philadelphia: Temple University, School of Business Administration, 1967. viii, 214 p. No index.

Sixteen nontechnical papers presented at a forum on systems management at Temple University on June 21-23, 1967. The forum was divided into the following topics: the challenge to systems management, systems management technology today, nondefense management application, technology development needs, and the research challenge. The purpose of the forum was to identify concepts and techniques of systems management.

Rosen, Judah Ben; Mangasarian, Olvi L.; and Ritter, K., eds. NONLINEAR PROGRAMMING. Foreword by J.B. Rosen. Publication No. 25 of the Mathematics Research Center, University of Wisconsin. New York: Academic Press, 1970. xii, 490 p. Index, pp. 487-90.

Twenty papers presented at a symposium on nonlinear programming held in Madison during May 4-6, 1970. Twelve papers are concerned primarily with computational algorithms; four papers are devoted to the theoretical aspects of nonlinear programming; and four papers represent applications in physics, statistics, and approximations. The areas covered include algorithms for nonlinear constraint problems, investigations of convergence scales, and the use of nonlinear programming for approximation. The contributors are I. Barrodale, R.H. Bartels, J.W. Daniel, R.J. Duffin, R. Fletcher, G.H. Golub, P. Huard, O. Krafft, C.E. Lemke, S.A. Lill, G.P. McCormick, R.R. Meyer, B. Mond, L.W. Neustadt, E. Polak, M.J.D. Powell, K. Ritter, F.D.K. Roberts, R.T. Rockafellar, M.A. Saunders, and G. Zoutendijk.

Ross, Miceal, ed. OPERATIONAL RESEARCH '72. New York: American Elsevier, 1973. xx, 732 p. Author Index, pp. 719-23. Subject Index, pp. 725-32.

Proceedings of the sixth International Federation of Operational Research Societies international conference on operational research held in Dublin, Ireland, during August 21-25, 1972. It includes three papers on critical path, operations research in environment problems, and operations in the USSR (Sir C. Goodeve, C.J. Hitch, and N.N. Moiseev); two papers read by representatives of allied organizations (IFIP and IFAC); and the contributions of eight workshop groups and general discussion forums; and twenty-eight national contributions dealing with such topics as news media, education, railroads, data processing, coal and power, health, communications, airlines, budgeting, and industry and defense. Three additional papers deal with recent developments of operations in the United States.

Saaty, Thomas L., and Weyl, F. Joachim, eds. THE SPIRIT AND THE USES OF THE MATHEMATICAL SCIENCES. New York: McGraw-Hill Book Co., 1969. x, 301 p.

Fifteen papers on the uses of mathematics in the sciences, grouped under the following headings: a basic form of creative thought, a medium for understanding nature, the challenge of living structures, and philosophical foundations of the mathematical mind. Included are the following articles: "Operation Research," by N. Dalkey; "Mathematics in Economic Analysis," by Wm. J. Baumol; and "Uses and Limitations of Mathematical Models in Social Sciences," by A. Rapoport.

Scarf, Herbert E.; Gilford, D[orothy].M.; and Shelly, M[aynard].W. II, eds. MULTISTAGE INVENTORY MODELS AND TECHNIQUES. Office of Naval Research Monographs on Mathematical Methods in Logistics. Stanford, Calif.: Stanford University Press, 1963. vii, 225 p.

A collection of seven papers on dynamic inventory problems and models which examine the distribution of inventories among a number of interrelated agencies.

Shubik, Martin, ed. GAME THEORY AND RELATED APPROACHES TO SOCIAL BEHAVIOR. Huntington, N.Y.: Krieger, 1964. xi, 390 p. Bibliography, pp. 363-76. Index, pp. 377-90.

An anthology of twenty-three essays (previously published) prefaced by the author's seventy-five-page introduction and followed by an annotated bibliography of 148 items. The essays are grouped under the following headings: political choice, power, and voting; bargaining, threats, and negotiations; and games (psychological, sociological, and political). Among the contributors are K.J. Arrow, R. Aumann, D. Black, H. Goldhamer, J. Harsanyi, M. Hausner, F. Ikle, A. Kaplan, D. Luce, O. Morgenstern, H. Raiffa, T. Schelling, L.S. Shapley, J. von Neumann, and K.V. Wilson.

_____. READINGS IN GAME THEORY AND POLITICAL BEHAVIOR. Editor's foreword by R.C. Snyder. Doubleday Short Studies in Political Science. Garden City, N.Y.: Doubleday & Co., 1954. xiv, 74 p. Selected Readings, pp. 73-74. No index.

A collection of ten nontechnical articles previously published in journals and elsewhere on the applications of game theory in political science. An introduction to the nature of game theory, by M. Shubik, is also included. Topics include military operations and games; rankings according to a preference scale; zero-sum two-person and n-person games with examples; international politics and game theory; the Colonel Blotto problem; and the unity of political science and economic science. The authors are K.J. Arrow, K.W. Deutsch, A. Kaplan, J. McDonald, J. Marshak, O. Morgenstern, J. von Neumann, M. Shubik, J.W. Tukey, and A. Wald. This collection is useful as a supplement to a first course in game theory.

Shuchman, Abraham, ed. SCIENTIFIC DECISION-MAKING IN BUSINESS: READINGS IN OPERATIONS RESEARCH FOR NONMATHEMATICIANS. New York: Holt, Rinehart and Winston, 1963. viii, 568 p. Index, pp. 562-68.

A collection of fifty-one articles that describe the aims, methods, and tools of management science without recourse to technical language or complex mathematical symbolism. The articles are grouped under the following headings: what is operations research?; the methodology of operations research: models and model building; the methodology of operations research: techniques; and some applications of operations research. It is useful as a supplemental reading text for courses in operations research and management science, and as a reference for executives who have little knowledge of mathematics and statistics.

Shuman, Larry J.; Speas, R. Dixon, Jr.; and Young, John P., eds. OPERATIONS RESEARCH IN HEALTH CARE: A CRITICAL ANALYSIS. Baltimore: Johns Hopkins University Press, 1975. xxvii, 433 p. Index, pp. 423-33.

A collection of sixteen heretofore unpublished articles on the philosophies and strategies of implementing operations research techniques in the health services field. The papers are grouped under the following three headings: the decision-maker's perspective (three articles); the researcher's perspective (three articles); and the state of the arts and potential (ten articles). There are no mathematics prerequisites for most of the papers. Topics include obstacles to the application of operations research; the influence of operations research; review of effectiveness measures; and the roles of industrial engineering, simulation, mathematical programming, cybernetics, stochastic processes, and information systems. The contributors are J.W. Bush, M.M. Chen, E.J. Connors, P.M. Densen, P.H. Diehr, A.O. Esogbue, L. Fisher, C.D. Flagle, J.R. Freeman, J. Goldman, R.L. Gue, Wm. J. Horvath, D. Howland, H.D. Kahn, I.M. King, R.A. Kronmal, E. Levine, L.R. Pondy, A. Sheldon, L.J. Shuman, R.D. Speas, Jr., D. Valinsky, H. Wolfe, J.P. Young, and J. Zaremba.

Smith, Walter L., and Wilkinson, William E., eds. CONGESTION THEORY. The University of North Carolina Monograph Series in Probability and Statistics. Chapel Hill: University of North Carolina Press, 1965. xv, 457 p. No index.

A collection of fourteen papers presented at the symposium on congestion theory held at the University of North Carolina on August 24-26, 1964. The papers treat such topics as the role of Green's function in congestion theory, networks of queues, divergent single server queues, Markovian queues, application of ballot theorems in the theory of queues, the use of the method of collective risks in queueing theory, and departure processes. The authors are D.R. Cox, D.P. Gaver, Jr., C.R. Heathcote, J. Keilson, J.F.C. Kingman, E.S. Page, F. Pollaczek, N.U. Prabhu, E. Reich,

J.Th. Runnenburg, T.L. Saaty, R. Syski, L. Takács, and G.H. Weiss.

Society for Industrial and Applied Mathematics, ed. STUDIES IN APPLIED MATHEMATICS. Vol. 3. Philadelphia: Society for Industrial and Applied Mathematics, 1969. v, 141 p. No index.

Ten papers presented at the symposium on applied probability and Monte Carlo methods and modern aspects of dynamics sponsored by the Air Force Office of Scientific Research at the 1967 national meeting of SIAM in Washington, D.C. Among the applications discussed are: molecular, linear transport of elementary particles, neutronics calculations, quantum statistics, classical fluids, the dynamics of variable stars, and rotating fluids. This book is useful primarily to applied mathematicians. The authors are D.L. Bunker, R.E. Christy, J.M. Cook, R.R. Coveyou, L.D. Fosdick, H.P. Greenspan, M. Leimdörfer, J.B. Parker, L. Verlet, and S.K. Zaremba.

_____. STUDIES IN OPTIMIZATION I. Philadelphia: Society for Industrial and Applied Mathematics, 1970. vi, 137 p. Abstract: The Development of Models of Ecological Processes, p. 137.

Ten papers presented at a symposium on optimization held in Toronto June 11-14, 1968, during the national meeting of SIAM. The purpose of the symposium was to bring together researchers in the fields of linear, nonlinear, and discrete and stochastic programming, and to present applications in the fields of economics, biology, and related fields.

Stilian, Gabriel, et al., eds. PERT: A NEW MANAGEMENT PLANNING AND CONTROL TECHNIQUE. Foreword by A. Appley. New York: American Management Association, 1962. 192 p. Bibliography, pp. 183-88. Index, pp. 189-92.

Fifteen essays heretofore unpublished on the methodology and concepts of systems which lend themselves to the PERT technique. The essays are grouped under the following headings: PERT and the manager, PERT theory, practical experience with PERT, and allied techniques. Three essays are written by G. Stilian and one each by the following authors: E.T. Alsaker, Wm. Bloom, E.O. Codier, W. Cosinuke, H.G. Francis, L.P. Harting, G.T. Hunter, A. McHugh, D.G. Malcolm, G.T. Mundoerff, and K.M. Tebo.

Taub, Abraham Haskel, ed. STUDIES IN APPLIED MATHEMATICS. Studies in Mathematics, vol. 7. Englewood Cliffs, N.J.: Prentice-Hall, 1971. xv, 217 p. Index, pp. 213-17.

Seven papers on mathematical model building in the physical

sciences. Among these is one paper by J. von Neuman which deals with model creation, mathematical analysis, and interpretation in the structure of galaxies, and another paper by D. Greenspan on the impact of computers on applied mathematics.

Thornley, Gail, ed. CRITICAL PATH ANALYSIS IN PRACTICE. New York: Tavistock Publications, 1968. xii, 152 p. Appendix 1: Glossary of Terms and Symbols, pp. 137-39. Appendix 2: Selected Reading List, pp. 140-42. Appendix 3: Survey of Computer Programs Available in Britain, pp. 143-48. Subject Index, pp. 149-51. Name Index, p. 152.

A collection of twenty papers on critical path analysis presented at meetings of the critical path analysis study group of the Operational Research Society in Britain during the period 1963-68. The papers are grouped under the following headings: review of basic CPA methods (three papers), early problems of implementation (three papers), further aspects of the technique (four papers), large networks (six papers), and alternative approaches to network problems (two papers). There are no mathematical prerequisites.

Tucker, A.W., and Luce, R. Duncan, eds. CONTRIBUTIONS TO THE THEORY OF GAMES. Vol. 4. Annals of Mathematics Studies, no. 40. Princeton, N.J.: Princeton University Press, 1959. viii, 453 p. A Bibliography of Game Theory, pp. 407-53.

A collection of nineteen papers on n-person games by the following authors: R.J. Aumann, B.R. Gelbaum, D.B. Gillies, J.H. Griesmer, H.M. Gurk, J.C. Harsanyi, J.R. Isbell, G.K. Kalisch, J.G. Kemeny, R.D. Luce, W.H. Mills, E.D. Nering, J. von Neumann, L.S. Shapley, M. Shubik, and W. Vickrey.

Van Rootselaar, B., ed. ANNALS OF SYSTEMS RESEARCH. Vol. 2, 1972. Publication of the Netherlands Society for Systems Research. Leiden, the Netherlands: H.E. Stenfert Kroese, 1973. vii, 139 p. No index.

Eight heretofore unpublished papers on systems analysis dealing with such topics as network and bond graphs in engineering modeling, analytical methods in information systems, a social planning model for a less developed economy, a general systems model concept, and the systems approach in political science, sociology, and biology. Familiarity with the techniques of operations research, and knowledge of calculus and set theory is assumed for some of the papers. The authors are S. Cohen, J.J. van Dixhoorn, A. Hovaguimian, G.P. Noordzij, L.U. de Sitter, N.W. de Smit, F.W. Umbach, and A.A. Vereen.

Veinott, Arthur F., Jr., ed. MATHEMATICAL STUDIES IN MANAGEMENT SCIENCE. Foreword by C.W. Churchman. New York: Macmillan, 1965. xiv, 481 p. No index.

A collection of thirty-seven reprints of articles from MANAGEMENT SCIENCE from 1956 to 1962 arranged under the following two main groups: deterministic decision models and stochastic decision models. Within the first group, the articles are classified under the following headings: transportation and network problems, topics in linear programming, quadratic programming, production and inventory control, and other models. Within the second group the articles are arranged under two headings: programming under uncertainty, and inventory models. Assumes knowledge of operations research techniques, advanced calculus, and matrix algebra.

Westcott, John Hugh, ed. AN EXPOSITION OF ADAPTIVE CONTROL. New York: Macmillan, 1962. vi, 135 p. Subject Index, p. 135.

Ten papers presented at a symposium held at the Imperial College of Science and Technology, London, in 1961. The topics of the papers include extremum-seeking regulators, quantization, perturbation measurement and control, optimum nonstationary filters, optimal trajectory problems, use of dynamic programming in optimization, and the design of a self-optimization control system. The contributors are P.F. Blackman, J.J. Florentin, S.M. Lyle, M.J. McCaim, D.Q. Mayne, J.D. Pearson, A.P. Roberts, B. McA. Sayers, D.G. Watts, and J.H. Westcott.

Williams, Thomas H., and Griffin, Charles H., eds. MANAGEMENT INFORMATION: A QUANTITATIVE ACCENT. The Willard J. Graham Series in Accounting. Homewood, Ill.: Richard D. Irwin, 1967. xv, 710 p. No index.

A collection of fifty-one articles which the editor feels will help bridge the gap between management science and managerial accounting. The articles, all previously published, are grouped under the following headings: the process of measurement (four articles), valuation of business resources (eight articles), analysis of the distribution function (seven articles), production planning and control (seven articles), inventory control (five articles), operations budgeting (six articles), capital budgeting (five articles), integrated planning models (six articles), and performance review (three articles). A background in elementary calculus, probability, and matrix algebra is helpful for understanding some of the articles.

Wismer, David A., ed. OPTIMIZATION METHODS FOR LARGE-SCALE SYSTEMS WITH APPLICATIONS. New York: McGraw-Hill Book Co., 1971. xii, 335 p.

Eight contributions, each one comprising one chapter, by different authors (J.D. Schoeffler contributed two chapters) on the theory and applications of large-scale systems. Proofs of theorems are omitted, and the theory-oriented chapters contain numerous examples and emphasize computational algorithms. Topics include multilevel systems and decomposition, static and dynamic system optimization, dynamic and trajectory decomposition, distributed and on-line multilevel systems, and aggregation. Assumes knowledge of the calculus of variations. The chapters conclude with references, and some chapters have appendixes. The contributors are M. Aoki, E.J. Bauman, J.D. Schoeffler, A.M. Geoffrion, G.B. Dantzig and M.M. Van Slyke, J.D. Pearson, and D.A. Wismer.

Wold, Herman O.A., ed. BIBLIOGRAPHY ON TIME SERIES AND STOCHASTIC PROCESSES. Cambridge, Mass.: MIT Press, 1966. xv, 516 p.

First published in 1965 for the International Statistics Institute by Oliver and Boyd (London), this bibliography contains about 10,000 entries arranged in the following three periods: before 1940, 1931-50, and 1951-59. The entries are annotated by a code system which provides six types of information: type of stochastic process, scientific nature of entry, group of problems, presence of empirical application, field of application, and language. This bibliography is part of the Teaching Aids Program of the International Statistical Institute (ISI) under sponsorship of UNESCO.

Wortman, Max S., Jr., and Luthans, Fred, eds. EMERGING CONCEPTS IN MANAGEMENT: PROCESS, BEHAVIORAL, QUANTITATIVE, AND SYSTEMS. New York: Macmillan, 1969. xiii, 1969. 462 p. Index, pp. 451-62.

A collection of forty-six reprints from journals which emphasize concepts and theories of management rather than specific techniques. The articles were written during the period 1964-68 and are grouped under the following headings: foundations of management, process approaches, behavioral approaches, quantitative approaches, and management in the future.

Zadeh, Lotfi A.; Neustadt, Lucien W.; and Balakrishnan, A.V., eds. COMPUTING METHODS IN OPTIMIZATION PROBLEMS--2. New York: Academic Press, 1969. ix, 393 p. No index.

A collection of thirty-one papers presented at an international conference on computing methods in optimization problems held in San Remo, Italy, on September 9-13, 1968. The conference emphasized computational methods, and included topics such as optimal control and trajectory problems; mathematical programming; economic, meteorology, biomedicine, and related areas; identification and inverse problems; computational aspects of decoding and information retrieval problems; and pattern recognition problems. Among the contributors are M. Aoki, M. Becker, J.M. Danskin, M.R. Hestenes, H.W. Kuhn, E. Polak, and L.W. Taylor.

Zelen, Marvin, ed. STATISTICAL THEORY OF RELIABILITY. Publication No. 9 of the Mathematics Research Center, University of Wisconsin. Madison: University of Wisconsin Press, 1963. xvii, 166 p. No index.

A collection of six papers presented at an advanced seminar which was sponsored by the Mathematics Research Center, University of Wisconsin, on May 8-10, 1962. The titles of the papers are as follows: "A Survey of Some Mathematical Models in the Theory of Reliability," by G.H. Weiss; "Redundancy for Reliability Improvement," by F. Proschan; "Maintenance and Replacement Policies," by R.E. Barlow; "Optimum Checking Procedures," by L.C. Hunter; "Confidence Limits for the Reliability of Complex Systems," by J.R. Rosenblatt; and "Problems in System Reliability Analysis," by Wm. Wolman.

APPENDIXES

Appendix A
METHODOLOGY

Ackoff, Russell L., with Gupta, S[hiv].K., and Minas, J.S. A CONCEPT OF
CORPORATE PLANNING. New York: Wiley-Interscience, 1970. ix, 158 p.
Appendix: Models, pp. 139-45. References and Bibliography, pp. 146-54.
Author Index, p. 155. Subject Index, pp. 157-58.

> An exposition of the objectives and logic of the corporate planning
> process. It stresses the questions: What should be done? Who
> should do it? and Why?

_____. SCIENTIFIC METHOD: OPTIMIZING APPLIED RESEARCH DECISIONS.
New York: Wiley, 1962. xii, 464 p. Appendix 1: Derivation of Min $M(d(y))$
Equation (sixteen) in chapter 8, pp. 447-49. Appendix 2: Analytical Derivation
of Decision Functions for a Simple Inventory Model, pp. 450-56. Author Index,
pp. 457-60. Subject Index, pp. 461-64.

> Presents the philosophy of the scientific method and optimization
> techniques with emphasis on planning the use of science in the
> pursuit of objectives without specifying the objectives. This book
> consists of fifteen chapters on definitions of problems and concepts,
> sampling, estimation, testing hypotheses, correlation, optimization,
> and organizing research. Chapters conclude with bibliographies.

Beer, Stafford. CYBERNETICS AND MANAGEMENT. New York: Wiley-
Interscience, 1959. xviii, 214 p. References, pp. 209-11. Index, pp. 212-14.

> An exposition of cybernetics for those interested in the problems of
> control. Topics include basic notions, the logical theory of cyber-
> netics, the biophysical theory of cybernetics, and the analogue
> theory of cybernetics.

_____. DECISION CONTROL: THE MEANING OF OPERATIONAL RESEARCH
AND MANAGEMENT CYBERNETICS. New York: Wiley-Interscience, 1966.
xii, 556 p. Index, pp. 551-56.

> A nonmathematical presentation of the role of science in manage-
> ment. It is divided into the following four parts: the nature of

operational research, the activity of operational research, the relevance of cybernetics, and outcomes.

Bellman, Richard E., and Smith, Charlene Paule. SIMULATIONS IN HUMAN SYSTEMS: DECISION-MAKING IN PSYCHOTHERAPY. Publications in Operations Research, no. 22. New York: Wiley-Interscience, 1973. xviii, 205 p. Author Index, pp. 197-201. Subject Index, pp. 202-5.

An exposition of decision making in human systems which assumes no previous familiarity with computers, simulation, psychotherapy, or systems theory. It discusses in a nonmathematical manner the use of mathematical concepts in decision making and it also provides an introduction to the technique of simulation. Topics include basic concepts of psychotherapy, taxonomy of systems, methodology. The book is aimed at a mixed audience, including engineers, mathematicians, educators, students of psychotherapy, and others interested in the analysis of any purposeful two-person interaction. The chapters conclude with notes and references.

Blalock, Hubert M., Jr. CAUSAL INFERENCES IN NONEXPERIMENTAL RESEARCH. Chapel Hill: University of North Carolina Press, 1964. xii, 193 p. Appendix: Some Related Approaches, pp. 191-93. Index, pp. 197-200.

A largely nonmathematical presentation of material on causal inferences appearing in philosophical, statistical, and social science literature. Chapters 1 to 3 present a discussion of concepts, such as experimental vs. nonexperimental inference, prediction vs. causal inference, multicollinearity, and the epistemological aspects of model building and causality. Chapter 4 deals with the distinction between correlation and regression coefficients with regard to grouping of observations, and chapter 5 gives various devices for dealing with "unclean data" so as to reduce or eliminate various types of confounding factors. Chapter 6 presents the summary and conclusions. Assumes knowledge of regression analysis.

Blin, J.M. PATTERNS AND CONFIGURATIONS IN ECONOMIC SCIENCE: A STUDY OF SOCIAL DECISION PROCESSES. Boston: Reidel, 1973. xv, 148 p. Bibliography, pp. 136-38. Appendixes, pp. 139-48. No index.

A pioneering effort in the application of pattern recognition theory, frequently used in electrical engineering and computer science, to economics and social theory. The five chapter titles are: "Welfare Economics and Public Decision Making," "Constitutional Choice and Majority Voting," "Optimization of Public Decisions in the Large: a Pattern Recognition Approach," "Algebraic Foundations of the Theory of Aggregation," and "Optimization of Public Decisions: New Results in the Theory of Aggregation."

Bose, Ashalata. THE METHODOLOGY AND SCOPE OF ECONOMICS. Calcutta, India: Bookland Private, 1970. 73 p. References, pp. 71-73.

A discussion of the scope and method of economics in the following three headings: positive-normative issues in economics with emphasis on ethico-social values, the scientific method as applied to economics, and mathematical and statistical methods in social accounting. The chapters conclude with references.

Box, George E.P., and Draper, Norman R. EVOLUTIONARY OPERATION: A STATISTICAL METHOD FOR PROCESS IMPROVEMENT. Wiley Series in Probability and Mathematical Statistics. New York: Wiley, 1969. xi, 237 p. Appendixes, pp. 196-215. Tables, pp. 218-26. References and Bibliography, pp. 227-31. Index, pp. 233-37.

A discussion of the philosophy and practice of evolutionary operations (EVOP) for chemists, engineers, foremen, and process superintendents responsible for operating industrial processes. Topics include the basic ideas, statistical concepts, factorial designs, and optimization. The four appendixes expand on the material, and knowledge of elementary statistics is assumed.

Braithwaite, Richard B. THEORY OF GAMES AS A TOOL FOR THE MORAL PHILOSOPHER. New York: Cambridge University Press, 1955. 76 p. Notes, pp. 59-76. Appendix: Algebraic Treatment of a General Two-person Collaboration Situation, pp. 59-76. No index.

A brief presentation of the philosophical aspects of the two-person game.

Churchman, Charles West. PREDICTION AND OPTIMAL DECISION: PHILOSOPHICAL ISSUES OF A SCIENCE OF VALUES. Prentice-Hall International Series in Management. Englewood Cliffs, N.J.: Prentice-Hall, 1961. xiv, 394 p. Index, pp. 385-94.

A presentation of the methodological problems of determining human values. The author discusses a wide range of topics, including utility, the teleology of measurement, probability, rational behavior, additivity of values, social groups, decision methods in science, and ethics. The chapters conclude with references.

Easton, Allan. COMPLEX MANAGERIAL DECISIONS INVOLVING MULTIPLE OBJECTIVES. Wiley Series in Management and Administration. New York: Wiley, 1973. xxii, 421 p. Appendix: Complex Cases Involving Multiple Objectives, pp. 363-413. Index, pp. 415-21.

An attempt to bridge the gap between theory and practice in decision making. It presents a number of decision/evaluation/rating models with variations and suggestions for use. The book is divided into the following four parts: introduction and background, decision elements and models, treatment of decision alternatives, and synthesis. Each chapter concludes with notes and references, a list of important concepts, overviews, questions for review and discussion, and suggested research topics. There are no mathematical prerequisites.

Elbing, Alvar O. BEHAVIORAL DECISIONS IN ORGANIZATIONS. Foreword by C.E. Summer. Glenview, Ill.: Scott, Foresman, 1970. xvi, 879 p. Readings on the Value Issue, pp. 790-870. Index, pp. 871-79.

Presents a step-by-step decision-making framework for those organizational problems which involve human factors. It is divided into two parts as follows: the model of the decision process, and the collection of readings. Many case studies are included, and there are no mathematics prerequisites.

Ellman, Michael. PLANNING PROBLEMS IN THE USSR: THE CONTRIBUTION OF MATHEMATICAL ECONOMICS TO THEIR SOLUTION, 1960-1971. University of Cambridge Department of Applied Economics Monograph 24. New York: Cambridge University Press, 1973. xx, 222 p. Glossary, pp. ix-xii. Dramatis Personae, pp. xiii-xviii. Important Dates, pp. xix-xx. Bibliography, pp. 191-217. Index, pp. 219-22.

A critical exposition of the contribution to overcoming the problems of the Soviet planning system made by the Soviet school of mathematical economists during the 1960s. Emphasis is given to the contribution of the schools of Kantorovich and Fedorenko to the problems of improving the planning and management of the Soviet economy. The main tool of analysis is linear programming.

_____. SOVIET PLANNING TODAY. PROPOSALS FOR AN OPTIMALLY FUNCTIONING ECONOMIC SYSTEM. University of Cambridge Department of Applied Economics Occasional Paper no. 25. New York: Cambridge University Press, 1971. xv, 219 p. References, pp. 194-219. No index.

This book sets forth the proposal of the Central Economic Mathematical Institute of the USSR Academy of Sciences for an optimally functioning economic system with recognition of capital intensity as a factor in price formation, the introduction of payments for the use of natural resources, and the development of wholesale trade. The techniques for planning include input-output analysis and linear programming. It is divided into the following three parts: chapters 1 to 5 explain the theory of the optimally functioning socialist economy. Chapters 6 and 7 describe some of the nonoptimalities of the existing planning system. Chapters 8 to 10 present the evaluation of the theory of the optimally functioning socialist economy as a guide to the solution of the problems of the Soviet economy. The topics are treated conceptually, and exposure to matrix algebra is helpful.

Fox, Karl A.; Sengupta, Jati K.; and Thorbecke, Erik. THE THEORY OF
QUANTITATIVE ECONOMIC POLICY WITH APPLICATIONS TO ECONOMIC
GROWTH, STABILIZATION AND PLANNING. 2d ed., rev. Studies in Mathematical and Managerial Economics, vol. 5. New York: American Elsevier,
1973. xviii, 620 p. Index, pp. 608-20.

A review of a wide range of economic models and economic problems, heretofore presented in isolation in journals and elsewhere,
from a policy point of view. It is divided into the following two
parts: the theory of quantitative economic policy; and applications to economic growth, stabilization, and planning. The first
part, which can be used as a text for a graduate-level course,
examines Tinbergen's approach to the theory of economic policy,
estimation of quantitative policy models, forecasting and decision
rules, programming aspects of economic policy models, control
theory approach with applications to optimal growth using Pontryagin's
maximum principle, optimal and adaptive control methods, and
decision making under risk and uncertainty using linear programming models. The second part, which is intended to supply motivation and empirical content to the first part, discusses the Klein-
Goldberger model (1955), the Brookings Quarterly Econometric
Model of the United States (1965), present status and prospects
for large-scale stabilization models of the United States as of
1971, models for regional growth and development planning, the
national planning models of the Netherlands, Japan, and France,
and the agricultural sector and economic development. Assumes
knowledge of econometrics, matrix algebra, and graduate economic
theory. Chapters conclude with references.

Georgescu-Roegen, Nicholas. THE ENTROPY LAW AND THE ECONOMIC
PROCESS. Cambridge, Mass.: Harvard University Press, 1971. xv, 475 p.
Appendixes, pp. 367-438. Index, pp. 439-57.

Develops the idea that economic behavior can be effectively described in terms of the entropy law of thermodynamics. After dis-
cussing the advantages of this approach to economics, the author
considers some general economic and social issues that can be effectively analyzed. The presentation is nonmathematical, but some
degree of mathematical maturity is helpful for comprehension. The
appendixes are devoted to the arithmetic continuum, ignorance and
information, a model for Boltzmann's H. theorem, and limitations
and extrapolations in biology.

Grubbstrom, Robert W. MARKET CYBERNETIC PROCESSES. Stockholm: Almqvist
and Wiksell, 1969. 358 p. References, pp. 348-58.

Presents decision processes of consumers, producers, and investors
in terms of cybernetic notions, while pointing out relationships
between mechanical and economic systems. The theory of automatic control is used to derive quantitative expressions for the
behavior of these three groups. Assumes knowledge of advanced
calculus.

471

Gupta, Syama Prasad. PLANNING MODELS IN INDIA: WITH PROJECTIONS TO 1975. Foreword by H.G. Johnson. Praeger Special Studies in International Economics and Development. New York: Praeger, 1971. xxv, 400 p. List of Tables, pp. xviii–xxiii. List of Abbreviations, p. xxv. Appendix A: A Feasibility and Consistency Check to the Fourth Plan, pp. 383–89. Bibliography, pp. 393–400. No index.

> A presentation of the author's experience in planning with the Indian Planning Commission in four parts. Part 1 identifies the explicit and implicit features of the Indian plans. Part 2 develops a multisectoral, intertemporal programming model for India containing both dynamic and recursive elements. Part 3 is devoted to the estimation of the technical and behavioral parameters of the model, and part 4 discusses the computational stages and economic implications of the model. The appendix demonstrates how the model can be revised in order to make it more operational. The models are based on regression theory rather than input-output analysis. The chapters conclude with notes and references.

Hamblin, Robert L.; Jacobsen, R. Brooke; and Miller, Jerry L. A MATHEMATICAL THEORY OF SOCIAL CHANGE. New York: Wiley-Interscience, 1973. xi, 237 p. References, pp. 217-27. Name Index, pp. 229-31. Subject Index, pp. 233-37.

> A text for graduate and undergraduate courses on social change in sociology, economics, public administration, geography, and political science. It presents a mathematical theory of social change and tests the theory using a wide range of empirical data. The main topics are types of diffusion, innovation theory, and socioeconomic development. Assumes knowledge of differential calculus.

Handy, Rollo. METHODOLOGY OF THE BEHAVIORAL SCIENCES: PROBLEMS AND CONTROVERSIES. American Lecture Series, no. 597. New York: Charles C. Thomas, 1964. xi, 182 p. Index, pp. 177-82.

> An intensive survey of the methodological issues in the behavioral science literature from about 1954 to 1964. It considers topics such as the role of mathematics and models, the scientific nature of various disciplines, the complexity of human behavior, and terminological problems.

Heal, Geoffrey M. THE THEORY OF ECONOMIC PLANNING. Advanced Textbooks in Economics, vol. 3. New York: American Elsevier, 1973. xiv, 409 p. Appendix A: Mathematical Appendix, pp. 347-88. Appendix B: A Fixed Proportions Technology, pp. 389-97. Bibliography, pp. 399-403. Index, pp. 405-9.

> An abstract presentation of the theoretical issues concerning the operation and properties of planned economic systems with little attention given to the behavior of individual units in the economy and the practical aspects such as input-output analysis and statistical

estimation. The two main issues considered are: short-term planning in which the price-guided procedures of the Lange-Arrow-Hurwicz and Malinvaud processes, as well as nonprice and mixed planning procedures, are considered; and long-term planning, in which preferences, objectives, characteristics, and existence problems of optimal planning are considered. Assumes knowledge of calculus.

Heesterman, A.R.G. ALLOCATION MODELS AND THEIR USE IN ECONOMIC PLANNING. Boston: Reidel, 1971. xiv, 203 p. Appendixes, pp. 184-200. Bibliography, pp. 201-3. No index.

An attempt to synthesize three different approaches to optimal planning: government, private enterprise, and welfare economics. Part 1 treats allocation, investment, and efficiency prices in input-output type models. Part 2 consists of an evaluation of individual projects by means of costing, discounted cash flow, and increasing returns to scale. Part 3 is on capita selecta on economic policy and takes up the distribution of output, opportunity cost and exchange price, and optimality conditions. The applications are in the fields of transport, education, and hospitals. The two appendixes are devoted to optimality conditions, such as the Kuhn-Tucker theorem for convex programming, and notation. Assumes knowledge of matrix algebra and differential calculus. The chapters conclude with questions and answers.

Hyvärinen, Lassi P. INFORMATION THEORY FOR SYSTEMS ENGINEERS. Econometrics and Operations Research, vol. 17. New York: Springer-Verlag, 1970. viii, 197 p. Appendix A: Fourier Transform and Related Concepts, pp. 139-51. Appendix B: Binary Logarithms and Entropies, p. 152. Appendix C: Computer Programs, pp. 152-57. Appendix D: Problems and Solutions, pp. 157-90. References, pp. 191-93. Subject Index, pp. 194-97.

A text for a course in information theory based on lectures given by the author at the IBM European Systems Research Institute in Geneva. It is an expanded and revised edition of the volume with the same title in the series, Lecture Notes in Operations Research and Mathematical Economics (1968), and it is divided into the following two parts: discrete information systems, and continuous information systems. Most of the problems in Appendix D are extensions of the text, and some require extensive numerical calculations by means of computer programs. Assumes knowledge of calculus, probability theory, and statistics.

Ivakhnenko, A.G., and Lapa, V.G. CYBERNETICS AND FORECASTING TECHNIQUES. Translated by Scripta Technica, and edited by R.N. McDonough. Modern Analytic and Computational Methods in Science and Mathematics. New York: American Elsevier, 1967. xxvii, 168 p. References, pp 163-66. Index, pp. 167-68.

Originally published in Russian in 1965, this book presents the
following three forecasting algorithms for large general-purpose
computers: forecasting of deterministic processes, i.e., extrapolation and interpolation; forecasting of stochastic processes based
on statistical forecasting theory; and forecasting based on adaptation
or learning of forecasting filters. The use of forecasting filters
for constructing a control system for periodic processes is also discussed. Examples are taken from the chemical industry, biology,
ocean turbulence processes, and forecasting of the relief of the
Dnieper river bottom. Assumes knowledge of advanced calculus
and probability theory.

Johnsen, Erik. STUDIES IN MULTIOBJECTIVE DECISION MODELS. Foreword
by C. Kihlstedt. Monograph no. 1. Lund, Sweden: Studentlitteratur, 1968.
628 p. Appendix A: Empirical Material for Chapter 2, pp. 567-92. References, pp. 593-612. Index, pp. 613-28.

An investigation of the problem of goal formulation in managerial
decision processes which draws heavily upon the works of many
researchers. It is a collection of cases from different scientific
disciplines. There are no mathematical prerequisites.

Kaufmann, Arnold. THE SCIENCE OF DECISION-MAKING: AN INTRODUC-
TION TO PRAXEOLOGY. Translated by R. Audley. World University Library.
New York: McGraw-Hill Book Co., 1968. 253 p. Bibliographical Refer-
ences, pp. 246-51. Index, pp. 252-53.

Translated from the French, this book develops a science of action
based on combinatorial methods rather than on continuous functions.
It considers many topics encountered in decision making, including
games, dynamic programming, cybernetics, simulation, sensitivity
analysis, and linear programming. The philosophical approach is
continuously emphasized. Assumes knowledge of college algebra.

Kibbee, Joel M.; Craft, Clifford J.; and Nanus, Burt. MANAGEMENT GAMES:
A NEW TECHNIQUE FOR EXECUTIVE DEVELOPMENT. Reinhold Management
Reference Series. New York: Reinhold, 1961. xii, 347 p. A Directory of
Management Games, pp 315-36. Bibliography, pp. 337-43. Index, pp. 345-47.

A nontechnical treatment of management games as an educational
technique rather than as a mathematical model. It is divided into
the following five parts: background: theory and practice, administration, game design, case studies, and references. There
are no mathematical prerequisites.

Klein, Lawrence R. AN ESSAY ON THE THEORY OF ECONOMIC PREDIC-
TION. Markham Economics Series. Chicago: Rand McNally, 1971. 140 p.
Index, pp. 135-40.

An essay based on the Yrjo Jahnsson lectures which the author

delivered at the University of Helsinki from April 1 to 5, 1968.
It deals with the measurement of prediction accuracy and prediction methodology, especially with regard to the improvement of
prediction accuracy in both linear and nonlinear econometric
models. The Wharton model forecasts of 1969 and new results on
statistical estimation to improve prediction are discussed. Assumes
knowledge of econometrics.

Koopmans, Tjalling C. THREE ESSAYS ON THE STATE OF ECONOMIC SCIENCE.
New York: McGraw-Hill Book Co., 1957. xi, 231 p. Index, pp. 221-31.

Presents mathematical and statistical concepts and tools used in
economic theory, economic observation, and measurement. The
manner of presentation is based on postulates, definitions, examples,
and theorems where intuitive and nonrigorous proofs are furnished.
Topics include convex sets, linear functions, Pareto optimality,
linear programming, input-output analysis, and production over
time.

Lange, Oskar Richard. POLITICAL ECONOMY. Vol. 1: GENERAL PROB-
LEMS. Translated by A.H. Walker. Elmsford, N.Y.: Pergamon Press, 1963.
xiv, 355 p. Index of Names, pp. 343-47. Subject Index, pp. 347-55.

Translated from the Polish, this book is a systematic monograph on
the general problems of political economy. Topics include the
subject matter of political economy, modes of production and
social formation, economic laws, the method of political economy,
the principle of economic rationality, the subjectivist and the
historical trends in political economy, and the social role of economic science. The presentation is nonmathematical except for
the eighteen-page appendix of chapter 5 which is devoted to the
mathematical foundations of programming.

_____. POLITICAL ECONOMY. Vol. 2. Edited by P.F. Knightsfield.
Translated by S.A. Klain and J. Stadler. Elmsford, N.Y.: Pergamon Press,
1971. xi, 249 p. Annex, pp. 203-21. Table of Contents to the Whole
Work and to Volume 2, pp. 222-39. Index of Names, pp. 241-43. Index
of Terms, pp. 244-49.

Originally published in Polish in 1968, volume 2 is a continuation
of the general problems of political economy begun in volume 1,
but from the point of view of methodology. It begins with a general discussion of the social process of production and reproduction,
and then develops production relationships, based on a matrix of
production techniques reflecting Marx's views on the renewal of
the means of production, labor productivity, and the equilibrium
conditions of reproduction. Assumes knowledge of differential and
integral calculus, and linear algebra. The mathematical appendix
to chapter 4 contains the equilibrium conditions of production.

———. WHOLES AND PARTS: A GENERAL THEORY OF SYSTEM BEHAVIOR.
Elmsford, N.Y.: Pergamon Press, 1965. v, 74 p.

A presentation of a system of behavior in terms of equations beginning with inputs and outputs and their linkage in the system. It
is useful as an introduction to a theory of social systems. Assumes
knowledge of calculus.

Lange, Oskar Richard, with Banasinski, A. INTRODUCTION TO ECONOMIC
CYBERNETICS. Translated by J. Stadler. Preface to English edition by A.
Zauberman. Elmsford, N.Y.: Pergamon Press, 1970. xv, 183 p. Bibliog-
raphy, pp. 175-77. Index of Names, pp. 179-80. Index of Terms, pp. 181-83.

Originally published in Polish in 1965, this book presents economic processes in terms of cybernetics and particularly its branch
called the theory of automatic control. The five chapter titles
are: "General Principles of Regulation and Control," "Cybernetic
Schemata of the Theory of Reproduction," "The Dynamics of Regulation Processes," "The Theory of Stability of Regulation Systems,"
and "A Generalization of the Theory of Regulation." Assumes
knowledge of matrix algebra and calculus.

Letwin, William. THE ORIGINS OF SCIENTIFIC ECONOMICS. Garden City,
N.Y.: Doubleday, 1964. viii, 345 p. Appendixes, pp. 249-323. Bibliography, pp. 325-30. Index, pp. 331-45.

An investigation of early scientific inquiry (including mathematical
inquiry) in economics in the following three parts: the old style,
including the inquiries of Sir Josiah Child and N. Bardon; the
new style, which deals with the scientific inquiry of the seven-
teeth century including those of J. Collins, Wm. Petty, J. Locke,
and D. North; and the legacy, including the economic theories
of the eighteenth century. The appendixes contain a description
of the work of J. Child, an estimate of the change in the money
supply in England during the period of recoinage, an early manu-
script by J. Locke, and a discussion of D. North's discourses upon
trade.

Machol, Robert E., and Gray, Paul, eds. RECENT DEVELOPMENTS IN IN-
FORMATION AND DECISION PROCESSES. New York: Macmillan, 1962.
x, 197 p. Index, pp. 195-97.

Twelve papers presented at a symposium on information and decision processes held at Purdue University in April 1961. The contributions are mainly exploratory in nature or reviews of developments in diverse fields. Topics include deferred decision theory,
Bayesian statistics, ergodic theory, entropy, mathematics of self-
organizing systems, theory of Markov chains, dynamic programming,
machine generation of mathematical theorems, and axioms of consumer behavior. Some of the papers assume knowledge of mathematical statistics. The contributors are R.E. Bellman, K.L. Chung,

P.A. Diamond, B. Dunham, R. Fridshal, H.H. Goode, T.C. Koopmans, S. Moriguti, J.H. North, H. Raiffa, H. Robbins, E. Samuel, L.J. Savage, M. Tribus, N. Wiener, and R. Williamson.

Magill, M.J.P. ON A GENERAL ECONOMIC THEORY OF MOTION. Lecture Notes in Operations Research and Mathematical Systems, vol. 36. New York: Springer-Verlag, 1970. vi, 95 p. Bibliography, pp. 91-95.

A development of a general economic theory of motion based on the theories of Ramsey, Fisher, von Neumann, and Samuelson, and on the methods of analytical mechanics introduced by Euler, Lagrange, and Hamilton. Provided certain assumptions are satisfied, the author claims that a general method of economic analysis is possible which renders unnecessary the subdivision of economics based upon the method of analysis. Assumes knowledge of advanced calculus, mathematical analysis, and differential equations.

Mennes, L.B.M. PLANNING ECONOMIC INTEGRATION AMONG DEVELOPING COUNTRIES. Rotterdam: Rotterdam University Press, 1973. xii, 153 p. Appendix A: Investments and Increases in Production, pp. 144-47. References, pp. 148-50. Index, pp. 151-53.

A quantitative approach to economic integration among developing countries from the point of view of planning.

Miller, David W., and Starr, Martin K. EXECUTIVE DECISIONS AND OPERATIONS RESEARCH. 2d ed., rev. Prentice-Hall International Series in Management. Englewood Cliffs, N.J.: Prentice-Hall, 1972. xvi, 607 p. Bibliography, pp. 589-94. Index, pp. 595-607.

An examination of the structure of decision problems from the viewpoint of an integrated theory of decisions using only elementary arithmetic. It is divided into the following five parts: organizations and decisions, the theory of decision, the nature of models, decision-problem paradigms, and the executive and operations research. The applications are in the areas of production, marketing, and finance. It is concerned with the role of operations research techniques, such as linear programming, in the decision process, rather than on their mathematical development. The chapters conclude with problems.

_____. THE STRUCTURE OF HUMAN DECISIONS. Englewood Cliffs, N.J.: Prentice-Hall, 1967. 179 p. Bibliography, pp. 173-74. Index, pp. 175-79.

An examination of the structure of decision problems by means of a special classification scheme. Topics include responsiblity for decisions, objectives of decisions, and the structure and analysis of decisions with illustrations from games and decision theory. There are no mathematical prerequisites. The specific techniques of operations research, cybernetics, and other new methodologies are not taken up. The chapters conclude with problems.

477

Moerman, P.A. METHODICAL TACTICAL PLANNING. Preface by P.A. Verheijen. Tilburg Studies on Economics, no. 4. Rotterdam: Rotterdam University Press, 1971. xv, 178 p. Appendixes, pp. 157-78. No index.

A synthesis of the views and techniques of mathematical operations research with applications to problems of commercial and production planning. It deals with problems such as how to translate the demand for a good into optimal levels of production and stocks with dynamic programming playing a partiularly important role. Assumes knowledge of calculus, matrix algebra, and probability theory. The appendixes contain a list of symbols, additional mathematical results, and flow diagrams for computer programs. The chapters conclude with references.

Montias, Michael. CENTRAL PLANNING IN POLAND. Yale Studies in Economics, 13. New Haven, Conn.: Yale University Press, 1962. xv, 410 p. Appendix A: Planning by Successive Approximation: Consistency Problems, pp. 335-48. Appendix B: The Efficiency Problem, pp. 349-70. Appendix C: Diagrams, pp. 371-74. Bibliography, pp. 375-92. Index of Names Cited, pp. 393-95. Subject Index, pp. 396-410.

A discussion of some of the theoretical and practical problems in the application of central planning in Poland between 1945 and 1961. Topics include the distribution of producer goods; employment, wages, and synthetic balances; origins and development of the price system; and the pricing of producer goods. This book is useful to model-builders and model-users because it sheds light on the problems of resource allocation in a planned economy and the contributions that econometric methods can make in solving these problems.

Morgenstern, Oskar. ON THE ACCURACY OF ECONOMIC OBSERVATIONS. 2d ed., rev. Princeton, N.J.: Princeton University Press, 1963. xiv, 322 p. Bibliography, pp. 307-15. Index, pp. 317-22.

An analysis of measurement errors of a broad range of economic statistics: foreign trade, prices, mining, agriculture, employment and unemployment, national income, and growth. One method of error measurement consists of the comparison of the same economic phenomenon with different sources of statistical data, and the second method is the revision over time of a given economic phenomenon prepared by the same source. Although the presentation is void of equations and mathematical notation, knowledge of statistical inference and some degree of mathematical maturity are helpful.

Morris, William T. MANAGEMENT SCIENCE IN ACTION. Irwin Series in Quantitative Analysis for Business. Homewood, Ill.: Richard D. Irwin, 1963. xi, 308 p. Index, pp. 303-8.

A nontechnical discussion of the role of science in business decisions.

National Academy of Science. THE MATHEMATICAL SCIENCES: A REPORT. Publication 1681. Washington, D.C.: National Academy of Sciences, 1968. xiv, 256 p. References, pp. 221-23. Appendixes, pp. 227-56.

A report on the present status and projected future needs, especially fiscal needs, of the mathematical sciences. Of particular interest is the section entitled "The State of the Mathematical Sciences," which includes a treatment of the uses of mathematics in several fields, including economics.

Nyblen, Goran. THE PROBLEM OF SUMMATION IN ECONOMIC SCIENCE: A METHODOLOGICAL STUDY WITH APPLICATIONS TO INTEREST, MONEY AND CYCLES. Lund, Sweden: C.W.K. Gleerup, 1951. xii, 289 p. List of References, pp. 283-87. Index of Names, pp. 288-89.

An investigation of two summation theorems: (1) von Neumann's and Morgenstern's theory of games which states that economic values are in general nonadditive; and (2) Arrow's theory, which states that it is impossible to aggregate many individual value schemes into one social welfare scheme. Chapters 1 to 4 deal with the methodology of the analysis, and chapters 5 to 7 are devoted to the application portion, taking up the empirical problems of interest rates, inflation, and business cycles. The problem of summation is viewed as being distinct from the problem of aggregation in economics. Assumes knowledge of elementary calculus.

Oparin, D. Ivanovich. MULTI-SECTOR ECONOMIC ACCOUNTS. Edited by K.J. Lancaster. Translated by P.F. Knightsfield. New York: Macmillan, 1963. ix, 70 p. Bibliography, pp. 67-68. Index, pp. 69-70.

Originally published in Russian, this book presents three models of expanding reproduction in an economy. Part 2 presents Karl Marx's scheme of expanding reproduction, part 3 deals with a model of production and consumption turnover of material goods in the course of their expanding reproduction, and part 1 presents the final model of expanding reproduction. Little mathematics is necessary for comprehension.

Papandreou, Andreas George. ECONOMICS AS A SCIENCE. New York: Lippincott, 1958. xi, 148 p. Index, pp. 147-48.

An examination of the nature of economics as a science and an exploration of its logical foundation. This methodological study makes extensive use of mathematical symbols and set theoretical concepts, and covers such topics as the concept of structure, the distinction between theories and models, and the extent of empirical relevance in economic laws. This book is useful as a supplement to courses in economic theory and mathematical economics.

Papandreou, Andreas George, and Zohar, Uri. PROJECT SELECTION FOR
NATIONAL PLANS. Vol. 1. National Planning and Socioeconomic Priori-
ties--A Two Volume Series. New York: Praeger, 1974. ix, 112 p. Appendix A: Income Generation, pp. 93-94. Appendix B: Profitability Analysis,
pp. 95-96. Appendix C: Breakeven Analysis, pp. 97-98. Appendix D: Internal Resource Utilization and Net Foreign Exchange Earnings, pp. 99-103.
Bibliography, pp. 105-7. Index, pp. 109-11. About the Author, p. 112.

A monograph which explores a method for evaluating the impact
of economic development projects in which there is a complete
social ranking of alternative states of the world. This method,
called the impact approach, goes beyond the cost-benefit technique
in three ways: (1) it deals with a multiplicity of social goals,
(2) it can evaluate projects that are large relative to the economy,
and (3) it takes into account the organizational structure of the
economy. This monograph is divided into the following two parts:
"The Impact Approach to Evaluation of Programs and Projects,"
and "Evaluation of Alternative Methods of Cashew Nut Processing
in Kenya: A Case Study." Assumes knowledge of matrix algebra.

Prior, Moody E. SCIENCE AND THE HUMANITIES. Evanston, III.: North-
western University Press, 1962. xii, 124 p. Notes, pp. 118-24.

A collection of five essays written by the author and presented in
the fall of 1951 at a three-day symposium on science and the humanities held in Aspen by the Markle Scholars in medicine.

Radner, Roy. NOTES ON THE THEORY OF ECONOMIC PLANNING. Training Seminar Series, 2. Athens, Greece: C. Serbinis Press, for the Center of
Economic Research, 1963. 140 p. Bibliography, pp. 137-40. No index.

A description of a general framework for the mathematical analysis of economic planning, and a presentation of a number of models
formulated in terms of real goods and services (rather than monetary magnitudes) that have been used in applications or that hold
promise of such use. Examples of models are: linear activity
analysis model of production possibilities, dynamic input-output
models, Cobb-Douglas production functions, and constant elasticity of substitution production functions. The properties of optimal paths of economic growth are considered with emphasis on the
role of shadow prices and interest rates as indicators of optimality.
Finally, explicit formulas are derived for optimal programs, both
for finite and infinite horizons, which show how certain aspects
of these programs, such as saving and consumption coefficients
and the rates of growth, depend upon the parameters of the production and social welfare functions. The cases considered include one commodity, two commodities with produced and primary
resources, and multicommodity with no primary resource. Assumes
knowledge of differential and integral calculus and matrix algebra.

Raisbeck, G. INFORMATION THEORY: AN INTRODUCTION FOR SCIEN-
TISTS AND ENGINEERS. Cambridge, Mass.: MIT Press, 1963. x, 105 p.
Bibliography, pp. 100-101. Index, pp. 103-5.

> An exposition of some ideas in information theory with applications
> in signal transmission and signal detection. It is not intended as
> a text or reference.

Roman, Daniel D. RESEARCH AND DEVELOPMENT MANAGEMENT: THE
ECONOMIES AND ADMINISTRATION OF TECHNOLOGY. New York: Appleton-
Century-Crofts, 1968. xv, 450 p. Selected Bibliography, pp. 434-42. In-
dex, pp. 443-50.

> A text for graduates and undergraduates on the operational charac-
> teristics, organization, and structure of research and development
> management.

Schlaifer, Robert. ANALYSIS OF DECISIONS UNDER UNCERTAINTY. New
York: McGraw-Hill Book Co., 1969. xvi, 729 p. Appendix: Sets, Functions, and Weighted Averages, pp. 642-52. Tables, pp. 654-719. Index,
pp. 721-29.

> An introduction to the logical analysis of the problems under un-
> certainty that arise in the practice of business administration. It
> is divided into the following three parts: foundations, the assess-
> ment of preferences and probabilities, and sampling and simulation.
> Numerous diagrams, graphs, and examples are used to illustrate
> the concepts, and the chapters conclude with detailed problems.

_____. COMPUTER PROGRAMS FOR ELEMENTARY DECISION MAKING.
Studies in Managerial Economics. Cambridge, Mass.: Graduate School of
Business Administration, Harvard University, 1971. xi, 247 p. No index.

> A presentation of the MANECON collection of computer programs
> on the analysis of decision problems under uncertainty of the order
> of complexity of those presented in the author's ANALYSIS OF
> DECISIONS UNDER UNCERTAINTY (see above) and the accompa-
> nying MANUAL OF CASES (McGraw-Hill, 1969). Most of the
> programs are used in the interactive mode, and many subroutines
> and functions are available to facilitate the writing of special
> programs whose diagrams are too complex to be analyzed by means
> of general purpose interaction programs. The MANECON collec-
> tion is divided into the following four parts: description of self-
> contained programs, description of the general purpose subpro-
> grams, mathematical notes on the programs and subprograms, and
> technical notes on the programs and subprograms. The collection
> is written in FORTRAN IV.

Shull, Fremont A.; Delbecq, André L.; and Cummins, Larry L. ORGANIZA-
TIONAL DECISION MAKING. New York: McGraw-Hill Book Co., 1970.
xvi, 320 p. Index, pp. 309-20.

A nonmathematical treatise on decision making from the psychological, small-group organizational, and econological perspectives.
The chapters conclude with bibliographies.

Simon, Julian L. BASIC RESEARCH METHODS IN SOCIAL SCIENCE: THE
ART OF EMPIRICAL INVESTIGATION. New York: Random House, 1969.
xiv, 525 p. Bibliography, pp. 481-99. Index, pp. 501-25.

An introduction to social-scientific research for students who have
never before studied the subject. It is divided into the following
five parts: the process of social-science research, the obstacles
to social-science knowledge, decisions and procedures, extracting
the meaning of data, and epilogue. The chapters conclude with
references.

Stone, Richard. THE ROLE OF MEASUREMENT IN ECONOMICS. University
of Cambridge Department of Applied Economics Monographs, 3. New York:
Cambridge University Press, 1951. vi, 85 p.

The Newmarch lectures given by the author during the academic
year 1948-49 at University College, London. Topics include estimation of population parameters, prediction and economic policy,
and the problem of aggregation. Knowledge of elementary econometrics is helpful.

Theil, Henri. ECONOMICS AND INFORMATION THEORY. Studies in Mathematical and Managerial Economics, vol. 7. New York: American Elsevier,
1967. xvii, 488 p. Bibliography, pp. 423-27. Tables, pp. 428-82. Index,
pp. 483-88.

A presentation of the problem of allocation proportions in economics using information theory. It is divided into the following
four parts: an introduction to information theory including such
topics as entropy, conditional entropy, and information inaccuracy;
applications to family households, index numbers, and consumer
demand theory; applications to the firm and international trade;
and the entropy of a continuous probability distribution with applications to forecasting evaluations. Information is defined in
terms of logarithms of density functions. Assumes knowledge of
linear algebra, calculus, and probability.

Tinbergen, Jan. ECONOMIC POLICY: PRINCIPLES AND DESIGN. Contributions to Economic Analysis, vol. 10. Chicago: Rand McNally, 1967.
xxviii, 276 p. Appendixes, pp. 225-73. Subject Index, pp. 274-75. Author
Index, p. 276.

A systematic treatment of the author's experience with policy problems in the Netherlands Central Planning Board. The objectives
are to describe the process of economic policy, to judge the consistency of its aims and means, to indicate the optimum policy for
attaining given aims, and to make suggestions regarding aims.

Twenty closed and open models and one group of economic models are used for illustrations. The appendixes consist of a survey of the models used in the presentation, problems which can be treated with the models, and detailed descriptions of the models. Assumes knowledge of elementary differential calculus and the rudiments of econometric model building.

_____. ON THE THEORY OF ECONOMIC POLICY. Contributions to Economic Analysis, vol. 1. New York: American Elsevier, 1952. viii, 78 p. No index.

Develops a number of simple mathematical models useful to economic policy using only the algebra of linear relations. The models are illustrated by examples used in government planning in the Netherlands Central Planning Office. The topics and concepts treated include variables, relations, and structure in quantitative policy problems; logical structure of the simplest quantitative policy problem (targets and instruments in equal numbers); the problem of inequality between number of targets and number of policy instruments; complications created by boundary conditions; effectiveness of political instruments; and a systematic survey of the characteristics of economic policy.

Tintner, Gerhard. METHODOLOGY OF MATHEMATICAL ECONOMICS AND ECONOMETRICS. Vol. 2. International Encyclopedia of Unified Science, no. 6. Chicago: University of Chicago Press, 1968. ix, 113 p. Bibliography, pp. 101-13. No index.

A collection of four essays (chapters) on methodological problems of mathematical economics, econometrics, and operations research aimed at the noneconomist as well as the economist. The essay titles are: "Introduction," "Mathematical Economics," "Econometrics," and "Welfare Economics and Economic Policy." Topics treated include the interpretation of probability statements; the superiority of nonstatic over static economic models; introduction to linear programming, game theory, and econometric terminology; the influence of ideological bias on economic thought; and the problem of ethics and social welfare. Assumes knowledge of differential calculus and matrix algebra.

Vidal, Pierre. NON-LINEAR SAMPLED-DATA SYSTEMS. Information and Systems Theory Series. New York: Gordon and Breach, 1969. xv, 346 p. Glossary of Principal Symbols Used, pp. xiii-xiv. Author Index, pp. 343-44. Subject Index, pp. 345-46.

A presentation of the theory of nonlinear sampled-data control systems based on investigations undertaken throughout the world. Topics include the Z-transform and discrete phase plane methods, methods using the describing function and signal-flow graphs, stability, and pulse-width and quantized sampled-data systems. The chapters conclude with references.

Wasson, Chester R. UNDERSTANDING QUANTITATIVE ANALYSIS. New York: Appleton-Century-Crofts, 1969. v, 263 p. A Selected Bibliography, p. 227. Greek Alphabet, p. 228. Appendix: Some Common Abbreviations and Conventions of Mathematical Language, pp. 229-32. Exercises, pp. 233-60. Index, pp. 261-63.

A nonmathematical exposition of the quantitative aspect of information with emphasis on the evaluation of uncertainty. It views numbers and symbols of quantity as a language, and considers such topics as Bayesian statistics, Bayesian decision and matrix decision models, network analysis, use of charts and averages, and simulation. The review exercises are designed to reinforce the reader's understanding of the topics. This book is useful to anyone desiring a basic knowledge of quantitative measurement.

Weddepohl, H.N. AXIOMATIC CHOICE MODELS AND DUALITY. Preface by J.J.J. Dalmulder. Tilburg Studies on Economics, no. 3. Rotterdam: Rotterdam University Press, 1970. x, 172 p. List of symbols, pp. ix-x. References, pp. 169-70. Index, pp. 171-72.

A development of the structure of rational choice in the following four parts: introduction, mathematical concepts for choice theory, choice models based on preferences and choice functions, and consumer choice theory. Assumes knowledge of set theory and point-set topology.

Wiener, Norbert. CYBERNETICS, OR CONTROL AND COMMUNICATION IN THE ANIMAL AND THE MACHINE. 2d ed., enl. New York: Wiley and MIT Press, 1961. xvii, 212 p. Index, pp. 205-12.

The second edition consists of the first edition, published in 1948, with two supplementary chapters on learning and self-reproducing machines, and brain waves and self-organizing systems. It is a largely nonmathematical exposition of the author's ideas and philosophical reflections on cybernetics. Topics include Newtonian and Bergsonian time, feedback and oscillation, computing machines and the nervous system, cybernetics and psychopathology, statistical mechanics, and information theory. Assumes familiarity with differential and integral calculus.

Williamson, Oliver E. CORPORATE CONTROL AND BUSINESS BEHAVIOR: AN INQUIRY INTO THE EFFECTS OF ORGANIZATION FORM OF ENTERPRISE BEHAVIOR. Englewood Cliffs, N.J.: Prentice-Hall, 1970. xii, 196 p. Bibliography, pp. 182-90. Index, pp. 191-96.

The first part (six chapters) deals with the phenomenon of managerial discretion, and the second part takes up organization innovation and its influence on goal pursuit and internal efficiency. Knowledge of calculus is helpful for understanding portions of the book. The chapters conclude with footnotes.

Young, John F. INFORMATION THEORY. New York: Wiley, 1971. v, 168 p. Appendix A: Probability, pp. 135-43. Appendix B: Formal Treatment of Probability, pp. 144-52. Bibliography, pp. 153-63. Index, pp. 165-68.

An elementary introduction to information theory requiring little mathematics and probability theory.

Zauberman, Alfred. ASPECTS OF PLANOMETRICS. London: Athlone Press, 1967. xiii, 318 p. Index of Subjects, pp. 303-12. Index of Names, pp. 313-18.

The historical development of Soviet mathematical economics with emphasis on the doctrinal significance of Soviet models which yield numerical solutions, such as linear programming models. Topics include planning techniques in pricing, profit, efficiency of investment, and efficiency of foreign trade. This book is a valuable reference on Soviet style mathematical economics.

Appendix B
MISCELLANEOUS

Aydelotte, William O.; Bogue, Allan G.; and Fogel, Robert William, eds.
THE DIMENSIONS OF QUANTITATIVE RESEARCH IN HISTORY. Quantitative Studies in History Series. Princeton, N.J.: Princeton University Press,
1972. ix, 435 p. Appendix, pp. 419-23. The Contributors, pp. 424-26.
Index, pp. 427-35.

> A collection of nine contributions (essays) on the application of
> mathematical methods to history. Five essays deal with American
> topics, two are on British subjects, and two on French topics.
> Four essays are sociological in nature, three are in political
> areas, and two in economic history. Topics include the social
> origins of grievances in the French revolution, expenditure levels
> in American cities, American land policy, social mobility of the
> religious and ethnic groups in Boston, and the analysis of Con-
> gressional elections. Knowledge of basic regression theory is
> helpful.

Bass, Frank M., et al. MATHEMATICAL MODELS AND METHODS IN MAR-
KETING. Irwin Series in Quantitative Analysis for Business. Homewood, Ill.:
Richard D. Irwin, 1961. xi, 545 p. Index, pp. 541-45.

> A collection of eighteen papers presented at an informal seminar
> of the Ford Foundation's Institute of Basic Mathematics for Appli-
> cations to Business held at Harvard University from September 1959
> to August 1960. The articles are grouped under the following
> five headings: introduction to the use of models (three papers),
> models of customer behavior (three papers), conceptual models for
> determining promotional effort (five papers), individual company
> models of promotional effort (four papers), and sales forecasting
> and inventory management (three papers). Assumes knowledge of
> differential and integral calculus.

Batchelor, James H. OPERATIONS RESEARCH: AN ANNOTATED BIBLIOG-
RAPHY. Vol. 1. Saint Louis: Saint Louis University Press, 1959. x, 865 p.
Index, pp. 781-866.

Contains an annotated listing of 3,840 items covering the years
up to but excluding 1958. The items are arranged in alphabetical
order by author or editor, or by institution, publication, or spon-
sor. Foreign language titles are followed by translations in English
and by English abstracts, which have been prepared to indicate
the nature of the article and the views and ideas of the authors.
In some cases abstracts are omitted where the title is adequate
to indicate the content or where the original is not available.
All papers from twelve journals, such as OPERATIONS RESEARCH
and MANAGEMENT SCIENCE, have been included, as well as
books, individual published papers, and reports of meetings on
operations research.

_____. OPERATIONS RESEARCH: AN ANNOTATED BIBLIOGRAPHY. Vol. 2.
Saint Louis: Saint Louis University Press, 1962. xi, 628 p. Abbreviations,
p. xi. Additional Entries to Volume 1, pp. 1-36. Index, pp. 545-626.
Corrections, pp. 627-28.

Contains an annotated listing of 2,528 items on operations research
and related subjects published for the years 1958 and 1959 and
350 items prior to 1958. Volumes 1 and 2 together contain 6,723
citations covering the years up to and including 1959.

Beckmann, Martin J. LOCATION THEORY. New York: American Elsevier,
1968. xii, 132 p. Author Index, pp. 127-28. Subject Index, pp. 129-32.

An exposition of the theory of location for graduates and undergraduates who have knowledge of differential and integral calculus. It concentrates on the theory of location of a single industry
and several interdependent activities, and develops a linear programming model of spatial equilibrium for both homogeneous and
heterogeneous goods. The presentation is intuitive, graphical,
and mathematical, with the mathematical development reserved
for notes at the ends of the chapters. The chapters also conclude
with selected references.

Berg, Robert L., ed. HEALTH STATUS INDEXES. Foreword by J.P. Cooney,
Jr. Chicago: Hospital Research and Educational Trust, 1973. xi, 262 p.
No index.

A collection of fifteen papers presented at a conference conducted
by the journal HEALTH SERVICES RESEARCH, and held in Tucson,
Arizona, on October 1-4, 1972. Topics include debility index
for long-term-care patients; a method for constructing proxy measures of health status; the G index for program priority; quantifying
the patient's preferences; techniques for the assessment of worth;
and guidelines for selecting a health status index. The participants
of the conference represent the fields of computer technology, education, economics, health care admininistration, industrial engineering, management science, psychometrics, sociology, and statistics.

Among the authors are R.L. Berg, J.W. Bush, M.K. Chen, B.E. Forst, D.C. Holloway, S. Katz, N.P. Kneppreth, M. Lerner, D.E. Skinner, G.A. Whitmore, and D.E. Yett.

Bernd, Joseph L., ed. MATHEMATICAL APPLICATIONS IN POLITICAL SCIENCE, IV. Charlottesville: University Press of Virginia, 1969. 83 p. No index.

A presentation of four technical articles presented at the 1968 summer institute on Mathematical Applications in Political Science at Virginia Polytechnic Institute. Articles 1 and 2 deal with methodology; article 3 employs spatial analysis to show the relationship of points of view in geometric space as a means of handling variables in a legislative context; and article 4 is a mathematical exploration of ecological regression.

Bos, Hendricus Cornelis. SPATIAL DISPERSION OF ECONOMIC ACTIVITY. New York: Gordon and Breach, 1965. ix, 99 p. List of symbols, pp. 96-97. Author Index, p. 98. Subject Index, p. 99.

A contribution to the understanding of the problem of space economics, that is, the spatial distribution of economic activity. The following special models are considered: dispersion of the production of a single industry, several vertically integrated industries and a large number of centers, several vertically integrated industries and different types of centers, several vertically integrated industries and a small number of centers, and industries producing final and intermediate products. The assumptions of the models are: agriculture and population are spread over a given area, the production of nonagricultural industries is characterized by indivisibilities leading to economies of scale, and the transport of goods and services gives rise to transportation costs. Assumes knowledge of college algebra and the rudiments of differential and integral calculus.

Brennan, Michael, ed. PATTERNS OF MARKET BEHAVIOR. Essays in Honor of Philip Taft. Providence, R.I.: Brown University Press, 1965. viii, 258 p. Bibliography of the Writings of Philip Taft, pp. 257-58.

A collection of thirteen essays in honor of Professor Philip Taft, a labor historian, on the occasion of his sixty-third birthday. The essays are grouped under the following headings: commodity markets, resource markets, money markets, and international markets. Assumes knowledge of matrix algebra and differential calculus. The authors are G.H. Borts, M.J. Brennan, P. Cagan, D. Carson, E.F. Denison, D. Gale, P. Hartland, E.S. Mills, H.P. Minsky, J.N. Morgan, M.B. Schopack, and J.L. Stein.

Brown, Robert Goodell. SMOOTHING, FORECASTING AND PREDICTION OF DISCRETE TIME SERIES. Prentice-Hall Quantitative Methods Series. Englewood Cliffs, N.J.: Prentice-Hall, 1963. 468 p. Appendix A: Regression,

Autocorrelation, and Spectral Analysis, pp. 387-401. Appendix B: The Z-transform, pp. 403-11. Appendix C: Samples of Time Series for Practice, pp. 413-34. Appendix D: Mathematical Tables, pp. 435-49. Bibliography, pp. 452-62. Glossary of Symbols, p. 463. Index, pp. 459-68.

A survey of the use of digital computers to compute forecasts of discrete time series. The primary concerns are the open-loop characteristics of one box that accepts current observations and delivers a forecast of the probability distribution from which future observations are drawn. The twenty-five chapters are grouped under the following headings: data, models, smoothing techniques, forecasting, error measurement and analysis, exploration of alternatives, and applications. Complete mathematical derivations are provided, with formulas and tables that are necessary to extend the range of coverage beyond the problems that are illustrated. Assumes knowledge of college algebra.

Bühlmann, Hans. MATHEMATICAL METHODS IN RISK THEORY. New York: Springer-Verlag, 1970. xii, 210 p. Appendix: The Generalized Riemann-Stieltjes Integral, pp. 201-5. Bibliography, pp. 206-8. Index, pp. 209-10.

A synthesis of modern scientific publications in the field of actuarial mathematics. The first part of the book presents mathematical models and includes such topics as the probability space of risk, the risk process, and the risk in the collective. The second part applies the models to such topics as premium calculation, the retention problem, and stability criteria (ruin criterion, dividend policy, and utility). It does not consider statistical estimation of the parameters of the models. Assumes knowledge of advanced calculus, mathematical analysis, and probability theory.

Campbell, Rita Ricardo. FOOD SAFETY REGULATION: A STUDY OF THE USE AND LIMITATIONS OF COST-BENEFIT ANALYSIS. AEI-Hoover Policy Studies, no. 12. Washington, D.C.: American Enterprise Institute; Stanford, Calif.: Hoover Institute, 1974. 59 p. Selected Bibliography, pp. 55-59. No index.

A study in cost-benefit analysis which focuses on the balance between costs and benefits and between biological risks and advantages involved in public policy decisions regulating the safety of the U.S. food supply. Topics include an analysis of both the individual's and society's risks and benefits; a case study involving the iron enrichment of flour and breads; and consumer involvement as an alternative to the growing complexity of governmental regulations. There are no mathematical prerequisites.

Carter, E. Eugene. PORTFOLIO ASPECTS OF CORPORATE CAPITAL BUDGETING: METHODS OF ANALYSIS, SURVEY OF APPLICATIONS, AND AN INTERACTIVE MODEL. Lexington, Mass.: D.C. Heath, 1974. xix, 222 p. Bibliography, pp. 205-13. Author and Corporation Index, pp. 217-19. Subject Index, pp. 221-22.

An analysis of the problem of portfolio selection by a corporation. After surveying the goals of corporation executives, the author presents a computer simulation model of a portfolio of projects in the interactive mode of computer usage. This simulation allows a manager to consider risk, and to study the portfolio for a company in a variety of dimensions (sales, earnings per share, cash flow, and net present value). The book concludes with the results of the model's usage by over one hundred executives in a laboratory setting. Knowledge of basic statistics and computer programming helpful.

Case Institute of Technology. A COMPREHENSIVE BIBLIOGRAPHY ON OPERATIONS RESEARCH. Foreword by D.B. Hertz. Publications in Operations Research, no. 8. New York: Wiley, 1963. Index Number 1: General Articles, pp. 355-59. Index Number 2: Bibliographies and Indexes, p. 360. Index Number 3: Entries by Technique, pp. 361-92; Entries by Type of Organization, pp. 392-403.

A continuation of a previous volume in the series which covered operations research literature through December 1956. The major operations research journals, JOURNAL OF THE OPERATIONS RESEARCH SOCIETY OF AMERICA, MANAGEMENT SCIENCE, OPERATIONS RESEARCH QUARTERLY, and NAVAL RESEARCH LOGISTICS QUARTERLY, have been covered in their entirety. In addition, articles from many other journals, American and foreign, have been included. The present volume contains some 5,500 new entries, most for the year 1958. The order of entries is alphabetical, and within a given letter, publications in English by cited author are listed first, followed by English publications without cited author, listed alphabetically by title. The final grouping within a given letter consists of foreign language publications, ordered in the same manner. Each entry has been given a ten-digit serial number to make case referencing more efficient. Each of the four appendixes consists of two parts: items in the previous volume (Publications in Operations Research, no. 4), and items in the present volume.

Cateora, Philip R., and Richardson, Lee, eds. READINGS IN MARKETING: THE QUALITATIVE AND QUANTITATIVE AREAS. New York: Appleton-Century-Crofts, 1967. xi, 462 p. No index.

A supplementary text for a principles course in marketing containing contributions in marketing made by both behavioral scientists and mathematicians. The forty-three reprints of papers previously published elsewhere are grouped under the following headings: scientific marketing management, science in marketing the consumer, marketing management, and marketing tomorrow. Among the techniques employed in the papers are linear programming, and Bayesian decision theory. Most of the papers require no mathematics prerequisites.

Chisholm, Roger K., and Whitaker, Gilbert R., Jr. FORECASTING METHODS. Homewood, III.: Richard D. Irwin, 1971. xi, 177 p. Index, pp. 175-77.

A technique-oriented book which can serve as a primary text in forecasting courses, a handbook for businessmen, or a reference for courses requiring a source of forecasting techniques. It discusses the following techniques: naive forecasting methods, polls and surveys, barometric or indicator, input-output analysis, and regression analysis. The methodology of forecasting is illustrated with simple examples. Portions of the 1963 Department of Commerce input-output table of the U.S. economy are included. Assumes knowledge of college algebra and basic statistics.

Churchill, Neil C.; Miller, Merton H.; and Trueblood, Robert M. AUDITING, MANAGEMENT GAMES, AND ACCOUNTING EDUCATION. Foreword by R.M. Cyert. Contributions to Management Education Series, vol. 2. Homewood, III.: Richard D. Irwin, 1964. x, 103 p. No index.

A complete and documented record of an experiment in which first-year graduate students in accounting were assigned the task of auditing the financial statements and control procedures of "firms" of advanced students who played the Carnegie Tech Management Game.

Corcoran, A. Wayne, and Mautz, Robert K. MATHEMATICAL APPLICATIONS IN ACCOUNTING. The Harbrace Series in Business and Economics. New York: Harcourt, Brace and World, 1968. xii, 249 p. Appendix, pp. 234-45. Index, pp. 247-49.

Primarily a supplement for the basic accounting courses (introductory, intermediate, advanced, cost, and managerial), or for courses in production management, managerial mathematics, industrial engineering, economics, finance, or operations research. The main topics are: time series mathematics with applications in discounting, compounding, mortgages, and sinking funds; calculus and probability with applications to learning curves and inventory control; and matrix algebra with applications to accounting and linear programming. The concepts are illustrated with numerous examples. The appendix contains tables such as the normal curve, logarithms, and interest tables. Assumes knowledge of high school algebra; all other mathematics is taken up as needed.

Dasgupta, Ajit K., and Pearce, D.W. COST-BENEFIT ANALYSIS: THEORY AND PRACTICE. New York: Barnes and Noble, 1972. 270 p. Bibliography, pp. 245-64. Index, pp. 265-70.

This book is divided into the following four parts: the objective function in cost-benefit analysis, accounting prices, decision formulas, and case studies. Topics include utility, costs, and benefits; Pareto optimality, compensation tests, and equity; social welfare functions; accounting price; external effects and public goods;

social rate of discount; formulas for project choice; and risk and
uncertainty. Assumes knowledge of high school algebra and graduate-level economic theory.

Dobrovolsky, Sergei P. THE ECONOMICS OF CORPORATE FINANCE. McGraw-
Hill Series in Finance. New York: McGraw-Hill, 1971. xii, 347 p. Index,
pp. 337-47.

A text for an advanced course in business finance for students who
have already taken an introductory course in this field and who
possess knowledge of calculus, and a first course in statistics.
Emphasis is on the theoretical principles rather than on the mathematical techniques for solving problems. It is divided into the following two parts: financial problems of the firm; and the financial system: capital markets, interest, and the flow of funds.
Topics include capital as a factor of production, profitability
measures, stock valuation, the cost of capital, depreciation as a
source of funds, the flow-of-funds accounts, theories of interest,
determinants of business investment, the term structure of interest
rates, and historical trends in capital formation in the United
States. Numerous examples are provided so that students with
relatively weak mathematical background can follow. Chapters
conclude with references.

Floud, Roderick. AN INTRODUCTION TO QUANTITATIVE METHODS FOR
HISTORIANS. Princeton, N.J.: Princeton University Press, 1973. ix, 220 p.
Bibliography, pp. 209-11. Logarithm Tables, pp. 212-15. Index, pp. 216-20.

An introduction to the skills required by the historian using quantitative evidence, and to the many other books on statistics, computing, econometrics, and mathematics which quantitative historians
may eventually need to read or consult. Topics include time series,
growth rates, trends, index numbers, sampling problems, and the
use of the computer.

_____, ed. ESSAYS IN QUANTITATIVE ECONOMIC HISTORY. New York:
Oxford University Press, for the Economic History Society, 1974. xii, 250 p.
Bibliography, p. 248. Index to Statistical Terms, pp. 249-50.

A collection of eleven articles, previously published in journals
and elsewhere, on the following aspects of economic history: pitfalls of historical statistics; geographical distribution of wealth in
England, 1334-1649; inflation and industrial growth; income inequality in England; statistical study of prices and production of
wheat in Great Britain, 1873-1914; British investments in Argentina,
1880-1914; postwar growth of the British economy; and English
bank deposits before 1844. The techniques of analysis are mainly
descriptive statistics, such as index numbers and frequency distributions. Included is an introduction by R. Floud. The contributors are D.K. Adie, E. Ames, J.P. Cooper, D. Felix, A.G.

Ford, C.C. Harris, Jr., R.C.O. Matthews, G. Ohlin, M. Olson, L.G. Sandberg, R.S. Schofield, and L. Soltow.

Francis, Jack Clark, and Archer, Stephen H. PORTFOLIO ANALYSIS. Prentice-Hall Foundations of Finance Series. Englewood Cliffs, N.J.: Prentice-Hall, 1971. xii, 268 p. Six Mathematical Appendixes, pp. 231-57. Subject Index, pp. 261-64. Name Index, pp. 267-68.

A text on the economic and mathematical tools for portfolio analysis and their applications. It is divided into the following parts: definitions and measurements, portfolio analysis, implications of portfolio analysis, and other issues. The appendixes cover summation, probability, simultaneous equations, quadratic equations, and computer programs for three-asset portfolio analysis.

Frank, Ronald E., and Green, Paul E. QUANTITATIVE METHODS IN MARKETING. Prentice-Hall Foundations of Marketing Series. Englewood Cliffs, N.J.: Prentice-Hall, 1967. v, 118 p. Selected Readings, p. 112. Index, pp. 113-18.

Presents the role of quantitative techniques in decision making with emphasis on the conceptual nature of the techniques and not on their computational aspects or on hypothesis-testing and estimation procedures. It considers four main topics: Bayesian decision theory and marketing analysis; experimental studies, which consist of the analysis of experimental-type data by the analysis of variance technique; observational studies, which consist of the analysis of nonexperimental data by factor analysis, discriminant analysis, and correlation and regression; and simulation for dealing with problems such as media scheduling and physical distribution. Assumes knowledge of basic statistics.

Goldman, Thomas A., ed. COST-EFFECTIVENESS ANALYSIS: NEW APPROACHES IN DECISION-MAKING. Preface by R.N. Grosse. Praeger Special Studies in U.S. Economic and Social Development. New York: Frederick A. Praeger, 1967. xvi, 231 p. No index.

Thirteen nontechnical papers presented at a symposium on cost-effectiveness sponsored by the Washington Operations Research Council in 1966. The papers dealt with many military applications of cost-effectiveness analysis. The nonmilitary applications are in the areas of poverty, such as job training programs; metropolitan transportation systems; and incentive contracting problems. The authors are R.S. Berg, S.M. Besen, A. Blumstein, E.R. Brussell, Wm. M. Capron, A.E. Fechter, A.C. Fisher, H.P. Hatry, J.F. Kain, M. Kamrass, J.D. McCullough, J.A. Navarro, Wm. A. Niskanen, E.S. Quade, and C.W. Scarborough.

Haggett, Peter, and Chorley, Richard J. NETWORK ANALYSIS IN GEOGRAPHY: EXPLORATIONS IN SPATIAL STRUCTURE. I. London: Edward Arnold,

1969. xii, 348 p. References, pp. 319-35. Further Reading, p. 337. Locational Index, pp. 338-41. General Index, pp. 342-48.

An exploration of the ways in which the analysis of a topologically distinct class of spatial structures--linear networks--might throw light on common geographical problems of morphometry, origin, growth, balance, and design. It is divided into the following three parts: spatial structures, evaluation of structures, and structural change. This book contains a detailed description of various types of networks and spatial designs, and changes in networks, and is useful to anyone interested in networks and their mathematical representation. Assumes knowledge of the rudiments of calculus.

Haley, Charles W., and Schall, Lawrence D. THE THEORY OF FINANCIAL DECISIONS. McGraw-Hill Series in Finance. New York: McGraw-Hill Book Co., 1973. xviii, 383 p. Bibliography, pp. 367-76. Index, pp. 377-83.

A text for a one-year course in finance at the undergraduate and graduate levels. Topics include: decisions under uncertainty, capital budgeting, probability and random variables, portfolio theory, single-period and multiperiod firm financial decision models, financing decisions in perfect markets, and firm investment in perfect and imperfect capital markets. Some proofs are provided, and numerous examples and illustrations are used to explain the concepts. Only special sections of the book and the chapter appendixes require knowledge of calculus for comprehension. The chapters conclude with suggested readings.

Hammond, James D., ed. ESSAYS IN THE THEORY OF RISK AND INSURANCE. Glenview, Ill.: Scott, Foresman and Co., 1968. 257 p. No index.

Fifteen articles previously published in journals and elsewhere emphasizing the foundations of risk and insurance grouped under the following three headings: the meaning and measurement of risk (four articles), risk-taking behavior (six articles), and risk and the insurance technique (five articles). Topics include subjective probabilities as a reaction to uncertainty, utility analysis of choice involving risk, game theory and insurance consumption, the legal definition of insurance, and government insurance and economic risk. Knowledge of elementary probability theory and basic statistics is assumed in many articles. The contributors are H.S. Deneberg, O.D. Dickerson, Wm. Feller, M. Friedman, D.B. Houston, M. Kaplan, S.A. Miller, J. Neggers, J. Niehans, F. Redlich, Y. Rim, L.J. Savage, G.L.S. Shackle, P. Slovic, and C.A. Williams, Jr.

Harberger, Arnold C. PROJECT EVALUATION: COLLECTED PAPERS. Markham Economics Series. Chicago: Markham Publishing Co., 1973. xii, 330 p. Index, pp. 325-30.

A collection of twelve essays written by the author on project appraisal, cost-benefit analysis, discount rates, social opportunity cost measurement, rate of return to capital, marginal cost pricing and social investment criteria, applications to transportation projects and irrigation dams, and capital assistance to developing nations. Nine of the essays are reprints of articles previously published in journals and elsewhere, and one article on irrigation dams was written in collaboration with L.G. Reca and J.A. Zapata.

Harberger, Arnold C., et al., eds. BENEFIT COST ANALYSIS 1971: AN ALDINE ANNUAL. Chicago: Aldine-Atherton, 1972. xxiv, 485 p. Index, pp. 475-85.

A collection of twenty-five articles originally published in journals and elsewhere grouped under the following headings: general principles for cost-benefit analysis (five articles), problems in the field of health and medical services (four articles), analyses of investment in human capital and income distribution (five articles), issues related to water resources (four articles), and miscellaneous topics: recreation, housing, transportation, and pollution (seven articles). A few articles assume knowledge of calculus, but most require no mathematics preparation. Among the contributors are Y. Barzel, V.R. Fuchs, W.I. Garms, A.C. Harberger, E.J. Mishan, C. Riordan, D.O. Sewell, B.A. Weisbrod, and D.F. Wood.

Harder, Theodor. INTRODUCTION TO MATHEMATICAL MODELS IN MARKET AND OPINION RESEARCH: WITH PRACTICAL APPLICATIONS, COMPUTING PROCEDURES, AND ESTIMATES OF COMPUTING REQUIREMENTS. Translated by P.H. Friedlander and E.H. Friedlander. Preface by R.F. Lazarsfeld. Boston: Reidel, 1969. ix, 194 p. Bibliography, pp. 193-94. No index.

Originally published in German in 1966, this book is an introduction to the quantitative evaluation of data in the fields of marketing and social science research from the standpoints of types of calculation, computing procedures, and magnitude of the computing effort needed. It utilizes the following mathematical concepts: multiple regression, first order difference equations, Markov chains, characteristic value problems of component analysis, factor analysis, combinatorial mathematics, linear programming, and matrix operations. Numerous examples are taken from the fields of advertising, market research, and from opinion research. Each of the four chapters concludes with selected references.

Haveman, Robert H.; Harberger, Arnold C.; Lynn, Laurence E., Jr.; Niskanen, William A.; Turvey, Ralph; and Zeckhauser, Richard, eds. With the Editorial Collaboration of Daniel Wisecarver. BENEFIT-COST AND POLICY ANALYSIS, 1973. Preface by R.H. Haveman and D. Wisecarver. An Aldine Annual on Forecasting, Decision-making, and Evaluation. Chicago: Aldine Publishing Co., 1974. xix, 524 p. No index.

The third annual volume concerned with cost-benefit and policy analysis. It consists of twenty-two papers, most of which are reprints from journals and elsewhere, grouped under the following two headings: the analysis and evaluation of public investment and resource management activities (thirteen papers), and the analysis and evaluation of social programs (nine papers). Topics include environmental control; the economics of flood alleviation; benefit-cost analysis of the Alaska pipeline and other natural resource problems; transportation; land use; income maintenance programs; health insurance; human investment; and regulation policies. A few papers assume knowledge of calculus. Among the authors are E.K. Browning, C.J. Cicchetti, M.S. Feldstein, J. Holahan, H.D. Jacoby, J.E. Reinhardt, D.C. Shoup, L.W. Weiss, and R.J. Zeckhauser.

Hester, Donald D., and Tobin, James, eds. FINANCIAL MARKETS AND ECONOMIC ACTIVITY. Cowles Foundation Monograph, 21. New York: Wiley, 1967. ix, 256 p. Appendix: Residuals from Baa Bond Adjustment Regression, pp. 244-45. Cumulative Author Index, pp. 247-50. Cumulative Subject Index, pp. 251-56.

A collection of six essays on the conditions of equilibrium in economy-wide financial markets. The titles of the essays are: "Commercial Banks as Creators of Money," by J. Tobin; "A Model of Bank Portfolio Selection," by R.C. Porter; "Financial Intermediaries and the Effectiveness of Monetary Controls," by J. Tobin and Wm. C. Brainard; "Financial Intermediaries and a Theory of Monetary Control," by Wm. C. Brainard; "Monetary Policy, Debt Management, and Interest Rates: a Quantitative Appraisal," by A.M. Okum; and "Determinants of Bond Yield Differentials; 1954 to 1959," by P.E. Sloane. Assumes knowledge of elementary calculus, probability theory, and econometrics.

_____. RISK AVERSION AND PORTFOLIO ANALYSIS. Cowles Foundation Monograph, 19. New York: Wiley, for the Cowles Foundation, 1967. ix, 180 p. Cumulative Author Index, pp. 171-74. Cumulative Subject Index, pp. 175-80.

A collection of seven essays on theoretical and empirical monetary economics by Yale graduate students and staff members of the Cowles Foundation. The essays are concerned, on the one hand, with attitudes of investors towards risk and average return and, on the other hand, with the opportunities which the market and the tax laws afford investors for purchasing less risk at the expense of expected return. The essay, "Stock Market Indices: A Principal Component Analysis," by G.L. Feeney and D.D. Hester, attempts to construct an index of stock prices which can be useful to investors. The other contributors are K. Hamada, S. Lepper, E.S. Phelps, R.N. Rosett, S. Royama, and J. Tobin.

_____. STUDIES IN PORTFOLIO BEHAVIOR. Cowles Foundation Monograph, 20. New York: Wiley, for the Cowles Foundation, 1967. ix, 258 p. Cumulative Author Index, pp. 249-52. Cumulative Subject Index, pp. 253-58.

A collection of six essays on empirical tests of hypotheses which attempt to explain quantitatively the portfolio behavior of households, nonfinancial corporations, banks, and life insurance companies. Monograph 20 is one of three (Monographs 19, 20, and 21) that bring together a total of nineteen essays on theoretical and empirical monetary economics written by recent Yale graduate students and staff members of the Cowles Foundation. The essays comprising Monograph 20 are "Consumer Expenditures and the Capital Account," by H.H. Watts and J.Tobin; "Consumer Debt and Spending: Some Evidence from Analysis of a Survey," by J. Tobin; "An Empirical Study of Cash, Securities, and Other Current Accounts of Large Corporations," by A.W. Heston; "An Empirical Examination of a Commercial Bank Loan Offer Function," by D.D. Hester; "An Empirical Model of Commercial Bank Portfolio Management," by J.L. Pierce; and "Life Insurance: the Experience of Four Companies," by L.S. Wehrle.

Hinote, Hubert. BENEFIT-COST ANALYSIS FOR WATER RESOURCE PROJECTS: A SELECTED ANNOTATED BIBLIOGRAPHY. Rev. ed. Knoxville, Tenn.: Center for Business and Economic Research, University of Tennessee, 1969. vi, 148 p. Author Index, pp. 143-45. Index to Government and other Agency Publications, pp. 146-48.

An annotated bibliography on major works (primarily journal articles) in benefit-cost analysis that have appeared in the period 1958-67. Emphasis is placed on works that fall into the following project purposes: flood control, navigation, water quality, recreation, and water supply. For each group, the literature is classified into the following basic steps in performing a benefit-cost analysis: definition, forecasting demand, benefit measurement or cost determination, evaluation techniques, and decision criteria. If a given work is devoted to two or more steps, it is listed in each step.

Ijiri, Yuji. THE FOUNDATIONS OF ACCOUNTING MEASUREMENT: A MATHEMATICAL, ECONOMIC, AND BEHAVIOR INQUIRY. Prentice-Hall International Series in Management. Englewood Cliffs, N.J.: Prentice-Hall, 1967. xvi, 235 p. Appendix A: An Introduction to Relational Systems, pp. 166-83. Appendix B: Functional Analysis of Aggregation, pp. 184-208. Bibliography, pp. 209-21. Index, pp. 223-35.

An inquiry into the measurement aspects of accounting and the foundations of accounting measurement. The presentations stress three basic inquiries: mathematical, economic, and behavioral. This book is useful to anyone interested in understanding the logical bases and uses or limitations of accounting. Assumes knowledge of high school algebra with the appendixes devoted to more advanced mathematical topics.

Jean, William H. THE ANALYTICAL THEORY OF FINANCE: A STUDY OF
THE INVESTMENT DECISION PROCESS OF THE INDIVIDUAL AND THE FIRM.
Foreword by Wm. Beranek. Holt, Rinehart and Winston Series in Finance.
New York: Holt, Rinehart and Winston, 1970. xiii, 206 p. Index, pp. 199-206.

A text for an advanced course in capital budgeting or finance for
graduates and undergraduates who have knowledge of calculus,
linear programming, and random variables. It is concerned primarily with project selection and the financial structure of the
firm, and the theory which is developed is based on investment
decisions under risk using the investor's utility function. Topics
include capital rationing models, capital structure of the firm,
portfolio theory, and the theory of security markets. Chapters
conclude with questions and references.

Jolson, Marvin A., and Hise, Richard T. QUANTITATIVE TECHNIQUES FOR
MARKETING DECISIONS. New York: Macmillan, 1973. xi, 238 p. Tables,
pp. 219-27. Answers to Selected Exercises, pp. 228-31. Index of Authors,
pp. 233-34. Subject Index, pp. 235-38.

A supplementary text for a course in marketing management based
on the following tools: Bayesian analysis, simulation and the
Markov process, linear programming, and differential calculus.
Illustrative examples and solved problems dominate the text. Assumes knowledge of college algebra, probability theory, and matrix
algebra. Each chapter contains about twelve exercises and a list
of references.

Kendall, Maurice George, ed. COST-BENEFIT ANALYSIS. New York: American
Elsevier, 1971. 328 p. No index.

A collection of twenty papers on cost-benefit analysis presented
at a symposium held in The Hague in July 1969 under the aegis
of the NATO Scientific Affairs Committee. The papers are grouped
under the following headings: opening ceremony and historical
review, health and community services, defense and research and
development, natural resources, transport, and investment problems.
Familiarity with linear algebra is helpful for some of the papers.
Among the authors are J.F. Boss, C. Bozon, D.J. Clough, P.F.
Gross, J.B. Heath, M.C. Heuston, R.G.N. Nicholson, A.H.
Pascal, B. Schwab, R. Turvey, and A. Williams.

King, William R. QUANTITATIVE ANALYSIS FOR MARKETING MANAGE-
MENT. New York: McGraw-Hill Book Co., 1967. xviii, 574 p. Appendix: Least Squares Analysis, pp. 565-66. Index, pp. 569-74.

A text on modern analytic marketing with emphasis on the role of
mathematical models. Topics include purchasing, pricing, advertising, promotional, distributional, marketing, and product decisions. Assumes knowledge of college algebra and calculus. The
chapters conclude with exercises and references.

Kotler, Philip. MARKETING DECISION MAKING: A MODEL BUILDING AP-
PROACH. New York: Holt, Rinehart and Winston, 1971. xi, 720 p. Appen-
dix: Mathematical Notes, pp. 683-706. Name Index, pp. 707-11. Subject
Index, pp. 713-20.

A systematic and self-contained theory of marketing analysis and
decision making which is divided into the following four parts:
macromarketing decision theory including a presentation of demand
laws, elasticity of demand, production functions, profit maximization by firms, the marketing mix problem, linear marketing effects
models, multiple products, and so forth; microeconomicing decision
theory models, consisting of distribution decision models, price decision models, sales force, and advertising decision models; models of
market behavior, including sales models for brand share sales, new
products, and established products; and from theory to practice,
including planning and control in marketing organization, implementing management science in marketing, and developing the
corporate marketing model. This book is useful as a text for a
course in marketing and as a reference for a mixed group, including marketing executives, operations researchers, economists, and
others. Assumes knowledge of linear programming, calculus, and
probability theory.

Latoné, Henry A., and Tuttle, Donald L. SECURITY ANALYSIS AND PORT-
FOLIO MANAGEMENT. New York: Ronald Press, 1970. xiv, 752 p. Abbreviations and Symbols, pp. 733-35. Index, pp. 736-52.

A text for a course in finance and investments for undergraduates
and graduates who have knowledge of descriptive statistics and
and college algebra. It is divided into the following four parts: characteristics of financial assets; security analysis: growth, capitali-
zation rates, and convergence; security analysis: forming probability belief; and portfolio management. Chapters conclude with
questions.

Leeflang, P.S.H. MATHEMATICAL MODELS IN MARKETING: A SURVEY,
THE STAGE OF DEVELOPMENT, SOME EXTENSIONS AND APPLICATIONS.
Leiden, the Netherlands: H.E. Stenfert Kroese B.V., 1974. vii, 213 p.
Some References to Marketing Simulation Models, p. 203. References, pp. 204-9.
Subject Index, pp. 210-13.

A text on the application of mathematical models to market research. It includes a survey, examination, and critical evaluation of existing consumer behavior, response, and policy models
used in marketing research; and some new response and policy
models based on a Markovian consumer behavior model. Assumes
knowledge of stochastic processes at the intermediate level.

Lefeber, Louis. ALLOCATION IN SPACE: PRODUCTION, TRANSPORT AND
INDUSTRIAL LOCATION. Contributions to Economic Analysis, vol. 14. New

York: American Elsevier, 1958. xv, 151 p. Appendix, pp. 135-47. Index, pp. 149-51.

A general equilibrium analysis of production and choice of industrial location, based on an extension of the neoclassical production analysis. The main tools employed are Lagrangian multipliers and mathematical programming. The appendix consists of an extension of the analysis to a complete general equilibrium analysis of production and consumption in space. Assumes knowledge of differential calculus and matrix algebra.

Levy, Haim, and Sarnat, Marshall. INVESTMENT AND PORTFOLIO ANALYSIS. Wiley Series in Finance. New York: Wiley, 1972. xvi, 604 p. Author Index, pp. 595-98. Subject Index, pp. 599-604.

A text for a course in investment analysis and portfolio selection for students with only a high school algebra background. It is divided into the following four parts: security analysis and the valuation of the firm, portfolio selection, applications of portfolio selection, and technical supplement. Topics include alternative measures of profits, investment decisions under uncertainty and proof of the expected utility principle follows from the von Neumann-Morgenstern axioms, the mean-variance criterion in investment choice, price determination in the stock market, the analysis of risk diversification by means of conglomerate mergers, and international diversification of investments. The more rigorous mathematical developments, including proofs of theorems, are reserved for chapter appendixes and the technical supplement. Assumes knowledge of differential and integral calculus and mathematical statistics. The chapters conclude with questions, problems, and selected references.

Lewis, Colin David. INDUSTRIAL FORECASTING TECHNIQUES. Foreword by N.A. Dudley. Modern Aids to Production Management. Brighton, England: Machinery Publishing, 1970. 118 p. Appendix A: The Cusum Technique of Monitoring, pp. 110-13. Index, pp. 115-18.

A description of some of the techniques in mathematical forecasting used by production planners. The methods considered are: linear additive trends; Holt's method and Brown's adaptive smoothing method (a computer method); linear ratio trend, using Muir's method; and combined linear and seasonal additive trend model, using the Holt method. The Box-Jenkins method is also considered. One chapter is devoted to a survey of computer programs currently available for routine forecasting. The techniques are illustrated by numerical examples and graphs, and the equations used for forecasting are relatively simple and require knowledge of only college algebra. The chapters conclude with references.

Loeckx, J.J.C. COMPUTABILITY AND DECIDABILITY: AN INTRODUCTION FOR STUDENTS OF COMPUTER SCIENCE. Lecture Notes in Economics and Mathematical Systems, vol. 68. New York: Springer-Verlag, 1972. 76 p. Appendix 1: Bibliographical Notes, p. 71. Appendix 2: List of the Most Important Notations, pp. 72-73. Appendix 3: List of the Most Important Concepts, pp. 74-76.

> An introduction to computability and decidability for the study of automata theory, formal language theory, and the theory of computing. The six chapter titles are: "Set and Functions," "Sets and Functions of Strings," "Computable Functions," "The Universal Turing Machine," "Some Functions Which Are Not Computable," and "Effectively Enumerable and Decidable Sets." Assumes knowledge of high school algebra and an introduction to computer programming.

McKean, Roland N. EFFICIENCY IN GOVERNMENT THROUGH SYSTEMS ANALYSIS: WITH EMPHASIS ON WATER RESOURCE DEVELOPMENT. Publications in Operations Research, no. 3. New York: Wiley, for the Rand Corporation, 1958. x, 336 p. Appendix on Possible Classifications of Expenditures by Programs, and Indicators of Performance, pp. 292-309. Bibliography, pp. 311-16. Published Rand Research, pp. 317-18. Index, pp. 319-36.

> A nonmathematical inquiry into the analysis of water-resource investments and civil government operations for use by a varied audience--cost-benefit analysts, operations researchers, governmental personnel, and all those who are concerned about economy in government. It begins with an analysis of methodological problems in the analysis of alternative actions using cost-benefit studies with illustrations from business, military planning, and water-resource development. Next, some special problems in the analysis of water-resource projects are examined and two case studies prepared by federal agencies are presented: the Green River watershed, and the Santa Maria project. The book concludes with suggestions for other uses of cost-benefit analysis, including health care.

Mao, James C.T. QUANTITATIVE ANALYSIS OF FINANCIAL DECISIONS. New York: Macmillan, 1969. xiv, 625 p. Tables, pp. 584-617. Index, pp. 619-25.

> A text for a graduate or undergraduate course in financial management with emphasis on theory rather than descriptive study of institutions and business practice. Topics include profit planning, financial budgeting, investment decision, cost of capital, valuation, shares, and working capital management. Assumes knowledge of calculus and matrix algebra. Each chapter contains review questions, exercises, and references.

Markowitz, Harry M. PORTFOLIO SELECTION: EFFICIENT DIVERSIFICATION OF INVESTMENTS. Cowles Foundation Monograph, no. 16. New Haven,

Conn.: Yale University Press, 1959. xiv, 351 p. Bibliography, pp. 305-7. Addendum, by M. Rubinstein, pp. 308-15. Appendixes: A, the Computation for Efficient Sets; B, A Simplex Method for Portfolio Selection; C, Alternative Axiom Systems for Expected Utility, pp. 316-47. Index, pp. 349-51.

A presentation of the mathematical techniques for the analysis of portfolios of securities which is aimed mainly at the nonmathematician. It is divided into the following four parts: the inputs, outputs, and objectives of a formal portfolio analysis; concepts and theorems necessary in portfolio analysis, particularly the statistical techniques of averages, variances, and covariances of return of a portfolio; a geometric analysis of efficient portfolios, the critical line method for deriving efficient portfolios, and the concept of semivariance as a measure of risk; and the theory of rational behavior and its application to the selection of portfolios. Assumes knowledge of college algebra, matrix algebra, and basic statistics. A few chapters conclude with exercises.

Martin, Michael J.C.; with Wells, W.B.; Coyle, R.G.; Rothwell, A.E.; Roberts, G.; Draper, D.H.; and Dension, R.A. MANAGEMENT SCIENCE AND URBAN PROBLEMS. Lexington, Mass.: D.C. Heath, 1974. xiv, 209 p. No index.

A description of the application of mathematical techniques and computers to several urban problems in the United Kingdom. Each project is described in two sections: a general description of the problem, including the costs and benefits of the recommended solution; and technical detailed descriptions of the mathematical model and computer techniques. The projects are: street lamp replacement policies; minimizing the cost of solid waste collection; evaluation of a water resource project sponsored by two public authorities; allocation of care to the elderly; and a comparative analysis of welfare services in England and Wales. The mathematical techniques are those used in cost-benefit analysis and economics, rather than the traditional operations research tools, and knowledge of college algebra should be sufficient for comprehension. Chapters conclude with notes and references.

Massy, William F.; Montgomery, David B.; and Morrison, Donald G. STOCHASTIC MODELS OF BUYING BEHAVIOR. Cambridge, Mass.: MIT Press, 1970. xiii, 464 p. Appendix: On Numerical Methods for Parameter Estimation, pp. 445-50. References, pp. 451-60. Index, pp. 461-64.

A standard reference on consumer behavior. After two introductory chapters on the meaning of stochastic models and on the estimation of their parameters, it takes up two main subjects: brand choice models, and purchase incidence models of buying behavior. The types of brand choice models are Bernoulli processes, Markov models, linear learning models, and probability diffusion models. The models used to explain whether a purchase is made (purchase incidence models) are exponential, logistic, and learning models.

A stochastic evolutionary adoption model is also developed. The book concludes with a model for predicting purchase behavior using simulation. Assumes knowledge of calculus, probability theory, and statistics.

Mattessich, Richard. ACCOUNTING AND ANALYTICAL METHODS: MEASURE-MENT AND PROJECTION OF INCOME AND WEALTH IN THE MICRO- AND MACRO-ECONOMY. Books in the Irwin Series in Accounting. Homewood, III.: Richard D. Irwin, 1964. xvii, 552 p. Appendix A: Set-theory and the Axiomatization of Accounting, pp. 437-65. Appendix B: Introduction to Matrix Algebra, pp. 466-77. Appendix C: Introduction to Linear Programming, pp. 479-96. Bibliography, pp. 497-524. Index of Authors, pp. 527-32. Index of Subjects, pp. 533-52.

A presentation of a unified view of accounting including new developments from the management sciences and their quantitative approaches. It is divided into the following three parts: on the essence and foundations of accounting; on valuation and hypotheses formulation; and on projection and planning for the future in which the accounting requirements for input-output analysis, linear programming, and other types of micro- and macro-models are discussed. Although it is written for academically trained accountants, it is also useful to economists, management scientists, engineers, and mathematicians in relating accounting to their own ideas.

Mishan, Edward J. COST-BENEFIT ANALYSIS: AN INFORMAL INTRODUC-TION. New Directions in Management and Economics. New York: Praeger, 1971. 364 p. Index, pp. 361-64.

Presents, by means of examples, some of the procedures and concepts that lie behind the techniques used in cost-benefit studies. No mathematics is used, other than the occasional resort to elementary algebra. It is divided into the following six parts: some simplified examples of cost-benefit studies, concepts of benefits and costs, external effects, investment criteria, uncertainty, and further notes. The chapters conclude with references and bibliographies.

_____. ECONOMICS FOR SOCIAL DECISIONS: ELEMENTS OF COST-BENEFIT ANALYSIS. Praeger University Series. New York: Praeger, 1973. 151 p. Index, pp. 148-51.

A nonmathematical introduction to the basic ideas of cost-benefit analysis rather than the specific techniques used for estimating the relevant magnitudes. It is divided into the following five parts: introduction, economic aspects of costs and benefits, external effects, investment criteria, and uncertainty. It was originally published in England under the title ELEMENTS OF COST-BENEFIT ANALYSIS (London: George Allen and Unwin, 1972). Topics include equity, consumer surplus, shadow pricing, externalities, investment criteria, and uncertainty. It is useful to engineers and statisticians who are called upon to contribute to the evaluation of a project.

Murdick, Robert G. MATHEMATICAL MODELS IN MARKETING. Scranton, Pa.: Intext Educational Publishers, 1971. ix, 293 p. No index.

Presents both brief and comprehensive marketing models based on twenty-six journal articles. The scholarly refinements have been omitted in the adaptations, and the student is referred to the original articles for such information as bibliographies and references. The revised articles, as well as the authors' contributions, are grouped into the following six parts: marketing strategy and product planning, customer behavior, forecasting demand and sales, selling and buying, logistics, and advertising and sales promotion. Topics include branch store planning, ruin model, method of search for new-product ideas, brand preferences and simple Markov processes, first-time purchase timing, spatial allocation of selling expense, bidding model, optimum order quantity, and expenditure policy for mail-order advertisers. Many of the articles assume knowledge of the techniques of operations research, such as linear programming, and calculus. Chapters conclude with problems on applications and analysis. Among the authors are A.C. Atkinson, F.M. Bass, Wm. Beranek, D.B. Hertz, P. Kotler, R.B. Maffei, F.M. Nicosia, J.A. Nordin, J.A. Parsons, J.B. Stewart, D.S. Tull, M.L. Vidale, and H.B. Wolfe.

Nicosia, Francesco M. CONSUMER DECISION PROCESSES: MARKETING AND ADVERTISING IMPLICATIONS. Foreword by H.A. Simon. Prentice-Hall Behavioral Sciences in Business Series. Englewood Cliffs, N.J.: Prentice-Hall, 1966. xviii, 284 p. Bibliography, pp. 247-72. Concept Index, pp. 275-77. Index of Authors, pp. 277-84.

An evaluation of various perspectives of the consumer in the fields of marketing, economics, and the behavioral sciences with emphasis on applications for decision making. The main tool employed is a system of differential equations in the micro-endogenous variables. Mathematics is used mainly in the final chapter which sets forth models of consumer decision processes.

Niskanen, William A.; Harberger, Arnold C.; Haveman, Robert H.; Turvey, Ralph; and Zeckhauser, Richard, eds. BENEFIT-COST AND POLICY DECISIONS, 1972. An Aldine Annual on Forecasting, Decision-making, and Evaluation. Chicago: Aldine Publishing Co., 1973. xiv, 535 p. Name Index, pp. 519-21. Subject Index, pp. 522-35.

Twenty-two of the best articles published during 1972 on the analysis of public allocation decisions grouped under the following headings: challenge and response (two), general theoretical contributions (four), the discount rate issue (three), evaluation of benefit and cost estimates (two), investments in people (six),

Miscellaneous

and investment in physical resources (five). Most of the articles require little mathematics preparation, although those dealing with the discount rate utilize differential and integral calculus. Among the contributors are S.N. Acharya, J.H. Dreze, R.H. Haveman, G.P. Jenkins, J.V. Krutilla, C.M. Lindsay, L. Merewitz, S. Paul, A. Sandmo, A.K. Sen, and A. Williams.

Nordhaus, William D. INVENTION, GROWTH, AND WELFARE: A THEORETICAL TREATMENT OF TECHNICAL CHANGE. MIT Monographs in Economics. Cambridge, Mass.: MIT Press, 1969. xiv, 168 p. Appendix A: Stability of Equilibrium and Research, pp. 140–48. Appendix B: Increasing Returns in a Competitive Industry, pp. 149–54. References, pp. 155–64. Index, pp. 165–68.

A largely nontechnical exploration of some of the problems raised by the economics of technology with the higher mathematics placed in the appendixes. It is divided into the following two parts: a theory of invention, which treats external economies, technology as a barrier to entry, the diffusion of technology in a growing economy, and the determination of the optimal royalty by an inventor and the optimal life of a patent; and invention in a growing economy, including the problems of invention in a general equilibrium framework with emphasis on technical change in the process of economic growth. Assumes knowledge of differential and integral calculus.

Novozhilov, V.V. PROBLEMS OF COST-BENEFIT ANALYSIS IN OPTIMAL PLANNING. Translated by H. McQuiston. White Plains, N.Y.: International Arts and Science Press, 1970. viii, 362 p. No index.

Originally published in Russian in 1967, this book uses cost-benefit analysis to analyze problems such as maximizing yields from capital investments, establishing principles for measuring production costs and benefits, measuring the results of labor in a socialistic economy, defining appropriate guidelines for planned price information, choosing between investment variants, determining the optimal life of capital equipment, and measuring labor productivity. While extensive use is made of mathematical tools and notation, little mathematics preparation is assumed. It is worthwhile reading for anyone interested in capital theory, economic systems, and the price mechanism.

Orgler, Yair E. ANALYTICAL METHODS IN LOAN EVALUATION. Foreword by P.M. Horvitz. Lexington, Mass.: D.C. Heath, 1975. xxii, 115 p. References, pp. 105–9. Index, pp. 111–15.

Applies quantitative techniques to loan selection and loan analysis, the latter consisting of two credit-scoring models for evaluating outstanding commercial and consumer loans, and a model for estimating the potential losses from an entire consumer loans portfolio held by a defaulted bank. Linear programming is used in developing the loan selection model; multivariate regression analysis is used in deriving the credit-scoting models; and a Markov chains process provides the basis for calculating the loss potential of a consumer loans portfolio. Useful to bank loan officers, examiners of bank regulatory agencies, bank auditors, and students of financial institutions.

Palomba, Giuseppe. A MATHEMATICAL INTERPRETATION OF THE BALANCE SHEET. New York: Augustus M. Kelley, 1968. 131 p. Index of Names, pp. 129-30. Table of Contents, p. 131.

A monograph on the logico-mathematical view of the problems created by the arbitrary choices that are made in the preparation of the balance sheet. These problems are: net capital as a function of time, rapid amortization, the net capital as a function of the balance sheet entries, distribution of the net capital among the various items of the assets and the ambivalent nature of profit, the dynamics of provisional costs and earnings and the theory of microcosmic perturbations, and problems of aggregation and of liquidity. Assumes knowledge of calculus and matrix algebra.

Peters, G.H. COST-BENEFIT ANALYSIS AND PUBLIC EXPENDITURES. 3d ed., rev. and enl. Eaton Paper 8. Westminister, England: Goron Pro-Print, for the Institute of Economic Affairs, 1973. 76 p. No index.

A brief nontechnical introduction to cost-benefit analysis as applied to public expenditures. The author explains the techniques of cost-benefit analysis and reviews its applications in water development, transportation, land conservation, urban development, regional planning, military defense, education, health, and the third London airport. There are no mathematics requirements.

Peterson, D.E., with Hardon, R.D. A QUANTITATIVE FRAMEWORK FOR FINANCIAL MANAGEMENT. Irwin Series in Quantitative Analysis for Business. Homewood, Ill.: Richard D. Irwin, 1969. xviii, 694 p. Appendixes, pp. 633-85. Index, pp. 689-94.

An intermediate-level text in financial management for graduates and undergraduates who have a background in mathematics comparable to Taro Yamane's MATHEMATICS FOR ECONOMISTS (Englewood Cliffs, N.J.: Prentice-Hall, 1968). It is divided into the following five parts: introduction, developing a short-run investment financing plan, developing a long-range enterprise investment plan, developing a long-range enterprise financing plan, and an integrated investment-financing plan. The appendixes consist of interest, logarithm, and probability tables. The main tool used in the analysis is mathematical programming. The chapters conclude with selected references and problems.

Rao, Ambar G. QUANTITATIVE THEORIES IN ADVERTISING. Publications in Operations Research, no. 21. New York: Wiley, 1970. x, 103 p. Appendix: Analysis of Factorial Designs, pp. 100-102. Index, p. 103.

A quantitative theory of advertising based on the author's research experience with a leading manufacturer of a packaged food product. The models developed use sales volume as a measure of advertising effectiveness, and they are applied to the study of advertising strategies for a variety of products, including those sold in a noninnovative market with many competing brands, as well as markets in which the number of competitors is small and the strategies of competition must be considered in formulating plans for selling any one brand. Other topics include defensive advertising policies with incomplete information, advertising over time, and advertising quality. Assumes knowledge of differential and integral calculus and matrix algebra. The chapters conclude with references.

Riley, Vera, and Gass, Saul I. LINEAR PROGRAMMING AND ASSOCIATED TECHNIQUES: A COMPREHENSIVE BIBLIOGRAPHY ON LINEAR, NONLINEAR, AND DYNAMIC PROGRAMMING. Rev. ed. Bibliographic Reference Series, no. 5. Baltimore: Johns Hopkins University Press, for the Operations Research Office of the Johns Hopkins University, 1959. x, 613 p. Appendix: Author Index, pp. 563-613.

A revised edition of a bibliography of some 500 references originally published in 1954. The revised edition contains references to over 1,000 items, and includes an intermixture of articles, books, monographs, documents, theses, conference proceedings, and so forth. It is divided into the following four parts: introduction, general theory, applications, and nonlinear and dynamic programming.

Rowney, Don Karl, and Graham, James Q., Jr., eds. QUANTITATIVE HISTORY: SELECTED READINGS IN THE QUANTITATIVE ANALYSIS OF HISTORICAL DATA. Homewood, Ill.: Dorsey Press, 1969. xiv, 488 p. Bibliography, pp. 473-79. Index, pp. 483-88.

Twenty-three essays arranged under the following headings: varieties of quantitative history; bureaucrats, deputies, and decision makers: studies in elite history; social history and social change; historical demography; "cliometrics," the new economic history; and voters and public: studies in legislative and electoral history.

Seal, Hilary I. STOCHASTIC THEORY OF A RISK BUSINESS. Wiley Series in Applied Probability and Statistics. New York: Wiley, 1969. xiii, 210 p. Appendix A: Renewal Processes, pp. 173-80. Appendix B: Numerical Inversion of Laplace Transforms, pp. 181-99. Glossary of Consistently Used Notation, pp. 201-4. Author Index, pp. 205-7. Subject Index, pp. 208-10.

A monograph on the mathematical and statistical theory of risk taking with emphasis drawn mainly from the insurance field. It is based on a survey of most of the literature; over 250 articles and books are cited in the references at the ends of the chapters, and an additional 100 references were read but not cited. Topics include distribution of aggregate claims, calculation of net premiums, the probability of ruin of a risk business, premium loading and reinsurance, and utility theory as applied to insurance. Assumes knowledge of at least one year of calculus and probability theory at the intermediate level.

Serck-Hanssen, Jan. OPTIMAL PATTERNS OF LOCATION. Contributions to Economic Analysis, vol. 66. New York: American Elsevier, 1970. xi, 235 p. Bibliography, pp. 229-31. Name Index, p. 232. Subject Index, pp. 233-35.

An analysis of optimal spatial distribution of firms based on Weber models for minimizing transportation costs, "programming models" with variables which are outputs of firms, and Lösch models which satisfy the second-order conditions for a minimum of total cost. Assumes knowledge of differential and integral calculus and matrix algebra.

Sharpe, William F. PORTFOLIO THEORY AND CAPITAL MARKETS. McGraw-Hill Series in Finance. New York: McGraw-Hill Book Co., 1970. xvi, 316 p. Supplement: The Mathematical Foundation, pp. 223-302. Bibliography, pp. 303-12. Index, pp. 313-16.

A summary and synthesis of some of the major contributions of the previous two decades in the fields of portfolio analysis and the theory of capital markets. Part 1 develops portfolio theory, covering the procedures for selecting investments. Part 2 deals with models of capital markets based on the assumption that investors act in accordance with the principles described in part 1. Part 3 is devoted to applications and extensions, and includes procedures for measuring the performance of portfolio-selection procedures, results of studies of mutual-fund performance, utility theory as it relates to portfolio theory, and state-preference theory as it relates to capital market theory. No mathematics beyond high school algebra is required. A supplement is provided for readers who have knowledge of elementary calculus.

Sheth, Jagdish N., ed. MODELS OF BUYER BEHAVIOR: CONCEPTUAL, QUANTITATIVE, AND EMPIRICAL. Series in Marketing Management. New York: Harper and Row, 1974. xiii, 441 p. Reference Bibliography, pp. 407-34. Index, pp. 435-41.

A supplement for advanced undergraduate and graduate courses in marketing. It consists of twenty-one chapters, mostly written individually by well-known scholars, arranged under the following headings: construct validation in model development (one chapter), models of innovative behavior and product adoption (five chapters), models of consumer typology and market segmentation (two chapters), and future (one chapter). Most of the chapters are nonmathematical, but several require knowledge of college algebra and matrix algebra for comprehension.

Simon, Leonard S., and Freimer, Marshall. ANALYTICAL MARKETING.
Foreword by Wm. F. Massy. The Harbrace Series in Business and Economics.
New York: Harcourt, Brace and World, 1970. xv, 346 p. Author Index,
pp. 339-42. Subject Index, pp. 343-46.

A text for an advanced marketing course emphasizing applications
of analytical tools and models for decision making. Topics include
mathematical programming, product and pricing decisions, distribution systems, sales force management, mass communications,
consumer purchase behavior, information for marketing decisions,
and four case studies. The chapters conclude with summaries.

Somers, G.G., and Wood, W.D., eds. COST-BENEFIT ANALYSIS OF MAN-
POWER POLICIES. Kingston, Ontario, Canada: Industrial Relations Centre,
Queens University, 1969. xix, 272 p. Index.

Proceedings of a North American Conference under the auspices
of the Center for Studies in Vocational and Technical Education,
the University of Wisconsin, and the Industrial Relations Centre,
Queens University, and sponsored by the Canadian Department of
Manpower and Immigration and the U.S. Department of Labor.
The first four papers are on the theoretical aspects of cost-benefit
analysis (B. Weisbrod, R. Judy, Wm. Dymond, and K.J. Arrow);
the next five papers deal with applications to manpower programs;
and the last paper, by N. Chamberlain, is a critique of the concept of human capital.

Stone, Bernell Kenneth. RISK, RETURN, AND EQUILIBRIUM: A GENERAL
SINGLE-PERIOD THEORY OF ASSET COLLECTION AND CAPITAL-MARKET
EQUILIBRIUM. Cambridge, Mass.: MIT Press, 1970. 150 p. Appendixes,
pp. 137-41. References, pp. 142-43. Index, pp. 145-50.

A supplement for graduate courses in economics and finance requiring knowledge of differential and integral calculus, and a
text in advanced finance courses dealing with risk, asset selection,
and capital markets. Topics include asset selection and market
equilibrium under a general two-parameter functional representation, mean variance models, risk-return relationships, the Sharpe
model and market equilibrium, and personal equilibrium. The appendixes are devoted to additional mathematical results.

Theil, Henri. LINEAR AGGREGATION OF ECONOMIC RELATIONS. Contributions to Economic Analysis, vol. 7. New York: American Elsevier, 1954.
xi, 205 p. Rules of Notation, pp. 199-200. Index, pp. 201-5.

Presents the problem of aggregation over individuals, commodities,
and time periods based on linear relations. The requirements for
"good" and "perfect" aggregation are also considered. The discussion is abstract and no empirical applications are presented.
Assumes knowledge of calculus and matrix algebra.

Theil, Henri; with Beerens, G.A.C.; DeLeeuw, C.G.; and Tilanus, C.B. AP-
PLIED ECONOMIC FORECASTING. Studies in Mathematical and Managerial
Economics, vol. 4. New York: American Elsevier, 1966. xxv, 474 p. In-
dex, pp. 471-74.

> A survey of the achievement of prediction methods in economics,
> partly at the theoretical but mainly at the empirical level. It is
> based primarily on the work of the Econometrics Institute of the
> Netherlands School of Economics on macromodels and input-output
> analysis. Advanced calculus and probability theory, or a course
> in econometrics, are essential for comprehension. Some of the
> chapters conclude with appendixes containing additional mathemat-
> ical results.

Thin, Tun. THEORY OF MARKETS. Cambridge, Mass.: Harvard University
Press, 1967. viii, 129 p. Appendixes, pp. 101-16. A Selected Bibliography,
pp. 117-18. Index, pp. 119-20.

> A study of pricing in perfect competition, perfect monopoly, and
> imperfect competition with emphasis given to oligopoly. The so-
> lutions offered by Cournot, Smithies, Chamberlain, Fellner, and
> Robinson are presented mathematically. The author's own solution
> to pricing in oligopoly is also presented. Assumes knowledge of
> calculus and matrix algebra.

Valentine, Jerome L., and Mennis, Edmund A. QUANTITATIVE TECHNIQUES
FOR FINANCIAL ANALYSIS. The Charter Financial Analysts Series. Homewood, Ill.: Richard D. Irwin, 1971. xiii, 284 p. Appendixes A to C,
pp. 265-75. Index, pp. 277-84.

> A book in finance for the practitioner which emphasizes the func-
> tions of tools, rather than elaborate computations and proofs. Topics
> include review of compound growth, rates of change in variables,
> hypothesis testing, analysis of variance, simple and multiple re-
> gression, linear programming with extensions to parametric, inte-
> ger, and nonlinear programming, and use of the computer. The
> appendixes consist of logarithms, interest tables, and the t, F,
> and Chi-square distributions.

Williams, Edward E., and Findlay, M. Chapman III. INVESTMENT ANALY-
SIS. Englewood Cliffs, N.J.: Prentice-Hall, 1974. xv, 476 p. Appendix
A: Financial Mathematics: Compound Interest, Present Values, and Yields,
pp. 399-409. Appendix B: Financial Statistics: Probability, Variance, and
Correlation Analysis, pp. 411-31. Appendix C: Tables, pp. 433-86. Index,
pp. 487-76.

> A comprehensive treatment of security analysis and portfolio theory
> for both students and practitioners who have knowledge of financial
> mathematics and statistics, although these subjects are reviewed in
> the appendixes. It is divided into the following four parts: the
> investment environment, security analysis, portfolio analysis, and

capital market efficiency. Topics include forecasting techniques, including regression, correlation, and discriminant analysis; analysis of fixed income securities; common stocks; basic analysis, valuation models, and growth and risk analysis; convertible and speculative securities; utility theory; capital market theory; and techniques of portfolio selection and elimination. Chapters conclude with problems and references.

Williamson, J. Peter. INVESTMENTS: NEW ANALYTIC TECHNIQUES. New Directions in Management and Economics. New York: Praeger, 1971. v, 325 p. Index, pp. 321-25.

A discussion of new, as well as established, techniques of investment analysis. Topics include the Markowitz portfolio selection model; multiindex models; Sharpe's single-index and linear models; stock-selection techniques, including the intrinsic value and leading indicator approaches; technical analysis, including the random walk theory; bond portfolio performance; investment mathematics; and the measurement of risk. Several chapters have appendixes devoted to the derivation of formulas, and one chapter is devoted to a review of the necessary statistical tools. Portions of the book require knowledge of elementary calculus.

Wolfe, J.N., ed. COST BENEFIT AND COST EFFECTIVENESS: STUDIES AND ANALYSIS. London: George Allen and Unwin, 1973. 236 p. Index, pp. 233-36.

Eleven papers on cost-benefit analysis initially presented at a conference of British economists held at the University of Edinburgh. The papers are grouped under the following headings: political economy (two papers), including the political acceptability and economic rationale of cost-benefit analysis; theory (three papers), including taxation, risk, and uncertainty in cost-benefit analysis; fields and areas (three papers), including cost-benefit analysis applied to airport problems and the evaluation of research and development; and frontiers (three papers), including the introduction of constraints in cost-benefit analysis and extensions of cost-benefit analysis into a general equilibrium setting. There is also an introductory paper by J.N. Wolfe. Knowledge of intermediate economic theory is helpful. The authors are M.E. Beesley, C. Blake, G. Corti, M.Q. Dalvi, C.D. Foster, J.B. Heath, P.M.S. Jones, N. Mansfield, W.N. Oulton, A. Peacock, H. Thomas, A.A. Walters, P. Watson, A. Williams, and J.N. Wolfe.

Wood, W.D., and Campbell, Harry F. COST-BENEFIT ANALYSIS AND THE ECONOMICS OF INVESTMENT IN HUMAN RESOURCES: AN ANNOTATED BIBLIOGRAPHY. Bibliography Series, no. 5. Kingston, Ontario: Industry Relations Centre, Queens University, 1970. vii, 211 p. Author Index, pp. 209-11.

An annotated listing of 389 articles, monographs, and books. The first four sections present the theoretical aspects of cost-benefit analysis, and the remaining four sections cover practical applications in schooling; training, retraining, and labor mobility; poverty and social welfare; and health. The annotations are lengthy (some are one page in length) and are divided into two parts: a heading which includes a brief introduction to the content of the article; and the abstract part which surveys the content of the article.

INDEXES

AUTHOR INDEX

This index is alphabetized letter by letter. In addition to authors, this index includes all editors, translators, compilers, and those who have contributed introductions and forewards to works cited in the text. It also includes authors of articles, books, and other works cited in the annotations.

Author Index

Aydelotte, William O. 487
Ayres, Frank, Jr. 6

B

Bacharach, M.O.L. 197
Bacharach, Michael 142, 157
Bacon, N.T. 97
Bagchi, Tapan Prasad 372
Baggaley, Andrew R. 6
Bailey, Martin J. 124
Bailey, Norman T. 6, 372
Baird, Robert N. 124
Bak, Thor A. 7
Baker, Bruce N. 310
Baker, Kenneth R. 310
Balakrishnan, A.V. 323, 372, 421-23, 463
Balinski, M.L. 423
Balintfy, Joseph L. 411
Ball, Richard J. 123, 174
Banasinski, A. 476
Banerji, Ranan B. 174
Bannink, R. 119
Banks, Jerry 258
Barish, Norman N. 424
Barlow, Richard E. 255, 464
Barna, Tibor 175
Barnetson, Paul 311
Barnett, Stephen 323
Barrett, Nancy Smith 92
Bar-Shalom, Y. 345
Barsov, A.S. 270
Bartholomew, D.J. 372
Bartlett, Maurice S. 373
Barton, Richard F. 397
Bashaw, W. Louis 7
Basil, Douglas C. 401
Bass, Frank M. 424, 487
Batchelor, James H. 487-88
Bate, R.R. 343
Bates, John 157
Batra, Raveendra N. 124
Batschelet, Edward 7
Battersby, Albert 311
Bauer, Friedrich L. 424
Bauman, E.J. 343
Baumol, William J. 93, 175
Baxter, Willard E. 8
Beach, Earl Francis 161

Beale, Evelyn Martin Landsdowne 270, 425
Beazer, William F. 397
Beckenbach, Edwin F. 175
Becker, Gary S. 93
Becker, Martin 323
Beckmann, Martin J. 270, 488
Beckmann, Peter 361
Beekman, John A. 373
Beer, Stafford 213, 467
Beerens, G.A.C. 511
Beggs, Robert I. 76
Beighey, D. Clyde 8
Beightler, Charles S. 358
Bell, D. 425
Bell, David John 425
Bell, Earl J. 276
Bellman, Richard E. 8-11, 228, 269, 270-71, 324-25, 351, 394, 400, 426, 468
Bello, Ignacio 11
Belsley, David A. 256
Beltrami, Edward J. 325
Belyayev, Yu. K. 259
Benavie, Arthur 12
Benes, Vaclav E. 311, 361
Ben-Israel, Adi 12
Bennion, Edward G. 213
Benson, O. 180
Bensoussan, Alain 326, 423, 427
Beranek, William 499
Berczi, Andrew 213
Berg, Robert L. 488
Berge, Claude 13, 272
Bergseth, F. Robert 50
Bergstrom, Abram R. 161
Berkeley, Edmund C. 13
Berman, Abraham 272
Bernd, Joseph L. 176, 188-89, 489
Berston, Hyman Maxwell 13
Bertele, Umberto 272
Beshers, James M. 176
Beveridge, Gordon S.G. 326
Bhagwati, Jagdish N. 176
Bharadwaj, Ranganath 142, 152
Bharucha-Reid, A.T. 374, 427
Bhat, U. Narayan 361, 374
Bicksler, James L. 176
Biegel, John E. 214
Bierman, Harold, Jr. 214

Author Index

Author Index

Author Index

Author Index

Author Index

Author Index

TITLE INDEX

This index is alphabetized letter by letter. It includes all titles,* in full, of books, monographs, and published reports cited in this text. Individual chapters, articles, and essay titles are not included.

Title Index

Title Index

Title Index

N

Title Index

SUBJECT INDEX

This index is alphabetized letter by letter. Emphasis is on key concepts and topics of general interest, as well as the use of techniques, such as linear programming, in specific areas of research. References to authors indicate that their theories or concepts are the subject of discussion. Underlined page numbers refer to main areas within the subject.

A

Accounting
  mathematical applications in 492, 498
  relation to input-output analysis and linear programming 504
Activity analysis 59, 64, 95, 192, 197, 217, 275, 290
  conferences on 192, 197
  in economic growth 197
  in economic planning 480
  in production 104, 109
Adaptive control 340, 359, 446, 462, 471
  conferences on 462
  dynamic programming approach to 272, 286, 324
  mathematics of 359
  sequential decision theory in 355
  statistical decision theory in 355
  systems 341, 342, 355, 357, 425
Adaptive processes 113, 306, 324
  in model construction 168
Adaptive programming 270
Additive functionals 379
Advertising, models of 508

Aggregation
  for cost and production functions 119
  in data analysis 189, 207
  in economic models 204
  in economics 128-29, 161, 173, 407, 468, 510
  of preference 118
Airline industry
  model building in 192
  use of linear programming in 294
Algebra 3, 49, 52
  business 24, 27
  college 14, 42
  in economics 24
  elementary 47
  linear 4, 9, 12, 14, 17, 19, 28, 49, 50, 57
  multilinear 53
ALGOL computer programs 338
Algorithms 17, 43
  for bang-bang control problems 287
  for continuous time control problems 287
  decomposition 186, 297
  in dynamic programming 284
  formulation of 38
  in integer programming 281, 286

Subject Index

Subject Index

Subject Index

Subject Index

Subject Index

Subject Index

Subject Index

Subject Index